STUDENT SOLUTIONS MANUAL

C. TRIMBLE AND ASSOCIATES

Introductory

MATHEMATICAL ANALYSIS

for Business, Economics, and the Life and Social Sciences

Twelfth Edition

Ernest F. Haeussler, Jr. | Richard S. Paul | Richard J. Wood

PEARSON
Prentice
Hall

Upper Saddle River, NJ 07458

Vice President and Editorial Director, Mathematics: Christine Hoag
Acquisitions Editor: Chuck Synovec
Print Supplement Editor: Joanne Wendelken
Senior Managing Editor: Linda Behrens
Assistant Managing Editor: Lynn Savino Wendel
Production Editor: Barbara Mack
Supplement Cover Manager: Paul Gourhan
Supplement Cover Designer: Victoria Colotta
Manufacturing Buyer: Ilene Kahn
Associate Director of Operations: Alexis Heydt-Long

© 2008 Pearson Education, Inc.
Pearson Prentice Hall
Pearson Education, Inc.
Upper Saddle River, NJ 07458

Pearson Prentice Hall™ is a trademark of Pearson Education, Inc.

The author and publisher of this book have used their best efforts in preparing this book. These efforts include the development, research, and testing of the theories and programs to determine their effectiveness. The author and publisher make no warranty of any kind, expressed or implied, with regard to these programs or the documentation contained in this book. The author and publisher shall not be liable in any event for incidental or consequential damages in connection with, or arising out of, the furnishing, performance, or use of these programs.

Printed in the United States of America

10 9 8 7 6 5 4 3 2 1

ISBN 13:   978-0-13-240424-2
ISBN 10:   0-13-240424-9

Pearson Education Ltd., *London*
Pearson Education Australia Pty. Ltd., *Sydney*
Pearson Education Singapore, Pte. Ltd.
Pearson Education North Asia Ltd., *Hong Kong*
Pearson Education Canada, Inc., *Toronto*
Pearson Educación de Mexico, S.A. de C.V.
Pearson Education—Japan, *Tokyo*
Pearson Education Malaysia, Pte. Ltd.

Table of Contents

Chapter 0   1

Chapter 1   21

Chapter 2   34

Chapter 3   57

Chapter 4   84

Chapter 5   102

Chapter 6   113

Chapter 7   149

Chapter 8   194

Chapter 9   218

Chapter 10   233

Chapter 11   246

Chapter 12   279

Chapter 13   306

Chapter 14   350

Chapter 15   396

Chapter 16   425

Chapter 17   433

Chapter 0

Problems 0.1

1. True; –13 is a negative integer.

3. False, because the natural numbers are 1, 2, 3, and so on.

5. True, because $5 = \dfrac{5}{1}$.

7. False, because $\sqrt{25} = 5,$ which is a positive integer.

9. False; we cannot divide by 0.

11. True

Problems 0.2

1. False, because 0 does not have a reciprocal.

3. False; the negative of 7 is –7 because $7 + (-7) = 0.$

5. False; $-x + y = y + (-x) = y - x.$

7. True; $\dfrac{x+2}{2} = \dfrac{x}{2} + \dfrac{2}{2} = \dfrac{x}{2} + 1.$

9. False; the left side is $5xy$, but the right side is $5x^2 y.$

11. distributive

13. associative

15. commutative and distributive

17. definition of subtraction

19. distributive

21. $2x(y - 7) = (2x)y - (2x)7 = 2xy - (7)(2x)$
$= 2xy - (7 \cdot 2)x = 2xy - 14x$

23. $(x + y)(2) = 2(x + y) = 2x + 2y$

25. $x[(2y + 1) + 3] = x[2y + (1 + 3)] = x[2y + 4]$
$= x(2y) + x(4) = (x \cdot 2)y + 4x = (2x)y + 4x$
$= 2xy + 4x$

27. $x(y - z + w) = x[(y - z) + w]$
$= x(y - z) + x(w) = x[y + (-z)] + xw$
$= x(y) + x(-z) + xw = xy - xz + xw$

29. $-6 + 2 = -4$

31. $7 - 2 = 5$

33. $-5 - (-13) = -5 + 13 = 8$

35. $(-2)(9) = -(2 \cdot 9) = -18$

37. $(-2)(-12) = 2(12) = 24$

39. $\dfrac{-1}{-\frac{1}{9}} = -1\left(-\dfrac{9}{1}\right) = 9$

41. $-7(x) = -(7x) = -7x$

43. $-[-6 + (-y)] = -(-6) - (-y) = 6 + y$

45. $-9 \div (-27) = \dfrac{-9}{-27} = \dfrac{9}{27} = \dfrac{9 \cdot 1}{9 \cdot 3} = \dfrac{1}{3}$

47. $2(-6 + 2) = 2(-4) = -8$

49. $(-2)(-4)(-1) = 8(-1) = -8$

51. $X(1) = X$

53. $4(5 + x) = 4(5) + 4(x) = 20 + 4x$

55. $0(-x) = 0$

57. $\dfrac{5}{1} = 5$

59. $\dfrac{3}{-2x} = \dfrac{3}{-(2x)} = -\dfrac{3}{2x}$

61. $\dfrac{a}{c}(3b) = \dfrac{a(3b)}{c} = \dfrac{3ab}{c}$

63. $\dfrac{-aby}{-ax} = \dfrac{-a \cdot by}{-a \cdot x} = \dfrac{by}{x}$

65. $\dfrac{2}{x} \cdot \dfrac{5}{y} = \dfrac{2 \cdot 5}{x \cdot y} = \dfrac{10}{xy}$

67. $\dfrac{5}{12} + \dfrac{3}{4} = \dfrac{5}{12} + \dfrac{9}{12} = \dfrac{5+9}{12} = \dfrac{14}{12} = \dfrac{2 \cdot 7}{2 \cdot 6} = \dfrac{7}{6}$

69. $\dfrac{4}{5} + \dfrac{6}{5} = \dfrac{4+6}{5} = \dfrac{10}{5} = 2$

71. $\dfrac{3}{2} - \dfrac{1}{4} + \dfrac{1}{6} = \dfrac{18}{12} - \dfrac{3}{12} + \dfrac{2}{12} = \dfrac{18-3+2}{12} = \dfrac{17}{12}$

73. $\dfrac{6}{\frac{x}{y}} = 6 \div \dfrac{x}{y} = 6 \cdot \dfrac{y}{x} = \dfrac{6y}{x}$

75. $\dfrac{\frac{-x}{y^2}}{\frac{z}{xy}} = -\dfrac{x}{y^2} \div \dfrac{z}{xy} = -\dfrac{x}{y^2} \cdot \dfrac{xy}{z} = -\dfrac{x^2}{yz}$

77. $\dfrac{0}{7} = 0$

79. $0 \cdot 0 = 0$

Problems 0.3

1. $(2^3)(2^2) = 2^{3+2} = 2^5 (= 32)$

3. $w^4 w^8 = w^{4+8} = w^{12}$

5. $\dfrac{x^3 x^5}{y^9 y^5} = \dfrac{x^{3+5}}{y^{9+5}} = \dfrac{x^8}{y^{14}}$

7. $\dfrac{(a^3)^7}{(b^4)^5} = \dfrac{a^{3 \cdot 7}}{b^{4 \cdot 5}} = \dfrac{a^{21}}{b^{20}}$

9. $(2x^2 y^3)^3 = 2^3 (x^2)^3 (y^3)^3$
$= 8x^{2 \cdot 3} y^{3 \cdot 3}$
$= 8x^6 y^9$

11. $\dfrac{x^9}{x^5} = x^{9-5} = x^4$

13. $\dfrac{(x^3)^6}{x(x^3)} = \dfrac{x^{3 \cdot 6}}{x^{1+3}} = \dfrac{x^{18}}{x^4} = x^{18-4} = x^{14}$

15. $\sqrt{25} = 5$

17. $\sqrt[7]{-128} = -2$

19. $\sqrt[4]{\dfrac{1}{16}} = \dfrac{\sqrt[4]{1}}{\sqrt[4]{16}} = \dfrac{1}{2}$

21. $(49)^{1/2} = \sqrt{49} = 7$

23. $9^{3/2} = \left(\sqrt{9}\right)^3 = (3)^3 = 27$

25. $(32)^{-2/5} = \dfrac{1}{(32)^{2/5}} = \dfrac{1}{\left(\sqrt[5]{32}\right)^2} = \dfrac{1}{(2)^2} = \dfrac{1}{4}$

27. $\left(\dfrac{1}{32}\right)^{4/5} = \left(\sqrt[5]{\dfrac{1}{32}}\right)^4 = \left(\dfrac{1}{2}\right)^4 = \dfrac{1}{16}$

29. $\sqrt{50} = \sqrt{25 \cdot 2} = \sqrt{25} \cdot \sqrt{2} = 5\sqrt{2}$

31. $\sqrt[3]{2x^3} = \sqrt[3]{2} \sqrt[3]{x^3} = x\sqrt[3]{2}$

33. $\sqrt{16x^4} = \sqrt{16}\sqrt{x^4} = 4x^2$

35. $2\sqrt{8} - 5\sqrt{27} + \sqrt[3]{128}$
$= 2\sqrt{4 \cdot 2} - 5\sqrt{9 \cdot 3} + \sqrt[3]{64 \cdot 2}$
$= 2 \cdot 2\sqrt{2} - 5 \cdot 3\sqrt{3} + 4\sqrt[3]{2}$
$= 4\sqrt{2} - 15\sqrt{3} + 4\sqrt[3]{2}$

37. $(9z^4)^{1/2} = \sqrt{9z^4} = \sqrt{3^2(z^2)^2} = \sqrt{3^2}\sqrt{(z^2)^2}$

$\quad = 3z^2$

39. $\left(\dfrac{27t^3}{8}\right)^{2/3} = \left(\left[\dfrac{3t}{2}\right]^3\right)^{2/3} = \left[\dfrac{3t}{2}\right]^2 = \dfrac{9t^2}{4}$

41. $\dfrac{a^5 b^{-3}}{c^2} = a^5 \cdot b^{-3} \cdot \dfrac{1}{c^2} = a^5 \cdot \dfrac{1}{b^3} \cdot \dfrac{1}{c^2} = \dfrac{a^5}{b^3 c^2}$

43. $5m^{-2}m^{-7} = 5m^{-2+(-7)} = 5m^{-9} = \dfrac{5}{m^9}$

45. $(3t)^{-2} = \dfrac{1}{(3t)^2} = \dfrac{1}{9t^2}$

47. $\sqrt[5]{5x^2} = (5x^2)^{1/5} = 5^{1/5}(x^2)^{1/5} = 5^{1/5}x^{2/5}$

49. $\sqrt{x} - \sqrt{y} = x^{1/2} - y^{1/2}$

51. $x^2\sqrt[4]{xy^{-2}z^3} = x^2(xy^{-2}z^3)^{1/4}$

$\quad = x^2 x^{1/4} y^{-2/4} z^{3/4}$

$\quad = \dfrac{x^{9/4} z^{3/4}}{y^{1/2}}$

53. $(2a - b + c)^{2/3} = \sqrt[3]{(2a - b + c)^2}$

55. $x^{-4/5} = \dfrac{1}{x^{4/5}} = \dfrac{1}{\sqrt[5]{x^4}}$

57. $3w^{-3/5} - (3w)^{-3/5} = \dfrac{3}{w^{3/5}} - \dfrac{1}{(3w)^{3/5}}$

$\quad = \dfrac{3}{\sqrt[5]{w^3}} - \dfrac{1}{\sqrt[5]{(3w)^3}} = \dfrac{3}{\sqrt[5]{w^3}} - \dfrac{1}{\sqrt[5]{27w^3}}$

59. $\dfrac{6}{\sqrt{5}} = \dfrac{6}{5^{1/2}} = \dfrac{6 \cdot 5^{1/2}}{5^{1/2} \cdot 5^{1/2}} = \dfrac{6\sqrt{5}}{5}$

61. $\dfrac{4}{\sqrt{2x}} = \dfrac{4}{(2x)^{1/2}} = \dfrac{4(2x)^{1/2}}{(2x)^{1/2}(2x)^{1/2}} = \dfrac{4\sqrt{2x}}{2x}$

$\quad = \dfrac{2\sqrt{2x}}{x}$

63. $\dfrac{1}{\sqrt[3]{3x}} = \dfrac{1}{(3x)^{1/3}} = \dfrac{1(3x)^{2/3}}{(3x)^{1/3}(3x)^{2/3}} = \dfrac{\sqrt[3]{(3x)^2}}{3x}$

$\quad = \dfrac{\sqrt[3]{9x^2}}{3x}$

65. $\dfrac{\sqrt{12}}{\sqrt{3}} = \sqrt{\dfrac{12}{3}} = \sqrt{4} = 2$

67. $\dfrac{\sqrt[5]{2}}{\sqrt[4]{a^2 b}} = \dfrac{\sqrt[5]{2}}{a^{2/4}b^{1/4}} = \dfrac{\sqrt[5]{2} \cdot a^{1/2}b^{3/4}}{a^{1/2}b^{1/4} \cdot a^{1/2}b^{3/4}}$

$\quad = \dfrac{2^{1/5}a^{1/2}b^{3/4}}{ab} = \dfrac{2^{4/20}a^{10/20}b^{15/20}}{ab}$

$\quad = \dfrac{(2^4 a^{10}b^{15})^{1/20}}{ab} = \dfrac{\sqrt[20]{16a^{10}b^{15}}}{ab}$

69. $2x^2 y^{-3}x^4 = 2x^6 y^{-3} = \dfrac{2x^6}{y^3}$

71. $\dfrac{\sqrt{243}}{\sqrt{3}} = \sqrt{\dfrac{243}{3}} = \sqrt{81} = 9$

73. $\dfrac{2^0}{(2^{-2}x^{1/2}y^{-2})^3} = \dfrac{1}{2^{-6}x^{3/2}y^{-6}} = \dfrac{2^6 y^6}{x^{3/2}}$

$\quad = \dfrac{64y^6 \cdot x^{1/2}}{x^{3/2} \cdot x^{1/2}} = \dfrac{64y^6 x^{1/2}}{x^2}$

75. $\sqrt[3]{x^2 yz^3}\sqrt[3]{xy^2} = \sqrt[3]{(x^2 yz^3)(xy^2)} = \sqrt[3]{x^3 y^3 z^3}$

$\quad = xyz$

77.
$$3^2(32)^{-2/5} = 3^2(2^5)^{-2/5}$$
$$= 3^2(2^{-2})$$
$$= 3^2 \cdot \frac{1}{2^2}$$
$$= \frac{9}{4}$$

79. $(2x^{-1}y^2)^2 = 2^2 x^{-2} y^4 = \dfrac{4y^4}{x^2}$

81.
$$\sqrt{x}\sqrt{x^2 y^3}\sqrt{xy^2} = x^{1/2}(x^2 y^3)^{1/2}(xy^2)^{1/2}$$
$$= x^{1/2}(xy^{3/2})(x^{1/2}y) = x^2 y^{5/2}$$

83. $\dfrac{(ab^{-3}c)^8}{(a^{-1}c^2)^{-3}} = \dfrac{a^8 b^{-24} c^8}{a^3 c^{-6}} = \dfrac{a^5 c^{14}}{b^{24}}$

85.
$$\frac{(x^2)^3}{x^4} \div \left[\frac{x^3}{(x^3)^2}\right]^2 = \frac{x^6}{x^4} \div \frac{(x^3)^2}{(x^6)^2}$$
$$= x^2 \div \frac{x^6}{x^{12}} = x^2 \div x^{6-12} = x^2 \div x^{-6}$$
$$= x^2 \div \frac{1}{x^6} = x^2 \cdot x^6 = x^8$$

87. $-\dfrac{8s^{-2}}{2s^3} = -\dfrac{4}{s^3 s^2} = -\dfrac{4}{s^5}$

89.
$$(3x^3 y^2 \div 2y^2 z^{-3})^4 = \left(\frac{3x^3 y^2}{2y^2 z^{-3}}\right)^4$$
$$= \left(\frac{3x^3 z^3}{2}\right)^4$$
$$= \frac{(3x^3 z^3)^4}{(2)^4}$$
$$= \frac{3^4 x^{12} z^{12}}{2^4}$$
$$= \frac{81x^{12} z^{12}}{16}$$

Problems 0.4

1. $8x - 4y + 2 + 3x + 2y - 5 = 11x - 2y - 3$

3. $8t^2 - 6s^2 + 4s^2 - 2t^2 + 6 = 6t^2 - 2s^2 + 6$

5. $\sqrt{a} + 2\sqrt{3b} - \sqrt{c} + 3\sqrt{3b}$
$$= \sqrt{a} + 5\sqrt{3b} - \sqrt{c}$$

7. $6x^2 - 10xy + \sqrt{2} - 2z + xy - 4$
$$= 6x^2 - 9xy - 2z + \sqrt{2} - 4$$

9. $\sqrt{x} + \sqrt{2y} - \sqrt{x} - \sqrt{3z} = \sqrt{2y} - \sqrt{3z}$

11. $9x + 9y - 21 - 24x + 6y - 6 = -15x + 15y - 27$

13. $5x^2 - 5y^2 + xy - 3x^2 - 8xy - 28y^2$
$$= 2x^2 - 33y^2 - 7xy$$

15. $2\{3[3x^2 + 6 - 2x^2 + 10]\} = 2\{3[x^2 + 16]\}$
$$= 2\{3x^2 + 48\} = 6x^2 + 96$$

17. $-5(8x^3 + 8x^2 - 2(x^2 - 5 + 2x))$
$$= -5(8x^3 + 8x^2 - 2x^2 + 10 - 4x)$$
$$= -5(8x^3 + 6x^2 - 4x + 10)$$
$$= -40x^3 - 30x^2 + 20x - 50$$

19. $x^2 + (4+5)x + 4(5) = x^2 + 9x + 20$

21. $(w+2)(w-5) = w^2 + (-5+2)x + 2(-5)$
$$= w^2 - 3w - 10$$

23. $(2x)(5x) + [(2)(2) + (3)(5)]x + 3(2)$
$$= 10x^2 + 19x + 6$$

25. $X^2 + 2(X)(2Y) + (2Y)^2 = X^2 + 4XY + 4Y^2$

27. $x^2 - 2(5)x + 5^2 = x^2 - 10x + 25$

29. $\left(\sqrt{3x}\right)^2 + 2\left(\sqrt{3x}\right)(5) + (5)^2$

$\quad = 3x + 10\sqrt{3x} + 25$

31. $(2s)^2 - 1^2 = 4s^2 - 1$

33. $x^2(x+4) - 3(x+4)$

$\quad = x^3 + 4x^2 - 3x - 12$

35. $x^2(3x^2 + 2x - 1) - 4(3x^2 + 2x - 1)$

$\quad = 3x^4 + 2x^3 - x^2 - 12x^2 - 8x + 4$

$\quad = 3x^4 + 2x^3 - 13x^2 - 8x + 4$

37. $x\{2(x^2 - 2x - 35) + 4[2x^2 - 12x]\}$

$\quad = x\{2x^2 - 4x - 70 + 8x^2 - 48x\}$

$\quad = x\{10x^2 - 52x - 70\}$

$\quad = 10x^3 - 52x^2 - 70x$

39. $x(3x + 2y - 4) + y(3x + 2y - 4)$

$\quad\quad\quad + 2(3x + 2y - 4)$

$\quad = 3x^2 + 2xy - 4x + 3xy + 2y^2$

$\quad\quad - 4y + 6x + 4y - 8$

$\quad = 3x^2 + 2y^2 + 5xy + 2x - 8$

41. $(2a)^3 + 3(2a)^2(3) + 3(2a)(3)^2 + (3)^3$

$\quad = 8a^3 + 36a^2 + 54a + 27$

43. $(2x)^3 - 3(2x)^2(3) + 3(2x)(3)^2 - 3^3$

$\quad = 8x^3 - 36x^2 + 54x - 27$

45. $\dfrac{z^2}{z} - \dfrac{18z}{z} = z - 18$

47. $\dfrac{6x^5}{2x^2} + \dfrac{4x^3}{2x^2} - \dfrac{1}{2x^2} = 3x^3 + 2x - \dfrac{1}{2x^2}$

49. $x+5 \overline{\smash{)}\,x^2 + 5x - 3}$ quotient x

$\quad\quad\quad \underline{x^2 + 5x}$

$\quad\quad\quad\quad\quad\quad -3$

Answer: $x + \dfrac{-3}{x+5}$

51. $x+2 \overline{\smash{)}\,3x^3 - 2x^2 + x - 3}$ quotient $3x^2 - 8x + 17$

$\quad\quad\quad \underline{3x^3 + 6x^2}$

$\quad\quad\quad\quad\quad -8x^2 + x$

$\quad\quad\quad\quad\quad \underline{-8x^2 - 16x}$

$\quad\quad\quad\quad\quad\quad\quad 17x - 3$

$\quad\quad\quad\quad\quad\quad\quad \underline{17x + 34}$

$\quad\quad\quad\quad\quad\quad\quad\quad -37$

Answer: $3x^2 - 8x + 17 + \dfrac{-37}{x+2}$

53. $x+2 \overline{\smash{)}\,x^3 + 0x^2 + 0x + 0}$ quotient $x^2 - 2x + 4$

$\quad\quad\quad \underline{x^3 + 2x^2}$

$\quad\quad\quad\quad -2x^2 + 0$

$\quad\quad\quad\quad \underline{-2x^2 - 4x}$

$\quad\quad\quad\quad\quad\quad 4x + 0$

$\quad\quad\quad\quad\quad\quad \underline{4x + 8}$

$\quad\quad\quad\quad\quad\quad\quad -8$

Answer: $x^2 - 2x + 4 - \dfrac{8}{x+2}$

55. $3x+2 \overline{\smash{)}\,3x^2 - 4x + 3}$ quotient $x - 2$

$\quad\quad\quad \underline{3x^2 + 2x}$

$\quad\quad\quad\quad -6x + 3$

$\quad\quad\quad\quad \underline{-6x - 4}$

$\quad\quad\quad\quad\quad\quad 7$

Answer: $x - 2 + \dfrac{7}{3x+2}$

Problems 0.5

1. $2(ax + b)$

3. $5x(2y + z)$

5. $4bc(2a^3 - 3ab^2 d + b^3 cd^2)$

7. $z^2 - 7^2 = (z + 7)(z - 7)$

9. $(p + 3)(p + 1)$

11. $(4x)^2 - 3^2 = (4x + 3)(4x - 3)$

13. $(a + 7)(a + 5)$

15. $x^2 + 2(3)(x) + 3^2 = (x + 3)^2$

17. $5(x^2 + 5x + 6)$
$= 5(x + 3)(x + 2)$

19. $3(x^2 - 1^2) = 3(x + 1)(x - 1)$

21. $6y^2 + 13y + 2 = (6y + 1)(y + 2)$

23. $2s(6s^2 + 5s - 4) = 2s(3s + 4)(2s - 1)$

25. $u^{3/5} v(u^2 - 4v^2) = u^{3/5} v(u + 2v)(u - 2v)$

27. $2x(x^2 + x - 6) = 2x(x + 3)(x - 2)$

29. $[2(2x + 1)]^2 = 2^2 (2x + 1)^2$
$= 4(2x + 1)^2$

31. $x(x^2 y^2 - 14xy + 49)$
$= x[(xy)^2 - 2(xy)(7) + 7^2]$
$= x(xy - 7)^2$

33. $x(x^2 - 4) + 2(4 - x^2)$
$= x(x^2 - 4) - 2(x^2 - 4)$
$= (x^2 - 4)(x - 2)$
$= (x + 2)(x - 2)(x - 2)$
$= (x + 2)(x - 2)^2$

35. $y^2(y^2 + 8y + 16) - (y^2 + 8y + 16)$
$= (y^2 + 8y + 16)(y^2 - 1)$
$= (y + 4)^2 (y + 1)(y - 1)$

37. $b^3 + 4^3 = (b + 4)(b^2 - 4(b) + 4^2)$
$= (b + 4)(b^2 - 4b + 16)$

39. $(x^3)^2 - 1^2 = (x^3 + 1)(x^3 - 1)$
$= (x + 1)(x^2 - x + 1)(x - 1)(x^2 + x + 1)$

41. $(x + 3)^2 (x - 1)[(x + 3) + (x - 1)]$
$= (x + 3)^2 (x - 1)[2x + 2]$
$= (x + 3)^2 (x - 1)[2(x + 1)]$
$= 2(x + 3)^2 (x - 1)(x + 1)$

43. $[P(1 + r)] + [P(1 + r)]r = [P(1 + r)](1 + r)$
$= P(1 + r)^2$

45. $(x^2)^2 - 4^2 = (x^2 + 4)(x^2 - 4)$
$= (x^2 + 4)(x + 2)(x - 2)$

47. $(y^4)^2 - 1^2 = (y^4 + 1)(y^4 - 1)$
$= (y^4 + 1)(y^2 + 1)(y^2 - 1)$
$= (y^4 + 1)(y^2 + 1)(y + 1)(y - 1)$

49. $(X^2 + 5)(X^2 - 1) = (X^2 + 5)(X + 1)(X - 1)$

51. $y(x^4 - 2x^2 + 1) = y(x^2 - 1)^2$
$= y[(x + 1)(x - 1)]^2$
$= y(x + 1)^2 (x - 1)^2$

Problems 0.6

1. $\dfrac{a^2-9}{a^2-3a} = \dfrac{(a-3)(a+3)}{a(a-3)} = \dfrac{a+3}{a}$

3. $\dfrac{x^2-9x+20}{x^2+x-20} = \dfrac{(x-5)(x-4)}{(x+5)(x-4)} = \dfrac{x-5}{x+5}$

5. $\dfrac{6x^2+x-2}{2x^2+3x-2} = \dfrac{(3x+2)(2x-1)}{(x+2)(2x-1)} = \dfrac{3x+2}{x+2}$

7. $\dfrac{y^2(-1)}{(y-3)(y+2)} = -\dfrac{y^2}{(y-3)(y+2)}$

9. $\dfrac{(ax-b)(c-x)}{(x-c)(ax+b)} = \dfrac{(ax-b)(-1)(x-c)}{(x-c)(ax+b)}$

$\qquad = \dfrac{(ax-b)(-1)}{ax+b}$

$\qquad = \dfrac{b-ax}{ax+b}$

11. $\dfrac{2(x-1)}{(x-4)(x+2)} \cdot \dfrac{(x+4)(x+1)}{(x+1)(x-1)}$

$\qquad = \dfrac{2(x-1)(x+4)(x+1)}{(x-4)(x+2)(x+1)(x-1)}$

$\qquad = \dfrac{2(x+4)}{(x-4)(x+2)}$

13. $\dfrac{X^2}{8} \cdot \dfrac{4}{X} = \dfrac{4X^2}{8X} = \dfrac{X}{2}$

15. $\dfrac{2m}{n^2} \cdot \dfrac{n^3}{6m} = \dfrac{2mn^3}{6mn^2} = \dfrac{n}{3}$

17. $\dfrac{4x}{3} \div 2x = \dfrac{4x}{3} \cdot \dfrac{1}{2x} = \dfrac{4x}{6x} = \dfrac{2}{3}$

19. $\dfrac{-9x^3}{1} \cdot \dfrac{3}{x} = \dfrac{-27x^3}{x} = -27x^2$

21. $\dfrac{x-3}{1} \cdot \dfrac{x-4}{(x-3)(x-4)} = \dfrac{x-3}{1} \cdot \dfrac{1}{x-3}$

$\qquad = \dfrac{x-3}{x-3}$

$\qquad = 1$

23. $\dfrac{10x^3}{(x+1)(x-1)} \cdot \dfrac{x+1}{5x} = \dfrac{10x^3(x+1)}{5x(x+1)(x-1)} = \dfrac{2x^2}{x-1}$

25. $\dfrac{(x+2)(x+5)}{(x+5)(x+1)} \cdot \dfrac{(x-4)(x+1)}{(x-4)(x+2)}$

$\qquad = \dfrac{x+2}{x+1} \cdot \dfrac{x+1}{x+2}$

$\qquad = \dfrac{(x+2)(x+1)}{(x+1)(x+2)}$

$\qquad = 1$

27. $\dfrac{(2x+3)(2x-3)}{(x+4)(x-1)} \cdot \dfrac{(1+x)(1-x)}{2x-3}$

$\qquad = \dfrac{(2x+3)(2x-3)(1+x)(1-x)}{(x+4)(x-1)(2x-3)}$

$\qquad = \dfrac{(2x+3)(1+x)(-1)(x-1)}{(x+4)(x-1)}$

$\qquad = -\dfrac{(2x+3)(1+x)}{x+4}$

29. $\dfrac{x^2+5x+6}{x+3} = \dfrac{(x+3)(x+2)}{x+3} = x+2$

31. LCD $= 3t$

$\qquad \dfrac{2}{t} + \dfrac{1}{3t} = \dfrac{6}{3t} + \dfrac{1}{3t} = \dfrac{6+1}{3t} = \dfrac{7}{3t}$

33. $\text{LCD} = x^3 - 1$

$$1 - \frac{x^3}{x^3 - 1} = \frac{x^3 - 1}{x^3 - 1} - \frac{x^3}{x^3 - 1}$$
$$= \frac{x^3 - 1 - x^3}{x^3 - 1}$$
$$= \frac{-1}{x^3 - 1}$$
$$= \frac{1}{1 - x^3}$$

35. $\text{LCD} = (2x - 1)(x + 3)$

$$\frac{4}{2x - 1} + \frac{x}{x + 3}$$
$$= \frac{4(x + 3)}{(2x - 1)(x + 3)} + \frac{x(2x - 1)}{(x + 3)(2x - 1)}$$
$$= \frac{4(x + 3) + x(2x - 1)}{(2x - 1)(x + 3)} = \frac{2x^2 + 3x + 12}{(2x - 1)(x + 3)}$$

37. $\text{LCD} = (x - 3)(x + 1)(x + 3)$

$$\frac{1}{(x - 3)(x + 1)} + \frac{1}{(x + 3)(x - 3)}$$
$$= \frac{x + 3}{(x - 3)(x + 1)(x + 3)} + \frac{x + 1}{(x - 3)(x + 1)(x + 3)}$$
$$= \frac{(x + 3) + (x + 1)}{(x - 3)(x + 1)(x + 3)}$$
$$= \frac{2x + 4}{(x - 3)(x + 1)(x + 3)}$$
$$= \frac{2(x + 2)}{(x - 3)(x + 1)(x + 3)}$$

39. $\text{LCD} = (x - 1)(x + 5)$

$$\frac{4}{x - 1} - 3 + \frac{-3x^2}{-(x - 1)(x + 5)}$$
$$= \frac{4(x + 5)}{(x - 1)(x + 5)} - \frac{3(x - 1)(x + 5)}{(x - 1)(x + 5)}$$
$$\quad + \frac{3x^2}{(x - 1)(x + 5)}$$
$$= \frac{4x + 20 - 3(x^2 + 4x - 5) + 3x^2}{(x - 1)(x + 5)}$$
$$= \frac{35 - 8x}{(x - 1)(x + 5)}$$

41. $\left(1 + \frac{1}{x}\right)^2 = \left(\frac{x}{x} + \frac{1}{x}\right)^2 = \left(\frac{x + 1}{x}\right)^2$
$$= \frac{x^2 + 2x + 1}{x^2}$$

43. $\left(\frac{1}{x} - y\right)^{-1} = \left(\frac{1}{x} - \frac{xy}{x}\right)^{-1}$
$$= \left(\frac{1 - xy}{x}\right)^{-1}$$
$$= \frac{x}{1 - xy}$$

45. Multiplying the numerator and denominator of the given fraction by x gives $\frac{7x + 1}{5x}$.

47. Multiplying numerator and denominator by $2x(x + 2)$ gives
$$\frac{3(2x)(x + 2) - 1(x + 2)}{x(2x)(x + 2) + x(2x)} = \frac{(x + 2)[3(2x) - 1]}{2x^2[(x + 2) + 1]}$$
$$= \frac{(x + 2)(6x - 1)}{2x^2(x + 3)}.$$

49. $\text{LCD} = \sqrt[3]{x+h} \cdot \sqrt[3]{x}$

$$\frac{3}{\sqrt[3]{x+h}} - \frac{3}{\sqrt[3]{x}} = \frac{3\sqrt[3]{x}}{\sqrt[3]{x+h}\sqrt[3]{x}} - \frac{3\sqrt[3]{x+h}}{\sqrt[3]{x+h}\sqrt[3]{x}}$$

$$= \frac{3\left(\sqrt[3]{x} - \sqrt[3]{x+h}\right)}{\sqrt[3]{x+h}\sqrt[3]{x}}$$

51. $\dfrac{1}{2+\sqrt{3}} \cdot \dfrac{2-\sqrt{3}}{2-\sqrt{3}} = \dfrac{2-\sqrt{3}}{4-3} = 2-\sqrt{3}$

53. $\dfrac{\sqrt{2}}{\sqrt{3}-\sqrt{6}} \cdot \dfrac{\sqrt{3}+\sqrt{6}}{\sqrt{3}+\sqrt{6}}$

$$= \frac{\sqrt{2}\left(\sqrt{3}+\sqrt{6}\right)}{3-6} = \frac{\sqrt{6}+\sqrt{12}}{-3} = -\frac{\sqrt{6}+2\sqrt{3}}{3}$$

55. $\dfrac{2\sqrt{2}}{\sqrt{2}-\sqrt{3}} \cdot \dfrac{\sqrt{2}+\sqrt{3}}{\sqrt{2}+\sqrt{3}} = \dfrac{2\sqrt{2}\left(\sqrt{2}+\sqrt{3}\right)}{2-3}$

$$= \frac{4+2\sqrt{6}}{-1} = -4-2\sqrt{6}$$

57. $\dfrac{3}{t+\sqrt{7}} \cdot \dfrac{t-\sqrt{7}}{t-\sqrt{7}} = \dfrac{3t-3\sqrt{7}}{t^2-7}$

59. $\dfrac{5\left(2-\sqrt{3}\right)}{\left(2+\sqrt{3}\right)\left(2-\sqrt{3}\right)} - \dfrac{4\left(1+\sqrt{2}\right)}{\left(1-\sqrt{2}\right)\left(1+\sqrt{2}\right)}$

$$= \frac{5\left(2-\sqrt{3}\right)}{4-3} - \frac{4\left(1+\sqrt{2}\right)}{1-2}$$

$$= \frac{5\left(2-\sqrt{3}\right)}{1} - \frac{4\left(1+\sqrt{2}\right)}{-1}$$

$$= 5\left(2-\sqrt{3}\right) + 4\left(1+\sqrt{2}\right) = 4\sqrt{2} - 5\sqrt{3} + 14$$

Problems 0.7

1. $9x - x^2 = 0$

Set $x = 1$:

$9(1) - (1)^2 \overset{?}{=} 0$

$9 - 1 \overset{?}{=} 0$

$8 \neq 0$

Set $x = 0$:

$9(0) - (0)^2 \overset{?}{=} 0$

$0 - 0 \overset{?}{=} 0$

$0 = 0$

Thus, 0 satisfies the equation, but 1 does not.

3. $z + 3(z - 4) = 5; \dfrac{17}{4}, 4$

Set $z = \dfrac{17}{4}$:

$\dfrac{17}{4} + 3\left(\dfrac{17}{4} - 4\right) \overset{?}{=} 5$

$\dfrac{17}{4} + \dfrac{51}{4} - 12 \overset{?}{=} 5$

$5 = 5$

Set $z = 4$:

$4 + 3(4 - 4) \overset{?}{=} 5$

$4 + 0 \overset{?}{=} 5$

$4 \neq 5$

Thus, $\dfrac{17}{4}$ satisfies the equation, but 4 does not.

5. $x(6 + x) - 2(x + 1) - 5x = 4$

Set $x = -2$:

$(-2)(6 - 2) - 2(-2 + 1) - 5(-2) \overset{?}{=} 4$

$-2(4) - 2(-1) + 10 \overset{?}{=} 4$

$-8 + 2 + 10 \overset{?}{=} 4$

$4 = 4$

Set $x = 0$:

$0(6) - 2(1) - 5(0) \overset{?}{=} 4$

$-2 \neq 4$

Thus, -2 satisfies the equation, but 0 does not.

7. Adding 5 to both sides; equivalence guaranteed

9. Raising both sides to the third power; equivalence not guaranteed.

11. Dividing both sides by x; equivalence not guaranteed

13. Multiplying both sides by $x - 1$; equivalence not guaranteed

15. Multiplying both sides by $\dfrac{2x-3}{2x}$;

equivalence not guaranteed

17. $4x = 10$

$$x = \frac{10}{4} = \frac{5}{2}$$

19. $3y = 0$

$$y = \frac{0}{3} = 0$$

21. $-8x = 12 - 20$

$-8x = -8$

$$x = \frac{-8}{-8} = 1$$

23. $5x - 3 = 9$

$5x = 12$

$$x = \frac{12}{5}$$

25. $7x + 7 = 2(x + 1)$

$7x + 7 = 2x + 2$

$5x + 7 = 2$

$5x = -5$

$$x = \frac{-5}{5} = -1$$

27. $5(p-7) - 2(3p-4) = 3p$

$5p - 35 - 6p + 8 = 3p$

$-p - 27 = 3p$

$-27 = 4p$

$$p = -\frac{27}{4}$$

29. $\dfrac{x}{5} = 2x - 6$

$x = 5(2x - 6)$

$x = 10x - 30$

$30 = 9x$

$$x = \frac{30}{9} = \frac{10}{3}$$

31. $7 + \dfrac{4x}{9} = \dfrac{x}{2}$

Multiplying both sides by $9 \cdot 2$ gives

$9 \cdot 2 \cdot 7 + 2(4x) = 9(x)$

$126 + 8x = 9x$

$x = 126$

33. $r = \dfrac{4}{3}r - 5$

Multiplying both sides by 3 gives

$3r = 4r - 15$

$-r = -15$

$r = 15$

35. $3x + \dfrac{x}{5} - 5 = \dfrac{1}{5} + 5x$

Multiplying both sides by 5 gives

$15x + x - 25 = 1 + 25x$

$16x - 25 = 1 + 25x$

$-9x = 26$

$$x = -\frac{26}{9}$$

37. $\dfrac{2y-3}{4} = \dfrac{6y+7}{3}$

Multiplying both sides by 12 gives

$3(2y - 3) = 4(6y + 7)$

$6y - 9 = 24y + 28$

$-18y = 37$

$$y = -\frac{37}{18}$$

39. $w - \dfrac{w}{2} + \dfrac{w}{6} - \dfrac{w}{24} = 120$

Multiplying both sides by 24 gives

$24w - 12w + 4w - w = 2880$

$15w = 2880$

$$w = \frac{2880}{15} = 192$$

41. $\dfrac{x+2}{3} - \dfrac{2-x}{6} = x - 2$

Multiplying both sides by 6 gives
$2(x + 2) - (2 - x) = 6(x - 2)$
$2x + 4 - 2 + x = 6x - 12$
$3x + 2 = 6x - 12$
$2 = 3x - 12$
$14 = 3x$
$x = \dfrac{14}{3}$

43. $\dfrac{9}{5}(3 - x) = \dfrac{3}{4}(x - 3)$

Multiplying both sides by 20 gives
$36(3 - x) = 15(x - 3)$
$108 - 36x = 15x - 45$
$153 = 51x$
$x = 3$

45. $\dfrac{4}{3}(5x - 2) = 7[x - (5x - 2)]$

$4(5x - 2) = 21(x - 5x + 2)$
$20x - 8 = -84x + 42$
$104x = 50$
$x = \dfrac{50}{104} = \dfrac{25}{52}$

47. $\dfrac{5}{x} = 25$

Multiplying both sides by x gives
$5 = 25x$
$x = \dfrac{5}{25}$
$x = \dfrac{1}{5}$

49. Multiplying both sides by $3 - x$ gives $7 = 0$, which is false. Thus there is no solution, so the solution set is $\varnothing$.

51. $\dfrac{3}{5 - 2x} = \dfrac{7}{2}$

$3(2) = 7(5 - 2x)$
$6 = 35 - 14x$
$14x = 29$
$x = \dfrac{29}{14}$

53. $\dfrac{q}{5q - 4} = \dfrac{1}{3}$

$3q = 5q - 4$
$-2q = -4$
$q = 2$

55. $\dfrac{1}{p - 1} = \dfrac{2}{p - 2}$

$p - 2 = 2(p - 1)$
$p - 2 = 2p - 2$
$p = 0$

57. $\dfrac{1}{x} + \dfrac{1}{7} = \dfrac{3}{7}$

$\dfrac{1}{x} = \dfrac{3}{7} - \dfrac{1}{7}$
$\dfrac{1}{x} = \dfrac{2}{7}$
$x = \dfrac{7}{2}$

59. $\dfrac{3x - 2}{2x + 3} = \dfrac{3x - 1}{2x + 1}$

$(3x - 2)(2x + 1) = (3x - 1)(2x + 3)$
$6x^2 - x - 2 = 6x^2 + 7x - 3$
$1 = 8x$
$x = \dfrac{1}{8}$

61. $\dfrac{y-6}{y} - \dfrac{6}{y} = \dfrac{y+6}{y-6}$

Multiplying both sides by $y(y-6)$ gives

$(y-6)^2 - 6(y-6) = y(y+6)$

$y^2 - 12y + 36 - 6y + 36 = y^2 + 6y$

$y^2 - 18y + 72 = y^2 + 6y$

$72 = 24y$

$y = 3$

63. $\dfrac{-5}{2x-3} = \dfrac{7}{3-2x} + \dfrac{11}{3x+5}$

Multiplying both sides by $(2x-3)(3x+5)$ gives

$-5(3x+5) = -7(3x+5) + 11(2x-3)$

$-15x - 25 = -21x - 35 + 22x - 33$

$-15x - 25 = x - 68$

$-16x = -43$

$x = \dfrac{43}{16}$

65. $\dfrac{9}{x-3} = \dfrac{3x}{x-3}$

$9 = 3x$

$x = 3$

But the given equation is not defined for $x = 3$, so there is no solution. The solution set is $\varnothing$.

67. $\sqrt{x+5} = 4$

$\left(\sqrt{x+5}\right)^2 = 4^2$

$x + 5 = 16$

$x = 11$

69. $\sqrt{3x-4} - 8 = 0$

$\sqrt{3x-4} = 8$

$\left(\sqrt{3x-4}\right)^2 = (8)^2$

$3x - 4 = 64$

$3x = 68$

$x = \dfrac{68}{3}$

71. $\sqrt{\dfrac{x}{2} + 1} = \dfrac{2}{3}$

$\left(\sqrt{\dfrac{x}{2} + 1}\right)^2 = \left(\dfrac{2}{3}\right)^2$

$\dfrac{x}{2} + 1 = \dfrac{4}{9}$

$\dfrac{x}{2} = -\dfrac{5}{9}$

$x = 2\left(-\dfrac{5}{9}\right) = -\dfrac{10}{9}$

73. $\sqrt{4x-6} = \sqrt{x}$

$\left(\sqrt{4x-6}\right)^2 = \left(\sqrt{x}\right)^2$

$4x - 6 = x$

$3x = 6$

$x = 2$

75. $(x-5)^{3/4} = 27$

$\left[(x-5)^{3/4}\right]^{4/3} = 27^{4/3}$

$x - 5 = 81$

$x = 86$

77. $\sqrt{y} + \sqrt{y+2} = 3$

$\sqrt{y+2} = 3 - \sqrt{y}$

$\left(\sqrt{y+2}\right)^2 = \left(3 - \sqrt{y}\right)^2$

$y + 2 = 9 - 6\sqrt{y} + y$

$6\sqrt{y} = 7$

$\left(6\sqrt{y}\right)^2 = 7^2$

$36y = 49$

$y = \dfrac{49}{36}$

79. $\sqrt{z^2 + 2z} = 3 + z$

$\left(\sqrt{z^2 + 2z}\right)^2 = (3 + z)^2$

$z^2 + 2z = 9 + 6z + z^2$

$-9 = 4z$

$z = -\dfrac{9}{4}$

81. $I = Prt$

$r = \dfrac{I}{Pt}$

83. $p = 8q - 1$

$p + 1 = 8q$

$q = \dfrac{p+1}{8}$

85. $S = P(1 + rt)$

$S = P + Prt$

$S - P = r(Pt)$

$r = \dfrac{S - P}{Pt}$

87. $A = \dfrac{R[1 - (1+i)^{-n}]}{i}$

$R = \dfrac{Ai}{1 - (1+i)^{-n}}$

89. $r = \dfrac{d}{1 - dt}$

$r(1 - dt) = d$

$r - rdt = d$

$-rdt = -r + d$

$t = -\dfrac{d-r}{rd} = \dfrac{r-d}{rd}$

91. $r = \dfrac{2mI}{B(n+1)}$

$r(n+1) = \dfrac{2mI}{B}$

$n + 1 = \dfrac{2mI}{rB}$

$n = \dfrac{2mI}{rB} - 1$

93. $P = 2l + 2w$

$660 = 2l + 2(160)$

$660 = 2l + 320$

$340 = 2l$

$l = \dfrac{340}{2} = 170$

The length of the rectangle is 170 m.

95. $c = x + 0.0825x = 1.0825x$

97. $V = C\left(1 - \dfrac{n}{N}\right)$

$2000 = 3200\left(1 - \dfrac{n}{8}\right)$

$2000 = 3200 - 400n$

$400n = 1200$

$n = 3$

The furniture will have a value of $2000 after 3 years.

99. Bronwyn's weekly salary for working h hours is
$27h + 18$. Steve's weekly salary for working h hours is $35h$.

$\dfrac{1}{5}(27h + 18 + 35h) = 550$

$62h + 18 = 2750$

$62h = 2732$

$h = \dfrac{2732}{62} \approx 44.1$

They must each work 44 hours each week.

101. $y = \dfrac{1.4x}{1 + 0.09x}$

With $y = 10$ the equation is

$10 = \dfrac{1.4x}{1 + 0.09x}$

$10(1 + 0.09x) = 1.4x$

$10 + 0.9x = 1.4x$

$10 = 0.5x$

$x = 20$

The prey density should be 20.

103. $t = \dfrac{d}{r - c}$

$t(r - c) = d$

$tr - tc = d$

$tr - d = tc$

$c = \dfrac{tr - d}{t} = r - \dfrac{d}{t}$

105. $s = \sqrt{30\,fd}$

Set $s = 45$ and (for dry concrete) $f = 0.8$.

$45 = \sqrt{30(0.8)d}$

$45 = \sqrt{24d}$

$(45)^2 = \left(\sqrt{24d}\right)^2$

$2025 = 24d$

$d = \dfrac{2025}{24} = \dfrac{675}{8} = 84\dfrac{3}{8} \approx 84$ ft

107. Let e be Tom's expenses in Nova Scotia before the HST tax. Then the HST tax is $0.15e$ and the total receipts are $e + 0.15e = 1.15e$. The percentage of the total that is HST is

$\dfrac{0.15e}{1.15e} = \dfrac{0.15}{1.15} = \dfrac{15}{115} = \dfrac{3}{23}$ or approximately 13%.

109. $-\dfrac{1}{2}$ is a root.

111. 0 is a root.

Problems 0.8

1. $x^2 - 4x + 4 = 0$

$(x - 2)^2 = 0$

$x - 2 = 0$

$x = 2$

3. $t^2 - 8t + 15 = 0$

$(t - 3)(t - 5) = 0$

$t - 3 = 0$ or $t - 5 = 0$

$t = 3$ or $t = 5$

5. $x^2 - 2x - 3 = 0$

$(x - 3)(x + 1) = 0$

$x - 3 = 0$ or $x + 1 = 0$

$x = 3$ or $x = -1$

7. $u^2 - 13u = -36$

$u^2 - 13u + 36 = 0$

$(u - 4)(u - 9) = 0$

$u - 4 = 0$ or $u - 9 = 0$

$u = 4$ or $u = 9$

9. $x^2 - 4 = 0$

$(x - 2)(x + 2) = 0$

$x - 2 = 0$ or $x + 2 = 0$

$x = 2$ or $x = -2$

11. $t^2 - 5t = 0$

$t(t - 5) = 0$

$t = 0$ or $t - 5 = 0$

$t = 0$ or $t = 5$

13. $4x^2 + 1 = 4x$

$4x^2 - 4x + 1 = 0$

$(2x - 1)^2 = 0$

$2x - 1 = 0$

$x = \dfrac{1}{2}$

15. $v(3v-5) = -2$

$3v^2 - 5v = -2$

$3v^2 - 5v + 2 = 0$

$(3v-2)(v-1) = 0$

$3v - 2 = 0$ or $v - 1 = 0$

$v = \dfrac{2}{3}$ or $v = 1$

17. $-x^2 + 3x + 10 = 0$

$x^2 - 3x - 10 = 0$

$(x-5)(x+2) = 0$

$x - 5 = 0$ or $x + 2 = 0$

$x = 5$ or $x = -2$

19. $2p^2 = 3p$

$2p^2 - 3p = 0$

$p(2p - 3) = 0$

$p = 0$ or $2p - 3 = 0$

$p = 0$ or $p = \dfrac{3}{2}$

21. $x(x+4)(x-1) = 0$

$x = 0$ or $x + 4 = 0$ or $x - 1 = 0$

$x = 0$ or $x = -4$ or $x = 1$

23. $t^3 - 49t = 0$

$t(t^2 - 49) = 0$

$t(t+7)(t-7) = 0$

$t = 0$ or $t + 7 = 0$ or $t - 7 = 0$

$t = 0$ or $t = -7$ or $t = 7$

25. $6x^3 + 5x^2 - 4x = 0$

$x(6x^2 + 5x - 4) = 0$

$x(2x - 1)(3x + 4) = 0$

$x = 0$ or $2x - 1 = 0$ or $3x + 4 = 0$

$x = 0$ or $x = \dfrac{1}{2}$ or $x = -\dfrac{4}{3}$

27. $(x-3)(x^2 - 4) = 0$

$(x-3)(x-2)(x+2) = 0$

$x - 3 = 0$ or $x - 2 = 0$ or $x + 2 = 0$

$x = 3$ or $x = 2$ or $x = -2$

29. $p(p-3)^2 - 4(p-3)^3 = 0$

$(p-3)^2[p - 4(p-3)] = 0$

$(p-3)^2(12 - 3p) = 0$

$3(p-3)^2(4 - p) = 0$

$p - 3 = 0$ or $4 - p = 0$

$p = 3$ or $p = 4$

31. $x^2 + 2x - 24 = 0$

$a = 1, b = 2, c = -24$

$x = \dfrac{-b \pm \sqrt{b^2 - 4ac}}{2a}$

$= \dfrac{-2 \pm \sqrt{4 - 4(1)(-24)}}{2(1)}$

$= \dfrac{-2 \pm \sqrt{100}}{2}$

$= \dfrac{-2 \pm 10}{2}$

$x = \dfrac{-2 + 10}{2} = 4$ or $x = \dfrac{-2 - 10}{2} = -6$

33. $4x^2 - 12x + 9 = 0$

$a = 4, b = -12, c = 9$

$x = \dfrac{-b \pm \sqrt{b^2 - 4ac}}{2a}$

$= \dfrac{-(-12) \pm \sqrt{144 - 4(4)(9)}}{2(4)}$

$= \dfrac{12 \pm \sqrt{0}}{8}$

$= \dfrac{12 \pm 0}{8}$

$= \dfrac{3}{2}$

35. $p^2 - 2p - 7 = 0$

$a = 1, b = -2, c = -7$

$p = \dfrac{-b \pm \sqrt{b^2 - 4ac}}{2a}$

$= \dfrac{-(-2) \pm \sqrt{(-2)^2 - 4(1)(-7)}}{2(1)}$

$= \dfrac{2 \pm \sqrt{32}}{2}$

$= 1 \pm 2\sqrt{2}$

$p = 1 + 2\sqrt{2}$ or $p = 1 - 2\sqrt{2}$

37. $4 - 2n + n^2 = 0$

$n^2 - 2n + 4 = 0$

$a = 1, b = -2, c = 4$

$n = \dfrac{-b \pm \sqrt{b^2 - 4ac}}{2a}$

$= \dfrac{-(-2) \pm \sqrt{4 - 4(1)(4)}}{2(1)}$

$= \dfrac{2 \pm \sqrt{-12}}{2}$

no real roots

39. $4x^2 + 5x - 2 = 0$

$a = 4, b = 5, c = -2$

$x = \dfrac{-b \pm \sqrt{b^2 - 4ac}}{2a}$

$= \dfrac{-5 \pm \sqrt{25 - 4(4)(-2)}}{2(4)}$

$= \dfrac{-5 \pm \sqrt{57}}{8}$

$x = \dfrac{-5 + \sqrt{57}}{8}$ or $x = \dfrac{-5 - \sqrt{57}}{8}$

41. $0.02w^2 - 0.3w = 20$

$0.02w^2 - 0.3w - 20 = 0$

$a = 0.02, b = -0.3, c = -20$

$w = \dfrac{-b \pm \sqrt{b^2 - 4ac}}{2a}$

$= \dfrac{-(-0.3) \pm \sqrt{0.09 - 4(0.02)(-20)}}{2(0.02)}$

$= \dfrac{0.3 \pm \sqrt{1.69}}{0.04}$

$= \dfrac{0.3 \pm 1.3}{0.04}$

$w = \dfrac{0.3 + 1.3}{0.04} = \dfrac{1.6}{0.04} = 40$ or

$w = \dfrac{0.3 - 1.3}{0.04} = \dfrac{-1.0}{0.04} = -25$

43. $2x^2 + 4x = 5$

$2x^2 + 4x - 5 = 0$

$a = 2, b = 4, c = -5$

$x = \dfrac{-b \pm \sqrt{b^2 - 4ac}}{2a}$

$= \dfrac{-4 \pm \sqrt{16 - 4(2)(-5)}}{2(2)}$

$= \dfrac{-4 \pm \sqrt{56}}{4}$

$= \dfrac{-4 \pm 2\sqrt{14}}{4}$

$= \dfrac{-2 \pm \sqrt{14}}{2}$

$x = \dfrac{-2 + \sqrt{14}}{2}$ or $x = \dfrac{-2 - \sqrt{14}}{2}$

45. $(x^2)^2 - 5(x^2) + 6 = 0$

Let $w = x^2$. Then

$w^2 - 5w + 6 = 0$

$(w - 3)(w - 2) = 0$

$w = 3, 2$

Thus $x^2 = 3$ or $x^2 = 2$, so $x = \pm\sqrt{3}, \pm\sqrt{2}$.

47. $3\left(\dfrac{1}{x}\right)^2 - 7\left(\dfrac{1}{x}\right) + 2 = 0$

Let $w = \dfrac{1}{x}$. Then

$3w^2 - 7w + 2 = 0$
$(3w - 1)(w - 2) = 0$
$$w = \dfrac{1}{3},\ 2$$

Thus, $x = 3, \dfrac{1}{2}$.

49. $(x^{-2})^2 - 9(x^{-2}) + 20 = 0$

Let $w = x^{-2}$. Then

$w^2 - 9w + 20 = 0$
$(w - 5)(w - 4) = 0$
$w = 5,\ 4$

Thus, $\dfrac{1}{x^2} = 5$ or $\dfrac{1}{x^2} = 4$, so $x^2 = \dfrac{1}{5}$ or

$x^2 = \dfrac{1}{4}$. $x = \pm\dfrac{\sqrt{5}}{5}, \pm\dfrac{1}{2}$.

51. $(X - 5)^2 + 7(X - 5) + 10 = 0$

Let $w = X - 5$. Then

$w^2 + 7w + 10 = 0$
$(w + 2)(w + 5) = 0$
$w = -2, -5$
Thus, $X - 5 = -2$ or $X - 5 = -5$, so $X = 3, 0$.

53. $\left(\dfrac{1}{x-2}\right)^2 - 12\left(\dfrac{1}{x-2}\right) + 35 = 0$

Let $w = \dfrac{1}{x-2}$, then

$w^2 - 12w + 35 = 0$
$(w - 7)(w - 5) = 0$
$w = 7,\ 5$

Thus, $\dfrac{1}{x-2} = 7$ or $\dfrac{1}{x-2} = 5$.

$x = \dfrac{15}{7}, \dfrac{11}{5}$.

55. $x^2 = \dfrac{x+3}{2}$

$2x^2 = x + 3$
$2x^2 - x - 3 = 0$
$(2x - 3)(x + 1) = 0$

Thus, $x = \dfrac{3}{2}, -1$.

57. $\dfrac{3}{x-4} + \dfrac{x-3}{x} = 2$

Multiplying both sides by the LCD, $x(x - 4)$, gives
$3x + (x - 3)(x - 4) = 2x(x - 4)$
$3x + x^2 - 7x + 12 = 2x^2 - 8x$
$x^2 - 4x + 12 = 2x^2 - 8x$
$0 = x^2 - 4x - 12$
$0 = (x - 6)(x + 2)$
Thus, $x = 6, -2$.

59. $\dfrac{3x+2}{x+1} - \dfrac{2x+1}{2x} = 1$

Multiplying both sides by the LCD, $2x(x + 1)$, gives
$2x(3x + 2) - (2x + 1)(x + 1) = 2x(x + 1)$
$6x^2 + 4x - (2x^2 + 3x + 1) = 2x^2 + 2x$
$4x^2 + x - 1 = 2x^2 + 2x$
$2x^2 - x - 1 = 0$
$(2x + 1)(x - 1) = 0$

Thus, $x = -\dfrac{1}{2}, 1$.

61. $\dfrac{2}{r-2} - \dfrac{r+1}{r+4} = 0$

Multiplying both sides by the LCD, $(r - 2)(r + 4)$, gives
$2(r + 4) - (r - 2)(r + 1) = 0$
$2r + 8 - (r^2 - r - 2) = 0$
$-r^2 + 3r + 10 = 0$
$r^2 - 3r - 10 = 0$
$(r - 5)(r + 2) = 0$
Thus, $r = 5, -2$.

63. $\dfrac{t+1}{t+2}+\dfrac{t+3}{t+4}=\dfrac{t+5}{t^2+6t+8}$

Multiplying both sides by the LCD,
$(t + 2)(t + 4)$, gives

$(t+1)(t+4)+(t+3)(t+2)=t+5$

$t^2+5t+4+t^2+5t+6=t+5$

$2t^2+10t+10=t+5$

$2t^2+9t+5=0$

$a = 2, b = 9, c = 5$

$t=\dfrac{-b\pm\sqrt{b^2-4ac}}{2a}$

$=\dfrac{-9\pm\sqrt{81-4(2)(5)}}{2(2)}$

$=\dfrac{-9\pm\sqrt{41}}{4}$

Thus $t=\dfrac{-9+\sqrt{41}}{4},\dfrac{-9-\sqrt{41}}{4}.$

65. $\dfrac{2}{x^2-1}-\dfrac{1}{x(x-1)}=\dfrac{2}{x^2}$

Multiplying both sides by the LCD,
$x^2(x+1)(x-1),$ gives

$2x^2-x(x+1)=2(x+1)(x-1)$

$2x^2-x^2-x=2x^2-2$

$x^2-x=2x^2-2$

$0=x^2+x-2$

$0=(x+2)(x-1)$

$x = -2$ or $x = 1$

But $x = 1$ does not check. The solution is –2.

67. $\left(\sqrt{2x-3}\right)^2=(x-3)^2$

$2x-3=x^2-6x+9$

$0=x^2-8x+12$

$0=(x-6)(x-2)$

$x = 6$ or $x = 2$

Only $x = 6$ checks.

69. $(q+2)^2=\left(2\sqrt{4q-7}\right)^2$

$q^2+4q+4=16q-28$

$q^2-12q+32=0$

$(q-4)(q-8)=0$

Thus, $q = 4, 8.$

71. $\sqrt{z+3}=\sqrt{3z}+1$

$\left(\sqrt{z+3}\right)^2=\left(\sqrt{3z}+1\right)^2$

$z+3=3z+2\sqrt{3z}+1$

$-2z+2=2\sqrt{3z}$

$-z+1=\sqrt{3z}$

$(-z+1)^2=\left(\sqrt{3z}\right)^2$

$z^2-2z+1=3z$

$z^2-5z+1=0$

$a = 1, b = -5, c = 1$

$z=\dfrac{-b\pm\sqrt{b^2-4ac}}{2a}$

$=\dfrac{-(-5)\pm\sqrt{(-5)^2-4(1)(1)}}{2(1)}$

$=\dfrac{5\pm\sqrt{21}}{2}$

Only $z=\dfrac{5-\sqrt{21}}{2}$ checks.

73. $\sqrt{x}+1=\sqrt{2x+1}$

$\left(\sqrt{x}+1\right)^2=\left(\sqrt{2x+1}\right)^2$

$x+2\sqrt{x}+1=2x+1$

$2\sqrt{x}=x$

$\left(2\sqrt{x}\right)^2=x^2$

$4x=x^2$

$0=x^2-4x$

$0=x(x-4)$

Thus, $x = 0, 4.$

75. $\left(\sqrt{x+3}+1\right)^2 = \left(3\sqrt{x}\right)^2$

$x+3+2\sqrt{x+3}+1 = 9x$

$2\sqrt{x+3} = 8x-4$

$\sqrt{x+3} = 4x-2$

$\left(\sqrt{x+3}\right)^2 = (4x-2)^2$

$x+3 = 16x^2 -16x+4$

$0 = 16x^2 -17x+1$

$0 = (16x-1)(x-1)$

$x = \dfrac{1}{16}$ or $x = 1$

Only $x = 1$ checks.

77. $x = \dfrac{-(-2.7) \pm \sqrt{(-2.7)^2 - 4(0.04)(8.6)}}{2(0.04)}$

≈ 64.15 or 3.35

79. Let l be the length of the picture, then its width is $l - 2$.

$l(l - 2) = 48$

$l^2 - 2l - 48 = 0$

$(l - 8)(l + 6) = 0$

$l - 8 = 0$ or $l + 6 = 0$

$l = 8$ or $l = -6$

Since length cannot be negative, $l = 8$. The width of the picture is

$l - 2 = 8 - 2 = 6$ inches.

The dimensions of the picture are 6 inches by 8 inches.

81. $\overline{M} = \dfrac{Q(Q+10)}{44}$

$44\overline{M} = Q^2 + 10Q$

$0 = Q^2 + 10Q - 44\overline{M}$

From the quadratic formula with $a = 1$, $b = 10$, $c = -44\overline{M}$,

$Q = \dfrac{-10 \pm \sqrt{100 - 4(1)(-44\overline{M})}}{2(1)}$

$= \dfrac{-10 + 2\sqrt{25 + 44\overline{M}}}{2}$

$= -5 \pm \sqrt{25 + 44\overline{M}}$

Thus, $-5 + \sqrt{25 + 44\overline{M}}$ is a root.

83. $\dfrac{A}{A+12}d = \dfrac{A+1}{24}d$.

Dividing both sides by d and then multiplying both sides by $24(A + 12)$ gives

$24A = (A + 12)(A + 1)$

$24A = A^2 + 13A + 12$

$0 = A^2 - 11A + 12$

From the quadratic formula,

$A = \dfrac{11 \pm \sqrt{121 - 48}}{2} = \dfrac{11 \pm \sqrt{73}}{2}$.

$A = \dfrac{11 + \sqrt{73}}{2} \approx 10$ or $A = \dfrac{11 - \sqrt{73}}{2} \approx 1$.

The doses are the same at 1 year and 10 years.

$c = d$ in Cowling's rule when $\dfrac{A+1}{24} = 1$,

which occurs when $A = 23$. Thus, adulthood is achieved at age 23 according to Cowling's rule. $c = d$ in Young's rule when

$\dfrac{A}{A+12} = 1$, which is never true. Thus, adulthood is never reached according to Young's rule.

Young's rule prescribes less than Cowling's for ages less than one year and greater than 10 years. Cowling's rule prescribes less for ages between 1 and 10.

85. a. When the object strikes the ground, h must be 0, so

$$0 = 39.2t - 4.9t^2 = 4.9t(8 - t)$$

$t = 0$ or $t = 8$

The object will strike the ground 8 s after being thrown.

b. Setting $h = 68.2$ gives

$$68.2 = 39.2t - 4.9t^2$$

$$4.9t^2 - 39.2t + 68.2 = 0$$

$$t = \frac{39.2 \pm \sqrt{(-39.2)^2 - 4(4.9)(68.2)}}{2(4.9)}$$

$$\approx \frac{39.2 \pm 14.1}{9.8}$$

$t \approx 5.4$ s or $t \approx 2.6$ s.

87. By a program, roots are 1.5 and 0.75.
Algebraically:

$$8x^2 - 18x + 9 = 0$$

$$(2x - 3)(4x - 3) = 0$$

Thus, $2x - 3 = 0$ or $4x - 3 = 0$.

So $x = \dfrac{3}{2} = 1.5$ or $x = \dfrac{3}{4} = 0.75$.

89. By a program, there are no real roots.

91. $(\pi t - 4)^2 = 4.1t - 3$

$$\pi^2 t^2 - 8\pi t + 16 = 4.1t - 3$$

$$\pi^2 t^2 + (-8\pi - 4.1)t + 19 = 0$$

Roots: 1.999, 0.963

Mathematical Snapshot Chapter 0

1.
```
LinReg
 y=ax+b
 a=7.221004E-4
 b=.0060813684
 r²=.999988045
 r=.9999940225
```

3.
```
QuadReg
 y=ax²+bx+c
 a=-3.226931E-9
 b=1.0165234E-5
 c=-.0055906575
 R²=.9922351962
```

The results agree.

Chapter 1

Problems 1.1

1. Let w be the width and $2w$ be the length of the plot.

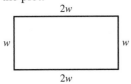

Then area = 800.

$(2w)w = 800$

$2w^2 = 800$

$w^2 = 400$

$w = 20$ ft

Thus the length is 40 ft, so the amount of fencing needed is $2(40) + 2(20) = 120$ ft.

3. Let n = number of ounces in each part. Then we have

$4n + 5n = 145$

$9n = 145$

$n = 16\frac{1}{9}$

Thus there should be $4\left(16\frac{1}{9}\right) = 64\frac{4}{9}$

ounces of A and $5\left(16\frac{1}{9}\right) = 80\frac{5}{9}$ ounces of B.

5. Let n = number of ounces in each part. Then we have

$2n + 1n = 16$

$3n = 16$

$n = \frac{16}{3}$

Thus the turpentine needed is

$(1)n = \frac{16}{3} = 5\frac{1}{3}$ ounces.

7. Let w = width (in meters) of pavement. The remaining plot for flowers has dimensions $8 - 2w$ by $4 - 2w$.

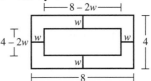

Thus

$(8 - 2w)(4 - 2w) = 12$

$32 - 24w + 4w^2 = 12$

$4w^2 - 24w + 20 = 0$

$w^2 - 6w + 5 = 0$

$(w - 1)(w - 5) = 0$

Hence $w = 1, 5$. But $w = 5$ is impossible since one dimension of the original plot is 4 m. Thus the width of the pavement should be 1 m.

9. Let q = number of tons for \$560,000 profit.

Profit = Total Revenue – Total Cost

$560,000 = 134q - (82q + 120,000)$

$560,000 = 52q - 120,000$

$680,000 = 52q$

$\dfrac{680,000}{52} = q$

$q \approx 13,076.9 \approx 13,077$ tons.

11. Let x = amount at 6% and

$20,000 - x$ = amount at $7\frac{1}{2}\%$.

$x(0.06) + (20,000 - x)(0.075) = 1440$

$-0.015x + 1500 = 1440$

$-0.015x = -60$

$x = 4000$, so $20,000 - x = 16,000$. Thus the investment should be \$4000 at 6% and

\$16,000 at $7\frac{1}{2}\%$.

13. Let p = selling price. Then profit = $0.2p$.
selling price = cost + profit
$p = 3.40 + 0.2p$
$0.8p = 3.40$
$$p = \frac{3.40}{0.8} = \$4.25$$

15. Following the procedure in Example 6 we obtain
$$3,000,000(1+r)^2 = 3,245,000$$
$$(1+r)^2 = \frac{649}{600}$$
$$1+r = \pm\sqrt{\frac{649}{600}}$$
$$r = -1 \pm \sqrt{\frac{649}{600}}$$
$$r \approx -2.04 \text{ or } 0.04$$
We choose $r \approx 0.04 = 4\%$.

17. Let n = number of room applications sent out.
$0.95n = 76$
$$n = \frac{76}{0.95} = 80$$

19. Let s = monthly salary of deputy sheriff.
$0.30s = 200$
$$s = \frac{200}{0.30}$$
$$\text{Yearly salary} = 12s = 12\left(\frac{200}{0.30}\right) = \$8000$$

21. Let q = number of cartridges sold to break even.
total revenue = total cost
$21.95q = 14.92q + 8500$
$7.03q = 8500$
$q \approx 1209.10$
1209 cartridges must be sold to approximately break even.

23. Let v = total annual vision-care expenses (in dollars) covered by program. Then
$35 + 0.80(v - 35) = 100$
$0.80v + 7 = 100$
$0.80v = 93$
$v = \$116.25$

25. Revenue
= (number of units sold)(price per unit)
Thus
$$400 = q\left[\frac{80-q}{4}\right]$$
$$1600 = 80q - q^2$$
$$q^2 - 80q + 1600 = 0$$
$$(q-40)^2 = 0$$
$$q = 40 \text{ units}$$

27. Let q = required number of units. We equate incomes under both proposals.
$2000 + 0.50q = 25,000$
$0.50q = 23,000$
$q = 46,000$ units

29. Let n = number of \$20 increases. Then at the rental charge of $400 + 20n$ dollars per unit, the number of units that can be rented is $50 - 2n$. The total of all monthly rents is $(400 + 20n)(50 - 2n)$, which must equal 20,240.
$$20,240 = (400 + 20n)(50 - 2n)$$
$$20,240 = 20,000 + 200n - 40n^2$$
$$40n^2 - 200n + 240 = 0$$
$$n^2 - 5n + 6 = 0$$
$$(n-2)(n-3) = 0$$
$$n = 2, 3$$
Thus the rent should be either
$400 + 2(\$20) = \$440$ or $400 + 3(\$20)$
= \$460.

31. $10,000 = 800p - 7p^2$

$7p^2 - 800p + 10,000 = 0$

$p = \dfrac{800 \pm \sqrt{640,000 - 280,000}}{14}$

$= \dfrac{800 \pm \sqrt{360,000}}{14} = \dfrac{800 \pm 600}{14}$

For $p > 50$ we choose

$p = \dfrac{800 + 600}{14} = \$100.$

33. To have supply = demand,

$2p - 10 = 200 - 3p$

$5p = 210$

$p = 42$

35. Let w = width (in ft) of enclosed area. Then length of enclosed area is

$300 - w - w = 300 - 2w.$

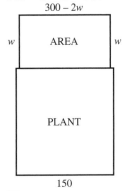

Thus

$w(300 - 2w) = 11,200$

$2w(150 - w) = 11,200$

$w(150 - w) = 5600$

$0 = w^2 - 150w + 5600$

$0 = (w - 80)(w - 70)$

Hence $w = 80, 70$. If $w = 70$, then length is $300 - 2w = 300 - 2(70) = 160$. Since the building has length of only 150 ft, we reject $w = 70$. If $w = 80$, then length is $300 - 2w = 300 - 2(80) = 140$. Thus the dimensions are 80 ft by 140 ft.

37. Original volume = $(10)(5)(2) = 100$ cm^3
Volume cut from bar = $0.28(100) = 28$ cm^3
Volume of new bar = $100 - 28 = 72$ cm^3
Let x = number of centimeters that the length and width are each reduced. Then

$(10 - x)(5 - x)2 = 72$

$(10 - x)(5 - x) = 36$

$x^2 - 15x + 50 = 36$

$x^2 - 15x + 14 = 0$

$(x - 1)(x - 14) = 0$

$x = 1$ or 14

Because of the length and width of the original bar, we reject $x = 14$ and choose $x = 1$. The new bar has length $10 - x = 10 - 1 = 9$ cm and width is $5 - x = 5 - 1 = 4$ cm.

39. Let x = amount of loan. Then the amount actually received is $x - 0.16x$. Hence,

$x - 0.16x = 195,000$

$0.84x = 195,000$

$x \approx 232,142.86$

To the nearest thousand, the loan amount is $232,000. In the general case, the amount received from a loan of L with a compensating balance of $p\%$ is $L - \dfrac{p}{100}L.$

$L - \dfrac{p}{100}L = E$

$\dfrac{100 - p}{100}L = E$

$L = \dfrac{100E}{100 - p}$

41. Let n = number of acres sold. Then $n + 20$ acres were originally purchased at a cost of $\dfrac{7200}{n + 20}$ each. The price of each acre sold was $30 + \left[\dfrac{7200}{n + 20}\right]$. Since the revenue from selling n acres is $7200 (the original cost of the parcel), we have

$$n\left[30 + \frac{7200}{n+20}\right] = 7200$$

$$n\left[\frac{30n + 600 + 7200}{n+20}\right] = 7200$$

$$n(30n + 600 + 7200) = 7200(n + 20)$$

$$30n^2 + 7800n = 7200n + 144,000$$

$$30n^2 + 600n - 144,000 = 0$$

$$n^2 + 20n - 4800 = 0$$

$$(n + 80)(n - 60) = 0$$

$n = 60$ acres (since $n > 0$), so 60 acres were sold.

43. Let q = number of units of B and
$q + 25$ = number of units of A produced.

Each unit of B costs $\dfrac{1000}{q}$, and each unit of

A costs $\dfrac{1500}{q+25}$. Therefore,

$$\frac{1500}{q+25} = \frac{1000}{q} + 2$$

$$1500q = 1000(q + 25) + 2(q)(q + 25)$$

$$0 = 2q^2 - 450q + 25,000$$

$$0 = q^2 - 225q + 12,500$$

$$0 = (q - 100)(q - 125)$$

$q = 100$ or $q = 125$

If $q = 100$, then $q + 25 = 125$; if $q = 125$,
$q + 25 = 150$. Thus the company produces
either 125 units of A and 100 units of B, or
150 units of A and 125 units of B.

Principles in Practice 1.2

1. $200 + 0.8S \geq 4500$
$0.8S \geq 4300$
$S \geq 5375$
He must sell at least 5375 products per
month.

Problems 1.2

1. $3x > 12$

$$x > \frac{12}{3}$$

$$x > 4$$

$(4, \infty)$

3. $5x - 11 \leq 9$

$$5x \leq 20$$

$$x \leq 4$$

$(-\infty, 4]$

5. $-4x \geq 2$

$$x \leq \frac{2}{-4}$$

$$x \leq -\frac{1}{2}$$

$$\left(-\infty, -\frac{1}{2}\right]$$

7. $5 - 7s > 3$
$-7s > -2$

$$s < \frac{2}{7}$$

$$\left(-\infty, \frac{2}{7}\right)$$

9. $3 < 2y + 3$
$0 < 2y$
$0 < y$
$y > 0$
$(0, \infty)$

11. $x + 5 \le 3 + 2x$
$-x \le -2$
$x \ge 2$
$[2, \infty)$

13. $3(2 - 3x) > 4(1 - 4x)$
$6 - 9x > 4 - 16x$
$7x > -2$
$x > -\dfrac{2}{7}$
$\left(-\dfrac{2}{7}, \infty\right)$

15. $2(4x - 2) > 4(2x + 1)$
$8x - 4 > 8x + 4$
$-4 > 4$, which is false for all x.
Thus the solution set is $\varnothing$.

17. $x + 2 < \sqrt{3} - x$
$2x < \sqrt{3} - 2$
$x < \dfrac{\sqrt{3} - 2}{2}$
$\left(-\infty, \dfrac{\sqrt{3} - 2}{2}\right)$

19. $\dfrac{5}{6}x < 40$
$5x < 240$
$x < 48$
$(-\infty, 48)$

21. $\dfrac{9y + 1}{4} \le 2y - 1$
$9y + 1 \le 8y - 4$
$y \le -5$
$(-\infty, -5]$

23. $-3x + 1 \le -3(x - 2) + 1$
$-3x + 1 \le -3x + 7$
$1 \le 7$, which is true for all x. The solution is
$-\infty < x < \infty$.
$(-\infty, \infty)$

25. $\dfrac{1 - t}{2} < \dfrac{3t - 7}{3}$
$3(1 - t) < 2(3t - 7)$
$3 - 3t < 6t - 14$
$-9t < -17$
$t > \dfrac{17}{9}$
$\left(\dfrac{17}{9}, \infty\right)$

27. $2x + 13 \ge \dfrac{1}{3}x - 7$
$6x + 39 \ge x - 21$
$5x \ge -60$
$x \ge -12$
$(-12, \infty)$

29. $\dfrac{2}{3}r < \dfrac{5}{6}r$
$4r < 5r$
$0 < r$
$r > 0$
$(0, \infty)$

25

31. $\dfrac{y}{2} + \dfrac{y}{3} > y + \dfrac{y}{5}$

$15y + 10y > 30y + 6y$

$25y > 36y$

$0 > 11y$

$0 > y$

$y < 0$

$(-\infty, 0)$

33. $0.1(0.03x + 4) \geq 0.02x + 0.434$

$0.003x + 0.4 \geq 0.02x + 0.434$

$-0.017x \geq 0.034$

$x \leq -2$

$(-\infty, -2]$

35. $12(50) < S < 12(150)$

$600 < S < 1800$

37. The measures of the acute angles of a right triangle sum to 90°. If x is the measure of one acute angle, the other angle has measure $90 - x$.

$x < 3(90 - x) + 10$

$x < 270 - 3x + 10$

$4x < 280$

$x < 70$

The measure of the angle is less than 70°.

Problems 1.3

1. Let q = number of units sold.

Profit > 0

Total revenue − Total cost > 0

$20q - (15q + 600{,}000) > 0$

$5q - 600{,}000 > 0$

$5q > 600{,}000$

$q > 120{,}000$

Thus at least 120,001 units must be sold.

3. Let x = number of miles driven per year.

If the auto is leased, the annual cost is

$12(420) + 0.06x$.

If the auto is purchased, the annual cost is

$4700 + 0.08x$. We want Rental cost $\leq$ Purchase cost.

$12(420) + 0.06x \le 4700 + 0.08x$

$5040 + 0.06x \le 4700 + 0.08x$

$340 \le 0.02x$

$17{,}000 \le x$

The number of miles driven per year must be at least 17,000.

5. Let q be the number of magazines printed. Then the cost of publication is $0.55q$. The number of magazines sold is $0.90q$. The revenue from dealers is $(0.60)(0.90q)$. If fewer than 30,000 magazines are sold, the only revenue is from the sales to dealers, while if more than 30,000 are sold, there are advertising revenues of $0.10(0.60)(0.90q - 30{,}000)$. Thus,

$$\text{Revenue} = \begin{cases} 0.6(0.9)q & \text{if} \quad 0.9q \le 30{,}000 \\ 0.6(0.9)q + 0.1(0.6)(0.9q - 30{,}000) & \text{if} \quad 0.9q > 30{,}000 \end{cases}$$

$$= \begin{cases} 0.54q & q \le 33{,}333 \\ 0.594q - 1800 & q > 33{,}333 \end{cases}$$

$\text{Profit} = \text{Revenue} - \text{Cost}$

$$= \begin{cases} 0.54q - 0.55q & q \le 33{,}333 \\ 0.594q - 1800 - 0.55q & q > 33{,}333 \end{cases}$$

$$= \begin{cases} -0.01q & q \le 33{,}333 \\ 0.044q - 1800 & q > 33{,}333 \end{cases}$$

Clearly, the profit is negative if fewer than 33,334 magazines are sold.

$0.044q - 1800 \ge 0$

$\quad 0.044q \ge 1800$

$\qquad q \ge 40{,}910$

Thus, at least 40,910 magazines must be printed in order to avoid a loss.

7. Let x = amount at $6\frac{3}{4}\%$ and $30{,}000 - x$ = amount at 5%. Then

interest at $6\frac{3}{4}\%$ + interest at 5% $\ge$ interest at $6\frac{1}{2}\%$

$x(0.0675) + (30{,}000 - x)(0.05) \ge (0.065)(30{,}000)$

$0.0175x + 1500 \ge 1950$

$0.0175x \ge 450$

$x \ge 25{,}714.29$

Thus at least \$25,714.29 must be invested at $6\frac{3}{4}\%$.

9. Let q be the number of units sold this month at \$4.00 each. Then $2500 - q$ will be sold at \$4.50 each. Then

Total revenue $\ge 10{,}750$

$4q + 4.5(2500 - q) \ge 10{,}750$

$-0.5q + 11{,}250 \ge 10{,}750$

$500 \ge 0.5q$

$1000 \ge q$

The maximum number of units that can be sold this month is 1000.

11. For $t < 40$, we want
income on hourly basis
  $>$ income on per-job basis
  $9t > 320 + 3(40 - t)$
  $9t > 440 - 3t$
  $12t > 440$
  $t > 36.7$ hr

13. Let $x =$ accounts receivable. Then

Acid test ratio $= \dfrac{450,000 + x}{398,000}$

$1.3 \le \dfrac{450,000 + x}{398,000}$

$517,400 \le 450,000 + x$
$x \ge 67,400$

The company must have at least \$67,400 in accounts receivable.

Principles in Practice 1.4

1. $|w - 22| \le 0.3$

Problems 1.4

1. $|-13| = 13$

3. $|8 - 2| = |6| = 6$

5. $\left|2\left(-\dfrac{7}{2}\right)\right| = |-7| = 7$

7. $|x| < 4$, $-4 < x < 4$

9. Because $2 - \sqrt{5} < 0$,
$\left|2 - \sqrt{5}\right| = -\left(2 - \sqrt{5}\right) = \sqrt{5} - 2$.

11. a. $|x - 7| < 3$

  b. $|x - 2| < 3$

  c. $|x - 7| \le 5$

  d. $|x - 7| = 4$

  e. $|x + 4| < 2$

  f. $|x| < 3$

  g. $|x| > 6$

  h. $|x - 105| < 3$

  i. $|x - 850| < 100$

13. $|p_1 - p_2| \le 9$

15. $|x| = 7$
$x = \pm 7$

17. $\left|\dfrac{x}{5}\right| = 7$
$\dfrac{x}{5} = \pm 7$
$x = \pm 35$

19. $|x - 5| = 8$
$x - 5 = \pm 8$
$x = 5 \pm 8$
$x = 13$ or $x = -3$

21. $|5x - 2| = 0$
$5x - 2 = 0$
$x = \dfrac{2}{5}$

23. $|7 - 4x| = 5$
$7 - 4x = \pm 5$
$-4x = -7 \pm 5$
$-4x = -2$ or -12
$x = \dfrac{1}{2}$ or $x = 3$

25. $|x| < M$

$-M < x < M$

$(-M, M)$

Note that $M > 0$ is required.

27. $\left|\dfrac{x}{4}\right| > 2$

$\dfrac{x}{4} < -2$ or $\dfrac{x}{4} > 2$

$x < -8$ or $x > 8$, so the solution is $(-\infty, -8) \cup (8, \infty)$.

29. $|x + 9| < 5$

$-5 < x + 9 < 5$

$-14 < x < -4$

$(-14, -4)$

31. $\left|x - \dfrac{1}{2}\right| > \dfrac{1}{2}$

$x - \dfrac{1}{2} < -\dfrac{1}{2}$ or $x - \dfrac{1}{2} > \dfrac{1}{2}$

$x < 0$ or $x > 1$

$(-\infty, 0) \cup (1, \infty)$

33. $|5 - 8x| \le 1$

$-1 \le 5 - 8x \le 1$

$-6 \le -8x \le -4$

$\dfrac{3}{4} \ge x \ge \dfrac{1}{2}$, which may be rewritten as

$\dfrac{1}{2} \le x \le \dfrac{3}{4}$.

The solution is $\left[\dfrac{1}{2}, \dfrac{3}{4}\right]$.

35. $\left|\dfrac{3x-8}{2}\right| \ge 4$

$\dfrac{3x-8}{2} \le -4$ or $\dfrac{3x-8}{2} \ge 4$

$3x - 8 \le -8$ or $3x - 8 \ge 8$

$3x \le 0$ or $3x \ge 16$

$x \le 0$ or $x \ge \dfrac{16}{3}$

The solution is $(-\infty, 0] \cup \left[\dfrac{16}{3}, \infty\right)$.

37. $|d - 35.2 \text{ m}| \le 20 \text{ cm or } |d - 35.2| \le 0.20$

39. $|x - \mu| > h\sigma$

Either $x - \mu < -h\sigma$, or $x - \mu > h\sigma$. Thus either $x < \mu - h\sigma$ or $x > \mu + h\sigma$, so the solution is $(-\infty, \mu - h\sigma) \cup (\mu + h\sigma, \infty)$.

Problems 1.5

1. The bounds of summation are 12 and 17; the index of summation is t.

3. $\displaystyle\sum_{i=1}^{7} 6i$

$= 6(1) + 6(2) + 6(3) + 6(4) + 6(5) + 6(6) + 6(7)$

$= 6 + 12 + 18 + 24 + 30 + 36 + 42$

$= 168$

5. $\displaystyle\sum_{k=3}^{9} (10k + 16)$

$= [10(3)+16] + [10(4)+16] + [10(5)+16] + [10(6)+16] + [10(7)+16] + [10(8)+16] + [10(9)+16]$

$= 46 + 56 + 66 + 76 + 86 + 96 + 106$

$= 532$

7. $36 + 37 + 38 + 39 + \cdots + 60 = \displaystyle\sum_{i=36}^{60} i$

9. $5^3 + 5^4 + 5^5 + 5^6 + 5^7 + 5^8 = \displaystyle\sum_{i=3}^{8} 5^i$

11. $2 + 4 + 8 + 16 + 32 + 64 + 128 + 256 = \displaystyle\sum_{i=1}^{8} 2^i$

13. $\displaystyle\sum_{k=1}^{43} 10 = 10 \sum_{k=1}^{43} 1 = 10(43) = 430$

15. $\displaystyle\sum_{k=1}^{n}\left(5\cdot\frac{1}{n}\right) = \left(5\cdot\frac{1}{n}\right)\sum_{k=1}^{n} 1 = \left(5\cdot\frac{1}{n}\right)(n) = 5$

17. $\displaystyle\sum_{k=51}^{100} 10k = 10\sum_{i=1}^{50}(i+50)$

$\displaystyle = 10\sum_{i=1}^{50} i + (10)(50)\sum_{i=1}^{50} 1$

$\displaystyle = 10\cdot\frac{50(51)}{2} + 500(50) = 12,750 + 25,000$

$= 37,750$

19. $\displaystyle\sum_{k=1}^{20}(5k^2 + 3k) = 5\sum_{k=1}^{20} k^2 + 3\sum_{k=1}^{20} k$

$\displaystyle = 5\cdot\frac{20(21)(41)}{6} + 3\frac{20(21)}{2}$

$= 5(2870) + 3(210) = 14,980$

21. $\displaystyle\sum_{k=51}^{100} k^2 = \sum_{i=1}^{50}(i+50)^2$

$\displaystyle = \sum_{i=1}^{50}(i^2 + 100i + 2500)$

$\displaystyle = \sum_{i=1}^{50} i^2 + 100\sum_{i=1}^{50} i + 2500\sum_{i=1}^{50} 1$

$\displaystyle = \frac{50(51)(101)}{6} + 100\frac{50(51)}{2} + 2500(50)$

$= 42,925 + 127,500 + 125,000 = 295,425$

23. $\displaystyle\sum_{k=1}^{10}\left\{\left[4 - \left(\frac{2k}{10}\right)^2\right]\left(\frac{2}{10}\right)\right\}$

$\displaystyle = \frac{1}{5}\sum_{k=1}^{10}\left(4 - \frac{1}{25}k^2\right)$

$\displaystyle = \frac{1}{5}(4)\sum_{k=1}^{10} 1 - \frac{1}{5}\left(\frac{1}{25}\right)\sum_{k=1}^{10} k^2$

$\displaystyle = \frac{4}{5}(10) - \frac{1}{125}\cdot\frac{10(11)(21)}{6} = 8 - \frac{1}{125}\cdot 385$

$\displaystyle = 8 - \frac{77}{25} = \frac{123}{25} = 4\frac{23}{25}$

25. $\displaystyle\sum_{k=1}^{n}\left\{\left[5 - \left(\frac{3}{n}\cdot k\right)^2\right]\frac{3}{n}\right\}$

$\displaystyle = \frac{3}{n}\sum_{k=1}^{n}\left(5 - \frac{9}{n^2}k^2\right)$

$\displaystyle = \frac{3}{n}(5)\sum_{k=1}^{n} 1 - \frac{3}{n}\left(\frac{9}{n^2}\right)\sum_{k=1}^{n} k^2$

$\displaystyle = \frac{15}{n}(n) - \frac{27}{n^3}\cdot\frac{n(n+1)(2n+1)}{6}$

$\displaystyle = 15 - \frac{9(n+1)(2n+1)}{2n^2}$

Chapter 1 Review Problems

1. $5x - 2 \geq 2(x - 7)$
$5x - 2 \geq 2x - 14$
$3x \geq -12$
$x \geq -4$
$[-4, \infty)$

3. $-(5x + 2) < -(2x + 4)$
$-5x - 2 < -2x - 4$
$-3x < -2$
$x > \dfrac{2}{3}$
$\left(\dfrac{2}{3}, \infty\right)$

5. $3p(1-p) > 3(2+p) - 3p^2$

$3p - 3p^2 > 6 + 3p - 3p^2$

$0 > 6$, which is false for all x. The solution set is $\varnothing$.

7. $\dfrac{x+5}{3} - \dfrac{1}{2} \le 2$

$2(x+5) - 3(1) \le 6(2)$

$2x + 10 - 3 \le 12$

$2x \le 5$

$x \le \dfrac{5}{2}$

$\left(-\infty, \dfrac{5}{2} \right]$

9. $\dfrac{1}{4}s - 3 \le \dfrac{1}{8}(3 + 2s)$

$2s - 24 \le 3 + 2s$

$0 \le 27$, which is true for all s. Thus $-\infty < s < \infty$, or $(-\infty, \infty)$.

11. $|3 - 2x| = 7$

$3 - 2x = 7$	or $3 - 2x = -7$
$-2x = 4$	or $-2x = -10$
$x = -2$	or $x = 5$

13. $|2z - 3| < 5$

$-5 < 2z - 3 < 5$

$-2 < 2z < 8$

$-1 < z < 4$

$(-1, 4)$

15. $|3 - 2x| \ge 4$

$3 - 2x \ge 4$	or $3 - 2x \le -4$
$-2x \ge 1$	or $-2x \le -7$
$x \le -\dfrac{1}{2}$	or $x \ge \dfrac{7}{2}$

The solution is $\left(-\infty, -\dfrac{1}{2} \right] \cup \left[\dfrac{7}{2}, \infty \right)$.

17. $\displaystyle\sum_{i=3}^{7} i^3 = \sum_{i=1}^{7} i^3 - \sum_{i=1}^{2} i^3$

$= \dfrac{7^2(8)^2}{4} - \dfrac{2^2(3)^2}{4}$

$= 784 - 9$

$= 775$

This uses Equation (1.9). By Equation (1.8),

$\displaystyle\sum_{i=3}^{7} i^3 = \sum_{i=1}^{5} (i+2)^3.$

19. Let x be the number of issues with a decline, and

$x + 48$ be the number of issues with an increase. Then

$x + (x + 48) = 1132$

$2x = 1084$

$x = 542$

21. Let q units be produced at A and $10,000 - q$ at B.

Cost at A + Cost at B $\le 117,000$

$[5q + 30,000] + [5.50(10,000 - q) + 35,000]$

$\le 117,000$

$-0.5q + 120,000 \le 117,000$

$-0.5q \le -3000$

$q \ge 6000$

Thus at least 6000 units must be produced at plant A.

23. Let c = operating costs

$\dfrac{c}{236,460} < 0.90$

$c < \$212,814$

Mathematical Snapshot Chapter 1

1. Here $m = 120$ and $M = 2\dfrac{1}{2}(60) = 150$. For

LP, $r = 2$, so the first t minutes take up $\dfrac{t}{2}$ of the 120 available minutes. For SP, $r = 1$, so the remaining $150 - t$ minutes take up

$\dfrac{150 - t}{1}$ of the 120 available.

$$\frac{t}{2} + \frac{150 - t}{1} = 120$$
$$t + 300 - 2t = 240$$
$$-t = -60$$
$$t = 60$$

Switch after 1 hour.

3. Use the reasoning in Exercise 1, with M unknown and $m = 120$.

$$\frac{t}{2} + \frac{M - t}{1} = 120$$
$$t + 2M - 2t = 240$$
$$-t = 240 - 2M$$
$$t = 2M - 240$$

The switch should be made after $2M - 240$ minutes.

5.
```
solve(X/12+(1080
-X)/20-74,X,450)
                600
```
$x = 600$

```
solve(X/15+(1590
-X)/24-74,X,450)
                310
■
```
$x = 310$

7. The first t minutes use $\dfrac{t}{R}$ of the m available minutes, the remaining $M - t$ minutes use $\dfrac{M - t}{r}$ of the m available.

$$\frac{t}{R} + \frac{M - t}{r} = m$$
$$\frac{t}{R} + \frac{M}{r} - \frac{t}{r} = m$$
$$t\left(\frac{1}{R} - \frac{1}{r}\right) = m - \frac{M}{r}$$
$$t\left(\frac{r - R}{rR}\right) = \frac{mr - M}{r}$$
$$t = \frac{R(mr - M)}{r - R}$$

Chapter 2

Principles in Practice 2.1

1. a. The formula for the area of a circle is πr^2, where r is the radius.

$$a(r) = \pi r^2$$

b. The domain of $a(r)$ is all real numbers.

c. Since a radius cannot be negative or zero, the domain for the function, in context, is $r > 0$.

3. a. If the price is \$18.50 per large pizza, $p = 18.5$.

$$18.5 = 26 - \frac{q}{40}$$

$$-7.5 = -\frac{q}{40}$$

$$300 = q$$

At a price of \$18.50 per large pizza, 300 pizzas are sold each week.

b. If 200 large pizzas are being sold each week, $q = 200$.

$$p = 26 - \frac{200}{40}$$

$$p = 26 - 5$$

$$p = 21$$

The price is \$21 per pizza if 200 large pizzas are being sold each week.

c. To double the number of large pizzas sold, use $q = 400$.

$$p = 26 - \frac{400}{40}$$

$$p = 26 - 10$$

$$p = 16$$

To sell 400 large pizzas each week, the price should be \$16 per pizza.

Problems 2.1

1. The functions are not equal because $f(x) \geq 0$ for all values of x, while $g(x)$ can be less than 0. For example,

$$f(-2) = \sqrt{(-2)^2} = \sqrt{4} = 2 \text{ and}$$

$g(-2) = -2$, thus $f(-2) \neq g(-2)$.

3. The functions are not equal because they have different domains. $h(x)$ is defined for all non-zero real numbers, while $k(x)$ is defined for all real numbers.

5. The denominator is zero when $x = 0$. Any other real number can be used for x.
Answer: all real numbers except 0

7. For $\sqrt{x-3}$ to be real, $x - 3 \geq 0$, so $x \geq 3$.
Answer: all real numbers ≥ 3

9. Any real number can be used for z.
Answer: all real numbers

11. We exclude values of x where

$$2x + 7 = 0$$
$$2x = -7$$
$$x = -\frac{7}{2}$$

Answer: all real numbers except $-\dfrac{7}{2}$

13. We exclude values of y for which $y^2 - 4y + 4 = 0$. $y^2 - 4y + 4 = (y-2)^2$, so we exclude values of y for which $y - 2 = 0$, thus $y = 2$.
Answer: all real numbers except 2.

15. We exclude all values of s for which

$$2s^2 - 7s - 4 = 0$$
$$(s - 4)(2s + 1) = 0$$
$$s = 4, -\frac{1}{2}$$

Answer: all real numbers except 4 and $-\dfrac{1}{2}$

17. $f(x) = 2x + 1$
$f(0) = 2(0) + 1 = 1$
$f(3) = 2(3) + 1 = 7$
$f(-4) = 2(-4) + 1 = -7$

19. $G(x) = 2 - x^2$

$G(-8) = 2 - (-8)^2 = 2 - 64 = -62$

$G(u) = 2 - u^2$

$G(u^2) = 2 - (u^2)^2 = 2 - u^4$

21. $\gamma(u) = 2u^2 - u$

$\gamma(-2) = 2(-2)^2 - (-2) = 8 + 2 = 10$

$\gamma(2v) = 2(2v)^2 - (2v) = 8v^2 - 2v$

$\gamma(x + a) = 2(x + a)^2 - (x + a)$
$\qquad = 2x^2 + 4ax + 2a^2 - x - a$

23. $f(x) = x^2 + 2x + 1$

$f(1) = 1^2 + 2(1) + 1 = 1 + 2 + 1 = 4$

$f(-1) = (-1)^2 + 2(-1) + 1 = 1 - 2 + 1 = 0$

$f(x + h) = (x + h)^2 + 2(x + h) + 1$
$\qquad = x^2 + 2xh + h^2 + 2x + 2h + 1$

25. $k(x) = \dfrac{x - 7}{x^2 + 2}$

$k(5) = \dfrac{5 - 7}{5^2 + 2} = -\dfrac{2}{27}$

$k(3x) = \dfrac{3x - 7}{(3x)^2 + 2} = \dfrac{3x - 7}{9x^2 + 2}$

$k(x + h) = \dfrac{(x + h) - 7}{(x + h)^2 + 2} = \dfrac{x + h - 7}{x^2 + 2xh + h^2 + 2}$

27. $f(x) = x^{4/3}$

$f(0) = 0^{4/3} = 0$

$f(64) = 64^{4/3} = \left(\sqrt[3]{64}\right)^4 = (4)^4 = 256$

$f\left(\dfrac{1}{8}\right) = \left(\dfrac{1}{8}\right)^{4/3} = \left(\sqrt[3]{\dfrac{1}{8}}\right)^4 = \left(\dfrac{1}{2}\right)^4 = \dfrac{1}{16}$

29. $f(x) = 4x - 5$

    a. $f(x + h) = 4(x + h) - 5 = 4x + 4h - 5$

    b. $\dfrac{f(x+h) - f(x)}{h} = \dfrac{(4x + 4h - 5) - (4x - 5)}{h} = \dfrac{4h}{h} = 4$

31. $f(x) = x^2 + 2x$

    a. $f(x + h) = (x + h)^2 + 2(x + h) = x^2 + 2xh + h^2 + 2x + 2h$

    b. $\dfrac{f(x+h) - f(x)}{h} = \dfrac{(x^2 + 2xh + h^2 + 2x + 2h) - (x^2 + 2x)}{h}$

        $= \dfrac{2xh + h^2 + 2h}{h} = 2x + h + 2$

33. $f(x) = 3 - 2x + 4x^2$

    a. $f(x + h) = 3 - 2(x + h) + 4(x + h)^2$

               $= 3 - 2x - 2h + 4(x^2 + 2xh + h^2)$

    b. $\dfrac{f(x+h) - f(x)}{h} = \dfrac{3 - 2x - 2h + 4x^2 + 8xh + 4h^2 - (3 - 2x + 4x^2)}{h}$

                   $= \dfrac{-2h + 8xh + 4h^2}{h}$

                   $= -2 + 8x + 4h$

35. $f(x) = \dfrac{1}{x}$

    a. $f(x + h) = \dfrac{1}{x + h}$

    b. $\dfrac{f(x+h) - f(x)}{h} = \dfrac{\frac{1}{x+h} - \frac{1}{x}}{h} = \dfrac{\frac{x - (x+h)}{x(x+h)}}{h} = \dfrac{-h}{x(x+h)h} = -\dfrac{1}{x(x+h)}$

37.
$$\frac{f(3+h)-f(3)}{h} = \frac{[5(3+h)+3]-[5(3)+3]}{h}$$
$$= \frac{[15+5h+3]-[15+3]}{h}$$
$$= \frac{18+5h-18}{h}$$
$$= \frac{5h}{h}$$
$$= 5$$

39. $9y - 3x - 4 = 0$

The equivalent form $y = \dfrac{3x+4}{9}$ shows that

for each input x there is exactly one output,

$\dfrac{3x+4}{9}$. Thus y is a function of x. Solving

for x gives $x = \dfrac{9y-4}{3}$. This shows that for

each input y there is exactly one output,

$\dfrac{9y-4}{3}$. Thus x is a function of y.

41. $y = 7x^2$

For each input x, there is exactly one output
$7x^2$. Thus y is a function of x. Solving for x

gives $x = \pm\sqrt{\dfrac{y}{7}}$. If, for example, $y = 7$, then

$x = \pm 1$, so x is not a function of y.

43. Yes, because corresponding to each input r

there is exactly one output, πr^2.

45. Weekly excess of income over expenses is
$6500 - 4800 = 1700$.
After t weeks the excess accumulates to
$1700t$. Thus the value of V of the business at
the end of t weeks is given by
$V = f(t) = 25,000 + 1700t$.

47. Yes; for each input q there corresponds
exactly one output, $1.25q$, so P is a function
of q. The dependent variable is P and the
independent variable is q.

49. The function can be written as $q = 48p$.
At \$8.39 per pound, the coffee house will
supply $q = 48(8.39) = 402.72$ pounds per
week.
At \$19.49 per pound, the coffee house will
supply $q = 48(19.49) = 935.52$ pounds per
week. The amount the coffee house supplies
increases as the price increases.

51. a.
$$f(1000) = \frac{\left(\sqrt[3]{1000}\right)^4}{2500}$$
$$= \frac{10^4}{2500}$$
$$= \frac{10,000}{2500}$$
$$= 4$$

b.
$$f(2000) = \frac{\left[\sqrt[3]{1000(2)}\right]^4}{2500} = \frac{\left(10\sqrt[3]{2}\right)^4}{2500}$$
$$= \frac{10,000\sqrt[3]{2^4}}{2500} = 4\sqrt[3]{2^3 \cdot 2} = 8\sqrt[3]{2}$$

c.
$$f(2I_0) = \frac{(2I_0)^{4/3}}{2500} = \frac{2^{4/3}I_0^{4/3}}{2500}$$
$$= 2\sqrt[3]{2}\left[\frac{I_0^{4/3}}{2500}\right] = 2\sqrt[3]{2}f(I_0)$$

Thus $f(2I_0) = 2\sqrt[3]{2}f(I_0)$, which
means that doubling the intensity
increases the response by a factor of
$2\sqrt[3]{2}$.

53. a. Domain: 3000, 2900, 2300, 2000
$f(2900) = 12, f(3000) = 10$

b. Domain: 10, 12, 17, 20
$g(10) = 3000, g(17) = 2300$

55. a. -5.13

b. 2.64

c. -17.43

57. a. 7.89

  b. 63.85

  c. 1.21

Principles in Practice 2.2

1. a. Let n = the number of visits and $p(n)$ be the premium amount.
$p(n) = 125$

  b. The premiums do not change regardless of the number of doctor visits.

  c. This is a constant function.

3. The price for n pairs of socks is given by
$$c(n) = \begin{cases} 3.5n & 0 \le n \le 5 \\ 3n & 5 < n \le 10 \\ 2.75n & 10 < n \end{cases}.$$

Problems 2.2

1. yes

3. no

5. yes

7. no

9. all real numbers

11. all real numbers

13. a. 3

  b. 7

15. a. 7

  b. 1

17. $f(x) = 8$
$f(2) = 8$
$f(t + 8) = 8$
$f\left(-\sqrt{17}\right) = 8$

19. $F(10) = 1$
$F\left(-\sqrt{3}\right) = -1$
$F(0) = 0$
$F\left(-\dfrac{18}{5}\right) = -1$

21. $G(8) = 8 - 1 = 7$
$G(3) = 3 - 1 = 2$
$G(-1) = 3 - (-1)^2 = 2$
$G(1) = 3 - (1)^2 = 2$

23. $6! = 6 \cdot 5 \cdot 4 \cdot 3 \cdot 2 \cdot 1 = 720$

25. $(4 - 2)! = 2! = 2 \cdot 1 = 2$

27. $\dfrac{n!}{(n-1)!} = \dfrac{n \cdot (n-1)!}{(n-1)!} = n$

29. Let i = the passenger's income and $c(i)$ = the cost for the ticket.
$c(i) = 4.5$
This is a constant function.

31. a. $C = 850 + 3q$

  b. $1600 = 850 + 3q$
$750 = 3q$
$q = 250$

33. The cost for buying n tickets is
$$c(n) = \begin{cases} 9.5n & 0 \le n < 12 \\ 8.75n & 12 \le n \end{cases}$$

35. $P(2) = \dfrac{3!\left(\frac{1}{4}\right)^2 \left(\frac{3}{4}\right)^1}{2!(1!)} = \dfrac{6\left(\frac{1}{16}\right)\left(\frac{3}{4}\right)}{2(1)} = \dfrac{9}{64}$

37. a. all T such that $30 \le T \le 39$

b. $f(30) = \dfrac{1}{24}(30) + \dfrac{11}{4} = \dfrac{5}{4} + \dfrac{11}{4} = \dfrac{16}{4} = 4$

$f(36) = \dfrac{1}{24}(36) + \dfrac{11}{4} = \dfrac{6}{4} + \dfrac{11}{4} = \dfrac{17}{4}$

$f(39) = \dfrac{4}{3}(39) - \dfrac{175}{4} = 52 - \dfrac{175}{4} = \dfrac{33}{4}$

39. a. 1182.74

 b. 4985.27

 c. 252.15

41. a. 2.21

 b. 9.98

 c. −14.52

Principles in Practice 2.3

1. The customer's price is
$(c \circ s)(x) = c(s(x)) = c(x+3) = 2(x+3)$
$= 2x + 6$

Problems 2.3

1. $f(x) = x + 3$, $g(x) = x + 5$

 a. $(f + g)(x) = f(x) + g(x)$
  $= (x+3) + (x+5)$
  $= 2x + 8$

 b. $(f + g)(0) = 2(0) + 8 = 8$

 c. $(f - g)(x) = f(x) - g(x)$
  $= (x+3) - (x+5)$
  $= -2$

 d. $(fg)(x) = f(x)g(x)$
  $= (x+3)(x+5)$
  $= x^2 + 8x + 15$

 e. $(fg)(-2) = (-2)^2 + 8(-2) + 15 = 3$

 f. $\dfrac{f}{g}(x) = \dfrac{f(x)}{g(x)} = \dfrac{x+3}{x+5}$

 g. $(f \circ g)(x) = f(g(x))$
  $= f(x+5)$
  $= (x+5) + 3$
  $= x + 8$

 h. $(f \circ g)(3) = 3 + 8 = 11$

 i. $(g \circ f)(x) = g(f(x))$
  $= g(x+3)$
  $= (x+3) + 5$
  $= x + 8$

 j. $(g \circ f)(3) = 3 + 8 = 11$

3. $f(x) = x^2 + 1$, $g(x) = x^2 - x$

 a. $(f + g)(x) = f(x) + g(x)$
  $= (x^2 + 1) + (x^2 - x)$
  $= 2x^2 - x + 1$

 b. $(f - g)(x) = f(x) - g(x)$
  $= (x^2 + 1) - (x^2 - x)$
  $= x + 1$

 c. $(f - g)\left(-\dfrac{1}{2}\right) = -\dfrac{1}{2} + 1 = \dfrac{1}{2}$

 d. $(fg)(x) = f(x)g(x)$
  $= (x^2 + 1)(x^2 - x)$
  $= x^4 - x^3 + x^2 - x$

 e. $\dfrac{f}{g}(x) = \dfrac{f(x)}{g(x)} = \dfrac{x^2 + 1}{x^2 - x}$

 f. $\dfrac{f}{g}\left(-\dfrac{1}{2}\right) = \dfrac{\left(-\frac{1}{2}\right)^2 + 1}{\left(-\frac{1}{2}\right)^2 - \left(-\frac{1}{2}\right)} = \dfrac{5}{3}$

g. $(f \circ g)(x) = f(g(x))$

$$= f(x^2 - x)$$
$$= (x^2 - x)^2 + 1$$
$$= x^4 - 2x^3 + x^2 + 1$$

h. $(g \circ f)(x) = g(f(x))$

$$= g(x^2 + 1)$$
$$= (x^2 + 1)^2 - (x^2 + 1)$$
$$= x^4 + x^2$$

i. $(g \circ f)(-3) = (-3)^4 + (-3)^2 = 90$

5. $f(g(2)) = f(4 - 4) = f(0) = 0 + 6 = 6$
$g(f(2)) = g(12 + 6) = g(18) = 4 - 36 = -32$

7. $(F \circ G)(t) = F(G(t))$

$$= F\left(\frac{2}{t-1}\right)$$

$$= \left(\frac{2}{t-1}\right)^2 + 7\left(\frac{2}{t-1}\right) + 1$$

$$= \frac{4}{(t-1)^2} + \frac{14}{t-1} + 1$$

$(G \circ F)(t) = G(F(t))$

$$= G(t^2 + 7t + 1)$$

$$= \frac{2}{(t^2 + 7t + 1) - 1}$$

$$= \frac{2}{t^2 + 7t}$$

9. $(f \circ g)(v) = f(g(v))$

$$= f\left(\sqrt{v+2}\right)$$

$$= \frac{1}{\left(\sqrt{v+2}\right)^2 + 1}$$

$$= \frac{1}{v + 2 + 1}$$

$$= \frac{1}{v + 3}$$

$(g \circ f)(v) = g(f(v))$

$$= g\left(\frac{1}{v^2 + 1}\right)$$

$$= \sqrt{\frac{1}{v^2 + 1} + 2}$$

$$= \sqrt{\frac{1 + 2(v^2 + 1)}{v^2 + 1}}$$

$$= \sqrt{\frac{2v^2 + 3}{v^2 + 1}}$$

11. Let $g(x) = 11x$ and $f(x) = x - 7$. Then
$h(x) = g(x) - 7 = f(g(x))$

13. Let $g(x) = x^2 - 2$ and $f(x) = \dfrac{1}{x}$. Then

$$h(x) = \frac{1}{x^2 - 2} = \frac{1}{g(x)} = f(g(x))$$

15. Let $g(x) = \dfrac{x^2 - 1}{x + 3}$ and $f(x) = \sqrt[4]{x}$.

Then $h(x) = \sqrt[4]{g(x)} = f(g(x))$.

17. a. The revenue is \$9.75 per pound of coffee sold, so $r(x) = 9.75x$.

   b. The expenses are $e(x) = 4500 + 4.25x$.

   c. Profit = revenue − expenses.
   $(r - e)(x) = 9.75x - (4500 + 4.25x)$
   $= 5.5x - 4500$.

19. $(g \circ f)(m) = g(f(m))$

$$= g\left(\frac{40m - m^2}{4}\right)$$

$$= 40\left(\frac{40m - m^2}{4}\right)$$

$$= 10(40m - m^2)$$

$$= 400m - 10m^2$$

This represents the total revenue received when the total output of m employees is sold.

21. a.   14.05

   b.   1169.64

23. a.   194.47

   b.   0.29

Problems 2.4

1. $f^{-1}(x) = \dfrac{x}{3} - \dfrac{7}{3}$

3. $F^{-1}(x) = 2x + 14$

5. $r(A) = \sqrt{\dfrac{A}{\pi}}$

7. $f(x) = 5x + 12$ is one-to-one, for if $f(x_1) = f(x_2)$ then $5x_1 + 12 = 5x_2 + 12$, so $5x_1 = 5x_2$ and thus $x_1 = x_2$.

9. $h(x) = (5x + 12)^2$, for $x \geq -\dfrac{5}{12}$, is one-to-one.

If $h(x_1) = h(x_2)$ then

$(5x_1 + 12)^2 = (5x_2 + 12)^2$.

Since $x \geq -\dfrac{5}{12}$ we have $5x + 12 \geq 0$, and

thus $(5x_1 + 12)^2 = (5x_2 + 12)^2$ only if

$5x_1 + 12 = 5x_2 + 12$, and hence $x_1 = x_2$.

11. The inverse of $f(x) = (4x - 5)^2$ for $x \geq \dfrac{5}{4}$

is $f^{-1}(x) = \dfrac{\sqrt{x}}{4} + \dfrac{5}{4}$, so to find the solution,

we find $f^{-1}(23)$.

$$f^{-1}(23) = \frac{\sqrt{23}}{4} + \frac{5}{4}$$

The solution is $x = \dfrac{\sqrt{23}}{4} + \dfrac{5}{4}$.

13. From $p = \dfrac{1,200,000}{q}$, we get

$q = \dfrac{1,200,000}{p}$. Since $q > 0$, p is also

greater than 0, so q as a function of p is

$$q = q(p) = \frac{1,200,000}{p}, \, p > 0.$$

$$p(q(p)) = p\left(\frac{1,200,000}{p}\right)$$

$$= \frac{1,200,000}{\frac{1,200,000}{p}}$$

$$= 1,200,000 \cdot \frac{p}{1,200,000}$$

$$= p$$

Similarly, $q(p(q)) = q$, so the functions are inverses.

Principles in Practice 2.5

1. Let $y =$ the amount of money in the account. Then, after one month, $y = 7250 - (1 \cdot 600) = \$6650$, and after two months $y = 7250 - (2 \cdot 600) = \$6050$. Thus, in general, if we let $x =$ the number of months during which Rachel spends from this account, $y = 7250 - 600x$. To identify the x-intercept, we set $y = 0$ and solve for x.

$y = 7250 - 600x$
$0 = 7250 - 600x$
$600x = 7250$

$$x = 12\frac{1}{12}$$

The x-intercept is $\left(12\frac{1}{12}, 0\right)$.

Therefore, after 12 months and approximately 2.5 days Rachel will deplete her savings. To identify the y-intercept, we set $x = 0$ and solve for y.
$y = 7250 - 600x$
$y = 7250 - 600(0)$
$y = 7250$
The y-intercept is $(0, 7250)$.
Therefore, before any months have gone by, Rachel has $7250 in her account.

3. The formula relating distance, time, and speed is $d = rt$, where d is the distance, r is the speed, and t is the time. Let $x =$ the time spent biking (in hours). Then, $12x =$ the distance traveled. Brett bikes $12 \cdot 2.5 = 30$ miles and then turns around and bikes the same distance back to the rental shop. Therefore, we can represent the distance from the turn-around point at any time x as $|30 - 12x|$. Similarly, the distance from the rental shop at any time x can be represented by the function $y = 30 - |30 - 12x|$.

x	0	1	2	2.5	3	4	5
y	0	12	24	30	24	12	0

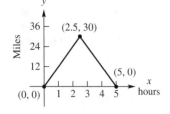

Problems 2.5

1.

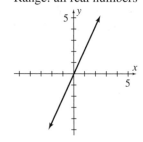

3. **a.** $f(0) = 1, f(2) = 2, f(4) = 3, f(-2) = 0$

   b. Domain: all real numbers

   c. Range: all real numbers

   d. $f(x) = 0$ for $x = -2$. So a real zero is -2.

5. **a.** $f(0) = 0, f(1) = 1, f(-1) = 1$

   b. Domain: all real numbers

   c. Range: all nonnegative real numbers

   d. $f(x) = 0$ for $x = 0$. So a real zero is 0.

7. $y = 2x$
   If $y = 0$, then $x = 0$. If $x = 0$, then $y = 0$.
   Intercept: $(0, 0)$
   y is a function of x. One-to-one.
   Domain: all real numbers
   Range: all real numbers

9. $y = 3x - 5$

If $y = 0$, then $0 = 3x - 5$, $x = \dfrac{5}{3}$.

If $x = 0$, then $y = -5$. Intercepts: $\left(\dfrac{5}{3}, 0\right)$,

$(0, -5)$
y is a function of x. One-to-one.
Domain: all real numbers
Range: all real numbers

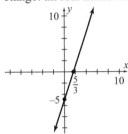

11. $y = x^4$

If $y = 0$, then $0 = x^4$, $x = 0$. If $x = 0$, then $y = 0$.
Intercept: $(0, 0)$
y is a function of x. Not one-to-one.
Domain: all real numbers
Range: all real numbers ≥ 0

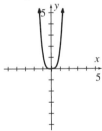

13. $x = 0$

If $y = 0$, then $x = 0$. If $x = 0$, then y can be any real number. Intercepts: every point on y-axis
y is not a function of x.

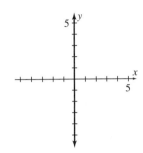

15. $y = x^3$

If $y = 0$, then $0 = x^3$, $x = 0$. If $x = 0$, then $y = 0$.
Intercept: $(0, 0)$. y is a function of x. One-to-one.
Domain: all real numbers
Range: all real numbers

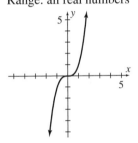

17. $x = -|y|$

If $y = 0$, then $x = 0$. If $x = 0$, then $0 = -|y|$, $y = 0$. Intercept: $(0, 0)$
y is not a function of x.

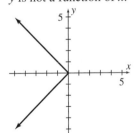

19. $2x + y - 2 = 0$

If $y = 0$, then $2x - 2 = 0$, $x = 1$. If $x = 0$, then $y - 2 = 0$, $y = 2$. Intercepts: $(1, 0)$, $(0, 2)$
Note that $y = 2 - 2x$. y is a function of x.
One-to-one.
Domain: all real numbers
Range: all real numbers

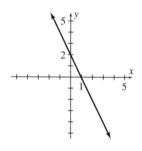

21. $s = f(t) = 4 - t^2$

If $s = 0$, then $0 = 4 - t^2$
$0 = (2 + t)(2 - t)$
$t = \pm 2$. If $t = 0$, then $s = 4$.
Intercepts: $(2, 0)$, $(-2, 0)$, $(0, 4)$
Domain: all real numbers
Range: all real numbers ≤ 4

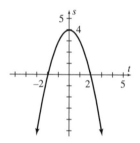

23. $y = h(x) = 3$
Because y cannot be 0, there is no x-intercept. If $x = 0$, then $y = 3$.
Intercept: $(0, 3)$
Domain: all real numbers
Range: 3

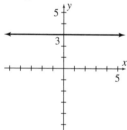

25. $y = h(x) = x^2 - 4x + 1$

If $y = 0$, then $0 = x^2 - 4x + 1$, and by the quadratic formula, $x = \dfrac{4 \pm \sqrt{12}}{2} = 2 \pm \sqrt{3}$. If $x = 0$, then $y = 1$. Intercepts:
$(2 \pm \sqrt{3}, 0), (0, 1)$
Domain: all real numbers
Range: all real numbers ≥ -3

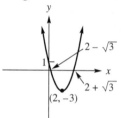

27. $f(t) = -t^3$

If $f(t) = 0$, then $0 = -t^3$, $t = 0$.
If $t = 0$, then $f(t) = 0$. Intercept: $(0, 0)$
Domain: all real numbers
Range: all real number

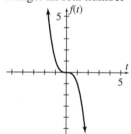

29. $s = f(t) = \sqrt{t^2 - 9}$

Note that for $\sqrt{t^2 - 9}$ to be a real number, $t^2 - 9 \geq 0$, so $t^2 \geq 9$, and $|t| \geq 3$. If $s = 0$, then $0 = \sqrt{t^2 - 9}$, $0 = t^2 - 9$, or $t = \pm 3$.
Because $|t| \geq 3$, we know $t \neq 0$, so no s-intercept exists.
Intercepts: $(-3, 0)$, $(3, 0)$
Domain: all real numbers $t \leq -3$ and ≥ 3
Range: all real numbers ≥ 0

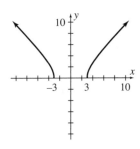

31. $f(x) = |2x - 1|$

If $f(x) = 0$, then $0 = |2x - 1|$, $2x - 1 = 0$, so

$x = \dfrac{1}{2}$.

If $x = 0$, then $f(x) = |-1| = 1$.

Intercepts: $\left(\dfrac{1}{2}, 0\right), (0, 1)$

Domain: all real numbers
Range: all real numbers ≥ 0

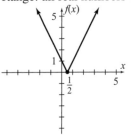

33. $F(t) = \dfrac{16}{t^2}$

If $F(t) = 0$, then $0 = \dfrac{16}{t^2}$, which has no

solution. Because $t \neq 0$, there is no vertical-axis intercept. No intercepts
Domain: all nonzero real numbers
Range: all positive real numbers

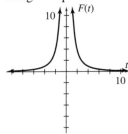

35. Domain: all real numbers ≥ 0
Range: all real numbers $1 \leq c < 8$

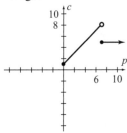

37. Domain: all real numbers
Range: all real numbers ≥ 0

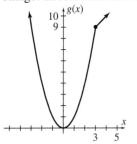

39. From the vertical-line test, the graphs that represent functions of x are (a), (b), and (d).

41. Let $y =$ the amount that is owed and let $x =$ the number of monthly payments made. Then, the amount Tara owes is represented by the equation $y = 2400 - 275x$.
To determine the x-intercept, we set $y = 0$ and solve for x.

$y = 2400 - 275x$
$0 = 2400 - 275x$
$275x = 2400$
$x = 8\dfrac{8}{11}$

The x-intercept is $\left(8\dfrac{8}{11}, 0\right)$. Therefore, Tara

will have paid off her debt after 9 months.
To determine the y-intercept, we set $x = 0$ and solve for y.
$y = 2400 - 275x$
$y = 2400 - 275(0)$
$y = 2400$
The y-intercept is $(0, 2400)$. Therefore, before any payments are made, Tara owes $2400.

43. As price increases, quantity supplied increases; p is a function of q.

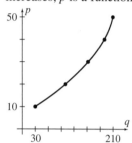

45.

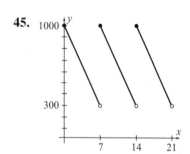

47. 0.39

49. −0.61, −0.04

51. −1.12

53. −1.70, 0

55.

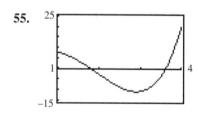

    a. maximum value of $f(x)$: 19.60

    b. minimum value of $f(x)$: −10.86

57.

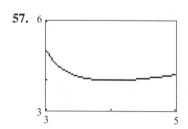

    a. maximum value of $f(x)$: 5

    b. minimum value of $f(x)$: 4

59.

    a. maximum value of $f(x)$: 28

    b. range: $(-\infty, 28]$

    c. real zeros: −4.02, 0.60

61.

    a. maximum value of $f(x)$: 34.21

    b. minimum value of $f(x)$: 18.68

    c. range: $[18.68, 34.21]$

    d. no intercept

Problems 2.6

1. $y = 5x$
Intercepts: If $y = 0$, then $5x = 0$, or $x = 0$; if $x = 0$, then $y = 5 \cdot 0 = 0$.
Testing for symmetry gives:

x-axis:	$-y = 5x$
	$y = -5x$
y-axis:	$y = 5(-x) = -5x$
origin:	$-y = 5(-x)$
	$y = 5x$

line $y = x$: (a, b) on graph, then $b = 5a$, and

$$a = \frac{1}{5}b \neq 5b \text{ for all } b, \text{ so } (b, a) \text{ is}$$

not on the graph.

Answer: $(0, 0)$; symmetry about origin

3. $2x^2 + y^2 x^4 = 8 - y$

Intercepts: If $y = 0$, then

$2x^2 = 8$, $x^2 = 4$, or $x = \pm 2$;

if $x = 0$, then $0 = 8 - y$, so $y = 8$.

Testing for symmetry gives:

x-axis: $2x^2 + (-y)^2 x^4 = 8 - (-y)$

$\qquad\qquad 2x^2 + y^2 x^4 = 8 + y$

y-axis: $2(-x)^2 + y^2(-x)^4 = 8 - y$

$\qquad\qquad 2x^2 + y^2 x^4 = 8 - y$

origin: $2(-x)^2 + (-y)^2(-x)^4 = 8 - (-y)$

$\qquad\qquad 2x^2 + y^2 x^4 = 8 + y$

line $y = x$: (a, b) on graph, then

$\qquad\qquad 2a^2 + b^2 a^4 = 8 - b$, but

$\qquad\qquad 2b^2 + a^2 b^4 = 8 - a$ will not

necessarily be true, so (b, a) is not on the graph.

Answer: $(\pm 2, 0)$, $(0, 8)$; symmetry about y-axis

5. $16x^2 - 9y^2 = 25$

Intercepts: If $y = 0$, then $16x^2 = 25$,

$$x^2 = \frac{25}{16}, \text{ so } x = \pm\frac{5}{4};$$

if $x = 0$, then $-9y^2 = 25$, $y^2 = -\frac{25}{9}$, which

has no real root.

Testing for symmetry gives:

x-axis: $16x^2 - 9(-y)^2 = 25$

$\qquad\qquad 16x^2 - 9y^2 = 25$

y-axis: $16(-x)^2 - 9y^2 = 25$

$\qquad\qquad 16x^2 - 9y^2 = 25$

origin: Since the graph has symmetry about x- and y-axes, there is also symmetry about the origin.

line $y = x$: (a, b) on graph, then

$\qquad\qquad 16a^2 - 9b^2 = 25$, and

$$a^2 = \frac{1}{16}(9b^2 + 25). \ (b, a) \text{ on}$$

graph, then $16b^2 - 9a^2 = 25$ and

$$a^2 = \frac{1}{9}(16b^2 - 25)$$

$$\neq \frac{1}{16}(9b^2 + 25)$$

for all b, so (b, a) and (a, b) are not always both on the graph.

Answer: $\left(\pm\frac{5}{4}, 0\right)$; symmetry about x-axis, y-axis, and origin.

7. $x = -2$

Intercepts: If $y = 0$, then $x = -2$; because $x \neq 0$, there is no y-intercept.

Testing for symmetry gives:

x-axis: $x = -2$

y-axis: $-x = -2$

$\qquad\qquad x = 2$

origin: $-x = -2$

$\qquad\qquad x = 2$

line $y = x$: (a, b) on graph, then $a = -2$, but b can be any value, so $(b, a) = (b, -2)$ is not necessarily on the graph.

Answer: $(-2, 0)$; symmetry about x-axis

9. $x = -y^{-4}$

Intercepts: Because $y \neq 0$, there is no x-intercept; if $x = 0$, then $0 = -\frac{1}{y^4}$, which

has no solution.

Testing for symmetry gives:

x-axis: $x = -(-y)^{-4}$

$\qquad\qquad x = -y^{-4}$

y-axis: $-x = -y^{-4}$

$\qquad\qquad x = y^{-4}$

origin:　　$-x = -(-y)^{-4}$

　　　　　　$x = y^{-4}$

line $y = x$:　(a, b) on graph, then $a = -b^{-4}$

　　　　and $b = (-a)^{-1/4} \neq -a^{-4}$ for all

　　　　a, so (b, a) is not on the graph.

Answer: no intercepts; symmetry about x-axis

11.　$x - 4y - y^2 + 21 = 0$

Intercepts: If $y = 0$, then $x + 21 = 0$, so $x = -21$; if $x = 0$, then $-4y - y^2 + 21 = 0$,

$y^2 + 4y - 21 = 0$, $(y + 7)(y - 3) = 0$, so $y = -7$ or $y = 3$.

Testing for symmetry gives:

x-axis:　　$x - 4(-y) - (-y)^2 + 21 = 0$

　　　　　　$x + 4y - y^2 + 21 = 0$

y-axis:　　$(-x) - 4y - y^2 + 21 = 0$

　　　　　　$-x - 4y - y^2 + 21 = 0$

origin:　　$(-x) - 4(-y) - (-y)^2 + 21 = 0$

　　　　　　$-x + 4y - y^2 + 21 = 0$

line $y = x$:　(a, b) on graph, then

　　　　$a - 4b - b^2 + 21 = 0$ and

　　　　$a = b^2 + 4b - 21$, but

　　　　$b = a^2 + 4a - 21$ will not necessarily be true, so (b, a) is not on the graph.

Answer: $(-21, 0)$, $(0, -7)$, $(0, 3)$; no symmetry

13.　$y = f(x) = \dfrac{x^3 - 2x^2 + x}{x^2 + 1}$

Intercepts: If $y = 0$, then

$\dfrac{x^3 - 2x^2 + x}{x^2 + 1} = \dfrac{x(x-1)^2}{x^2 + 1} = 0$, so $x = 0, 1$;

if $x = 0$, then $y = 0$.

Testing for symmetry gives:

x-axis:　　Because f is not the zero function, there is no x-axis symmetry

y-axis:　　$y = \dfrac{(-x)^3 - 2(-x)^2 + (-x)}{(-x)^2 + 1}$

　　　　　　$y = \dfrac{-x^3 - 2x^2 - x}{x^2 + 1}$

origin:　　$-y = \dfrac{(-x)^3 - 2(-x)^2 + (-x)}{(-x)^2 + 1}$

　　　　　　$y = \dfrac{x^3 + 2x^2 + x}{x^2 + 1}$

line $y = x$:　(a, b) on graph, then

　　　　$b = \dfrac{a^3 - 2a^2 + a}{a^2 + 1}$, but

　　　　$a = \dfrac{b^3 - 2b^2 + b}{b^2 + 1}$ is not

necessarily true, so (b, a) is not on the graph.

Answer: $(1, 0)$, $(0, 0)$; no symmetry of the given types

15.　$y = \dfrac{3}{x^3 + 8}$

Intercepts: If $y = 0$, then $\dfrac{3}{x^3 + 8} = 0$, which

has no solution; if $x = 0$, then $y = \dfrac{3}{8}$.

Testing for symmetry gives:

x-axis:　　$-y = \dfrac{3}{x^3 + 8}$

　　　　　　$y = -\dfrac{3}{x^3 + 8}$

y-axis:　　$y = \dfrac{3}{(-x)^3 + 8}$

　　　　　　$y = \dfrac{3}{-x^3 + 8}$

origin:　　$-y = \dfrac{3}{(-x)^3 + 8}$

　　　　　　$-y = \dfrac{3}{-x^3 + 8}$

　　　　　　$y = \dfrac{3}{x^3 - 8}$

line $y = x$: (a, b) on graph, then $b = \dfrac{3}{a^3 + 8}$

and $a = \sqrt[3]{\dfrac{3}{b} - 8} \neq \dfrac{3}{b^3 + 8}$ for all

b, so (b, a) is not on the graph.

Answer: $\left(0, \dfrac{3}{8}\right)$; no symmetry of the given

types

17. $3x + y^2 = 9$

Intercepts: If $y = 0$, then $3x = 9$, so $x = 3$;
if $x = 0$, then $y^2 = 9$, so $y = \pm 3$.
Testing for symmetry gives:

x-axis: $\quad 3x + (-y)^2 = 9$
$\qquad\qquad 3x + y^2 = 9$

y-axis: $\quad 3(-x) + y^2 = 9$
$\qquad\qquad -3x + y^2 = 9$

origin: $\quad 3(-x) + (-y)^2 = 9$
$\qquad\qquad -3x + y^2 = 9$

line $y = x$: (a, b) on graph, then $3a + b^2 = 9$

and $a = \dfrac{1}{3}(9 - b^2)$, but

$b = \dfrac{1}{3}(9 - a^2)$ will not

necessarily be true, so (b, a) is
not on the graph.

Answer: $(3, 0)$, $(0, \pm 3)$; symmetry about
x-axis

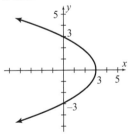

19. $y = f(x) = x^3 - 4x$

Intercepts: If $y = 0$, then $x^3 - 4x = 0$,
$x(x + 2)(x - 2) = 0$, so $x = 0$ or $x = \pm 2$; if
$x = 0$, then $y = 0$.
Testing for symmetry gives:

x-axis: $\quad$ Because f is not the zero
$\qquad\qquad$ function, there is no x-axis
$\qquad\qquad$ symmetry.

y-axis: $\quad y = (-x)^3 - 4(-x)$
$\qquad\qquad y = -x^3 + 4x$

origin: $\quad -y = (-x)^3 - 4(-x)$
$\qquad\qquad y = x^3 - 4x$

line $y = x$: (a, b) on graph, then
$\qquad\qquad b = a^3 - 4a$, but $a = b^3 - 4b$ will
$\qquad\qquad$ not necessarily be true, so (b, a)
$\qquad\qquad$ is not on the graph.

Answer: $(0, 0)$, $(\pm 2, 0)$; symmetry about
origin.

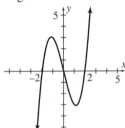

21. $|x| - |y| = 0$

Intercepts: If $y = 0$, then $|x| = 0$, so $x = 0$; if
$x = 0$, then $-|y| = 0$, so $y = 0$.
Testing for symmetry gives:

x-axis: $\quad |x| - |-y| = 0$
$\qquad\qquad |x| - |y| = 0$

y-axis: $\quad |-x| - |y| = 0$
$\qquad\qquad |x| - |y| = 0$

origin: $\quad$ Since there is symmetry about
$\qquad\qquad$ the x- and y-axes, symmetry
$\qquad\qquad$ about origin exists.

line $y = x$: (a, b) on graph, then $|a| - |b| = 0$,

thus $|a| = |b|$, and $|b| - |a| = 0$, so

(b, a) is on the graph.

Answer: $(0, 0)$; symmetry about x-axis, y-axis, origin, line $y = x$.

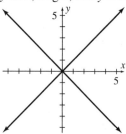

23. $9x^2 + 4y^2 = 25$

Intercepts: If $y = 0$, then $9x^2 = 25$,

$x^2 = \dfrac{25}{9}$, so

$x = \pm\dfrac{5}{3}$; if $x = 0$, then $4y^2 = 25$, so

$y = \pm\dfrac{5}{2}$.

Testing for symmetry gives:

x-axis: $9x^2 + 4(-y)^2 = 25$

$9x^2 + 4y^2 = 25$

y-axis: $9(-x)^2 + 4y^2 = 25$

$9x^2 + 4y^2 = 25$

origin: Since there is symmetry about x- and y-axes, symmetry about origin exists.

line $y = x$: (a, b) on graph, then $9a^2 + 4b^2 = 25$

and $b^2 = \dfrac{1}{4}(25 - 9a^2)$. (b, a) on

graph, then $9b^2 + 4a^2 = 25$ and

$b^2 = \dfrac{1}{9}(25 - 4a^2)$, so (a, b) and

(b, a) are not always both on the graph.

Answer: $\left(\pm\dfrac{5}{3}, 0\right), \left(0, \pm\dfrac{5}{2}\right)$; symmetry about x-axis, y-axis, origin

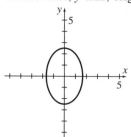

25.

$y = f(x) = 5 - 1.96x^2 - \pi x^4$. Replacing x by $-x$ gives $y = 5 - 1.96(-x)^2 - \pi(-x)^4$ or $y = 5 - 1.96x^2 - \pi x^4$, which is equivalent to original equation. Thus the graph is symmetric about the y-axis.

a. Intercepts: $(\pm0.99, 0)$, $(0, 5)$

b. Maximum value of $f(x)$: 5

c. Range: $(-\infty, 5]$

27.

Problems 2.7

1.

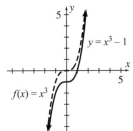

3.

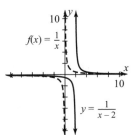

5.

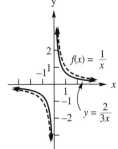

7.

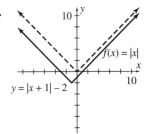

9.

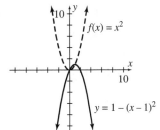

11.

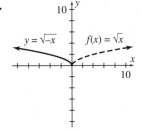

13. Translate 3 units to the left, stretch vertically away from the *x*-axis by a factor of 2, reflect about the *x*-axis, and move 2 units upward.

15. Reflect about the *y*-axis and translate 5 units downward.

17.

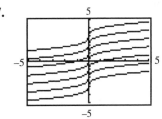

Compared to the graph for $k = 0$, the graphs for $k = 1, 2$, and 3 are vertical shifts upward of 1, 2, and 3 units, respectively. The graphs for $k = -1, -2$, and -3 are vertical shifts downward of 1, 2, and 3 units, respectively.

19.

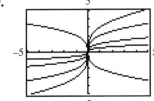

Compared to the graph for $k = 1$, the graphs for $k = 2$ and 3 are vertical stretches away from the x-axis by factors of 2 and 3, respectively. The graph for $k = \dfrac{1}{2}$ is a vertical shrinking toward the x-axis.

Chapter 2 Review Problems

1. Denominator is 0 when
$$x^2 - 6x + 5 = 0$$
$$(x-1)(x-5) = 0$$
$$x = 1, 5$$
Domain: all real numbers except 1 and 5.

3. all real numbers

5. For $\sqrt{x}$ to be real, x must be nonnegative. For the denominator $x - 1$ to be different from 0, x cannot be 1. Both conditions are satisfied by all nonnegative numbers except 1.
Domain: all nonnegative real numbers except 1.

7. $f(x) = 3x^2 - 4x + 7$

$f(0) = 3(0)^2 - 4(0) + 7 = 7$

$f(-3) = 3(-3)^2 - 4(-3) + 7$
$\qquad = 27 + 12 + 7$
$\qquad = 46$

$f(5) = 3(5)^2 - 4(5) + 7 = 75 - 20 + 7 = 62$

$f(t) = 3t^2 - 4t + 7$

9. $G(x) = \sqrt[4]{x-3}$

$G(3) = \sqrt[4]{3-3} = \sqrt[4]{0} = 0$

$G(19) = \sqrt[4]{19-3} = \sqrt[4]{16} = 2$

$G(t+1) = \sqrt[4]{(t+1)-3} = \sqrt[4]{t-2}$

$G(x^3) = \sqrt[4]{x^3 - 3}$

11. $h(u) = \dfrac{\sqrt{u+4}}{u}$

$h(5) = \dfrac{\sqrt{5+4}}{5} = \dfrac{\sqrt{9}}{5} = \dfrac{3}{5}$

$h(-4) = \dfrac{\sqrt{-4+4}}{-4} = \dfrac{0}{-4} = 0$

$$h(x) = \frac{\sqrt{x+4}}{x}$$

$$h(u-4) = \frac{\sqrt{(u-4)+4}}{u-4} = \frac{\sqrt{u}}{u-4}$$

13. $f(4) = 4 + 16 = 20$
$f(-2) = -3$
$f(0) = -3$
$f(1)$ is not defined.

15. a. $f(x + h) = 3 - 7(x + h) = 3 - 7x - 7h$

   b. $\dfrac{f(x+h)-f(x)}{h} = \dfrac{(3-7x-7h)-(3-7x)}{h} = \dfrac{-7h}{h} = -7$

17. a. $f(x+h) = 4(x+h)^2 + 2(x+h) - 5 = 4x^2 + 8xh + 4h^2 + 2x + 2h - 5$

   b. $\dfrac{f(x+h)-f(x)}{h} = \dfrac{(4x^2 + 8xh + 4h^2 + 2x + 2h - 5) - (4x^2 + 2x - 5)}{h}$

$$= \frac{8xh + 4h^2 + 2h}{h}$$

$$= 8x + 4h + 2$$

19. $f(x) = 3x - 1$, $g(x) = 2x + 3$

   a. $(f + g)(x) = f(x) + g(x) = (3x - 1) + (2x + 3) = 5x + 2$

   b. $(f + g)(4) = 5(4) + 2 = 22$

   c. $(f - g)(x) = f(x) - g(x) = (3x - 1) - (2x + 3) = x - 4$

   d. $(fg)(x) = f(x)g(x) = (3x-1)(2x+3) = 6x^2 + 7x - 3$

   e. $(fg)(1) = 6(1)^2 + 7(1) - 3 = 10$

   f. $\dfrac{f}{g}(x) = \dfrac{f(x)}{g(x)} = \dfrac{3x-1}{2x+3}$

   g. $(f \circ g)(x) = f(g(x)) = f(2x+3) = 3(2x+3) - 1 = 6x + 8$

   h. $(f \circ g)(5) = 6(5) + 8 = 38$

   i. $(g \circ f)(x) = g(f(x)) = g(3x-1) = 2(3x-1) + 3 = 6x + 1$

21. $f(x) = \dfrac{1}{x^2}$, $g(x) = x + 1$

$(f \circ g)(x) = f(g(x)) = f(x+1) = \dfrac{1}{(x+1)^2}$

$(g \circ f)(x) = g(f(x))$

$\qquad = g\left(\dfrac{1}{x^2}\right)$

$\qquad = \dfrac{1}{x^2} + 1$

$\qquad = \dfrac{1+x^2}{x^2}$

23. $f(x) = \sqrt{x+2}$, $g(x) = x^3$

$(f \circ g)(x) = f(g(x)) = f(x^3) = \sqrt{x^3 + 2}$

$(g \circ f)(x) = g(f(x))$

$\qquad = g\left(\sqrt{x+2}\right)$

$\qquad = \left(\sqrt{x+2}\right)^3$

$\qquad = (x+2)^{3/2}$

25. $y = 3x - x^3$

Intercepts: If $y = 0$, then $0 = 3x - x^3$,

$x(3 - x^2) = 0$, $x = 0, \pm\sqrt{3}$.

If $x = 0$, then $y = 0$.

Testing for symmetry gives:

x-axis: $\quad -y = 3x - x^3$

$\qquad\qquad y = -3x + x^3$, which is not the
original equation.

y-axis: $\quad y = 3(-x) - (-x)^3$

$\qquad\qquad y = -3x + x^3$

origin: $\quad -y = 3(-x) - (-x)^3$

$\qquad\qquad y = 3x - x^3$, which is the
original equation.

line $y = x$: (a, b) on graph, then $b = 3a - a^3$,

but $a = 3b - b^3$ is not necessarily
true, so (b, a) is not on the graph.

Answer: $(0, 0)$, $\left(\pm\sqrt{3}, 0\right)$; symmetry about
origin

27. $y = 9 - x^2$

Intercepts: If $y = 0$, then

$0 = 9 - x^2 = (3+x)(3-x)$, or $x = \pm 3$

If $x = 0$, then $y = 9$.

Testing for symmetry gives:

x-axis: $\quad -y = 9 - x^2$

$\qquad\qquad y = -9 + x^2$, which is not the
original equation.

y-axis: $\quad y = 9 - (-x)^2$

$\qquad\qquad y = 9 - x^2$, which is the original
equation.

origin: $\quad -y = 9 - (-x)^2$

$\qquad\qquad y = -9 + x^2$, which is not the
original equation.

line $y = x$: (a, b) on graph, then $b = 9 - a^2$

$\qquad\qquad$ and $a = \pm\sqrt{9-b} \neq 9 - b^2$ for all

$\qquad\qquad b$, so (b, a) is not on the graph.

Answer: $(0, 9)$, $(\pm 3, 0)$; symmetry about
y-axis.

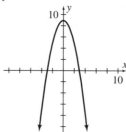

29. $G(u) = \sqrt{u+4}$

If $G(u) = 0$, then $0 = \sqrt{u+4}$.

$0 = u + 4$,

$u = -4$

If $u = 0$, then $G(u) = \sqrt{4} = 2$.

Intercepts: $(0, 2)$, $(-4, 0)$

Domain: all real numbers u such that

$u \geq -4$

Range: all real numbers ≥ 0

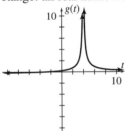

31. $y = g(t) = \dfrac{2}{|t-4|}$

If $y = 0$, then $0 = \dfrac{2}{|t-4|}$, which has no

solution. If $t = 0$, then $y = \dfrac{2}{4} = \dfrac{1}{2}$.

Intercept: $\left(0, \dfrac{1}{2}\right)$

Domain: all real numbers t such that $t \neq 4$
Range: all real numbers > 0

33. Domain: all real numbers.
Range: all real numbers ≤ 2

35.

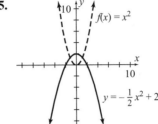

37. From the vertical-line test, the graphs that represent functions of x are (a) and (c).

39.

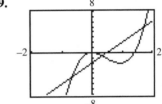

$-0.67; 0.34, 1.73$

41.

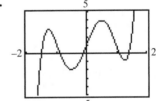

$-1.50, -0.88, -0.11, 1.09, 1.40$

43.

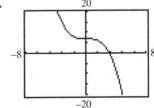

a. $(-\infty, \infty)$

b. $(1.92, 0), (0, 7)$

45. $k = 0, 2, 4$

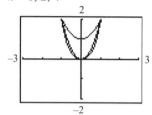

$k = 1, 3$

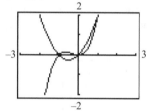

    a.   0, 2, 4

    b.   none

Mathematical Snapshot Chapter 2

  1. $f(23,000) = 1510 + 0.15(23,000 - 15,100)$
$$= 2695$$
     The tax on \$23,000 is \$2695.

  3. $f(290,000) = 42,170 + 0.33(290,000 - 188,450)$
$$= 75,681.5$$
     The tax on \$290,000 is \$75,681.50.

  5.   Answers may vary.

  7. $g(x) = x - f(x)$

$$= \begin{cases} x - 0.10x & \text{if} \quad 0 \le x \le 15,100 \\ x - [1510 + 0.15(x - 15,100)] & \text{if} \quad 15,100 < x \le 61,300 \\ x - [8440 + 0.25(x - 61,300)] & \text{if} \quad 61,300 < x \le 123,700 \\ x - [24,040 + 0.28(x - 123,700)] & \text{if} \quad 123,700 < x \le 188,450 \\ x - [42,170 + 0.33(x - 188,450)] & \text{if} \quad 188,450 < x \le 336,550 \\ x - [91,043 + 0.35(x - 336,550)] & \text{if} \quad x > 336,550 \end{cases}$$

$$= \begin{cases} 0.90x & \text{if} \quad 0 \le x \le 15,100 \\ 0.85x + 755 & \text{if} \quad 15,100 < x \le 61,300 \\ 0.75x + 6885 & \text{if} \quad 61,300 < x \le 123,700 \\ 0.72x + 10,596 & \text{if} \quad 123,700 < x \le 188,450 \\ 0.67x + 20,018.50 & \text{if} \quad 188,450 < x \le 336,550 \\ 0.65x + 26,749.50 & \text{if} \quad x > 336,550 \end{cases}$$

Chapter 3

Principles in Practice 3.1

1. Let x = the time (in years) and let y = the selling price. Then,
   In 1991: $x_1 = 1991$ and $y_1 = 32{,}000$
   In 1994: $x_2 = 1994$ and $y_2 = 26{,}000$
   The slope is
   $$m = \frac{y_2 - y_1}{x_2 - x_1}$$
   $$= \frac{26{,}000 - 32{,}000}{1994 - 1991}$$
   $$= \frac{-6000}{3}$$
   $$= -2000$$
   The car depreciated $2000 per year.

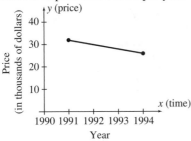

3. A linear function relating Fahrenheit temperature to Celsius temperature can be found by using the point-slope form of an equation of a line.
   $$m = \frac{F_2 - F_1}{C_2 - C_1} = \frac{77 - 41}{25 - 5} = \frac{36}{20} = \frac{9}{5}$$
   $$F - F_1 = m(C - C_1)$$
   $$F - 41 = \frac{9}{5}(C - 5)$$
   $$F - 41 = \frac{9}{5}C - 9$$
   $$F = \frac{9}{5}C + 32$$

5. $$F = \frac{9}{5}C + 32$$
   $$5(F) = 5\left(\frac{9}{5}C + 32\right)$$
   $$5F = 9C + 160$$
   $$0 = 9C - 5F + 160$$
   Thus, $9C - 5F + 160 = 0$ is a general linear form of $F = \frac{9}{5}C + 32$.

7. Right angles are formed by perpendicular lines. The slopes of the sides of the triangle are:
   $$\overline{AB}\left\{ m = \frac{0 - 0}{6 - 0} = \frac{0}{6} = 0 \right.$$
   $$\overline{BC}\left\{ m = \frac{7 - 0}{7 - 6} = \frac{7}{1} = 7 \right.$$
   $$\overline{AC}\left\{ m = \frac{7 - 0}{7 - 0} = \frac{7}{7} = 1 \right.$$

   Since none of the slopes are negative reciprocals of each other, there are no perpendicular lines. Therefore, the points do not define a right triangle.

Problems 3.1

1. $$m = \frac{10 - 1}{7 - 4} = \frac{9}{3} = 3$$

3. $$m = \frac{-3 - (-2)}{8 - 6} = \frac{-1}{2} = -\frac{1}{2}$$

5. The difference in the x-coordinates is $5 - 5 = 0$, so the slope is undefined.

7. $$m = \frac{-2 - (-2)}{4 - 5} = \frac{0}{-1} = 0$$

9.
$$y - 7 = -5[x - (-1)]$$
$$y - 7 = -5(x + 1)$$
$$y - 7 = -5x - 5$$
$$5x + y - 2 = 0$$

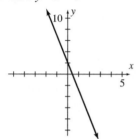

11.
$$y - 5 = -\frac{1}{4}[x - (-2)]$$
$$4(y - 5) = -(x + 2)$$
$$4y - 20 = -x - 2$$
$$x + 4y - 18 = 0$$

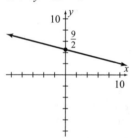

13. $m = \dfrac{4 - 1}{1 - (-6)} = \dfrac{3}{7}$

$$y - 4 = \frac{3}{7}(x - 1)$$
$$7(y - 4) = 3(x - 1)$$
$$7y - 28 = 3x - 3$$
$$3x - 7y + 25 = 0$$

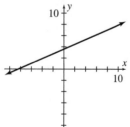

15. $m = \dfrac{-8 - (-4)}{-2 - (-3)} = \dfrac{-4}{1} = -4$

$$y - (-4) = -4[x - (-3)]$$
$$y + 4 = -4x - 12$$
$$4x + y + 16 = 0$$

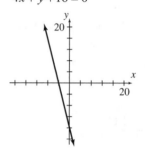

17.
$$y = 2x + 4$$
$$2x - y + 4 = 0$$

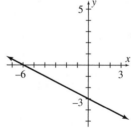

19.
$$y = -\frac{1}{2}x - 3$$
$$2y = 2\left(-\frac{1}{2}x - 3\right)$$
$$2y = -x - 6$$
$$x + 2y + 6 = 0$$

21. A horizontal line has the form $y = b$. Thus $y = -3$, or $y + 3 = 0$.

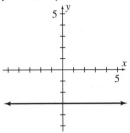

23. A vertical line has the form $x = a$. Thus $x = 2$, or $x - 2 = 0$.

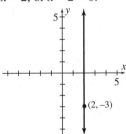

25. $y = 4x - 6$ has the form $y = mx + b$, where $m = 4$ and $b = -6$.

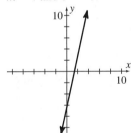

27. $3x + 5y - 9 = 0$
$$5y = -3x + 9$$
$$y = -\frac{3}{5}x + \frac{9}{5}$$
$$m = -\frac{3}{5},\ b = \frac{9}{5}$$

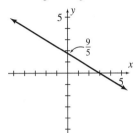

29. $x = -5$ is a vertical line. Thus the slope is undefined. There is no y-intercept.

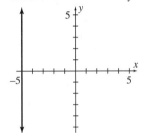

31. $y = 3x$
$y = 3x + 0$
$m = 3,\ b = 0$

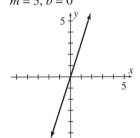

33. $y = 3$
$y = 0x + 3$
$m = 0, b = 3$

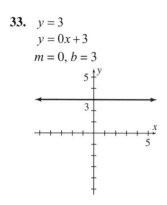

35. $2x = 5 - 3y$, or $2x + 3y - 5 = 0$ (general form)

$3y = -2x + 5$, or $y = -\dfrac{2}{3}x + \dfrac{5}{3}$ (slope-intercept form)

37. $4x + 9y - 5 = 0$ is a general form.

$9y = -4x + 5$, or $y = -\dfrac{4}{9}x + \dfrac{5}{9}$ (slope-intercept form)

39.
$$-\dfrac{x}{2} + \dfrac{2y}{3} = -4\dfrac{3}{4}$$
$$12\left(-\dfrac{x}{2} + \dfrac{2y}{3}\right) = 12\left(-\dfrac{19}{4}\right)$$
$$-6x + 8y = -57$$
$$6x - 8y - 57 = 0 \text{ (general form)}$$
$$-8y = -6x + 57$$
$$y = \dfrac{3}{4}x - \dfrac{57}{8} \text{ (slope}$$
$$\text{-intercept form)}$$

41. The lines $y = 7x + 2$ and $y = 7x - 3$ have the same slope, 7. Thus they are parallel.

43. The lines $y = 5x + 2$ and $-5x + y - 3 = 0$ (or $y = 5x + 3$) have the same slope, 5. Thus they are parallel.

45. The line $x + 3y + 5 = 0$ $\left(\text{or } y = -\dfrac{1}{3}x - \dfrac{5}{3}\right)$

has slope $m_1 = -\dfrac{1}{3}$ and the line $y = -3x$ has

slope $m_2 = -3$. Since $m_1 \neq m_2$ and

$m_1 \neq -\dfrac{1}{m_2}$, the lines are neither parallel nor

perpendicular.

47. The line $y = 3$ is horizontal and the line

$x = -\dfrac{1}{3}$ is vertical, so the lines are

perpendicular.

49. The line $3x + y = 4$ (or $y = -3x + 4$) has slope
$m_1 = -3$, and the line $x - 3y + 1 = 0$

$\left(\text{or } y = \dfrac{1}{3}x + \dfrac{1}{3}\right)$ has slope $m_2 = \dfrac{1}{3}$. Since

$m_2 = -\dfrac{1}{m_1}$, the lines are perpendicular.

51. The slope of $y = -\dfrac{x}{4} - 2$ is $-\dfrac{1}{4}$, so the

slope of a line parallel to it must also be

$-\dfrac{1}{4}$. An equation of the desired line is

$$y - 1 = -\dfrac{1}{4}(x - 1) \text{ or } y = -\dfrac{1}{4}x + \dfrac{5}{4}.$$

53. $y = 2$ is a horizontal line. A line parallel to it has the form $y = b$. Since the line must pass through $(2, 1)$ its equation is $y = 1$.

55. The slope of $y = 3x - 5$ is 3, so the slope of a

line perpendicular to it must have slope $-\dfrac{1}{3}$.

An equation of the desired line is

$$y - 4 = -\dfrac{1}{3}(x - 3), \text{ or } y = -\dfrac{1}{3}x + 5.$$

57. $y = -3$ is a horizontal line, so the perpendicular line must be vertical with equation of the form $x = a$. Since that line passes through $(5, 2)$, its equation is $x = 5$.

59. The line $2x + 3y + 6 = 0$ has slope $-\dfrac{2}{3}$, so the slope of a line parallel to it must also be $-\dfrac{2}{3}$. An equation of the desired line is

$$y - (-5) = -\frac{2}{3}[x - (-7)], \text{ or } y = -\frac{2}{3}x - \frac{29}{3}.$$

61. $(1, 2), (-3, 8)$

$$m = \frac{8-2}{-3-1} = \frac{6}{-4} = -\frac{3}{2}$$

Point-slope form: $y - 2 = -\dfrac{3}{2}(x-1)$. When the x-coordinate is 5,

$$y - 2 = -\frac{3}{2}(5-1)$$
$$y - 2 = -\frac{3}{2}(4)$$
$$y - 2 = -6$$
$$y = -4$$

Thus the point is $(5, -4)$.

63. Let $x =$ the time (in years) and $y =$ the price per share. Then,

In 1988: $x_1 = 1988$ and $y_1 = 37$

In 1998: $x_2 = 1998$ and $y_2 = 8$

The slope is

$$m = \frac{8-37}{1998-1988} = \frac{-29}{10} = -2.9$$

The stock price dropped an average of $2.90 per year.

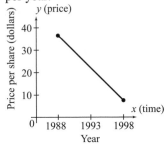

65. The owner's profits increased as a function of time. Let $x =$ the time (in years) and let $y =$ the profit (in dollars). The given points are $(x_1, y_1) = (0, -100,000)$ and $(x_2, y_2) = (5, 40,000)$.

$$m = \frac{y_2 - y_1}{x_2 - x_1} = \frac{40,000 - (-100,000)}{5-0}$$
$$= \frac{140,000}{5}$$
$$= 28,000$$

Using the point-slope form with $m = 28,000$ and $(x_1, y_1) = (0, -100,000)$ gives

$$y - y_1 = m(x - x_1)$$
$$y - (-100,000) = 28,000(x - 0)$$
$$y + 100,000 = 28,000x$$
$$y = 28,000x - 100,000$$

67. A general linear form of $d = 184 + t$ is $-t + d - 184 = 0$.

69. The slopes of the sides of the figure are:

$$\overline{AB} \left\{ m = \frac{4-0}{0-0} = \frac{4}{0} = \text{ undefined (vertical)} \right.$$

$$\overline{CD} \left\{ m = \frac{7-3}{2-2} = \frac{4}{0} = \text{ undefined (vertical)} \right.$$

$$\overline{AC} \left\{ m = \frac{3-0}{2-0} = \frac{3}{2} \right.$$

$$\overline{BD} \left\{ m = \frac{7-4}{2-0} = \frac{3}{2} \right.$$

Since $\overline{AB}$ is parallel to $\overline{CD}$ and $\overline{AC}$ is parallel to $\overline{BC}$, $ABCD$ is a parallelogram.

71. The line has slope 59.82 and passes through $(6, 1128.50)$. Thus
$C - 1128.50 = 59.82(T - 6)$ or
$C = 59.82T + 769.58$.

73.

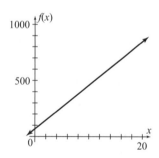

The graph of the equation $y = -0.9x - 7.3$
shows that when $x = 0$, $y = 7.3$. Thus, the
y-intercept is 7.3.

75. The slope is 7.1.

Principles in Practice 3.2

1. Let $x =$ the number of skis that are produced
and let $y =$ the number of boots that are
produced. Then, the equation
$8x + 14y = 1000$ describes all possible
production levels of the two products.

3. Answers may vary, but two possible points
are (0, 60) and (2, 140).
$f(x) = 40x + 60$

x	$f(x)$
0	60
2	140

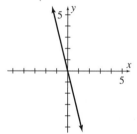

5. Let $y = f(x) = $ a linear function that describes
the value of the necklace after x years. The
problem states that $f(3) = 360$ and
$f(7) = 640$. Thus,
$(x_1, y_1) = (3, 360)$ and $(x_2, y_2) = (7, 640)$.
The slope is

$$m = \frac{y_2 - y_1}{x_2 - x_1} = \frac{640 - 360}{7 - 3} = \frac{280}{4} = 70$$

Using the point-slope form with $m = 70$ and
$(x_1, y_1) = (3, 360)$ gives

$$y - y_1 = m(x - x_1)$$
$$y - 360 = 70(x - 3)$$
$$y = f(x) = 70x + 150$$

Problems 3.2

1. $y = f(x) = -4x = -4x + 0$ has the form
$f(x) = ax + b$ where $a = -4$ (the slope) and
$b = 0$ (the vertical-axis intercept).

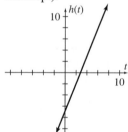

3. $h(t) = 5t - 7$ has the form $h(t) = at + b$ with a
$= 5$ (the slope) and $b = -7$ (the vertical-axis
intercept).

5. $h(q) = \dfrac{2 - q}{7} = \dfrac{2}{7} - \dfrac{1}{7}q$ has the form

$h(q) = aq + b$ where $a = -\dfrac{1}{7}$ (the slope) and

$b = \dfrac{2}{7}$ (the vertical-axis intercept).

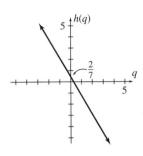

7. $f(x) = ax + b = 4x + b$. Since $f(2) = 8$,
$8 = 4(2) + b$, $8 = 8 + b$, $b = 0 \Rightarrow f(x) = 4x$.

9. Let $y = f(x)$. The points $(1, 2)$ and $(-2, 8)$ lie
on the graph of f. $m = \dfrac{8-2}{-2-1} = -2$. Thus
$y - 2 = -2(x - 1)$, so
$y = -2x + 4 \Rightarrow f(x) = -2x + 4$.

11. $f(x) = ax + b = -\dfrac{2}{3}x + b$. Since

$f\left(-\dfrac{2}{3}\right) = -\dfrac{2}{3}$,

we have

$-\dfrac{2}{3} = -\dfrac{2}{3}\left(-\dfrac{2}{3}\right) + b$

$b = -\dfrac{2}{3} - \dfrac{4}{9} = -\dfrac{10}{9}$,

so $f(x) = -\dfrac{2}{3}x - \dfrac{10}{9}$.

13. Let $y = f(x)$. The points $(-2, -1)$ and $(-4, -3)$
lie on the graph of f. $m = \dfrac{-3+1}{-4+2} = 1$. Thus
$y + 1 = 1(x + 2)$, so $y = x + 1 \Rightarrow f(x) = x + 1$.

15. The points $(40, 12.75)$ and $(25, 18.75)$ lie on
the graph of the equation, which is a line.
$m = \dfrac{18.75 - 12.75}{25 - 40} = -\dfrac{2}{5}$. Hence an equation
of the line is $p - 12.75 = -\dfrac{2}{5}(q - 40)$, which

can be written $p = -\dfrac{2}{5}q + 28.75$. When

$q = 37$, then $p = -\dfrac{2}{5}(37) + 28.75 = \$13.95$.

17. The line passes through $(3000, 940)$ and
$(2200, 740)$, so $m = \dfrac{740 - 940}{2200 - 3000} = 0.25$.
Then
$p - 740 = 0.25(q - 2200)$ or
$p = 0.25q + 190$.

19. The line passing through $(10, 40)$ and $(20, 70)$ has slope $\dfrac{70 - 40}{20 - 10} = 3$, so an equation
for the line is
$c - 40 = 3(q - 10)$
$c = 3q + 10$
If $q = 35$, then
$c = 3(35) + 10 = 105 + 10 = \$115$.

21. If $x =$ the number of kilowatt hours used in a
month, then $f(x) =$ the total monthly
charges for x kilowatt hours of electricity. If
$f(x)$ is a linear function it has the form
$f(x) = ax + b$. The problem states that
$f(380) = 51.65$. Since 12.5 cents are charged
per kilowatt hour used, $a = 0.125$.
$f(x) = ax + b$
$51.65 = 0.125(380) + b$
$51.65 = 47.5 + b$
$4.15 = b$
Hence, $f(x) = 0.125x + 4.15$ is a linear
function that describes the total monthly
charges for any number of kilowatt hours x.

23. Each year the value decreases by
$0.10(1800)$. After t years the total decrease
is $0.10(1800)t$. Thus
$v = 1800 - 0.10(1800)t$
$v = -180t + 1800$
The slope is -180.

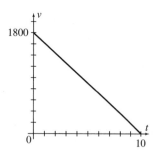

25. The line has slope 45,000 and passes through (5, 960,000). Thus
$y - 960{,}000 = 45{,}000(x - 5)$ or
$y = f(x) = 45{,}000x + 735{,}000$.

27. If $x =$ the number of hours of service, then $f(x) =$ the price of x hours of service. Let $y = f(x)$. $f(1) = 159$ and $f(3) = 287$, so (1, 159) and (3, 287) lie on the graph of f
which has slope $a = \dfrac{287 - 159}{3 - 1} = 64$. Using
(1, 159), we get $y - 159 = 64(x - 1)$ or
$y = 64x + 95$, so $f(x) = 64x + 95$.

29. At \$200/ton, x tons cost $200x$, and at \$2000/acre, y acres cost $2000y$. Hence the required equation is
$200x + 2000y = 20{,}000$, which can be written as $x + 10y = 100$.

31. a. $m = \dfrac{100 - 65}{100 - 56} = \dfrac{35}{44}$

$y - 100 = \dfrac{35}{44}(x - 100)$

$y = \dfrac{35}{44}x - \dfrac{3500}{44} + 100$

$y = \dfrac{35}{44}x + \dfrac{225}{11}$

b. $62 = \dfrac{35}{44}x + \dfrac{225}{11}$

$\dfrac{35}{44}x = 62 - \dfrac{225}{11}$

$x = \dfrac{1828}{35} \approx 52.2$

52.2 is the lowest passing score on original scale.

33. $p = f(t) = at + b$, $f(5) = 0.32$,
$a = \text{slope} = 0.059$.

a. $p = f(t) = 0.059t + b$. Since $f(5) = 0.32$,
$0.32 = 0.059(5) + b$, $0.32 = 0.295 + b$,
so $b = 0.025$. Thus $p = 0.059t + 0.025$.

b. When $t = 9$, then
$p = 0.059(9) + 0.025 = 0.556$.

35. a. $m = \dfrac{t_2 - t_1}{c_2 - c_1} = \dfrac{80 - 68}{172 - 124} = \dfrac{12}{48} = \dfrac{1}{4}$.

$t - 68 = \dfrac{1}{4}(c - 124)$, $t - 68 = \dfrac{1}{4}c - 31$,

or $t = \dfrac{1}{4}c + 37$.

b. Since c is the number of chirps per minute, then $\dfrac{1}{4}c$ is the number of chirps in $\dfrac{1}{4}$ minute or 15 seconds. Thus from part (a), to estimate temperature add 37 to the number of chirps in 15 seconds.

Principles in Practice 3.3

1. In the quadratic function
$y = P(x) = -x^2 + 2x + 399$, $a = -1$, $b = 2$,
$c = 399$. Since $a < 0$, the parabola opens downward. The x-coordinate of the vertex is
$-\dfrac{b}{2a} = -\dfrac{2}{2(-1)} = 1$.
The y-coordinate of the vertex is

$P(1) = -\left(1^2\right) + 2(1) + 399 = 400$. Thus, the vertex is (1, 400). Since $c = 399$, the y-intercept is (0, 399). To find the x-intercepts we set $y = p(x) = 0$.

$0 = -x^2 + 2x + 399$

$0 = -\left(x^2 - 2x - 399\right)$

$0 = -(x + 19)(x - 21)$

Thus, the x-intercepts are (−19, 0) and (21, 0).

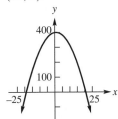

If the model is correct, this is not a good business, since it will lose money if more than 21 minivans are sold.

3. If we express the revenue r as a function of the quantity produced q, we obtain

$r = pq$

$r = (6 - 0.003q)q$

$r = 6q - 0.003q^2$

We note that this is a quadratic function with $a = -0.003$, $b = 6$, and $c = 0$. Since $a < 0$, the graph of the function is a parabola that opens downward, and r is maximum at the vertex (q, r).

$q = -\dfrac{b}{2a} = -\dfrac{6}{2(-0.003)} = 1000$

$r = 6(1000) - 0.003(1000)^2 = 3000$

Thus, the maximum revenue that the manufacturer can receive is $3000, which occurs at a production level of 1000 units.

Problems 3.3

1. $f(x) = 5x^2$ has the form

$f(x) = ax^2 + bx + c$ where $a = 5$, $b = 0$, and $c = 0 \Rightarrow$ quadratic.

3. $g(x) = 7 - 6x$ cannot be put in the form $g(x) = ax^2 + bx + c$ where $a \neq 0 \Rightarrow$ not quadratic.

5. $h(q) = (3 - q)^2 = 9 - 6q + q^2$ has form $h(q) = aq^2 + bq + c$ where $a = 1$, $b = -6$, and $c = 9 \Rightarrow$ quadratic.

7. $f(s) = \dfrac{s^2 - 9}{2} = \dfrac{1}{2}s^2 - \dfrac{9}{2}$ has the form

$f(s) = as^2 + bs + c$ where $a = \dfrac{1}{2}$, $b = 0$,

and $c = -\dfrac{9}{2} \Rightarrow$ quadratic.

9. $y = f(x) = -4x^2 + 8x + 7$

$a = -4$, $b = 8$, $c = 7$

  a. Vertex occurs when

$x = -\dfrac{b}{2a} = -\dfrac{8}{2(-4)} = 1$. When $x = 1$,

then $y = f(1) = -4(1)^2 + 8(1) + 7 = 11$.

Vertex: (1, 11)

  b. $a = -4 < 0$, so the vertex corresponds to the highest point.

11. $y = x^2 + x - 6$

$a = 1$, $b = 1$, $c = -6$

  a. $c = -6$. Thus the y-intercept is −6.

  b. $x^2 + x - 6 = (x - 2)(x + 3) = 0$, so $x = 2, -3$.

x-intercepts: 2, −3

  c. $-\dfrac{b}{2a} = -\dfrac{1}{2}$

$f\left(-\dfrac{1}{2}\right) = \left(-\dfrac{1}{2}\right)^2 - \dfrac{1}{2} - 6 = -\dfrac{25}{4}$

Vertex: $\left(-\dfrac{1}{2}, -\dfrac{25}{4}\right)$

13.　$y = f(x) = x^2 - 6x + 5$

　　$a = 1, b = -6, c = 5$

　　Vertex: $-\dfrac{b}{2a} = \dfrac{-6}{2 \cdot 1} = 3$

　　$f(3) = 3^2 - 6(3) + 5 = -4$

　　Vertex = $(3, -4)$

　　y-intercept: $c = 5$

　　x-intercepts:

　　$x^2 - 6x + 5 = (x-1)(x-5) = 0$, so $x = 1, 5$.

　　Range: all $y \geq -4$

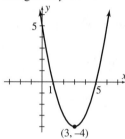

15.　$y = g(x) = -2x^2 - 6x$

　　$a = -2, b = -6, c = 0$

　　Vertex: $-\dfrac{b}{2a} = -\dfrac{-6}{2(-2)} = -\dfrac{6}{4} = -\dfrac{3}{2}$

　　$f\left(\dfrac{-3}{2}\right) = -2\left(\dfrac{-3}{2}\right)^2 - 6\left(\dfrac{-3}{2}\right) = \dfrac{-9}{2} + 9 = \dfrac{9}{2}$

　　Vertex: $\left(-\dfrac{3}{2}, \dfrac{9}{2}\right)$

　　y-intercept: $c = 0$

　　x-intercepts: $-2x^2 - 6x = -2x(x+3) = 0$, so $x = 0, -3$.

　　Range: all $y \leq \dfrac{9}{2}$

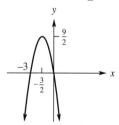

17.　$s = h(t) = t^2 + 6t + 9$

　　$a = 1, b = 6, c = 9$

　　Vertex: $-\dfrac{b}{2a} = -\dfrac{6}{2 \cdot 1} = -3$

　　$h(-3) = (-3)^2 + 6(-3) + 9 = 0$

　　Vertex = $(-3, 0)$

　　s-intercept: $c = 9$

　　t-intercepts: $t^2 + 6t + 9 = (t+3)^2 = 0$, so $t = -3$.

　　Range: all $s \geq 0$

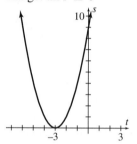

19.　$y = f(x) = -9 + 8x - 2x^2$

　　$a = -2, b = 8, c = -9$

　　Vertex: $-\dfrac{b}{2a} = -\dfrac{8}{2(-2)} = 2$

　　$f(2) = -9 + 8(2) - 2(2)^2 = -1$

　　Vertex = $(2, -1)$

　　y-intercept: $c = -9$

　　x-intercepts: Because the parabola opens downward ($a < 0$) and the vertex is below the x-axis, there is no x-intercept.

　　Range: $y \leq -1$

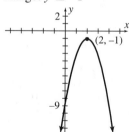

21. $t = f(s) = s^2 - 8s + 14$

$a = 1, b = -8, c = 14$

Vertex: $-\dfrac{b}{2a} = -\dfrac{-8}{2 \cdot 1} = 4$

$f(4) = 4^2 - 8(4) + 14 = -2$

Vertex $= (4, -2)$

t-intercept: $c = 14$

s-intercepts: Solving $s^2 - 8s + 14 = 0$ by the quadratic formula:

$s = \dfrac{-(-8) \pm \sqrt{(-8)^2 - 4(1)(14)}}{2(1)}$

$= \dfrac{8 \pm \sqrt{8}}{2} = \dfrac{8 \pm 2\sqrt{2}}{2} = 4 \pm \sqrt{2}$

Range: all $t \geq -2$

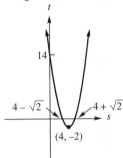

23. $f(x) = 49x^2 - 10x + 17$

Since $a = 49 > 0$, the parabola opens upward and $f(x)$ has a minimum value that occurs when $x = -\dfrac{b}{2a} = -\dfrac{-10}{2 \cdot 49} = \dfrac{5}{49}$. The minimum value is

$f\left(\dfrac{5}{49}\right) = 49\left(\dfrac{5}{49}\right)^2 - 10\left(\dfrac{5}{49}\right) + 17 = \dfrac{808}{49}$.

25. $f(x) = 4x - 50 - 0.1x^2$

Since $a = -0.1 < 0$, the parabola opens downward and $f(x)$ has a maximum value that occurs when

$x = -\dfrac{b}{2a} = -\dfrac{4}{2(-0.1)} = 20$. The maximum

value is

$f(20) = 4(20) - 50 - 0.1(20)^2 = -10$.

27. $f(x) = x^2 - 2x + 4$

$a = 1, b = -2, c = 4$

$v = -\dfrac{b}{2a} = -\dfrac{-2}{2(1)} = 1$

The restricted function is

$g(x) = x^2 - 2x + 4$,

$x \geq 1$. From the quadratic formula applied to $x^2 - 2x + 4 - y = 0$, we get

$x = \dfrac{2 \pm \sqrt{4 - 4(1)(4 - y)}}{2(1)} = 1 \pm \sqrt{1 - (4 - y)}$

So the inverse of $g(x)$ is

$g^{-1}(x) = 1 + \sqrt{x - 3}$,

$x \geq 3$.

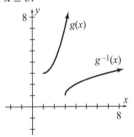

29. If we express the revenue r as a function of the quantity produced q, we obtain

$r = pq$

$r = (200 - 5q)q$

$r = 200q - 5q^2$

This is a quadratic function with $a = -5$, $b = 200$, and $c = 0$. Since $a < 0$, the graph of the function is a parabola that opens downward, and r is maximum at the vertex (q, r).

$q = -\dfrac{b}{2a} = -\dfrac{200}{2(-5)} = 20$

$r = 200(20) - 5(20)^2 = 2000$

Thus, the maximum revenue that the manufacturer can receive is $2000, which occurs at a production level of 20 units.

31. If we express the revenue r as a function of the quantity produced q, we obtain

$r = pq$

$r = (2400 - 6q)q$

$r = 2400q - 6q^2$

This is a quadratic function with $a = -6$, $b = 2400$, and $c = 0$. Since $a < 0$, the graph of the function is a parabola that opens downward, and r is maximum at the vertex (q, r).

$q = -\dfrac{b}{2a} = -\dfrac{2400}{2(-6)} = 200$

$r = 2400(200) - 6(200)^2 = 240,000$

Thus, the maximum revenue that the manufacturer can receive is $240,000, which occurs at a production level of 200 units.

33. In the quadratic function

$P(x) = -x^2 + 18x + 144,$

$a = -1$, $b = 18$, and $c = 144$. Since $a < 0$, the graph of the function is a parabola that opens downward. The x-coordinate of the vertex is $-\dfrac{b}{2a} = -\dfrac{18}{2(-1)} = 9$. The

y-coordinate of the vertex is

$P(9) = -\left(9^2\right) + 18(9) + 144 = 225$. Thus, the

vertex is (9, 225). Since $c = 144$, the y-intercept is (0, 144). To find the x-intercepts, let $y = P(x) = 0$.

$0 = -x^2 + 18x + 144$

$0 = -\left(x^2 - 18x - 144\right)$

$0 = -(x - 24)(x + 6)$

Thus, the x-intercepts are (24, 0) and (–6, 0).

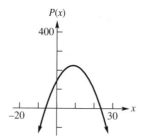

35. $f(P) = -\dfrac{1}{50}P^2 + 2P + 20$, where

$0 \le P \le 100$. Because $a = -\dfrac{1}{50} < 0$, $f(P)$

has a maximum value that occurs at the vertex.

$-\dfrac{b}{2a} = -\dfrac{2}{2\left(-\frac{1}{50}\right)} = 50$. The maximum

value of $f(P)$ is

$f(50) = \dfrac{-1}{50}(50)^2 + 2(50) + 20 = 70$ grams.

37. $h(t) = -16t^2 + 85t + 22$

Since $a = -16 < 0$, $h(t)$ has a maximum value that occurs at the vertex where

$t = -\dfrac{b}{2a} = -\dfrac{85}{2(-16)} \approx 2.7$ sec. When

$t = 2.7$, then

$h(t) = -16(2.7)^2 + 85(2.7) + 22$
$= 134.86$ feet.

39. In the quadratic function

$h(t) = -16t^2 + 80t + 16$, $a = -16$, $b = 80$, and

$c = 16$. Since $a < 0$, the graph of the function is a parabola that opens downward. The x-coordinate of the vertex is

$-\dfrac{b}{2a} = -\dfrac{80}{2(-16)} = \dfrac{5}{2}$.

The y-coordinate of the vertex is

$h\left(\dfrac{5}{2}\right) = -16\left(\dfrac{5}{2}\right)^2 + 80\left(\dfrac{5}{2}\right) + 16 = 116$

Thus, the vertex is $\left(\dfrac{5}{2}, 116\right)$. Since $c = 16$,

the y-intercept is $(0, 16)$. To find the x-intercepts, we let $y = h(t) = 0$.

$$0 = -16t^2 + 80t + 16$$

$$t = \frac{-b \pm \sqrt{b^2 - 4ac}}{2a}$$

$$= \frac{-80 \pm \sqrt{80^2 - 4(-16)(16)}}{2(-16)}$$

$$= \frac{-80 \pm \sqrt{7424}}{-32} = \frac{5 \pm \sqrt{29}}{2}$$

Thus, the x-intercepts are $\left(\dfrac{5 + \sqrt{29}}{2}, 0\right)$ and

$\left(\dfrac{5 - \sqrt{29}}{2}, 0\right)$.

41. Since the total length of fencing is 500, the side opposite the highway has length $500 - 2x$. The area A is given by
    $$A = x(500 - 2x) = 500x - 2x^2,$$
    which is quadratic with $a = -2 < 0$. Thus A is maximum when $x = -\dfrac{500}{2(-2)} = 125$. Then the side opposite the highway is $500 - 2x = 500 - 2(125) = 250$. Thus the dimensions are 125 ft by 250 ft.

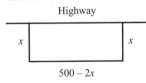

43. $(1.11, 2.88)$

45. **a.**   none

    b.   one

    c.   two

47. 4.89

Principles in Practice 3.4

1. Let x = the number invested at 9% and let y = the amount invested at 8%. Then, the problem states
   $$\begin{cases} x + y = 200{,}000, \\ 0.09x + 0.08y = 17{,}200. \end{cases}$$
   We eliminate x by multiplying the first equation by -0.09 and then adding
   $$\begin{cases} -0.09x - 0.09y = -18{,}000, \\ 0.09x + 0.08y = 17{,}200. \end{cases}$$
   $$-0.01y = -800,$$
   $$y = 80{,}000.$$
   Therefore,
   $$\begin{cases} x = 120{,}000, \\ y = 80{,}000. \end{cases}$$
   Thus, \$120,000 is invested at 9% and \$80,000 is invested at 8%.

3. Let A = the number of fish of species A, and let B = the number of fish of species B. Then, the number of milligrams of the first supplement that will be consumed is $15A + 20B = 100{,}000$. The number of milligrams of the second supplement that will be consumed is $30A + 40B = 200{,}000$.
   $$\begin{cases} 15A + 20B = 100{,}000, \\ 30A + 40B = 200{,}000. \end{cases}$$
   We multiply the second equation by $-\dfrac{1}{2}$ and then add.
   $$\begin{cases} 15A + 20B = 100{,}000, \\ -15A - 20B = -100{,}000, \end{cases}$$
   $$0 = 0$$

Thus, there are infinitely many solutions of the form $A = \dfrac{20,000}{3} - \dfrac{4}{3}r$, $B = r$, where $0 \le r \le 5000$.

Problems 3.4

1. $\begin{cases} x + 4y = 3, & (1) \\ 3x - 2y = -5. & (2) \end{cases}$

From Eq. (1), $x = 3 - 4y$. Substituting in Eq. (2) gives
$3(3 - 4y) - 2y = -5$
$9 - 12y - 2y = -5$
$-14y = -14,$
or $y = 1 \Rightarrow x = 3 - 4y = 3 - 4(1) = -1.$
Thus $x = -1$, $y = 1$.

3. $\begin{cases} 3x - 4y = 13, & (1) \\ 2x + 3y = 3. & (2) \end{cases}$

Multiplying Eq. (1) by 3 and Eq. (2) by 4 gives
$\begin{cases} 9x - 12y = 39, \\ 8x + 12y = 12. \end{cases}$
Adding gives
$17x = 51$
$x = 3$
From Eq. (2) we have
$2(3) + 3y = 3$
$3y = -3$
$y = -1$
Thus $x = 3$, $y = -1$.

5. $\begin{cases} u + v = 5 \\ u - v = 7 \end{cases}$

From the first equation, $v = 5 - u$.
Substituting in the second equation gives
$u - (5 - u) = 7$
$\quad 2u - 5 = 7$
$\qquad 2u = 12$
or $u = 6$ so $v = 5 - u = 5 - 6 = -1$.
Thus, $u = 6$, $v = -1$.

7. $\begin{cases} x - 2y = -7, & (1) \\ 5x + 3y = -9. & (2) \end{cases}$

From Eq. (1), $x = 2y - 7$. Substituting in Eq. (2) gives
$5(2y - 7) + 3y = -9$
$13y = 26$
$y = 2 \Rightarrow x = 2y - 7 = 2(2) - 7 = -3.$
Thus $x = -3$, $y = 2$.

9. $\begin{cases} 4x - 3y - 2 = 3x - 7y, \\ x + 5y - 2 = y + 4. \end{cases}$

Simplifying, we have
$\begin{cases} x + 4y = 2, \\ x + 4y = 6. \end{cases}$
Subtracting the second equation from the first gives $0 = -4$, which is never true. Thus there is no solution.

11. $\begin{cases} \dfrac{2}{3}x + \dfrac{1}{2}y = 2, \\[2mm] \dfrac{3}{8}x + \dfrac{5}{6}y = -\dfrac{11}{2}. \end{cases}$

Clearing fractions gives the system
$\begin{cases} 4x + 3y = 12, \\ 9x + 20y = -132. \end{cases}$
Multiplying the first equation by 9 and the second equation by -4 gives
$\begin{cases} 36x + 27y = 108, \\ -36x - 80y = 528. \end{cases}$
Adding gives
$-53y = 636$
$y = -12$
From $4x + 3y = 12$, we have
$4x + 3(-12) = 12$
$4x = 48 \Rightarrow x = 12.$ Thus $x = 12$, $y = -12$.

13. $\begin{cases} 5p + 11q = 7, & (1) \\ 10p + 22q = 33. & (2) \end{cases}$

Multiplying Eq. (1) by –2 gives
$\begin{cases} -10p - 22q = -14, \\ 10p + 22q = 33. \end{cases}$
Adding gives 0 = 19, which is never true, so the system has no solution.

15. $\begin{cases} 2x + y + 6z = 3, & (1) \\ x - y + 4z = 1, & (2) \\ 3x + 2y - 2z = 2. & (3) \end{cases}$

Adding Eq. (1) and (2), and adding 2 times Eq. (2) to Eq. (3) gives
$\begin{cases} 3x + 10z = 4, \\ 5x + 6z = 4. \end{cases}$
Multiplying the first equation by 5 and the second equation by –3 gives
$\begin{cases} 15x + 50z = 20, \\ -15x - 18z = -12. \end{cases}$

Adding gives $32z = 8$, or $z = \dfrac{1}{4}$. From

$3x + 10z = 4$, we have

$3x + 10\left(\dfrac{1}{4}\right) = 4$

$3x = \dfrac{3}{2}$

$x = \dfrac{1}{2}$

From $2x + y + 6z = 3$, we have

$2\left(\dfrac{1}{2}\right) + y + 6\left(\dfrac{1}{4}\right) = 3$

$y = \dfrac{1}{2}$

Therefore $x = \dfrac{1}{2}$, $y = \dfrac{1}{2}$, $z = \dfrac{1}{4}$.

17. $\begin{cases} x + 4y + 3z = 10 \\ 4x + 2y - 2z = -2 \\ 3x - y + z = 11 \end{cases}$

From the third equation, $y = 3x + z - 11$.
Substituting in the first two equations gives
$\begin{cases} x + 4(3x + z - 11) + 3z = 10 \\ 4x + 2(3x + z - 11) - 2z = -2 \end{cases}$

or
$\begin{cases} 13x + 7z = 54 \\ 10x = 20 \end{cases}$

From the last equation we have $x = 2$.
Thus $13(2) + 7z = 54$, and $7z = 28$, hence $z = 4$.
Substitute these two values to solve for y:
$y = 3(2) + 4 - 11 = -1$
Therefore, $x = 2$, $y = -1$, $z = 4$.

19. $\begin{cases} x - 2z = 1, & (1) \\ y + z = 3. & (2) \end{cases}$

From Eq. (1), $x = 1 + 2z$; from Eq. (2),
$y = 3 - z$. Setting $z = r$ gives the parametric solution $x = 1 + 2r$, $y = 3 - r$, $z = r$, where r is any real number.

21. $\begin{cases} x - y + 2z = 0, & (1) \\ 2x + y - z = 0 & (2) \\ x + 2y - 3z = 0 & (3) \end{cases}$

Adding Eq. (1) to Eq. (3) gives
$\begin{cases} x - y + 2z = 0, \\ 2x + y - z = 0 \\ 2x + y - z = 0 \end{cases}$

We can ignore the third equation because the second equation can be used to reduce it to 0 = 0. We have
$\begin{cases} x - y + 2z = 0, \\ 2x + y - z = 0. \end{cases}$
Adding the first equation to the second gives
$3x + z = 0$

$x = -\dfrac{1}{3}z$

Substituting in the first equation we have

$$-\frac{1}{3}z - y + 2z = 0$$

$$y = \frac{5}{3}z$$

Letting $z = r$ gives the parametric solution

$x = -\frac{1}{3}r$, $y = \frac{5}{3}r$, $z = r$, where r is any real number.

23. $\begin{cases} 2x + 2y - z = 3, & (1) \\ 4x + 4y - 2z = 6. & (2) \end{cases}$

Multiplying Eq. (2) by $-\frac{1}{2}$ gives

$\begin{cases} 2x + 2y - z = 3, \\ -2x - 2y + z = -3. \end{cases}$

Adding the first equation to the second equation gives

$\begin{cases} 2x + 2y - z = 3, \\ 0 = 0. \end{cases}$

Solving the first equation for x, we have

$x = \frac{3}{2} - y + \frac{1}{2}z$. Letting $y = r$ and $z = s$

gives the parametric solution

$x = \frac{3}{2} - r + \frac{1}{2}s$, $y = r$, $z = s$, where r and s are any real numbers.

25. Let x = number of gallons of 20% solution and y = number of gallons of 35% solution. Then

$\begin{cases} x + y = 800, & (1) \\ 0.20x + 0.35y = 0.25(800). & (2) \end{cases}$

From Eq. (1), $y = 800 - x$. Substituting in Eq. (2) gives

$0.20x + 0.35(800 - x) = 0.25(800)$

$-0.15x + 280 = 200$

$-0.15x = -80$

$x = \frac{1600}{3} \approx 533.3$

$y = 800 - x = 800 - \frac{1600}{3} = \frac{800}{3} \approx 266.7.$

Thus 533.3 gal of 20% solution and 266.7 gal of 35% solution must be mixed.

27. Let C = the number of pounds of cotton, let P = the number of pounds of polyester, and let N = the number of pounds of nylon. If the final blend will cost \$3.25 per pound to make, then $4C + 3P + 2N = 3.25$. Furthermore, if we use the same amount of nylon as polyester to prepare, say, 1 pound of fabric, then $N = P$ and $C + P + N = 1$. Thus, the system of equations is

$\begin{cases} 4C + 3P + 2N = 3.25, \\ C + P + N = 1, \\ N = P. \end{cases}$

Simplifying gives

$\begin{cases} 4C + 5N = 3.25, \\ C + 2N = 1, \\ N = P. \end{cases}$

$\begin{cases} N = 0.25, \\ C = 0.5, \\ P = 0.25. \end{cases}$

Thus, each pound of the final fabric will contain 0.25 lb each of nylon and polyester, and 0.5 lb of cotton.

29. Let p = speed of airplane in still air and w = wind speed. Now convert the time into minutes and solve the system

$\begin{cases} p + w = \dfrac{900}{175} \\ p - w = \dfrac{900}{206}, \end{cases}$

Thus $2p = \dfrac{900}{175} + \dfrac{900}{206} = \dfrac{36}{7} + \dfrac{450}{103}$

$p = \dfrac{3429}{721}$ miles per minute

$w = \dfrac{279}{721}$ miles per minute

Multiplying by 60 to get miles per hour we have $p \approx 285$ and $w \approx 23.2$

Plane speed in still air is about 285 mph and wind speed is about 23.2 mph.

31. Let x = number of early American units and y = number of Contemporary units. The fact that 20% more of early American styles are sold than Contemporary styles means that
$x = y + 0.20y$
$x = 1.20y$
An analysis of profit gives
$250x + 350y = 130,000$. Thus we have the system
$$\begin{cases} x = 1.20y, & (1) \\ 250x + 350y = 130,000. & (2) \end{cases}$$
Substituting $1.20y$ for x in Eq. (2) gives
$250(1.20y) + 350y = 130,000$
$300y + 350y = 130,000$
$650y = 130,000$
$y = 200$
Thus $x = 1.20y = 1.20(200) = 240$.
Therefore 240 units of early American and 200 units of Contemporary must be sold.

33. Let x = number of calculators produced at Exton, and y = number of calculators produced at Whyton. The total cost of Exton is $7.50x + 7000$, and the total cost at Whyton is $6.00y + 8800$. Thus $7.50x + 7000 = 6.00y + 8800$. Also,
$x + y = 1500$. This gives the system
$$\begin{cases} x + y = 1500, & (1) \\ 7.50x + 7000 = 6.00y + 8800. & (2) \end{cases}$$
From Eq. (1), $y = 1500 - x$. Substituting in Eq. (2) gives
$7.50x + 7000 = 6.00(1500 - x) + 8800$
$7.50x + 7000 = 9000 - 6x + 8800$
$13.5x = 10,800$
$x = 800$
Thus $y = 1500 - x = 1500 - 800 = 700$.
Therefore 800 calculators must be made at the Exton plant and 700 calculators at the Whyton plant.

35. Let x = rate on first \$100,000 and y = rate on sales over \$100,000. Then
$$\begin{cases} 100,000x + 75,000y = 8500, & (1) \\ 100,000x + 180,000y = 14,800. & (2) \end{cases}$$
Subtracting Eq. (1) from Eq. (2) gives
$105,000y = 6300$
$y = 0.06$

Substituting in Eq. (1) gives
$100,000x + 75,000(0.06) = 8500$
$100,000x + 4500 = 8500$, $100,000x = 4000$,
or $x = 0.04$. Thus the rate is 4% on the first \$100,000 and 6% on the remainder.

37. Let x = number of loose-filled boxes and y = number of boxes of clam-shells that will be filled. Then $8y$ clam-shells will be used. This will take $20x + 2.2(8y)$ pounds of peaches.
$$\begin{cases} x = y & (1) \\ 20x + 17.6y = 3600 & (2) \end{cases}$$
Substitute $x = y$ in Eq. (2).
$20x + 17.6x = 3600$
$37.6x = 3600$
$x \approx 95.74$
$y = x \approx 95.74$
Thus, 95 boxes will be loose-filled and $8(95) = 760$ clam-shells will be used, for a total of 190 boxes.

39. Let c = number of chairs company makes, r = number of rockers, and l = number of chaise lounges.
Wood used: $(1)c + (1)r + (1)l = 400$
Plastic used: $(1)c + (1)r + (2)l = 600$
Aluminum used: $(2)c + (3)r + (5)l = 1500$
Thus we have the system
$$\begin{cases} c + r + l = 400, & (1) \\ c + r + 2l = 600, & (2) \\ 2c + 3r + 5l = 1500. & (3) \end{cases}$$
Subtracting Eq. (1) from Eq. (2) gives $l = 200$. Adding -2 times Eq. (1) to Eq. (3) gives $r + 3l = 700$, from which
$r + 3(200) = 700$,
$r = 100$
From Eq. (1) we have $c + 100 + 200 = 400$, or $c = 100$. Thus 100 chairs, 100 rockers and 200 chaise lounges should be made.

41. Let x = number of skilled workers employed,
y = number of semiskilled workers employed,
z = number of shipping clerks employed.
Then we have the system

$$\begin{cases} \text{number of workers:} & x+y+z=70, & (1) \\ \text{wages:} & 16x+9.5y+10z=725 & (2) \\ \text{semiskilled:} & y=2x & (3) \end{cases}$$

From the last equation, $y=2x$ so substitute into the first two equations:

$$\begin{cases} x+2x+z=70 \\ 16x+9.5(2x)+10z=725 \end{cases}$$

or

$$\begin{cases} 3x+z=70 \\ 35x+10z=725 \end{cases}$$

Adding -10 times the first equation to the second gives:

$5x=25$

$x=5$

So $y=2x=10$

$z=70-3x=70-15=55$

The company should hire 5 skilled workers, 10 semiskilled workers, and 55 shipping clerks.

45. $x=3, y=2$

47. $x=8.3, y=14.0$

Problems 3.5

In the following solutions, any reference to Eq. (1) or Eq. (2) refers to the first or second equation, respectively, in the given system.

1. From Eq. (2), $y=3-2x$. Substituting in Eq. (1) gives

$$3-2x=x^2-9$$

$$0=x^2+2x-12$$

$$x=\frac{-b\pm\sqrt{b^2-4ac}}{2a}$$

$$=\frac{-2\pm\sqrt{2^2-4(1)(-12)}}{2(1)}$$

$$=\frac{-2\pm\sqrt{52}}{2}$$

$$=-1\pm\sqrt{13}$$

From $y=3-2x$, if $x=-1+\sqrt{13}$, then $y=5-2\sqrt{13}$; if $x=-1-\sqrt{13}$, then $y=5+2\sqrt{13}$.

There are two solutions: $x=-1+\sqrt{13}$, $y=5-2\sqrt{13}$; $x=-1-\sqrt{13}$, $y=5+2\sqrt{13}$.

3. From Eq. (2), $q = p - 1$. Substituting in Eq. (1) gives

$$p^2 = 5 - (p - 1)$$

$$p^2 + p - 6 = 0$$

$$(p + 3)(p - 2) = 0$$

Thus $p = -3, 2$. From $q = p - 1$, if $p = -3$, we have $q = -3 - 1 = -4$; if $p = 2$, then $q = 2 - 1 = 1$. There are two solutions: $p = -3, q = -4; p = 2, q = 1$.

5. Substituting $y = x^2$ into $x = y^2$ gives

$$x = x^4, \ x^4 - x = 0$$

$$x\left(x^3 - 1\right) = 0$$

Thus $x = 0, 1$. From $y = x^2$, if $x = 0$, then $y = 0^2 = 0$; $x = 1$, then $y = 1^2 = 1$. There are two solutions: $x = 0, y = 0; x = 1, y = 1$.

7. Substituting $y = x^2 - 2x$ in Eq. (1) gives

$$x^2 - 2x = 4x - x^2 + 8$$

$$2x^2 - 6x - 8 = 0$$

$$x^2 - 3x - 4 = 0$$

$$(x - 4)(x + 1) = 0$$

Thus $x = 4, -1$. From $y = x^2 - 2x$, if $x = 4$, then we have $y = 4^2 - 2(4) = 8$; if $x = -1$, then $y = (-1)^2 - 2(-1) = 3$. There are two solutions: $x = 4, y = 8; x = -1, y = 3$.

9. Substituting $p = \sqrt{q}$ in Eq. (2) gives

$\sqrt{q} = q^2$. Squaring both sides gives

$$q = q^4$$

$$q^4 - q = 0$$

$$q\left(q^3 - 1\right) = 0$$

Thus $q = 0, 1$. From $p = \sqrt{q}$, if $q = 0$, then $p = \sqrt{0} = 0$; if $q = 1$, then $p = \sqrt{1} = 1$. There are two solutions: $p = 0, q = 0; p = 1, q = 1$.

11. Replacing x^2 by $y^2 + 13$ in Eq. (2) gives

$$y = \left(y^2 + 13\right) - 15$$

$$y^2 - y - 2 = 0$$

$$(y - 2)(y + 1) = 0$$

Thus $y = 2, -1$. If $y = 2$, then $x^2 = y^2 + 13 = 2^2 + 13 = 17$, so $x = \pm\sqrt{17}$. If $y = -1$, then $x^2 = y^2 + 13 = (-1)^2 + 13 = 14$, so $x = \pm\sqrt{14}$. The system has four solutions: $x = \sqrt{17}, y = 2; \ x = -\sqrt{17}, y = 2; \ x = \sqrt{14}, y = -1; \ x = -\sqrt{14}, y = -1$.

13. From Eq. (1), $y = x - 1$. Substituting in Eq. (2) gives

$$x - 1 = 2\sqrt{x + 2}$$

$$(x - 1)^2 = 4(x + 2)$$

$$x^2 - 2x + 1 = 4x + 8$$

$$x^2 - 6x - 7 = 0$$

$$(x + 1)(x - 7) = 0$$

Thus $x = -1$ or 7. From $y = x - 1$, if $x = -1$, then $y = -2$; if $x = 7$, then $y = 6$. However, from Eq. (2), $y \geq 0$. The only solution is $x = 7, y = 6$.

15. We can write the following system of equations.

$$\begin{cases} y = 0.01x^2 + 0.01x + 7, \\ y = 0.01x + 8.0. \end{cases}$$

By substituting $0.01x + 8.0$ for y in the first equation and simplifying, we obtain

$$0.01x + 8.0 = 0.01x^2 + 0.01x + 7$$

$$0 = 0.01x^2 - 1$$

$$0 = (0.1x + 1)(0.1x - 1)$$

$$x = -10 \quad \text{or} \quad x = 10$$

If $x = -10$ then $y = 7.9$, and if $x = 10$ then $y = 8.1$. The rope touches the streamer twice, 10 feet away from center on each side at $(-10, 7.9)$ and $(10, 8.1)$.

17. The system has 3 solutions.

19. $x = -1.3$, $y = 5.1$

21. $x = 1.76$

23. $x = -1.46$

Problems 3.6

1. Equating p-values gives

$$\frac{4}{100}q + 3 = -\frac{6}{100} + 13$$

$$\frac{10}{100}q = 10$$

$$q = 100$$

$$p = \frac{4}{100}(100) + 3 = 7$$

Thus, the equilibrium point is (100, 7).

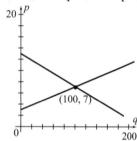

3. $\begin{cases} 35q - 2p + 250 = 0, & (1) \\ 65q + p - 537.5 = 0. & (2) \end{cases}$

Multiplying Eq. (2) by 2 and adding equations gives
$165q - 825 = 0$
$q = 5$
From Eq. (2),
$65(5) + p - 537.5 = 0$
$p = 212.50$
Thus the equilibrium point is (5, 212.50).

5. Equating p-values:

$$2q + 20 = 200 - 2q^2$$

$$2q^2 + 2q - 180 = 0$$

$$q^2 + q - 90 = 0$$

$$(q + 10)(q - 9) = 0$$

Thus $q = -10$, 9. Since $q \geq 0$, choose $q = 9$.
Then $p = 2q + 20 = 2(9) + 20 = 38$. The equilibrium point is (9, 38).

7. Equating p-values gives $20 - q = \sqrt{q + 10}$.
Squaring both sides gives

$$400 - 40q + q^2 = q + 10$$

$$q^2 - 41q + 390 = 0$$

$$(q - 26)(q - 15) = 0$$

Thus $q = 26$, 15. If $q = 26$, then
$p = 20 - q = 20 - 26 = -6$. But p cannot be negative. If $q = 15$, then
$p = 20 - q = 20 - 15 = 5$. The equilibrium point is (15, 5).

9. Letting $y_{TR} = y_{TC}$ gives $4q = 2q + 5000$, or $q = 2500$ units.

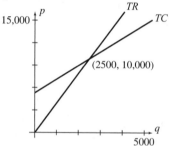

11. Letting $y_{TR} = y_{TC}$ gives
$0.05q = 0.85q + 600$
$-0.80q = 600$
$q = -750$, which is negative. Thus one cannot break even at any level of production.

13. Letting $y_{TR} = y_{TC}$ gives

$$90 - \frac{900}{q + 3} = 1.1q + 37.3$$

$$90(q + 3) - 900 = (1.1q + 37.3)(q + 3)$$

$$90q + 270 - 900 = 1.1q^2 + 40.6q + 111.9$$

$$1.1q^2 - 49.4q + 741.9 = 0$$

$$q = \frac{-b \pm \sqrt{b^2 - 4ac}}{2a}$$

$$= \frac{49.4 \pm \sqrt{(-49.4)^2 - 4(1.1)(741.9)}}{2(1.1)}$$

$$= \frac{49.4 \pm \sqrt{-824}}{2.2}$$

There are no real solutions, therefore one cannot break even at any level of production.

15. $\begin{cases} 3q - 200p + 1800 = 0, & (1) \\ 3q + 100p - 1800 = 0. & (2) \end{cases}$

    a. Subtracting Eq. (2) from Eq. (1) gives
$$-300p + 3600 = 0$$
$$p = \$12$$

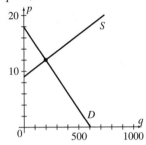

    b. Before the tax, the supply equation is
$$3q - 200p + 1800 = 0$$
$$-200p = -3q - 1800$$
$$p = \frac{3}{200}q + 9$$

    After the tax, the supply equation is
$$p = \frac{3}{200}q + 9 + 0.27$$
$$p = \frac{3}{200}q + 9.27$$

    This equation can be written
$-3q + 200p - 1854 = 0$, and the new
system to solve is
$$\begin{cases} -3q + 200p - 1854 = 0, \\ 3q + 100p - 1800 = 0. \end{cases}$$
Adding gives
$$300p - 3654 = 0$$
$$\Rightarrow p = \frac{3654}{300} = \$12.18.$$

17. Since profit = total revenue − total cost, then
$4600 = 8.35q - (2116 + 7.20q)$. Solving
gives $4600 = 1.15q - 2116$
$$1.15q = 6716$$
$$q = \frac{6716}{1.15} = 5840 \text{ units}$$
For a loss (negative profit) of \$1150, we
solve $-1150 = 8.35q - (2116 + 7.20q)$. Thus

$$-1150 = 1.15q - 2116$$
$$1.15q = 966$$
$$q = 840 \text{ units}$$
To break even, we have $y_{TR} = y_{TC}$, or
$$8.35q = 2116 + 7.20q$$
$$1.15q = 2116$$
$$q = 1840 \text{ units}$$

19. Let q = break-even quantity. Since total
revenue is $5q$, we have $5q = 200,000$, which
yields $q = 40,000$. Let c be the variable cost
per unit. Then at the break even point,
Tot. Rev. = Tot. Cost
$$= \text{Variable Cost + Fixed Cost.}$$
Thus
$$200,000 = 40,000c + 40,000$$
$$160,000 = 40,000c$$
$$c = \$4.$$

21. $y_{TC} = 3q + 1250$: $y_{TR} = 60\sqrt{q}$. Letting
$y_{TR} = y_{TC}$ gives
$$60\sqrt{q} = 3q + 1250$$
$$20\sqrt{q} = q + \frac{1250}{3}$$
Squaring gives
$$400q = q^2 + \frac{2500}{3}q + \left(\frac{1250}{3}\right)^2$$
$$q^2 + \frac{1300}{3}q + \frac{1,562,500}{9} = 0$$
Using the quadratic formula,
$$q = \frac{-\frac{1300}{3} \pm \sqrt{\left(\frac{1300}{3}\right)^2 - 4(1)\left(\frac{1,562,500}{9}\right)}}{2},$$
which is not real. Thus total cost always
exceeds total revenue; there is no break-
even point.

23. After the subsidy the supply equation is
$$p = \left[\frac{8}{100}q + 50\right] - 1.50$$
$$p = \frac{8}{100}q + 48.50$$
The system to consider is

$$\begin{cases} p = \dfrac{8}{100}q + 48.50, \\[2mm] p = -\dfrac{7}{100}q + 65. \end{cases}$$

Equating p-values gives

$$\frac{8}{100}q + 48.50 = -\frac{7}{100}q + 65$$

$$\frac{15}{100}q = 16.5$$

$$q = 110$$

When $q = 110$, then

$$p = \frac{8}{100}q + 48.50 = \frac{8}{100}(110) + 48.50$$

$$= 8.8 + 48.50 = 57.30.$$

Thus the original equilibrium price decreases by \$0.70.

25. Equating q_A-values gives

$$7 - p_A + p_B = -3 + 4p_A - 2p_B$$
$$10 = 5p_A - 3p_B$$

Equating q_B-values gives

$$21 + p_A - p_B = -5 - 2p_A + 4p_B$$
$$26 = -3p_A + 5p_B$$

Now we solve

$$\begin{cases} 10 = 5p_A - 3p_B \\ 26 = -3p_A + 5p_B \end{cases}$$

Adding 3 times the first equation to 5 times the second equation gives

$$160 = 16p_B$$

$$p_B = 10$$

From $5p_A - 3p_B = 10$, $5p_A - 3(10) = 10$ or

$$p_A = 8.$$

Thus $p_A = 8$ and $p_B = 10$.

27. 2.4 and 11.3

Chapter 3 Review Problems

1. Solving $\dfrac{k-5}{3-2} = 4$ gives $k - 5 = 4$, $k = 9$.

3. $(-2, 3)$ and $(0, -1)$ lie on the line, so

$$m = \frac{-1-3}{0-(-2)} = -2. \text{ Slope-intercept form:}$$

$y = mx + b \Rightarrow y = -2x - 1$. A general form:
$2x + y + 1 = 0$.

5. $y - 4 = \dfrac{1}{2}(x - 10)$

$$y - 4 = \frac{1}{2}x - 5$$

$y = \dfrac{1}{2}x - 1$, which is slope-intercept form.

Clearing fractions, we have

$$2y = 2\left(\frac{1}{2}x - 1\right)$$

$$2y = x - 2$$

$x - 2y - 2 = 0$, which is a general form.

7. Slope of a horizontal line is 0. Thus
$y - 4 = 0[x - (-2)]$
$y - 4 = 0$,
so slope-intercept form is $y = 4$. A general form is $y - 4 = 0$.

9. The line $2y + 5x = 2$ $\left(\text{or } y = -\dfrac{5}{2}x + 1\right)$ has

slope $-\dfrac{5}{2}$, so the line perpendicular to it

has slope $\dfrac{2}{5}$. Since the y-intercept is -3, the

equation is $y = \dfrac{2}{5}x - 3$. A general form is

$2x - 5y - 15 = 0$.

In Problems 11–15, $m_1 = $ slope of first line, and $m_2 = $ slope of second line.

11. $x + 4y + 2 = 0$ $\left(\text{or } y = -\dfrac{1}{4}x - \dfrac{1}{2}\right)$ has slope

$m_1 = -\dfrac{1}{4}$ and $8x - 2y - 2 = 0$ (or $y = 4x - 1$)

has slope $m_2 = 4$. Since $m_1 = -\dfrac{1}{m_2}$, the

lines are perpendicular to each other.

13. $x - 3 = 2(y + 4)$ $\left(\text{or } y = \dfrac{1}{2}x - \dfrac{11}{2}\right)$ has slope

$m_1 = \dfrac{1}{2}$, and $y = 4x + 2$ has slope $m_2 = 4$.

Since $m_1 \neq m_2$ and $m_1 \neq -\dfrac{1}{m_2}$, the lines are

neither parallel nor perpendicular to each other.

15. $y = 3x + 5$ has slope 3, and $6x - 2y = 7$

$\left(\text{or } y = 3x - \dfrac{7}{2}\right)$ has slope 3. Since

$m_1 = m_2$, the lines are parallel.

17. $3x - 2y = 4$

$-2y = -3x + 4$

$y = \dfrac{3}{2}x - 2$

$m = \dfrac{3}{2}$

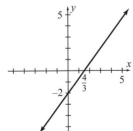

19. $4 - 3y = 0$

$-3y = -4$

$y = \dfrac{4}{3}$

$m = 0$

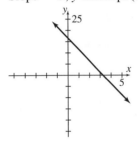

21. $y = f(x) = 17 - 5x$ has the linear form
$f(x) = ax + b$, where $a = -5$ and $b = 17$.
Slope $= -5$; y-intercept $(0, 17)$.

23. $y = f(x) = 9 - x^2$ has the quadratic form

$f(x) = ax^2 + bx + c$, where $a = -1$, $b = 0$

and $c = 9$.

Vertex: $-\dfrac{b}{2a} = -\dfrac{0}{2(-1)} = 0$

$f(0) = 9 - 0^2 = 9$

$\Rightarrow$ Vertex $= (0, 9)$

y-intercept: $c = 9$

x-intercepts: $9 - x^2 = (3 - x)(3 + x) = 0$, so

$x = 3, -3$.

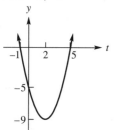

25. $y = h(t) = t^2 - 4t - 5$ has the quadratic form

$h(t) = at^2 + bt + c$, where $a = 1$, $b = -4$, and $c = -5$.

Vertex: $-\dfrac{b}{2a} = -\dfrac{-4}{2 \cdot 1} = 2$

$h(2) = 2^2 - 4(2) - 5 = -9$

$\Rightarrow$ Vertex = $(2, -9)$
y-intercept: $c = -5$
t-intercepts: $t^2 - 4t - 5 = (t-5)(t+1) = 0$
$\Rightarrow t = 5, -1$

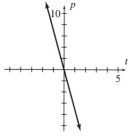

27. $p = g(t) = -7t$ has the linear form
$g(t) = at + b$, where $a = -7$ and $b = 0$.
Slope = -7; p-intercept $(0, 0)$

29. $y = F(x) = -\left(x^2 + 2x + 3\right) = -x^2 - 2x - 3$

has the quadratic form $F(x) = ax^2 + bx + c$,
where $a = -1$, $b = -2$, and $c = -3$

Vertex: $-\dfrac{b}{2a} = -\dfrac{-2}{2(-1)} = -1$

$F(-1) = -\left[(-1)^2 + 2(-1) + 3\right] = -2$

$\Rightarrow$ Vertex = $(-1, -2)$
y-intercept: $c = -3$
x-intercepts: Because the parabola opens downward ($a < 0$) and the vertex is below the x-axis, there is no x-intercept.

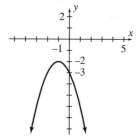

31. $\begin{cases} 2x - y = 6, & (1) \\ 3x + 2y = 5. & (2) \end{cases}$

From Eq. (1), $y = 2x - 6$. Substituting in Eq. (2) gives
$3x + 2(2x - 6) = 5$
$7x - 12 = 5$, $7x = 17$

$x = \dfrac{17}{7} \Rightarrow y = 2x - 6 = 2 \cdot \dfrac{17}{7} - 6 = -\dfrac{8}{7}$.

Thus $x = \dfrac{17}{7}$, $y = -\dfrac{8}{7}$.

33. $\begin{cases} 7x + 5y = 5 \\ 6x + 5y = 3 \end{cases}$

Subtracting the second equation from the first equation gives $x = 2$. Then

$7(2) + 5y = 5$, or $5y = -9$, so $y = -\dfrac{9}{5}$. Thus

$x = 2$, $y = -\dfrac{9}{5}$.

35. $\begin{cases} \dfrac{1}{4}x - \dfrac{3}{2}y = -4, & (1) \\ \dfrac{3}{4}x + \dfrac{1}{2}y = 8. & (2) \end{cases}$

Multiplying Eq. (2) by 3 gives

$\begin{cases} \dfrac{1}{4}x - \dfrac{3}{2}y = -4, \\ \dfrac{9}{4}x + \dfrac{3}{2}y = 24. \end{cases}$

Adding the first equation to the second gives

$\dfrac{5}{2}x = 20$

$x = 8$

From Eq. (1),

$\dfrac{1}{4}(8) - \dfrac{3}{2}y = -4$

$-\dfrac{3}{2}y = -6$

$y = 4$

Thus

$x = 8, \ y = 4.$

37. $\begin{cases} 3x - 2y + z = -2, & (1) \\ 2x + y + z = 1, & (2) \\ x + 3y - z = 3. & (3) \end{cases}$

Subtracting Eq. (2) from Eq. (1) and adding Eq. (2) to Eq. (3) gives

$\begin{cases} x - 3y = -3, \\ 3x + 4y = 4. \end{cases}$

Multiplying the first equation by -3 gives

$\begin{cases} -3x + 9y = 9, \\ 3x + 4y = 4. \end{cases}$

Adding the first equation to the second gives

$13y = 13$

$y = 1$

From the equation $x - 3y = -3$, we get

$x - 3(1) = -3$

$x = 0$

From $3x - 2y + z = -2$, we get

$3(0) - 2(1) + z = -2$

$z = 0$

Thus $x = 0, \ y = 1, \ z = 0.$

39. $\begin{cases} x^2 - y + 5x = 2, & (1) \\ x^2 + y = 3. & (2) \end{cases}$

From Eq. (2), $y = 3 - x^2$. Substituting in Eq. (1) gives

$x^2 - (3 - x^2) + 5x = 2$

$2x^2 + 5x - 5 = 0$

$x = \dfrac{-b \pm \sqrt{b^2 - 4ac}}{2a}$

$= \dfrac{-5 \pm \sqrt{5^2 - 4(2)(-5)}}{2(2)}$

$= \dfrac{-5 \pm \sqrt{65}}{4}$

Since $y = 3 - x^2$, if $x = \dfrac{-5 + \sqrt{65}}{4}$, then

$y = \dfrac{-21 + 5\sqrt{65}}{8}$; if $x = \dfrac{-5 - \sqrt{65}}{4}$, then

$y = \dfrac{-21 - 5\sqrt{65}}{8}.$

Thus, the two solutions are

$x = \dfrac{-5 + \sqrt{65}}{4}, \ y = \dfrac{-21 + 5\sqrt{65}}{8},$ and

$x = \dfrac{-5 - \sqrt{65}}{4}, \ y = \dfrac{-21 - 5\sqrt{65}}{8}.$

41. $\begin{cases} x + 2z = -2, & (1) \\ x + y + z = 5. & (2) \end{cases}$

From Eq. (1) we have $x = -2 - 2z$. Substituting in Eq. (2) gives

$-2 - 2z + y + z = 5$, so $y = 7 + z$. Letting $z = r$ gives the parametric solution $x = -2 - 2r, \ y = 7 + r, \ z = r$, where r is any real number.

43. $\begin{cases} x - y - z = 0, & (1) \\ 2x - 2y + 3z = 0. & (2) \end{cases}$

Multiplying Eq. (1) by –2 gives
$$\begin{cases} -2x + 2y + 2z = 0, \\ 2x - 2y + 3z = 0. \end{cases}$$
Adding the first equation to the second gives
$$\begin{cases} -2x + 2y + 2z = 0, \\ 5z = 0. \end{cases}$$
From the second equation, $z = 0$.
Substituting in Eq. (1) gives $x - y - 0 = 0$, so
$x = y$. Letting $y = r$ gives the parametric
solution $x = r$, $y = r$, $z = 0$, where r is any
real number.

45. $a = 1$ when $b = 2$; $a = 5$ when $b = 3$, so
$$m = \frac{a_2 - a_1}{b_2 - b_1} = \frac{5 - 1}{3 - 2} = \frac{4}{1} = 4.$$
Thus an equation relating a and b is
$$a - 1 = 4(b - 2)$$
$$a - 1 = 4b - 8$$
$$a - 4b = -7$$
When $b = 5$, then $a = 4b - 7 = 4(5) - 7 = 13$.

47. Slope is $\frac{-4}{3} \Rightarrow f(x) = ax + b = -\frac{4}{3}x + b$.
Since $f(1) = 5$,
$$5 = -\frac{4}{3}(1) + b$$
$$b = \frac{19}{3}$$
Thus $f(x) = -\frac{4}{3}x + \frac{19}{3}$.

49. $r = pq = (200 - 2q)q = 200q - 2q^2$, which
is a quadratic function with $a = -2$, $b = 200$,
$c = 0$. Since $a < 0$, r has a maximum value
when $q = -\frac{b}{2a} = -\frac{200}{-4} = 50$ units. If
$q = 50$, then
$r = [200 - 2(50)](50) = \$5000$.

51. $\begin{cases} 120p - q - 240 = 0, \\ 100p + q - 1200 = 0. \end{cases}$

Adding gives $220p - 1440 = 0$, or
$$p = \frac{1440}{220} \approx 6.55.$$

53. $y_{TR} = 16q$; $y_{TC} = 8q + 10,000$. Letting
$y_{TR} = y_{TC}$ gives
$$16q = 8q + 10,000$$
$$8q = 10,000$$
$$q = 1250$$
If $q = 1250$, then
$y_{TR} = 16(1250) = 20,000$. Thus the
break-even point is (1250, 20,000) or
1250 units, \$20,000.

55. Equating L-values gives
$$0.0183 - \frac{0.0042}{p} = 0.0005 + \frac{0.0378}{p}$$
$$0.0178 = \frac{0.042}{p}$$
$$0.0178p = 0.042$$
$$p \approx 2.36$$
The equilibrium pollution level is about
2.36 tons per square kilometer.

57. $x = 7.29$, $y = -0.78$

59. $x = 0.75$, $y = 1.43$

Mathematical Snapshot Chapter 3

1. $P_1(6000) = 39.99 + 0.45(6000 - 450)$
$$= 2537.49$$
$P_6(6000) = 199.99$
He loses $\$2537.49 - \$199.99 = \$2337.50$ by
using P_1.

3. The graph shows that P_3 and P_4 intersect
when the second branch of P_3 crosses the
first branch of P_4 Thus
$$79.99 + 0.35(t - 1350) = 99.99$$
$$t \approx 1407.14$$
P_3 is best for usage between 950 and
1407.14 minutes.

5. The graph shows that P_5 and P_6 intersect when the second branch of P_5 crosses the first branch of P_6 Thus

$$149.99 + 0.25(t - 4000) = 199.99$$
$$t = 4200$$

P_5 is best for usage between 2200 and 4200 minutes.

7. No; answers may vary.

Chapter 4

Principles in Practice 4.1

1. The shapes of the graphs are the same. The value of A scales the value of any point by A.

3. If V = the value of the car and r = the annual rate at which V depreciates, then after 1 year the value of the car is $V - rV = V(1 - r)$. Since $r = 0.15$, the factor by which V decreases for the first year is $1 - r = 1 - 0.15 = 0.85$. Similarly, after the second year the value of the car is $V(1-r) - r[V(1-r)] = V(1-r)^2$. Again, since $r = 0.15$, the multiplicative decrease for the second year is $(1-r)^2 = (1-0.15)^2 = 0.72$. This pattern will continue as shown in the table.

Year	Multiplicative Decrease	Expression
0	1	0.85^0
1	0.85	0.85^1
2	0.72	0.85^2
3	0.61	0.85^3

Thus, the depreciation is exponential with a base of $1 - r = 1 - 0.15 = 0.85$. If we graph the multiplicative decrease as a function of years, we obtain the following.

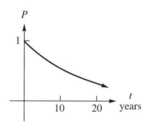

5. $S = P(1+r)^n$

$S = 2000(1+0.13)^5 = 2000(1.13)^5$

≈ 3684.87

The value of the investment after 5 years will be $3684.87. The interest earned over the first 5 years is $3684.87 - 2000 = \$1684.87$.

7. $P = e^{-0.06t} = \left(\dfrac{1}{e}\right)^{0.06t}$

Since $0 < \dfrac{1}{e} < 1$, the graph is that of an exponential function falling from left to right.

x	y
0	1
2	0.89
4	0.79
6	0.70
8	0.62
10	0.55

Problems 4.1

1.

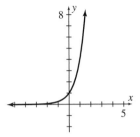

3.

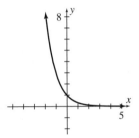

5.

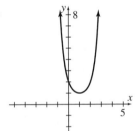

7.

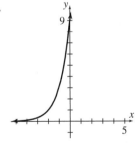

9.

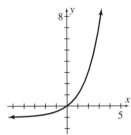

11.

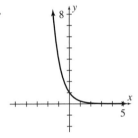

13. For the curves, the bases involved are 0.4, 2, and 5. For base 5, the curve rises from left to right, and in the first quadrant it rises faster than the curve for base 2. Thus the graph of $y = 5^x$ is B.

15. For 2015 we have $t = 20$, so
$$P = 125,000(1.11)^{\frac{20}{20}} = 125,000(1.11)^1$$
$$= 138,750.$$

17. With $c = \dfrac{1}{2}$, $P = 1 - \dfrac{1}{2}\left(\dfrac{1}{2}\right)^{n-1} = 1 - \left(\dfrac{1}{2}\right)^n$.

$n = 1$: $P = 1 - \left(\dfrac{1}{2}\right)^1 = 1 - \dfrac{1}{2} = \dfrac{1}{2}$

$n = 2$: $P = 1 - \left(\dfrac{1}{2}\right)^2 = 1 - \dfrac{1}{4} = \dfrac{3}{4}$

$n = 3$: $P = 1 - \left(\dfrac{1}{2}\right)^3 = 1 - \dfrac{1}{8} = \dfrac{7}{8}$

19. a. $4000(1.06)^7 \approx \$6014.52$

  b. $6014.52 - 4000 = \$2014.52$

21. a. $700(1.035)^{30} \approx \$1964.76$

b. $1964.76 - 700 = \$1264.76$

23. a. $3000\left(1+\dfrac{0.0875}{4}\right)^{64} \approx 11,983.37$

b. $11,983.37 - 3000 = \$8983.37$

25. a. $5000(1.0075)^{30} \approx \$6256.36$

b. $6256.36 - 5000 = \$1256.36$

27. a. $8000\left(1+\dfrac{0.0625}{365}\right)^{3(365)} \approx \$9649.69$

b. $9649.69 - 8000 = \$1649.69$

29. $6500\left(1+\dfrac{0.04}{4}\right)^{24} \approx \$8253.28$

31. a. $N = 400(1.05)^{t}$

b. When $t = 1$, then $N = 400(1.05)^{1} = 420.$

c. When $t = 4$, then
$N = 400(1.05)^{4} \approx 486.$

33. Let $P =$ the amount of plastic recycled and let
$r =$ the rate at which P increases each year. Then after the first year, the amount of plastic recycled, increases from P to $P + rP = P(1 + r)$, since $r = 0.3$, the factor by which P increases for the first year, is $1 + r = 1 + 0.3 = 1.3$. Similarly, during the second year, the amount of plastic recycled increases from $P(1 + r)$ to
$P(1 + r) + r[P(1+r)] = P(1+r)^{2}$. Again, since $r = 0.3$, the multiplicative increase for the second year is
$(1+r)^{2} = (1+0.3)^{2} = (1.3)^{2} = 1.69$. This pattern will continue as shown in the table.

Year	Multiplicative Increase	Expression
0	1	1.3^{0}
1	1.3	1.3^{1}
2	1.69	1.3^{2}
3	2.20	1.3^{3}

Thus, the increase in recycling is exponential with a
base $= 1 + r = 1 + 0.3 = 1.3$. If we graph the multiplicative increase as function of years, we obtaining the following.

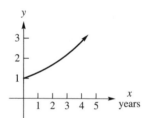

From the graph it appears that recycling will triple after about 4 years.

35. $P = 350,000(1-0.015)^{t} = 350,000(0.985)^{t}$, where P is the population after t years. When $t = 3$,
$P = 350,000(0.985)^{3} \approx 334,485.$

37. 4.4817

39. 0.4966

41.

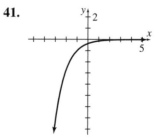

43. For $x = 3$, $P = \dfrac{e^{-3}3^3}{3!} \approx 0.2240$

45. $e^{kt} = \left(e^k\right)^t = b^t$, where $b = e^k$

47. a. When $t = 0$,
$N = 12e^{-0.031(0)} = 12 \cdot 1 = 12$.

    b. When $t = 10$,
$N = 12e^{-0.031(10)} = 12e^{-0.31} = 8.8$.

    c. When $t = 44$,
$N = 12e^{-0.031(44)} = 12e^{-1.364} \approx 3.1$.

    d. After 44 hours, approximately $\dfrac{1}{4}$ of the initial amount remains. Because $\dfrac{1}{4} = \left(\dfrac{1}{2}\right)\left(\dfrac{1}{2}\right)$, 44 hours corresponds to 2 half-lives. Thus the half-life is approximately 22 hours.

49. After one half-life, $\dfrac{1}{2}$ gram remains. After two half-lives, $\dfrac{1}{2} \cdot \dfrac{1}{2} = \left(\dfrac{1}{2}\right)^2 = \dfrac{1}{4}$ gram remains. Continuing in this manner, after n half-lives, $\left(\dfrac{1}{2}\right)^n$ gram remains. Because $\dfrac{1}{16} = \left(\dfrac{1}{2}\right)^4$, after 4 half-lives, $\dfrac{1}{16}$ gram remains. This corresponds to $4 \cdot 8 = 32$ years.

51. $f(x) = \dfrac{e^{-4}4^x}{x!}$

$f(2) = \dfrac{e^{-4}4^2}{2!} \approx 0.1465$

53.

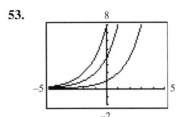

If $f(x) = 2^x$, then
$y = 2^a \cdot 2^x = 2^{x+a} = f(x+a)$. Thus, the graph of $y = 2^a \cdot 2^x$ is the graph of $y = 2^x$ shifted a units to the left.

55. 3.17

57. $300\left(\dfrac{4}{3}\right)^{4.1} \approx 976$

$300\left(\dfrac{4}{3}\right)^{4.2} \approx 1004$

4.2 minutes

59. The first integer t for which the graph of $P = 2500(1.043)^t$ lies on or above the horizontal line $P = 5000$ is 17.

Principles in Practice 4.2

1. If $16 = 2^t$ is the exponential form then $t = \log_2 16$ is the logarithmic form, where t represents the number of times the bacteria have doubled.

3. Let R = the amount of material recycled every year. If the amount being recycled increases by 50% every year, then the amount recycled at the end of y years is $R(1+r)^y = R(1+0.5)^y = R(1.5)^y$ Thus, the multiplicative increase in recycling at the end of y years is $(1.5)^y$. If we let x = the multiplicative increase, then $x = (1.5)^y$ and, in logarithmic form, $\log_{1.5} x = y$.

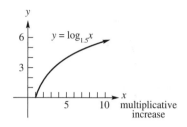

5. The equation $t(r) = \dfrac{\ln 4}{r}$ can be rewritten as

$r = \dfrac{\ln 4}{t(r)}$. When this equation is graphed we

find that the annual rate r needed to quadruple the investment in 10 years is approximately 13.9%. Alternatively, we can solve for r by setting $t(r) = 10$.

$r = \dfrac{\ln(4)}{t(r)}$

$r = \dfrac{\ln(4)}{10} \approx 0.139 \text{ or } \approx 13.9\%$

Problems 4.2

1. $\log 10{,}000 = 4$

3. $2^6 = 64$

5. $\ln 20.0855 = 3$

7. $e^{1.09861} = 3$

9.

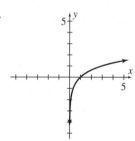

11.

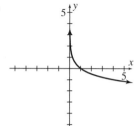

13.

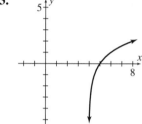

15.

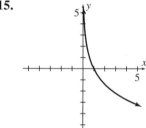

17. Because $6^2 = 36$, $\log_6 36 = 2$

19. Because $3^3 = 27$, $\log_3 27 = 3$

21. Because $7^1 = 7$, $\log_7 7 = 1$

23. Because $10^{-2} = 0.01$, $\log 0.01 = -2$

25. Because $5^0 = 1$, $\log_5 1 = 0$

27. Because $2^{-3} = \dfrac{1}{8}$, $\log_2 \dfrac{1}{8} = -3$

29. $3^4 = x$
$x = 81$

31. $5^3 = x$
$x = 125$

33. $10^{-1} = x$
$x = \dfrac{1}{10}$

35. $e^{-3} = x$

37. $x^3 = 8$
$x = 2$

39. $x^{-1} = \dfrac{1}{6}$
$x = 6$

41. $3^{-3} = x$
$x = \dfrac{1}{27}$

43. $x^2 = 6 - x$
$x^2 + x - 6 = 0$
$(x + 3)(x - 2) = 0$
The roots of this equation are –3 and 2. But since $x > 0$, we choose $x = 2$.

45. $2 + \log_2 4 = 3x - 1$
$2 + 2 = 3x - 1$
$5 = 3x$
$x = \dfrac{5}{3}$

47. $x^2 = 2x + 8$
$x^2 - 2x - 8 = 0$
$(x - 4)(x + 2) = 0$
The roots of this equation are 4 and –2. But since $x > 0$, we choose $x = 4$.

49. $e^{3x} = 2$
$3x = \ln 2$
$x = \dfrac{\ln 2}{3}$

51. $e^{2x-5} + 1 = 4$
$e^{2x-5} = 3$
$2x - 5 = \ln 3$
$x = \dfrac{5 + \ln 3}{2}$

53. 1.60944

55. 2.00013

57. If V = the value of the antique. If the value appreciates by 10% every year, then at the end of y years the value of the antique is $V(1+r)^y = V(1+0.10)^y = V(1.10)^y$. Thus, the multiplicative increase in value at the end of y years is $(1.10)^y$. If we let x = the multiplicative increase, then $x = (1.10)^y$, and, in logarithm form, $\log_{1.10} x = y$.

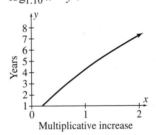

Multiplicative increase

59. $p = \log\left[10 + \dfrac{1980}{2}\right] = \log[10 + 990] = \log 1000$
$= 3$

61. a. If $t = k$, then $N = N_0\left(2^1\right) = 2N_0$

b. From part (a), $N = 2N_0$ when $t = k$. Thus k is the time it takes for the population to double.

c. $N_1 = N_0 2^{\frac{t}{k}}$

$\dfrac{N_1}{N_0} = 2^{\frac{t}{k}}$

$\dfrac{t}{k} = \log_2 \dfrac{N_1}{N_0}$

$t = k \log_2 \dfrac{N_1}{N_0}$

63. $T = \dfrac{\ln 2}{0.01920} \approx 36.1$ minutes

65.   From $\log_y x = 3$, $y^3 = x$; from $\log_z x = 2$,

$z^2 = x$. Thus $z^2 = y^3$ or $z = y^{\frac{3}{2}}$.

67.

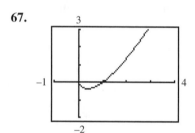

a.   (0, 1)

b.   [−0.37, ∞)

69.   For $y = e^x$, if $y = 3$, then $3 = e^x$ or $x = \ln 3$.

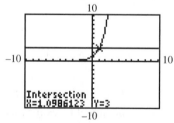

From the graph of $y = e^x$, when $y = 3$, then
$x = \ln 3 \approx 1.10$.

71.

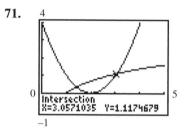

1.41, 3.06

Principles in Practice 4.3

1.   The magnitude (Richter Scale) of an

earthquake is given by $R = \log\left(\dfrac{I}{I_0}\right)$ where

I is the intensity of the earthquake and I_0 is
the intensity of a zero-level reference
earthquake. $\dfrac{I}{I_0} =$ how many times greater
the earthquake is than a zero-level
earthquake. Thus, when $\dfrac{I}{I_0} = 900,000$,

$R_1 = \log(900,000)$

When $\dfrac{I}{I_0} = 9000$

$R_2 = \log(9000)$
$R_1 - R_2 = \log(900,000) - \log 9000$

$= \log \dfrac{900,000}{9000} = \log 100 = \log 10^2$
$= 2 \log 10$
$= 2$
Thus, the two earthquakes differ by 2 on the
Richter scale.

Problems 4.3

1. $\log 30 = \log(2 \cdot 3 \cdot 5)$
$= \log 2 + \log 3 + \log 5$
$= a + b + c$

3. $\log \dfrac{2}{3} = \log 2 - \log 3 = a - b$

5. $\log \dfrac{8}{3} = \log 8 - \log 3 = \log 2^3 - \log 3$

$\qquad = 3\log 2 - \log 3 = 3a - b$

7. $\log 36 = \log(2 \cdot 3)^2 = 2\log(2 \cdot 3)$

$\qquad = 2(\log 2 + \log 3) = 2(a + b)$

9. $\log_2 3 = \dfrac{\log_{10} 3}{\log_{10} 2} = \dfrac{\log 3}{\log 2} = \dfrac{b}{a}$

11. $\log_7 7^{48} = 48$

13. $\log 0.0000001 = \log 10^{-7} = -7$

15. $\ln e^{5.01} = \log_e e^{5.01} = 5.01$

17. $\ln \dfrac{1}{e^2} = -\ln e^2 = -\log_e e^2 = -2$

19. $\log \dfrac{1}{10} + \ln e^3 = \log_{10} \dfrac{1}{10} + \log_e e^3$

$\qquad\qquad = -1 + 3$

$\qquad\qquad = 2$

21. $\ln\left[x(x+1)^2 \right] = \ln x + \ln(x+1)^2$

$\qquad = \ln x + 2\ln(x+1)$

23. $\ln \dfrac{x^2}{(x+1)^3} = \ln x^2 - \ln(x+1)^3$

$\qquad = 2\ln x - 3\ln(x+1)$

25. $\ln\left(\dfrac{x+1}{x+2} \right)^4 = 4\ln \dfrac{x+1}{x+2}$

$\qquad\qquad = 4[\ln(x+1) - \ln(x+2)]$

27. $\ln \dfrac{x}{(x+1)(x+2)} = \ln x - \ln[(x+1)(x+2)]$

$\qquad = \ln x - [\ln(x+1) + \ln(x+2)]$

$\qquad = \ln x - \ln(x+1) - \ln(x+2)$

29. $\ln \dfrac{\sqrt{x}}{(x+1)^2(x+2)^3}$

$\qquad = \ln x^{\frac{1}{2}} - \ln\left[(x+1)^2(x+2)^3 \right]$

$\qquad = \dfrac{1}{2}\ln x - \left[\ln(x+1)^2 + \ln(x+2)^3 \right]$

$\qquad = \dfrac{1}{2}\ln x - [2\ln(x+1) + 3\ln(x+2)]$

$\qquad = \dfrac{1}{2}\ln x - 2\ln(x+1) - 3\ln(x+2)$

31. $\ln\left[\dfrac{1}{x+2} \sqrt[5]{\dfrac{x^2}{x+1}} \right] = \ln\left[\dfrac{1}{x+2}\left(\dfrac{x^2}{x+1} \right)^{\frac{1}{5}} \right]$

$\qquad = \ln \dfrac{x^{\frac{2}{5}}}{(x+2)(x+1)^{\frac{1}{5}}}$

$\qquad = \ln x^{\frac{2}{5}} - \ln\left[(x+2)(x+1)^{\frac{1}{5}} \right]$

$\qquad = \dfrac{2}{5}\ln x - \left[\ln(x+2) + \ln(x+1)^{\frac{1}{5}} \right]$

$\qquad = \dfrac{2}{5}\ln x - \ln(x+2) - \dfrac{1}{5}\ln(x+1)$

33. $\log (6 \cdot 4) = \log 24$

35. $\log_2 \dfrac{2x}{x+1}$

37. $5\log_2 10 + 2\log_2 13 = \log_2 10^5 + \log_2 13^2$

$\qquad\qquad\qquad\qquad\quad = \log_2(10^5 \cdot 13^2)$

39. $\log 100 + \log(1.05)^{10} = \log\left[100(1.05)^{10} \right]$

41. $e^{4\ln 3 - 3\ln 4} = e^{\ln 3^4 - \ln 4^3}$

$\qquad = e^{\ln\left(\frac{3^4}{4^3} \right)} = \dfrac{3^4}{4^3} = \dfrac{81}{64}$

43. $\log_6 54 - \log_6 9 = \log_6 \dfrac{54}{9} = \log_6 6 = 1$

45. $e^{\ln(2x)} = 5$

$2x = 5$

$x = \dfrac{5}{2}$

47. $10^{\log x^2} = 4$

$x^2 = 4$

$x = \pm 2$

49. From the change of base formula with $b = 2$, $m = 2x + 1$, and $a = e$, we have

$$\log_2(2x+1) = \frac{\log_e(2x+1)}{\log_e 2} = \frac{\ln(2x+1)}{\ln 2}$$

51. From the change of base formula with $b = 3$, $m = x^2 + 1$, and $a = e$, we have

$$\log_3\left(x^2+1\right) = \frac{\log_e\left(x^2+1\right)}{\log_e 3} = \frac{\ln\left(x^2+1\right)}{\ln 3}.$$

53. $e^{\ln z} = 7e^y$

$z = 7e^y$

$\dfrac{z}{7} = e^y$

$y = \ln \dfrac{z}{7}$

55. $C = B + E$

$C = B\left(1 + \dfrac{E}{B}\right)$

$\ln C = \ln\left[B\left(1 + \dfrac{E}{B}\right)\right]$

$\ln C = \ln B + \ln\left(1 + \dfrac{E}{B}\right)$

57. $y = \log_6 x = \dfrac{\ln x}{\ln 6}$

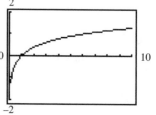

59. By the change of base formula,

$\log x = \dfrac{\ln x}{\ln 10}$. Thus the graphs of $y = \log x$

and $y = \dfrac{\ln x}{\ln 10}$ are identical.

Principles in Practice 4.4

1. Let $x =$ the number and let $y =$ the unknown exponent. Then

$x \cdot 32^y = x \cdot 4^{(3y-9)}$

$32^y = 4^{(3y-9)}$

$\log 32^y = \log 4^{(3y-9)}$

$y \log 32 = (3y - 9) \log 4$

$y \log 32 = 3y \log 4 - 9 \log 4$

$y(\log 32 - 3 \log 4) = -9 \log 4$

$y = \dfrac{-9 \log 4}{\log \frac{32}{4^3}} = \dfrac{-18 \log 2}{\log \frac{1}{2}} = \dfrac{-18 \log 2}{-\log 2}$

$y = 18$

Thus, Greg used 32 to the power of 18.

3. The magnitude (Richter Scale) of an

earthquake is given by $R = \log\left(\dfrac{I}{I_0}\right)$ where

I is the intensity of the earthquake and I_0 is the intensity of a zero-level reference

earthquake. $\dfrac{I}{I_0} =$ how many times greater

the earthquake is than a zero-level earthquake.

$R_1 = \log(675,000)$

$R_2 = \log\left(\dfrac{I}{I_0}\right)$

Since $R_1 - 4 = R_2$

$\log(675,000) - 4 = \log\left(\dfrac{I}{I_0}\right)$

$\log\left(6.75 \times 10^5\right) - 4 = \log\left(\dfrac{I}{I_0}\right)$

$\log 6.75 + 5\log 10 - 4 = \log\left(\dfrac{I}{I_0}\right)$

$1.829 = \log\left(\dfrac{I}{I_0}\right)$

$10^{1.829} = \dfrac{I}{I_0}$

$67.5 = \dfrac{I}{I_0}$

Thus, the other earthquake is 67.5 times as intense as a zero-level earthquake.

Problems 4.4

1. $\log(3x + 2) = \log(2x + 5)$

$\quad 3x + 2 = 2x + 5$

$\quad\quad x = 3$

3. $\log 7 - \log(x - 1) = \log 4$

$\quad \log\dfrac{7}{x-1} = \log 4$

$\quad \dfrac{7}{x-1} = 4$

$\quad 7 = 4x - 4$

$\quad 4x = 11$

$\quad x = \dfrac{11}{4} = 2.75$

5. $\ln(-x) = \ln\left(x^2 - 6\right)$

$\quad -x = x^2 - 6$

$\quad x^2 + x - 6 = 0$

$\quad (x + 3)(x - 2) = 0$

$\quad x = -3 \text{ or } x = 2$

However, $x = -3$ is the only value that satisfies the original equation.

$x = -3$

7. $e^{2x}e^{5x} = e^{14}$

$\quad e^{7x} = e^{14}$

$\quad 7x = 14$

$\quad x = 2$

9. $(81)^{4x} = 9$

$\quad (3^4)^{4x} = 3^2$

$\quad 3^{16x} = 3^2$

$\quad 16x = 2$

$\quad x = \dfrac{2}{16} = \dfrac{1}{8} = 0.125$

11. $e^{2x} = 9$

$\quad (e^x)^2 = 3^2$

$\quad e^x = 3$

$\quad x = \ln 3 \approx 1.099$

13. $2e^{5x+2} = 17$

$\quad e^{5x+2} = \dfrac{17}{2}$

$\quad 5x + 2 = \ln\left(\dfrac{17}{2}\right)$

$\quad 5x = \ln\left(\dfrac{17}{2}\right) - 2$

$\quad x = \dfrac{1}{5}\left[\ln\left(\dfrac{17}{2}\right) - 2\right] \approx 0.028$

15. $10^{\frac{4}{x}} = 6$

$\dfrac{4}{x} = \log 6$

$x = \dfrac{4}{\log 6} \approx 5.140$

17. $\dfrac{5}{10^{2x}} = 7$

$10^{2x} = \dfrac{5}{7}$

$2x = \log \dfrac{5}{7}$

$x = \dfrac{\log\left(\frac{5}{7}\right)}{2} \approx -0.073$

19. $2^x = 5$

$\ln 2^x = \ln 5$
$x \ln 2 = \ln 5$

$x = \dfrac{\ln 5}{\ln 2} \approx 2.322$

21. $7^{3x-2} = 5$

$\ln 7^{3x-2} = \ln 5$
$(3x - 2)\ln 7 = \ln 5$

$3x - 2 = \dfrac{\ln 5}{\ln 7}$

$3x = \dfrac{\ln 5}{\ln 7} + 2$

$x = \dfrac{\frac{\ln 5}{\ln 7} + 2}{3} \approx 0.942$

23. $2^{-\frac{2x}{3}} = \dfrac{4}{5}$

$\ln 2^{-\frac{2x}{3}} = \ln \dfrac{4}{5}$

$-\dfrac{2x}{3}\ln 2 = \ln \dfrac{4}{5}$

$-\dfrac{2x}{3} = \dfrac{\ln\left(\frac{4}{5}\right)}{\ln 2}$

$x = -\dfrac{3\ln\left(\frac{4}{5}\right)}{2\ln 2} \approx 0.483$

25. $(4)5^{3-x} - 7 = 2$

$5^{3-x} = \dfrac{9}{4}$

$\ln 5^{3-x} = \ln \dfrac{9}{4}$

$(3 - x)\ln 5 = \ln \dfrac{9}{4}$

$3 - x = \dfrac{\ln\left(\frac{9}{4}\right)}{\ln 5}$

$x = 3 - \dfrac{\ln\left(\frac{9}{4}\right)}{\ln 5} \approx 2.496$

27. $\log(x - 3) = 3$

$10^3 = x - 3$

$x = 10^3 + 3 = 1003$

29. $\log_4(9x - 4) = 2$

$4^2 = 9x - 4$

$9x = 4^2 + 4$

$x = \dfrac{4^2 + 4}{9} = \dfrac{20}{9} \approx 2.222$

31. $\log(3x - 1) - \log(x - 3) = 2$

$$\log \frac{3x - 1}{x - 3} = 2$$

$$10^2 = \frac{3x - 1}{x - 3}$$

$$100(x - 3) = 3x - 1$$

$$97x = 299$$

$$x = \frac{299}{97} \approx 3.082$$

33. $$\log_2(5x + 1) = 4 - \log_2(3x - 2)$$

$$\log_2(5x + 1) + \log_2(3x - 2) = 4$$

$$\log[(5x + 1)(3x - 2)] = 4$$

$$(5x + 1)(3x - 2) = 2^4$$

$$15x^2 - 7x - 2 = 16$$

$$15x^2 - 7x - 18 = 0$$

$x \approx 1.353$ or $x \approx -0.887$

However, $x \approx 1.353$ is the only value that satisfies the original equation.

$x \approx 1.353$

35. $\log_2\left(\dfrac{2}{x}\right) = 3 + \log_2 x$

$$\log_2\left(\frac{2}{x}\right) - \log_2 x = 3$$

$$\log_2 \frac{\frac{2}{x}}{x} = 3$$

$$\log_2 \frac{2}{x^2} = 3$$

$$2^3 = \frac{2}{x^2}$$

$$x^2 = \frac{1}{4}$$

$$x = \pm\frac{1}{2}$$

However, $x = \dfrac{1}{2}$ is the only value that satisfies the original equation.

$$x = \frac{1}{2} = 0.5$$

37. $\log S = \log 12.4 + 0.26 \log A$

$\log S = \log 12.4 + \log A^{0.26}$

$\log S = \log \left[12.4 A^{0.26} \right]$

$S = 12.4 A^{0.26}$

39. a. When $t = 0$,

$Q = 100e^{-0.035(0)} = 100e^0$

$= 100 \cdot 1 = 100.$

b. If $Q = 20$, then $20 = 100e^{-0.035t}$.
Solving for t gives

$\dfrac{20}{100} = e^{-0.035t}$

$\dfrac{1}{5} = e^{-0.035t}$

$\ln \dfrac{1}{5} = -0.035t$

$-\ln 5 = -0.035t$

$t = \dfrac{\ln 5}{0.035} \approx 46$

41. If $P = 1,500,000$, then

$1,500,000 = 1,000,000(1.02)^t$. Solving for t gives

$\dfrac{1,500,000}{1,000,000} = (1.02)^t$

$1.5 = (1.02)^t$

$\ln 1.5 = \ln(1.02)^t$

$\ln 1.5 = t \ln 1.02$

$t = \dfrac{\ln 1.5}{\ln 1.02} \approx 20.5$

43. $q = 80 - 2^p$

$2^p = 80 - q$

$\log 2^p = \log(80 - q)$

$p \log 2 = \log(80 - q)$

$p = \dfrac{\log(80 - q)}{\log 2}$

When $q = 60$, then $p = \dfrac{\log 20}{\log 2} \approx 4.32$.

45. $q = 1000 \left(\dfrac{1}{2} \right)^{0.8^t}$

$\log q = \log 1000 + \log \left(\dfrac{1}{2} \right)^{0.8^t}$

$\log q = 3 + 0.8^t \log \dfrac{1}{2}$

$\log q = 3 + 0.8^t (-\log 2)$

$\log(q) - 3 = 0.8^t (-\log 2)$

Thus

$0.8^t = \dfrac{\log(q) - 3}{-\log 2} = \dfrac{3 - \log q}{\log 2}$

$t \log(0.8) = \log \left(\dfrac{3 - \log q}{\log 2} \right)$.

$t = \dfrac{\log \left(\frac{3 - \log q}{\log 2} \right)}{\log(0.8)}$

$$y = Ab^{a^x}$$
$$\log y = \log A + \log b^{a^x}$$
$$\log y = \log A + a^x \log b$$
$$\log y - \log A = a^x \log b$$
$$a^x = \frac{\log y - \log A}{\log b}$$
$$\log a^x = \log\left(\frac{\log y - \log A}{\log b}\right)$$
$$x \log a = \log\left(\frac{\log y - \log A}{\log b}\right)$$
$$x = \frac{\log\left(\frac{\log y - \log A}{\log b}\right)}{\log a}$$

The previous solution was the special case

$y = q$, $A = 1000$, $b = \dfrac{1}{2}$, $a = 0.8$, and $x = t$.

47. $\log_2 x = 5 - \log_2(x+4)$ is equivalent to
$0 = 5 - \log_2(x+4) - \log_2 x$, or
$0 = 5 - \dfrac{\ln(x+4)}{\ln 2} - \dfrac{\ln x}{\ln 2}$. Thus the solutions
of the original equation are the zeros of the
function $y = 5 - \dfrac{\ln(x+4)}{\ln 2} - \dfrac{\ln x}{\ln 2}$.

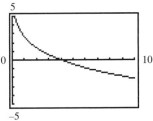

From the graph of this function, the only
zero is $x = 4$. Thus 4 is the only solution of
the original equation.

49.

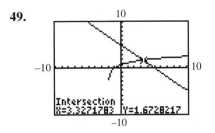

3.33

Chapter 4 Review Problems

1. $\log_3 243 = 5$

3. $81^{\frac{1}{4}} = 3$

5. $\ln 54.598 = 4$

7. Because $5^3 = 125$, $\log_5 125 = 3$

9. Because $3^{-4} = \dfrac{1}{81}$, $\log_3 \dfrac{1}{81} = -4$

11. Because $\left(\dfrac{1}{3}\right)^{-2} = 3^2 = 9$, $\log_{\frac{1}{3}} 9 = -2$

13. $5^x = 625$
    $x = 4$

15. $2^{-5} = x$
    $x = \dfrac{1}{2^5} = \dfrac{1}{32}$

17. $\ln(2x+3) = 0$
    $e^0 = 2x+3$
    $1 = 2x+3$
    $2x = -2$
    $x = -1$

19. $\log 8000 = \log(2 \cdot 10)^3 = 3\log(2 \cdot 10)$
    $= 3(\log 2 + \log 10) = 3(a+1)$

21. $3\log 7 - 2\log 5 = \log 7^3 - \log 5^2 = \log \dfrac{7^3}{5^2}$

23. $2\ln x + \ln y - 3\ln z = \ln x^2 + \ln y - \ln z^3 = \ln x^2 y - \ln z^3 = \ln \dfrac{x^2 y}{z^3}$

25. $\dfrac{1}{2}\log_2 x + 2\log_2 x^2 - 3\log_2(x+1) - 4\log_2(x+2)$

$= \log_2 x^{\frac{1}{2}} + \log_2 \left(x^2\right)^2 - \left[\log_2(x+1)^3 + \log_2(x+2)^4\right] = \log_2\left(x^{\frac{1}{2}}x^4\right) - \log_2\left[(x+1)^3(x+2)^4\right]$

$= \log_2 \dfrac{x^{\frac{9}{2}}}{(x+1)^3(x+2)^4}$

27. $\ln \dfrac{x^3 y^2}{z^{-5}} = \ln x^3 y^2 - \ln z^{-5}$

$= \ln x^3 + \ln y^2 - \ln z^{-5}$

$= 3\ln x + 2\ln y + 5\ln z$

29. $\ln \sqrt[3]{xyz} = \ln(xyz)^{\frac{1}{3}} = \dfrac{1}{3}\ln(xyz)$

$= \dfrac{1}{3}(\ln x + \ln y + \ln z)$

31. $\ln\left[\dfrac{1}{x}\sqrt{\dfrac{y}{z}}\right] = \ln \dfrac{\left(\dfrac{y}{z}\right)^{1/2}}{x} = \ln\left(\dfrac{y}{z}\right)^{\frac{1}{2}} - \ln x$

$= \dfrac{1}{2}\ln \dfrac{y}{z} - \ln x = \dfrac{1}{2}(\ln y - \ln z) - \ln x$

33. $\log_3(x+5) = \dfrac{\log_e(x+5)}{\log_e 3} = \dfrac{\ln(x+5)}{\ln 3}$

35. $\log_5 19 = \dfrac{\log_2 19}{\log_2 5} = \dfrac{4.2479}{2.3219} \approx 1.8295$

37. $\ln\left(16\sqrt{3}\right) = \ln 4^2 + \ln \sqrt{3} = 2\ln 4 + \dfrac{1}{2}\ln 3$

$= 2y + \dfrac{1}{2}x$

39. $10^{\log x} + \log 10^x + \log 10 = x + x + 1 = 2x + 1$

41. In exponential form, $y = e^{x^2 + 2}$.

43.

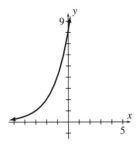

45. $\log(5x+1) = \log(4x+6)$
$5x+1 = 4x+6$
$x = 5$

47. $3^{4x} = 9^{x+1}$
$3^{4x} = \left(3^2\right)^{x+1}$
$3^{4x} = 3^{2(x+1)}$
$4x = 2(x+1)$
$4x = 2x+2$
$2x = 2$
$x = 1$

49. $\log x + \log(10x) = 3$
$\log x + \log 10 + \log x = 3$
$2\log(x) + 1 = 3$
$2\log(x) = 2$
$\log x = 1$
$x = 10^1 = 10$

51. $\ln(\log_x 3) = 2$
$\log_x 3 = e^2$
$x^{e^2} = 3$
$(x^{e^2})^{-e^2} = 3^{-e^2}$
$x^{e^2 \cdot -e^2} = 3^{-e^2}$
$x^1 = 3^{-e^2}$
$x = \dfrac{1}{3^{e^2}}$

53. $e^{3x} = 14$
$3x = \ln 14$
$x = \dfrac{\ln 14}{3} \approx 0.880$

55. $3\left(10^{x+4} - 3\right) = 9$
$10^{x+4} - 3 = 3$
$10^{x+4} = 6$
$x + 4 = \log 6$
$x = \log(6) - 4 \approx -3.222$

57. $4^{x+3} = 7$
$\ln 4^{x+3} = \ln 7$
$(x+3)\ln 4 = \ln 7$
$x + 3 = \dfrac{\ln 7}{\ln 4}$
$x = \dfrac{\ln 7}{\ln 4} - 3 \approx -1.596$

59. Quarterly rate $= \dfrac{0.06}{4} = 0.015$

$6\dfrac{1}{2}$ yr $= 26$ quarters

   a. $2600(1.015)^{26} \approx \$3829.04$

   b. $3829.04 - 2600 = \$1229.04$

61. $12\left(1\dfrac{1}{6}\%\right) = 14\%$

63. a. $P = 6000[1 + (-0.005)]^t$ or
       $P = 6000(0.995)^t$

   b. When $t = 10$, then
       $P = 6000(0.995)^{10} \approx 5707.$

65. $N = 10e^{-0.41t}$
   a. When $t = 0$, then
       $N = 10e^0 = 10 \cdot 1 = 10$ mg

   b. When $t = 2$, then
       $N = 10e^{-0.82} \approx 4.4$ mg

   c. When $t = 10$, then
       $N = 10e^{-4.1} \approx 0.2$ mg

d. $\dfrac{\ln 2}{0.41} \approx 1.7$

e. If $N = 1$, then $1 = 10e^{-0.41t}$. Solving for t gives

$$\frac{1}{10} = e^{-0.41t}$$

$$-0.41t = \ln \frac{1}{10} = -\ln 10$$

$$t = \frac{\ln 10}{0.41} \approx 5.6$$

67. $R = 10e^{-\frac{t}{40}}$

a. If $t = 20$, $R = 10e^{-\frac{20}{40}} = 10e^{-\frac{1}{2}} \approx 6$.

b. $5 = 10e^{-\frac{t}{40}}$, $\dfrac{1}{2} = e^{-\frac{t}{40}}$. Thus

$$-\frac{t}{40} = \ln \frac{1}{2} = -\ln 2$$

$$t = 40 \ln 2 \approx 28.$$

69. $T_t - T_e = \left(T_t - T_e\right)_o e^{-at}$

$$e^{-at} = \frac{T_t - T_e}{\left(T_t - T_e\right)_o}$$

$$-at = \ln \frac{T_t - T_e}{\left(T_t - T_e\right)_o}$$

$$a = -\frac{1}{t} \ln \frac{T_t - T_e}{\left(T_t - T_e\right)_o}$$

$$a = \frac{1}{t} \ln \frac{\left(T_t - T_e\right)_o}{T_t - T_e}$$

71.

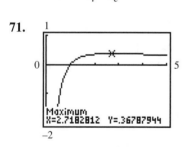

$(-\infty, 0.37]$

73.

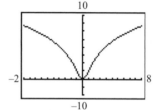

2.53

75. $y = \log_2\left(x^2 + 1\right) = \dfrac{\ln\left(x^2 + 1\right)}{\ln 2}$

77.

$$y = \frac{3^x}{9} = \frac{3^x}{3^2} = 3^{x-2}.$$

If $f(x) = 3^x$, then we have

$y = 3^{x-2} = f(x-2)$. Thus the graph of

$y = \dfrac{3^x}{9}$ is the graph of $y = 3^x$ shifted 2

units to the right.

Mathematical Snapshot Chapter 4

1. $T = \dfrac{P\left(1 - e^{-dkI}\right)}{e^{kI} - 1}$

  a. $T\left(e^{kI} - 1\right) = P\left(1 - e^{-dkI}\right)$

     $\dfrac{T\left(e^{kI} - 1\right)}{1 - e^{-dkI}} = P$ or $P = \dfrac{T\left(e^{kI} - 1\right)}{1 - e^{-dkI}}$

  b. $T\left(e^{kI} - 1\right) = P - Pe^{-dkI}$

     $Pe^{-dkI} = P - T\left(e^{kI} - 1\right)$

     $e^{-dkI} = \dfrac{P - T\left(e^{kI} - 1\right)}{P}$

     $-dkI = \ln\left[\dfrac{P - T\left(e^{kI} - 1\right)}{P}\right]$

     $d = -\dfrac{1}{kI}\ln\left[\dfrac{P - T\left(e^{kI} - 1\right)}{P}\right]$

     $d = \dfrac{1}{kI}\ln\left[\dfrac{P}{P - T\left(e^{kI} - 1\right)}\right]$

3. $P = 100,\ I = 4,\ d = 3,\ H = 8,\ k = \dfrac{\ln 2}{H} = \dfrac{\ln 2}{8}$

  a. $T = \dfrac{P\left(1 - e^{-dkI}\right)}{e^{kI} - 1} = \dfrac{100\left(1 - e^{-3\cdot\frac{\ln 2}{8}\cdot 4}\right)}{e^{\frac{\ln 2}{8}\cdot 4} - 1}$

     $= \dfrac{100\left(1 - \left[e^{\ln 2}\right]^{-\frac{3}{2}}\right)}{\left[e^{\ln 2}\right]^{\frac{1}{2}} - 1} = \dfrac{100\left(1 - 2^{-\frac{3}{2}}\right)}{2^{\frac{1}{2}} - 1} \approx 156$

  b. $R = P\left(1 - e^{-dkI}\right)$. From part (a),

     $P\left(1 - e^{-dkI}\right) = 100\left(1 - 2^{-\frac{3}{2}}\right)$. Thus

     $R = 100\left(1 - 2^{-\frac{3}{2}}\right) \approx 65.$

Chapter 5

Principles in Practice 5.1

1. Let $P = 518$ and let $n = 3(365) = 1095$.

$$S = P(1+r)^n$$

$$S = 518\left(1+\frac{r}{365}\right)^{1095}$$

By graphing S as a function of the nominal rate r, we find that when $r = 0.049$, $S = 600$. Thus, at the nominal rate of 4.9% compounded daily, the initial amount of $518 will grow to $600 after 3 years.

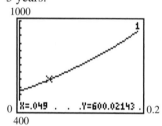

3. Let $n = 12$.

$$r_e = \left(1+\frac{r}{n}\right)^n - 1$$

$$r_e = \left(1+\frac{r}{12}\right)^{12} - 1$$

By graphing r_e as a function of r, we find that, when the nominal rate $r = 0.077208$ or 7.7208%, the effective rate $r_e = 0.08$ or 8%.

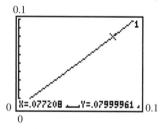

Problems 5.1

1. a. $6000(1.08)^8 \approx \$11,105.58$

b. $11,105.58 - 6000 = \$5105.58$

3. $(1.015)^2 - 1 \approx 0.030225$ or 3.023%

5. $\left(1+\dfrac{0.04}{365}\right)^{365} - 1 \approx 0.04081$ or 4.081%

7. a. A nominal rate compounded yearly is the same as the effective rate, so the effective rate is 10%.

b. $\left(1+\dfrac{0.10}{2}\right)^2 - 1 = 0.1025$ or 10.25%

c. $\left(1+\dfrac{0.10}{4}\right)^4 - 1 \approx 0.10381$ or 10.381%

d. $\left(1+\dfrac{0.10}{12}\right)^{12} - 1 \approx 0.10471$ or 10.471%

e. $\left(1+\dfrac{0.10}{365}\right)^{365} - 1 \approx 0.10516$ or 10.516%

9. Let r_e be the effective rate. Then

$$2000(1+r_e)^5 = 2950$$

$$(1+r_e)^5 = \frac{2950}{2000}$$

$$1+r_e = \sqrt[5]{\frac{2950}{2000}}$$

$$r_e = \sqrt[5]{\frac{2950}{2000}} - 1$$

$$r_e \approx 0.0808 \text{ or } 8.08\%.$$

11. From Example 6, the number of years, n, is given by $n = \dfrac{\ln 2}{\ln(1.09)} \approx 8.0$ years.

13. $6000(1.08)^7 \approx \$10,282.95$

15. $21,500(1.06)^{10} \approx \$38,503.23$

17. a. $(0.015)(12) = 0.18$ or 18%

   b. $(1.015)^{12} - 1 \approx 0.1956$ or 19.56%

19. The compound amount after the first four years is $2000(1.06)^4$. After the next four years the compound amount is

$\left[2000(1.06)^4\right](1.03)^8 \approx \$3198.54$.

21. 7.8% compounded semiannually is equivalent to an effective rate of

$(1.039)^2 - 1 = 0.079521$ or 7.9521%. Thus 8% compounded annually, which is the effective rate, is the better rate.

23. a. $\left(1 + \dfrac{0.0475}{360}\right)^{365} - 1 \approx 0.0493$ or 4.93%

   b. $\left(1 + \dfrac{0.0475}{365}\right)^{365} - 1 \approx 0.0486$ or 4.86%

25. Let r_e = effective rate.

$300,000 = 100,000\left(1 + r_e\right)^{10}$

$\left(1 + r_e\right)^{10} = 3$

$1 + r_e = \sqrt[10]{3}$

$r_e = \sqrt[10]{3} - 1 \approx 0.1161$ or 11.61%.

27. Let r = the required nominal rate.

$420\left(1 + \dfrac{r}{2}\right)^{28} = 1000$

$\left(1 + \dfrac{r}{2}\right)^{28} = \dfrac{1000}{420} = \dfrac{50}{21}$

$1 + \dfrac{r}{2} = \sqrt[28]{\dfrac{50}{21}}$

$r = 2\left[\sqrt[28]{\dfrac{50}{21}} - 1\right] \approx 0.0629$ or 6.29%

Problems 5.2

1. $6000(1.05)^{-20} \approx \$2261.34$

3. $4000(1.035)^{-24} \approx \$1751.83$

5. $9000\left(1 + \dfrac{0.08}{4}\right)^{-22} \approx \$5821.55$

7. $8000\left(1 + \dfrac{0.10}{12}\right)^{-60} \approx \$4862.31$

9. $10,000\left(1 + \dfrac{0.095}{365}\right)^{-4(365)} \approx \$6838.95$

11. $12,000\left(1 + \dfrac{0.053}{12}\right)^{-12} \approx \$11,381.89$

13. $27,000(1.03)^{-22} \approx \$14,091.10$

15. Let x be the payment 2 years from now. The equation of value at year 2 is

$x = 600(1.04)^{-2} + 800(1.04)^{-4}$

$x \approx \$1238.58$

17. Let x be the payment at the end of 6 years. The equation of value at year 6 is

$$2000(1.025)^4 + 4000(1.025)^2 + x = 5000(1.025) + 5000(1.025)^{-4}$$

$$x = 5000(1.025) + 5000(1.025)^{-4} - 2000(1.025)^4 - 4000(1.025)^2$$

$$x \approx \$3244.63.$$

19. a. $NPV = 8000(1.025)^{-6} + 10,000(1.025)^{-8} + 14,000(1.025)^{-12} - 25,000 \approx \$515.62$

b. Since $NPV > 0$, the investment is profitable.

21. We consider the value of each investment at the end of eight years. The savings account has a value of $10,000(1.03)^{16} \approx \$16,047.06$.

The business investment has a value of $16,000. Thus the better choice is the savings account.

23. $1000\left(1 + \dfrac{0.075}{4}\right)^{-80} \approx \$226.25$

25. Let r be the nominal discount rate, compounded quarterly. Then

$$4700 = 10,000\left(1 + \frac{r}{4}\right)^{-32}$$

$$4700 = \frac{10,000}{\left(1 + \frac{r}{4}\right)^{32}}$$

$$\left(1 + \frac{r}{4}\right)^{32} = \frac{10,000}{4700} = \frac{100}{47}$$

$$1 + \frac{r}{4} = \sqrt[32]{\frac{100}{47}}$$

$$r = 4\left[\sqrt[32]{\frac{100}{47}} - 1\right] \approx 0.0955 \text{ or } 9.55\%$$

Problems 5.3

1. $S = 4000e^{0.0625(6)} \approx \$5819.97$
   $5819.97 - 4000 = \$1819.97$

3. $P = 2500e^{-0.0675(8)} \approx \$1456.87$

5. $e^{0.04} - 1 \approx 0.0408$
   Answer: 4.08%

7. $e^{0.03} - 1 \approx 0.0305$
   Answer: 3.05%

9. $S = 100e^{0.045(2)} \approx \$109.42$

11. $P = 1,000,000e^{-0.05(5)} \approx \$778,800.78$

13. a. $25,000(1+0.035)^{25} = \$59,081$

  b. $P = 59,081e^{-(0.045)(25)} \approx \$19,181$

15. Effective rate $= e^r - 1$. Thus $0.05 = e^r - 1$,
  $e^r = 1.05$, $r = \ln 1.05 \approx 0.0488$.
  Answer: 4.88%

17. $3P = Pe^{0.07t}$
  $3 = e^{0.07t}$
  $0.07t = \ln 3$
  $t = \dfrac{\ln 3}{0.07} \approx 16$
  Answer: 16 years

19. The accumulated amounts under each option are:

  a. $1000e^{(0.035)(2)} \approx \$1072.51$

  b. $1020(1.0175)^4 \approx \$1093.30$

  c. $500e^{(0.035)(2)} + 500(1.0175)^4$
    $\approx 536.25 + 535.93 = \$1072.18$

21. a. $9000(1.0125)^4 \approx \$9458.51$

  b. After one year the accumulated amount of the investment is
    $10,000e^{0.055} \approx \$10,565.41$. The payoff for the loan (including interest) is $1000 + 1000(0.08) = \$1080$. The net return is $10,565.41 - 1080 = \$9485.41$. Thus, this strategy is better by $9485.41 - 9458.51 = \$26.90$.

Principles in Practice 5.4

1. Let $a = 64$ and let $r = \dfrac{3}{4}$. Then, the next five heights of the ball are $64\left(\dfrac{3}{4}\right)$,

  $64\left(\dfrac{3}{4}\right)^2$, $64\left(\dfrac{3}{4}\right)^3$, $64\left(\dfrac{3}{4}\right)^4$, $64\left(\dfrac{3}{4}\right)^5$, or

  48 ft, 36 ft, 27 ft, $20\dfrac{1}{4}$ ft, and $15\dfrac{3}{16}$ ft.

3. The total vertical distance traveled in the air after n bounces is equal to 2 times the sum of heights. If $a = 6$ and $r = \dfrac{2}{3}$, then when the ball hits the ground for the twelfth time, $n = 12$ and the distance traveled in the air is

$$2s = 2\left[\frac{a\left(1 - r^n\right)}{1-r}\right] = 2\left[\frac{6\left(1 - \left(\frac{2}{3}\right)^{12}\right)}{1 - \frac{2}{3}}\right]$$

  ≈ 35.72 meters

5. Let $R = 500$ and let $n = 72$. Then, the present value A of the annuity is given by

$$A = R\left(\frac{1 - (1+r)^{-n}}{r}\right) = 500\left(\frac{1 - (1+r)^{-72}}{r}\right)$$

By graphing A as a function of r, we find that when $r \approx 0.005167$, $A = 30,000$. Thus, if the present value of the annuity is $30,000, the monthly interest rate is 0.5167%, and the nominal rate is $12(0.005167) = 0.062$ or 6.2%.

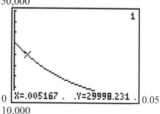

7. Let $r = \dfrac{0.048}{4} = 0.012$, and $n = 24$.

$$A = R\left(\frac{1-(1+r)^{-n}}{r}\right)$$

$$A = R\left(\frac{1-(1+0.012)^{-24}}{0.012}\right)$$

$$= R\left(\frac{1-(1.012)^{-24}}{0.012}\right)$$

By Graphing A as a function of R, we find that when $R = 723.03$, $A = 15,000$. Thus the monthly payment is \$723.03 if the present value of the annuity is \$15,000.

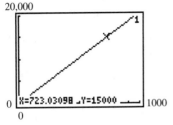

9. Let $R = 2000$ and let $r = 0.057$. Then, the value of the IRA at the end of 15 years, when $n = 15$, is given by

$$S = R\left(\frac{(1+r)^n - 1}{r}\right)$$

$$S = 2000\left(\frac{(1+0.057)^{15} - 1}{0.057}\right) \approx 45,502.06$$

Thus, at the end of 15 years the IRA will be worth \$45,502.06.

Problems 5.4

1. 64

$$64\left(\frac{1}{2}\right) = 32$$

$$64\left(\frac{1}{2}\right)^2 = 16$$

$$64\left(\frac{1}{2}\right)^3 = 8$$

$$64\left(\frac{1}{2}\right)^4 = 4$$

3. 100

$$100(1.02) = 102$$

$$100(1.02)^2 = 104.04$$

5. $s = \dfrac{\frac{4}{7}\left[1-\left(\frac{4}{7}\right)^5\right]}{1-\frac{4}{7}}$

$$= \frac{\frac{4}{7}\left[\frac{15,783}{16,807}\right]}{\frac{3}{7}}$$

$$= \frac{21,044}{16,807}$$

7. $s = \dfrac{1\left[1-(0.1)^6\right]}{1-0.1} = 1.11111$

9. $a_{\overline{35}\,|\,0.04} \approx 18.664613$

11. $s_{\overline{8}\,|\,0.0075} \approx 8.213180$

13. $600a_{\overline{6}\,|\,0.06} \approx 600(4.917324) \approx \$2950.39$

15. $2000a_{\overline{18}\,|\,0.02} \approx 2000(14.992031)$

$$\approx \$29,984.06$$

17. $800 + 800a_{\overline{11}|0.035} \approx 800 + 800(9.001551)$

$\approx \$8001.24$

19. $2000s_{\overline{36}|0.0125} \approx 2000(45.115505)$

$\approx \$90,231.01$

21. $5000s_{\overline{20}|0.07} \approx 5000(40.995492)$

$\approx \$204,977.46$

23. $1200\left(s_{\overline{13}|0.08} - 1\right) \approx 1200(21.495297 - 1)$

$\approx \$24,594.36$

25. $175a_{\overline{32}|\frac{0.04}{12}} - 25a_{\overline{8}|\frac{0.04}{12}}$

$\approx 175(30.304595) - 25(7.881321)$

$\approx \$5106.27$

27. $R = \dfrac{5000}{a_{\overline{12}|0.015}} \approx \dfrac{5000}{10.907505} \approx \$458.40$

29. a. $\left(50s_{\overline{48}|0.005}\right)(1.005)^{24}$

$\approx 50(54.097832)(1.005)^{24}$

$\approx \$3048.85$

   b. $3048.85 - 48(50) = \$648.85$

31. $R = \dfrac{48,000}{s_{\overline{10}|0.07}} \approx \dfrac{48,000}{13.816448} \approx \$3474.12$

33. The original annual payment is $\dfrac{25,000}{s_{\overline{10}|0.06}}$.

After six years the value of the fund is

$\dfrac{25,000}{s_{\overline{10}|0.06}} s_{\overline{6}|0.06}$.

This accumulates to

$\left[\dfrac{25,000}{s_{\overline{10}|0.06}} s_{\overline{6}|0.06}\right](1.07)^4.$

Let x be the amount of the new payment.

$xs_{\overline{4}|0.07}$

$= 25,000 - \left[\dfrac{25,000}{s_{\overline{10}|0.06}} s_{\overline{6}|0.06}(1.07)^4\right]$

$x = \dfrac{25,000 - \left[\dfrac{25,000}{s_{\overline{10}|0.06}} s_{\overline{6}|0.06}(1.07)^4\right]}{s_{\overline{4}|0.07}}$

$x \approx \dfrac{25,000 - \left[\dfrac{25,000}{13.180795}(6.975319)(1.07)^4\right]}{4.439943}$

$x \approx \$1725$

35. $s_{\overline{60}|0.017} = \dfrac{(1.017)^{60} - 1}{0.017} \approx 102.91305$

37. $750a_{\overline{480}|0.0135} = 750\left[\dfrac{1 - (1.0135)^{-480}}{0.0135}\right]$

$\approx 55,466.57$

39. $R = \dfrac{3000}{s_{\overline{20}|0.01375}} = \dfrac{3000(0.01375)}{(1.01375)^{20} - 1}$

$\approx \$131.34$

41. $200,000 + 200,000a_{\overline{19}|0.10}$

$= 200,000 + 200,000\left[\dfrac{1 - (1.10)^{-19}}{0.10}\right]$

$\approx \$1,872,984.02$

43. For the first situation, the compound amount is

$$\left[2000\left(s_{\overline{11}|0.07}-1\right)\right](1.07)^{30}$$

$$=2000\left[\frac{(1.07)^{11}-1}{0.07}-1\right](1.07)^{30}$$

$\approx \$225,073,$
so the net earnings are
$225,073 - 20,000 = \$205,073.$
For the second situation, the compound amount is

$$2000\left(s_{\overline{31}|0.07}-1\right)=2000\left[\frac{(1.07)^{31}-1}{0.07}-1\right]\approx\$202,146,$$

so the net earnings are $202,146 - 60,000 = \$142,146.$

45. $40,000\dfrac{1-e^{-0.04(5)}}{0.04}\approx\$181,269.25$

Problems 5.5

1. $R=\dfrac{8000}{a_{\overline{36}|\frac{0.14}{12}}}\approx\dfrac{8000}{29.258904}\approx\$273.42$

3. $R=\dfrac{8000}{a_{\overline{36}|\frac{0.04}{12}}}\approx\dfrac{8000}{33.870766}\approx\$236.19$

Finance charge $= 36(236.19) - 8000 = \$502.84$

5. a. $R=\dfrac{7500}{a_{\overline{36}|\frac{0.04}{12}}}\approx\dfrac{7500}{33.870766}\approx\$221.43$

b. $7500\dfrac{0.04}{12}=\$25$

c. $221.43 - 25 = \$196.43$

7. $R=\dfrac{5000}{a_{\overline{4}|0.07}}\approx\dfrac{5000}{3.387211}\approx\$1476.14$

The interest for the first period is $(0.07)(5000) = \$350$, so the principal repaid at the end of that period is $1476.14 - 350 = \$1126.14$. The principal outstanding at the beginning of period 2 is $5000 - 1126.14 = \$3873.86$. The interest for period 2 is $(0.07)(3873.86) = \$271.17$, so the principal repaid at the end of that period is $1476.14 - 271.17 = \$1204.97$. The principal outstanding at beginning of period 3 is $3873.86 - 1204.97 = \$2668.89$. Continuing in this manner, we construct the following amortization schedule.

Period	Prin. Outs. at Beginning	Int. for Period	Pmt. at End	Prin. Repaid at End
1	5000.00	350.00	1476.14	1126.14
2	3873.86	271.17	1476.14	1204.97
3	2668.89	186.82	1476.14	1289.32
4	1379.57	96.57	1476.14	1379.57
Total		904.56	5904.56	5000.00

9. $R = \dfrac{900}{a_{\overline{5}|0.025}} \approx \dfrac{900}{4.645828} \approx \$193.72$

The interest for period 1 is $(0.025)(900) = \$22.50$, so the principal repaid at the end of that period is $193.72 - 22.50 = \$171.22$. The principal outstanding at the beginning of period 2 is $900 - 171.22 = \$728.78$. The interest for that period is $(0.025)(728.78) = \$18.22$, so the principal repaid at the end of that period is $193.72 - 18.22 = \$175.50$. The principal outstanding at the beginning of period 3 is $728.78 - 175.50 = \$553.28$. Continuing in this manner, we obtain the following amortization schedule. Note the adjustment in the final payment.

Period	Prin. Outs. at Beginning	Int. for Period	Pmt. at End	Prin. Repaid at End
1	900.00	22.50	193.72	171.22
2	728.78	18.22	193.72	175.50
3	553.28	13.83	193.72	179.89
4	313.39	9.33	193.72	184.39
5	189.00	4.73	193.73	189.00
Total		68.61	968.61	900.00

11. From Eq. (1),

$$n = \frac{\ln\left[\dfrac{100}{100-1000(0.02)}\right]}{\ln(1.02)} \approx 11.268 .$$

Thus the number of full payments is 11.

13. Each of the original payments is $\dfrac{18{,}000}{a_{\overline{15}\,|\,0.035}}$.

After two years the value of the remaining

payments is $\left[\dfrac{18{,}000}{a_{\overline{15}\,|\,0.035}}\right] a_{\overline{11}\,|\,0.035}$. Thus the

new semi-annual payment is

$\dfrac{18{,}000\,a_{\overline{11}\,|\,0.035}}{a_{\overline{15}\,|\,0.035}} \cdot \dfrac{1}{a_{\overline{11}\,|\,0.04}}$

$= \dfrac{18{,}000(9.001551)}{11.517411} \cdot \dfrac{1}{8.760477}$

$\approx \$1606.$

15. a. Monthly interest rate is $\dfrac{0.092}{12}$.

Monthly payment is

$\dfrac{245{,}000}{a_{\overline{300}\,|\,\frac{0.092}{12}}}$

$= 245{,}000 \left[\dfrac{\frac{0.092}{12}}{1-\left(1+\frac{0.092}{12}\right)^{-300}}\right]$

$\approx \$2089.69$

b. $245{,}000\left(\dfrac{0.092}{12}\right) = \$1878.33$

c. $2089.69 - 1878.33 = \$211.36$

d. $300(2089.69) - 245{,}000 = \$381{,}907$

17. $n = \dfrac{\ln\left[\dfrac{100}{100-2000(0.015)}\right]}{\ln 1.015} \approx 23.956.$ Thus the

number of full payments is 23.

19. Present value of mortgage payments is

$600\,a_{\overline{360}\,|\,\frac{0.076}{12}} = 600\left[\dfrac{1-\left(1+\frac{0.076}{12}\right)^{-360}}{\frac{0.076}{12}}\right]$

$\approx \$84{,}976.84$

This amount is 75% of the purchase price x.

$0.75x = 84{,}976.84$

$x = \$113{,}302.45 \approx \$113{,}302$

21. $\dfrac{25{,}000}{a_{\overline{60}\,|\,0.0125}} - \dfrac{25{,}000}{a_{\overline{60}\,|\,0.01}}$

$= 25{,}000\left[\dfrac{1}{a_{\overline{60}\,|\,0.0125}} - \dfrac{1}{a_{\overline{60}\,|\,0.01}}\right]$

$= 25{,}000\left[\dfrac{0.0125}{1-(1.0125)^{-60}} - \dfrac{0.01}{1-(1.01)^{-60}}\right]$

$\approx \$38.64$

Chapter 5 Review Problems

1. $s = 3 + 2 + 2\cdot\dfrac{2}{3} + \cdots + 3\left(\dfrac{2}{3}\right)^5$

$= \dfrac{3\left[1-\left(\frac{2}{3}\right)^6\right]}{1-\frac{2}{3}} = \dfrac{3\left[\frac{665}{729}\right]}{\frac{1}{3}} = \dfrac{665}{81}$

3. 8.2% compounded semiannually
corresponds to an effective rate of
$(1.041)^2 - 1 = 0.083681$ or 8.37%. Thus the
better choice is 8.5% compounded annually.

5. Let x be the payment at the end of 2 years. The equation of value at the end of year 2 is

$$1000(1.04)^4 + x = 1200(1.04)^{-4} + 1000(1.04)^{-8}$$

$$x = 1200(1.04)^{-4} + 1000(1.04)^{-8} - 1000(1.04)^4$$

$$\approx \$586.60$$

7. a. $A = 200a_{\overline{13}|0.04} \approx 200(9.985648)$

$$\approx \$1997.13$$

    b. $S = 200s_{\overline{13}|0.04} \approx 200(16.626838)$

$$\approx \$3325.37$$

9. $200s_{\overline{13}|\frac{0.08}{12}} - 200 \approx 200(13.532926) - 200$

$$\approx \$2506.59$$

11. $\dfrac{5000}{s_{\overline{5}|0.06}} \approx \dfrac{5000}{5.637093} \approx \$886.98$

13. Let x be the first payment. The equation of value now is

$$x + 2x(1.07)^{-3} = 500(1.05)^{-3} + 500(1.03)^{-8}$$

$$x\left[1 + 2(1.07)^{-3}\right] = 500(1.05)^{-3} + 500(1.03)^{-8}$$

$$x = \frac{500(1.05)^{-3} + 500(1.03)^{-8}}{1 + 2(1.07)^{-3}}$$

$$x \approx \$314.00$$

15. $R = \dfrac{15,000}{a_{\overline{5}|0.0075}} \approx \dfrac{15,000}{4.889440} \approx \$3067.84$

The interest for period 1 is $(0.0075)(15,000) = \$112.50$, so the principal repaid at the end of that period is $3067.84 - 112.50 = \$2955.34$. The principal outstanding at beginning of period 2 is $15,000 - 2955.34 = \$12,044.66$. The interest for period 2 is $0.0075(12,044.66) = \$90.33$, so the principal repaid at the end of that period is $3067.84 - 90.33 = \$2977.51$. Principal outstanding at the beginning of period 3 is $12,044.66 - 2977.51 = \$9067.15$. Continuing, we obtain the following amortization schedule. Note the adjustment in the final payment.

Period	Prin. Outs. at Beginning	Int. for Period	Pmt. at End	Prin. Repaid at End
1	15,000	112.50	3067.84	2955.34
2	12,044.66	90.33	3067.84	2977.51
3	9067.15	68.00	3067.84	2999.84
4	6067.31	45.50	3067.84	3022.34
5	3044.97	22.84	3067.81	3044.97
Total		339.17	15,339.17	15,000.00

17. The monthly payment is

$$\frac{11,000}{a_{\overline{48}\,|\,\frac{0.055}{12}}} = 11,000\left[\frac{\frac{0.055}{12}}{1-\left(1+\frac{0.055}{12}\right)^{-48}}\right] \approx \$255.82$$

The finance charge is $48(255.82) - 11,000 = \$1279.36$

Mathematical Snapshot Chapter 5

1. $\dfrac{0.085}{2} = 0.0425,$ thus $R = 0.0425(25,000) = 1062.50.$

$$P = 25,000(1.0825)^{-25} + 1062.50 \cdot \frac{1-(1.0825)^{-25}}{\sqrt{1.0825}-1}$$

$$\approx \$26,102.13$$

3. The normal yield curve assumes a stable economic climate. By contrast, if investors are expecting a drop in interest rates, and with it a drop in yields from future investments, they will gladly give up liquidity for long-term investment at current, more favorable, interest rates. T-bills, which force the investor to find a new investment in a short time, are correspondingly less attractive, and so prices drop and yields rise.

Chapter 6

Principles in Practice 6.1

1. There are 3 rows, one for each source. There are two columns, one for each raw material. Thus, the size of the matrix is 3×2. Alternatively, she could use a 2×3 matrix.

Problems 6.1

1. a. The size is the number of rows by the columns. Thus **A** is 2×3, **B** is 3×3, **C** is 3×2, **D** is 2×2, **E** is 4×4, **F** is 1×2, **G** is 3×1, **H** is 3×3, and **J** is 1×1.

b. A square matrix has the same number of rows as columns. Thus the square matrices are **B, D, E, H,** and **J**.

c. An upper triangular matrix is a square matrix where all entries below the main diagonal are zeros. Thus **H** and **J** are upper triangular. A lower triangular matrix is a square matrix where all entries above the main diagonal are zeros. Thus **D** and **J** are lower triangular.

d. A row vector (or row matrix) has only one row. Thus **F** and **J** are row vectors.

e. A column vector (or column matrix) has only one column. Thus **G** and **J** are column vectors.

3. a_{21} is the entry in the 2nd row and 1st column, namely 6.

5. a_{32} is the entry in the 3rd row and 2nd column, namely 4.

7. a_{44} is the entry in the 4th row and 4th column, namely 0.

9. The main diagonal entries are the entries on the diagonal extending from the upper left corner to the lower right corner. Thus the main diagonal entries are 7, 2, 1, 0.

11.
$$\begin{bmatrix} -2\cdot1+3\cdot1 & -2\cdot1+3\cdot2 & -2\cdot1+3\cdot3 & -2\cdot1+3\cdot4 & -2\cdot1+3\cdot5 \\ -2\cdot2+3\cdot1 & -2\cdot2+3\cdot2 & -2\cdot2+3\cdot3 & -2\cdot2+3\cdot4 & -2\cdot2+3\cdot5 \\ -2\cdot3+3\cdot1 & -2\cdot3+3\cdot2 & -2\cdot3+3\cdot3 & -2\cdot3+3\cdot4 & -2\cdot3+3\cdot5 \end{bmatrix} = \begin{bmatrix} 1 & 4 & 7 & 10 & 13 \\ -1 & 2 & 5 & 8 & 11 \\ -3 & 0 & 3 & 6 & 9 \end{bmatrix}$$

13. $12 \cdot 10 = 120$, so **A** has 120 entries. For a_{33}, $i = 3 = j$, so $a_{33} = 1$. Since $5 \neq 2$, $a_{52} = 0$. For $a_{10,\,10}$, $i = 10 = j$, so $a_{10,\,10} = 1$. Since $12 \neq 10$, $a_{12,\,10} = 0$.

15. A zero matrix is a matrix in which all entries are zeros.

a.
$$\begin{bmatrix} 0 & 0 & 0 & 0 \\ 0 & 0 & 0 & 0 \\ 0 & 0 & 0 & 0 \\ 0 & 0 & 0 & 0 \end{bmatrix}$$

b.
$$\begin{bmatrix} 0 & 0 & 0 & 0 & 0 & 0 \\ 0 & 0 & 0 & 0 & 0 & 0 \\ 0 & 0 & 0 & 0 & 0 & 0 \\ 0 & 0 & 0 & 0 & 0 & 0 \\ 0 & 0 & 0 & 0 & 0 & 0 \\ 0 & 0 & 0 & 0 & 0 & 0 \end{bmatrix}$$

17. $\mathbf{A}^T = \begin{bmatrix} 6 & -3 \\ 2 & 4 \end{bmatrix}^T = \begin{bmatrix} 6 & 2 \\ -3 & 4 \end{bmatrix}$

19. $\mathbf{A}^T = \begin{bmatrix} 1 & 3 & 7 & 3 \\ 3 & 2 & -2 & 0 \\ -4 & 5 & 0 & 1 \end{bmatrix}^T = \begin{bmatrix} 1 & 3 & -4 \\ 3 & 2 & 5 \\ 7 & -2 & 0 \\ 3 & 0 & 1 \end{bmatrix}$

21. **a.** **A** and **C** are diagonal matrices.

    b. All are them are triangular matrices.

23. $\mathbf{A}^T = \begin{bmatrix} 1 & 0 & -1 \\ 7 & 0 & 9 \end{bmatrix}^T = \begin{bmatrix} 1 & 7 \\ 0 & 0 \\ -1 & 9 \end{bmatrix}$

    $(\mathbf{A}^T)^T = \begin{bmatrix} 1 & 7 \\ 0 & 0 \\ -1 & 9 \end{bmatrix}^T = \begin{bmatrix} 1 & 0 & -1 \\ 7 & 0 & 9 \end{bmatrix} = \mathbf{A}$

25. Equating corresponding entries gives $6 = 6$, $2 = 2$, $x = 6$, $7 = 7$, $3y = 2$, and $2z = 7$. Thus $x = 6$, $y = \dfrac{2}{3}$, $z = \dfrac{7}{2}$.

27. Equating corresponding entries gives $2x = y$, $7 = 7$, $7 = 7$, and $2y = y$. Now $2y = y$ yields $y = 0$. Thus from $2x = y$ we get $2x = 0$, so $x = 0$. The solution is $x = 0$, $y = 0$.

29. **a.** From **J**, the entry in row 3 (super-duper) and column 2 (white) is 7. Thus in January, 7 white super-duper models were sold.

    b. From **F**, the entry in row 2 (deluxe) and column 3 (blue) is 3. Thus in February, 3 blue deluxe models were sold.

    c. The entries in row 1 (regular) and column 4 (purple) give the number of purple regular models sold. For **J** the entry is 2 and for **F** the entry is 4. Thus more purple regular models were sold in February.

    d. In both January and February, the deluxe blue models (row 2, column 3) sold the same number of units (3).

    e. In January a total of $0 + 1 + 3 + 5 = 9$ deluxe models were sold. In February a total of $2 + 3 + 3 + 2 = 10$ deluxe models were sold. Thus more deluxe models were sold in February.

    f. In January a total of $2 + 0 + 2 = 4$ red widgets were sold, while in February a total of $0 + 2 + 4 = 6$ red widgets were sold. Thus more red widgets were sold in February.

    g. Adding all entries in matrix **J** yields that a total of 38 widgets were sold in January.

31. By equating entries we find that x must satisfy $x^2 + 2000x = 2001$ and $\sqrt{x^2} = -x$. The second equation implies that $x < 0$. From the first equation, $x^2 + 2000x - 2001 = 0$, $(x + 2001)(x - 1) = 0$, so $x = -2001$.

33. $\begin{bmatrix} 3 & 1 & 1 \\ 1 & 7 & 4 \\ 4 & 3 & 1 \\ 2 & 6 & 2 \end{bmatrix}$

Principles in Practice 6.2

1. $T = J + F = \begin{bmatrix} 120 & 80 \\ 105 & 130 \end{bmatrix} + \begin{bmatrix} 110 & 140 \\ 85 & 125 \end{bmatrix}$

   $= \begin{bmatrix} 120+110 & 80+140 \\ 105+85 & 130+125 \end{bmatrix} = \begin{bmatrix} 230 & 220 \\ 190 & 255 \end{bmatrix}$

Problems 6.2

1. $\begin{bmatrix} 2 & 0 & -3 \\ -1 & 4 & 0 \\ 1 & -6 & 5 \end{bmatrix} + \begin{bmatrix} 2 & -3 & 4 \\ -1 & 6 & 5 \\ 9 & 11 & -2 \end{bmatrix} = \begin{bmatrix} 2+2 & 0+(-3) & -3+4 \\ -1+(-1) & 4+6 & 0+5 \\ 1+9 & -6+11 & 5+(-2) \end{bmatrix} = \begin{bmatrix} 4 & -3 & 1 \\ -2 & 10 & 5 \\ 10 & 5 & 3 \end{bmatrix}$

3. $\begin{bmatrix} 1 & 4 \\ -2 & 7 \\ 6 & 9 \end{bmatrix} - \begin{bmatrix} 6 & -1 \\ 7 & 2 \\ 1 & 0 \end{bmatrix} = \begin{bmatrix} 1-6 & 4-(-1) \\ -2-7 & 7-2 \\ 6-1 & 9-0 \end{bmatrix} = \begin{bmatrix} -5 & 5 \\ -9 & 5 \\ 5 & 9 \end{bmatrix}$

5. $2[2 \;\; -1 \;\; 3] + 4[-2 \;\; 0 \;\; 1] - 0[2 \;\; 3 \;\; 1]$
   $= [4 \;\; -2 \;\; 6] + [-8 \;\; 0 \;\; 4] - [0 \;\; 0 \;\; 0]$
   $= [4-8-0 \;\; -2+0-0 \;\; 6+4-0]$
   $= [-4 \;\; -2 \;\; 10]$

7. $\begin{bmatrix} 1 & 2 \\ 3 & 4 \end{bmatrix}$ has size 2×2, and $\begin{bmatrix} 7 \\ 2 \end{bmatrix}$ has size 2×1. Thus the sum is not defined.

9. $-6\begin{bmatrix} 2 & -6 & 7 & 1 \\ 7 & 1 & 6 & -2 \end{bmatrix} = \begin{bmatrix} -6 \cdot 2 & -6(-6) & -6 \cdot 7 & -6 \cdot 1 \\ -6 \cdot 7 & -6 \cdot 1 & -6 \cdot 6 & -6(-2) \end{bmatrix} = \begin{bmatrix} -12 & 36 & -42 & -6 \\ -42 & -6 & -36 & 12 \end{bmatrix}$

11. $\begin{bmatrix} 1 & -5 & 0 \\ -2 & 7 & 0 \\ 4 & 6 & 10 \end{bmatrix} + \frac{1}{5}\begin{bmatrix} 10 & 0 & 30 \\ 0 & 5 & 0 \\ 5 & 20 & 25 \end{bmatrix} = \begin{bmatrix} 1 & -5 & 0 \\ -2 & 7 & 0 \\ 4 & 6 & 10 \end{bmatrix} + \begin{bmatrix} 2 & 0 & 6 \\ 0 & 1 & 0 \\ 1 & 4 & 5 \end{bmatrix} = \begin{bmatrix} 3 & -5 & 6 \\ -2 & 8 & 0 \\ 5 & 10 & 15 \end{bmatrix}$

13. $-\mathbf{B} = -\begin{bmatrix} -6 & -5 \\ 2 & -3 \end{bmatrix} = (-1)\begin{bmatrix} -6 & -5 \\ 2 & -3 \end{bmatrix} = \begin{bmatrix} -1(-6) & -1(-5) \\ -1(2) & -1(-3) \end{bmatrix} = \begin{bmatrix} 6 & 5 \\ -2 & 3 \end{bmatrix}$

15. $2\mathbf{O} = 2\begin{bmatrix} 0 & 0 \\ 0 & 0 \end{bmatrix} = \begin{bmatrix} 2 \cdot 0 & 2 \cdot 0 \\ 2 \cdot 0 & 2 \cdot 0 \end{bmatrix} = \begin{bmatrix} 0 & 0 \\ 0 & 0 \end{bmatrix} = \mathbf{O}$

17. $3(2\mathbf{A} - 3\mathbf{B}) = 3\left\{ 2\begin{bmatrix} 2 & 1 \\ 3 & -3 \end{bmatrix} - 3\begin{bmatrix} -6 & -5 \\ 2 & -3 \end{bmatrix} \right\} = 3\left\{ \begin{bmatrix} 4 & 2 \\ 6 & -6 \end{bmatrix} - \begin{bmatrix} -18 & -15 \\ 6 & -9 \end{bmatrix} \right\} = 3\begin{bmatrix} 22 & 17 \\ 0 & 3 \end{bmatrix} = \begin{bmatrix} 66 & 51 \\ 0 & 9 \end{bmatrix}$

19. $3(\mathbf{A} - \mathbf{C})$ is a 2×2 matrix and 6 is a number. Therefore $3(\mathbf{A} - \mathbf{C}) + 6$ is not defined.

21. $2\mathbf{B} - 3\mathbf{A} + 2\mathbf{C} = 2\begin{bmatrix} -6 & -5 \\ 2 & -3 \end{bmatrix} - 3\begin{bmatrix} 2 & 1 \\ 3 & -3 \end{bmatrix} + 2\begin{bmatrix} -2 & -1 \\ -3 & 3 \end{bmatrix}$

$= \begin{bmatrix} -12 & -10 \\ 4 & -6 \end{bmatrix} - \begin{bmatrix} 6 & 3 \\ 9 & -9 \end{bmatrix} + \begin{bmatrix} -4 & -2 \\ -6 & 6 \end{bmatrix}$

$= \begin{bmatrix} -18 & -13 \\ -5 & 3 \end{bmatrix} + \begin{bmatrix} -4 & -2 \\ -6 & 6 \end{bmatrix} = \begin{bmatrix} -22 & -15 \\ -11 & 9 \end{bmatrix}$

23. $\dfrac{1}{2}\mathbf{A} - 2(\mathbf{B} + 2\mathbf{C}) = \dfrac{1}{2}\begin{bmatrix} 2 & 1 \\ 3 & -3 \end{bmatrix} - 2\left\{ \begin{bmatrix} -6 & -5 \\ 2 & -3 \end{bmatrix} + 2\begin{bmatrix} -2 & -1 \\ -3 & 3 \end{bmatrix} \right\}$

$= \begin{bmatrix} 1 & \frac{1}{2} \\ \frac{3}{2} & -\frac{3}{2} \end{bmatrix} - 2\left\{ \begin{bmatrix} -6 & -5 \\ 2 & -3 \end{bmatrix} + \begin{bmatrix} -4 & -2 \\ -6 & 6 \end{bmatrix} \right\}$

$= \begin{bmatrix} 1 & \frac{1}{2} \\ \frac{3}{2} & -\frac{3}{2} \end{bmatrix} - 2\begin{bmatrix} -10 & -7 \\ -4 & 3 \end{bmatrix} = \begin{bmatrix} 1 & \frac{1}{2} \\ \frac{3}{2} & -\frac{3}{2} \end{bmatrix} - \begin{bmatrix} -20 & -14 \\ -8 & 6 \end{bmatrix} = \begin{bmatrix} 21 & \frac{29}{2} \\ \frac{19}{2} & -\frac{15}{2} \end{bmatrix}$

25. $3(\mathbf{A} + \mathbf{B}) = 3\begin{bmatrix} -4 & -4 \\ 5 & -6 \end{bmatrix} = \begin{bmatrix} -12 & -12 \\ 15 & -18 \end{bmatrix}$

$3\mathbf{A} + 3\mathbf{B} = \begin{bmatrix} 6 & 3 \\ 9 & -9 \end{bmatrix} + \begin{bmatrix} -18 & -15 \\ 6 & -9 \end{bmatrix} = \begin{bmatrix} -12 & -12 \\ 15 & -18 \end{bmatrix}$

Thus $3(\mathbf{A} + \mathbf{B}) = 3\mathbf{A} + 3\mathbf{B}$.

27. $k_1\left(k_2\mathbf{A}\right) = k_1\begin{bmatrix} 2k_2 & k_2 \\ 3k_2 & -3k_2 \end{bmatrix} = \begin{bmatrix} 2k_1k_2 & k_1k_2 \\ 3k_1k_2 & -3k_1k_2 \end{bmatrix}$

$\left(k_1k_2\right)\mathbf{A} = \begin{bmatrix} 2k_1k_2 & k_1k_2 \\ 3k_1k_2 & -3k_1k_2 \end{bmatrix}$

Thus $k_1\left(k_2\mathbf{A}\right) = \left(k_1k_2\right)\mathbf{A}$.

29. $3\mathbf{A} + \mathbf{D}^{\mathrm{T}} = 3\begin{bmatrix} 1 & 2 \\ 0 & -1 \\ 7 & 0 \end{bmatrix} + \begin{bmatrix} 1 & 1 \\ 2 & 0 \\ -1 & 2 \end{bmatrix} = \begin{bmatrix} 3 & 6 \\ 0 & -3 \\ 21 & 0 \end{bmatrix} + \begin{bmatrix} 1 & 1 \\ 2 & 0 \\ -1 & 2 \end{bmatrix} = \begin{bmatrix} 4 & 7 \\ 2 & -3 \\ 20 & 2 \end{bmatrix}$

31. $2\mathbf{B}^{\mathrm{T}} - 3\mathbf{C}^{\mathrm{T}} = 2\begin{bmatrix} 1 & 4 \\ 3 & -1 \end{bmatrix} - 3\begin{bmatrix} 1 & 1 \\ 0 & 2 \end{bmatrix} = \begin{bmatrix} 2 & 8 \\ 6 & -2 \end{bmatrix} - \begin{bmatrix} 3 & 3 \\ 0 & 6 \end{bmatrix} = \begin{bmatrix} -1 & 5 \\ 6 & -8 \end{bmatrix}$

33. $\mathbf{C}^{\mathrm{T}} - \mathbf{D} = \begin{bmatrix} 1 & 0 \\ 1 & 2 \end{bmatrix}^{\mathrm{T}} - \begin{bmatrix} 1 & 2 & -1 \\ 1 & 0 & 2 \end{bmatrix}$ is impossible because $\mathbf{C}^{\mathrm{T}}$ and $\mathbf{D}$ are not of the same size.

35. $x\begin{bmatrix} 3 \\ 2 \end{bmatrix} - y\begin{bmatrix} -4 \\ 7 \end{bmatrix} = 3\begin{bmatrix} 2 \\ 4 \end{bmatrix}$

$\begin{bmatrix} 3x \\ 2x \end{bmatrix} - \begin{bmatrix} -4y \\ 7y \end{bmatrix} = \begin{bmatrix} 6 \\ 12 \end{bmatrix} = \begin{bmatrix} 3x+4y \\ 2x-7y \end{bmatrix} = \begin{bmatrix} 6 \\ 12 \end{bmatrix}$

Equating corresponding entries gives

$\begin{cases} 3x+4y=6 \\ 2x-7y=12 \end{cases}$

Multiply the first equation by 2 and the second equation by −3 to get

$\begin{cases} 6x+8y=12 \\ -6x+21y=-36 \end{cases}$

Now add the two equations to get
$29y=-24$

$y=-\dfrac{24}{29}$

Therefore

$3x=6-4\left(-\dfrac{24}{29}\right)=\dfrac{270}{29}$

$x=\dfrac{90}{29}$

The solution is $x=\dfrac{90}{29}$, $y=-\dfrac{24}{29}$.

37. $3\begin{bmatrix} x \\ y \end{bmatrix} - 3\begin{bmatrix} -2 \\ 4 \end{bmatrix} = 4\begin{bmatrix} 6 \\ -2 \end{bmatrix}$

$\begin{bmatrix} 3x+6 \\ 3y-12 \end{bmatrix} = \begin{bmatrix} 24 \\ -8 \end{bmatrix}$

$3x+6=24$, $3x=18$, or $x=6$.

$3y-12=-8$, $3y=4$, or $y=\dfrac{4}{3}$.

Thus $x=6$, $y=\dfrac{4}{3}$.

39. $\begin{bmatrix} 2 \\ 4 \\ 6 \end{bmatrix} + 2\begin{bmatrix} x \\ y \\ 4z \end{bmatrix} = \begin{bmatrix} -10 \\ -24 \\ 14 \end{bmatrix}$

$\begin{bmatrix} 2+2x \\ 4+2y \\ 6+8z \end{bmatrix} = \begin{bmatrix} -10 \\ -24 \\ 14 \end{bmatrix}$

$2+2x=-10$, $2x=-12$, or $x=-6$.

$4+2y=-24$, $2y=-28$, or $y=-14$.
$6+8z=14$, $8z=8$, or $z=1$.
Thus $x=-6$, $y=-14$, $z=1$.

41. $\mathbf{X}+\mathbf{Y} = \begin{bmatrix} 30 & 50 \\ 800 & 720 \\ 25 & 30 \end{bmatrix} + \begin{bmatrix} 15 & 25 \\ 960 & 800 \\ 10 & 5 \end{bmatrix}$

$= \begin{bmatrix} 30+15 & 50+25 \\ 800+960 & 720+800 \\ 25+10 & 30+5 \end{bmatrix} = \begin{bmatrix} 45 & 75 \\ 1760 & 1520 \\ 35 & 35 \end{bmatrix}$

43. $\mathbf{P}+0.1\mathbf{P} = [p_1 \ p_2 \ p_3]+[0.1p_1 \ 0.1p_2 \ 0.1p_3]$
$= [1.1p_1 \ 1.1p_2 \ 1.1p_3] = 1.1\mathbf{P}$

Thus P must be multiplied by 1.1.

45. $\begin{bmatrix} 15 & -4 & 26 \\ 4 & 7 & 30 \end{bmatrix}$

47. $\begin{bmatrix} -10 & 22 & 12 \\ 24 & 36 & -44 \end{bmatrix}$

Principles in Practice 6.3

1. Represent the value of each book by $[28 \ 22 \ 16]$ and the number of each book

by $\begin{bmatrix} 100 \\ 70 \\ 90 \end{bmatrix}$.

The total value is given by the following matrix product.

$[28 \ 22 \ 16]\begin{bmatrix} 100 \\ 70 \\ 90 \end{bmatrix} = [2800+1540+1440]$

$= [5780]$
The total value is \$5780.

3. First, write the equations with the variable terms on the left-hand side.

$$\begin{cases} y + \dfrac{8}{5}x = \dfrac{8}{5} \\ y + \dfrac{1}{3}x = \dfrac{5}{3} \end{cases}$$

Let $\mathbf{A} = \begin{bmatrix} 1 & \dfrac{8}{5} \\ 1 & \dfrac{1}{3} \end{bmatrix}$, $\mathbf{X} = \begin{bmatrix} y \\ x \end{bmatrix}$, and $\mathbf{B} = \begin{bmatrix} \dfrac{8}{5} \\ \dfrac{5}{3} \end{bmatrix}$.

Then the pair of lines is equivalent to the matrix equation $\mathbf{AX} = \mathbf{B}$ or $\begin{bmatrix} 1 & \dfrac{8}{5} \\ 1 & \dfrac{1}{3} \end{bmatrix} \begin{bmatrix} y \\ x \end{bmatrix} = \begin{bmatrix} \dfrac{8}{5} \\ \dfrac{5}{3} \end{bmatrix}$.

Problems 6.3

1. $c_{11} = 1(0) + 3(-2) + (-2)(3) = -12$

3. $c_{32} = 0(-2) + 4(4) + 3(1) = 19$

5. $c_{31} = 0(0) + 4(-2) + 3(3) = 1$

7. $\mathbf{A}$ is 2×3 and $\mathbf{E}$ is 3×2, so $\mathbf{AE}$ is 2×2; $2 \cdot 2 = 4$ entries.

9. $\mathbf{E}$ is 3×2 and $\mathbf{C}$ is 2×5, so $\mathbf{EC}$ is 3×5; $3 \cdot 5 = 15$ entries.

11. $\mathbf{F}$ is 2×3 and $\mathbf{B}$ is 3×1, so $\mathbf{FB}$ is 2×1; $2 \cdot 1 = 2$ entries.

13. $\mathbf{E}$ is 3×2, $\mathbf{E}^{\mathrm{T}}$ is 2×3, and $\mathbf{B}$ is 3×1, so $\mathbf{EE}^{\mathrm{T}}\mathbf{B}$ is 3×1; $3 \cdot 1 = 3$ entries.

15. $\mathbf{E}$ is 3×2. $\mathbf{F}$ is 2×3 and $\mathbf{B}$ is 3×1, so $\mathbf{FB}$ is 2×1. Thus $\mathbf{E(FB)}$ is 3×1; $3 \cdot 1 = 3$ entries.

17. An identity matrix is a square matrix (in this case 4×4) with 1's on the main diagonal and all other entries 0's.

$$\mathbf{I}_4 = \begin{bmatrix} 1 & 0 & 0 & 0 \\ 0 & 1 & 0 & 0 \\ 0 & 0 & 1 & 0 \\ 0 & 0 & 0 & 1 \end{bmatrix}$$

19. $\begin{bmatrix} 2 & -4 \\ 3 & 2 \end{bmatrix} \begin{bmatrix} 4 & 0 \\ -1 & 3 \end{bmatrix} = \begin{bmatrix} 2(4)+(-4)(-1) & 2(0)+(-4)(3) \\ 3(4)+2(-1) & 3(0)+2(3) \end{bmatrix} = \begin{bmatrix} 12 & -12 \\ 10 & 6 \end{bmatrix}$

21. $\begin{bmatrix} 2 & 0 & 3 \\ -1 & 4 & 5 \end{bmatrix} \begin{bmatrix} 1 \\ 4 \\ 7 \end{bmatrix} = \begin{bmatrix} 2(1)+0(4)+3(7) \\ -1(1)+4(4)+5(7) \end{bmatrix} = \begin{bmatrix} 23 \\ 50 \end{bmatrix}$

23. $\begin{bmatrix} 1 & 4 & -1 \\ 0 & 0 & 2 \\ -2 & 1 & 1 \end{bmatrix} \begin{bmatrix} 2 & 1 & 0 \\ 0 & -1 & 1 \\ 1 & 1 & 2 \end{bmatrix}$

$= \begin{bmatrix} 1(2)+4(0)+(-1)1 & 1(1)+4(-1)+(-1)(1) & 1(0)+4(1)+(-1)(2) \\ 0(2)+0(0)+2(1) & 0(1)+0(-1)+2(1) & 0(0)+0(1)+2(2) \\ -2(2)+1(0)+1(1) & -2(1)+1(-1)+1(1) & -2(0)+1(1)+1(2) \end{bmatrix} = \begin{bmatrix} 1 & -4 & 2 \\ 2 & 2 & 4 \\ -3 & -2 & 3 \end{bmatrix}$

25. $[1 \quad -2 \quad 5] \begin{bmatrix} 1 & 5 & -2 & -1 \\ 0 & 0 & 2 & 1 \\ -1 & 0 & 1 & -3 \end{bmatrix}$

$= [1+0-5 \quad 5+0+0 \quad -2-4+5 \quad -1-2-15]$

$= [-4 \quad 5 \quad -1 \quad -18]$

27. $\begin{bmatrix} 2 \\ 3 \\ -4 \\ 1 \end{bmatrix} [2 \quad 3 \quad -2 \quad 3] = \begin{bmatrix} 2(2) & 2(3) & 2(-2) & 2(3) \\ 3(2) & 3(3) & 3(-2) & 3(3) \\ -4(2) & -4(3) & -4(-2) & -4(3) \\ 1(2) & 1(3) & 1(-2) & 1(3) \end{bmatrix} = \begin{bmatrix} 4 & 6 & -4 & 6 \\ 6 & 9 & -6 & 9 \\ -8 & -12 & 8 & -12 \\ 2 & 3 & -2 & 3 \end{bmatrix}$

29. $3\left\{ \begin{bmatrix} -2 & 0 & 2 \\ 3 & -1 & 1 \end{bmatrix} + 2 \begin{bmatrix} -1 & 0 & 2 \\ 1 & 1 & -2 \end{bmatrix} \right\} \begin{bmatrix} 1 & 2 \\ 3 & 4 \\ 5 & 6 \end{bmatrix}$

$= 3\left\{ \begin{bmatrix} -2 & 0 & 2 \\ 3 & -1 & 1 \end{bmatrix} + \begin{bmatrix} -2 & 0 & 4 \\ 2 & 2 & -4 \end{bmatrix} \right\} \begin{bmatrix} 1 & 2 \\ 3 & 4 \\ 5 & 6 \end{bmatrix}$

$= 3\left\{ \begin{bmatrix} -4 & 0 & 6 \\ 5 & 1 & -3 \end{bmatrix} \right\} \begin{bmatrix} 1 & 2 \\ 3 & 4 \\ 5 & 6 \end{bmatrix} = \begin{bmatrix} -12 & 0 & 18 \\ 15 & 3 & -9 \end{bmatrix} \begin{bmatrix} 1 & 2 \\ 3 & 4 \\ 5 & 6 \end{bmatrix}$

$= \begin{bmatrix} -12(1)+0(3)+18(5) & -12(2)+0(4)+18(6) \\ 15(1)+3(3)+(-9)(5) & 15(2)+3(4)+(-9)(6) \end{bmatrix} = \begin{bmatrix} 78 & 84 \\ -21 & -12 \end{bmatrix}$

31. $\begin{bmatrix} 1 & 2 \\ 3 & 4 \end{bmatrix} \left\{ \begin{bmatrix} 2 & 0 & 1 \\ 1 & 0 & -2 \end{bmatrix} \begin{bmatrix} 1 & -2 \\ 2 & 1 \\ 3 & 0 \end{bmatrix} \right\} = \begin{bmatrix} 1 & 2 \\ 3 & 4 \end{bmatrix} \left\{ \begin{bmatrix} 2+0+3 & -4+0+0 \\ 1+0-6 & -2+0+0 \end{bmatrix} \right\}$

$= \begin{bmatrix} 1 & 2 \\ 3 & 4 \end{bmatrix} \begin{bmatrix} 5 & -4 \\ -5 & -2 \end{bmatrix} = \begin{bmatrix} 5-10 & -4-4 \\ 15-20 & -12-8 \end{bmatrix} = \begin{bmatrix} -5 & -8 \\ -5 & -20 \end{bmatrix}$

33. $\begin{bmatrix} 0 & 0 & 1 \\ 0 & 1 & 0 \\ 1 & 0 & 0 \end{bmatrix} \begin{bmatrix} x \\ y \\ z \end{bmatrix} = \begin{bmatrix} 0 \cdot x + 0 \cdot y + 1 \cdot z \\ 0 \cdot x + 1 \cdot y + 0 \cdot z \\ 1 \cdot x + 0 \cdot y + 0 \cdot z \end{bmatrix} = \begin{bmatrix} z \\ y \\ x \end{bmatrix}$

35. $\begin{bmatrix} 2 & 1 & 3 \\ 4 & 9 & 7 \end{bmatrix} \begin{bmatrix} x_1 \\ x_2 \\ x_3 \end{bmatrix} = \begin{bmatrix} 2x_1 + x_2 + 3x_3 \\ 4x_1 + 9x_2 + 7x_3 \end{bmatrix}$

37. $\mathbf{D} - \dfrac{1}{3}\mathbf{EI} = \mathbf{D} - \dfrac{1}{3}\mathbf{E} = \begin{bmatrix} 1 & 0 & 0 \\ 0 & 1 & 1 \\ 1 & 2 & 1 \end{bmatrix} - \dfrac{1}{3}\begin{bmatrix} 3 & 0 & 0 \\ 0 & 6 & 0 \\ 0 & 0 & 3 \end{bmatrix}$

$= \begin{bmatrix} 1 & 0 & 0 \\ 0 & 1 & 1 \\ 1 & 2 & 1 \end{bmatrix} - \begin{bmatrix} 1 & 0 & 0 \\ 0 & 2 & 0 \\ 0 & 0 & 1 \end{bmatrix}$

$= \begin{bmatrix} 0 & 0 & 0 \\ 0 & -1 & 1 \\ 1 & 2 & 0 \end{bmatrix}$

39. $3\mathbf{A} - 2\mathbf{BC} = 3\begin{bmatrix} 1 & -2 \\ 0 & 3 \end{bmatrix} - 2\begin{bmatrix} -2 & 3 & 0 \\ 1 & -4 & 1 \end{bmatrix}\begin{bmatrix} -1 & 1 \\ 0 & 3 \\ 2 & 4 \end{bmatrix}$

$= \begin{bmatrix} 3 & -6 \\ 0 & 9 \end{bmatrix} - 2\begin{bmatrix} 2+0+0 & -2+9+0 \\ -1+0+2 & 1-12+4 \end{bmatrix}$

$= \begin{bmatrix} 3 & -6 \\ 0 & 9 \end{bmatrix} - \begin{bmatrix} 4 & 14 \\ 2 & -14 \end{bmatrix} = \begin{bmatrix} -1 & -20 \\ -2 & 23 \end{bmatrix}$

41. $3\mathbf{I} - \dfrac{2}{3}\mathbf{FE} = 3\mathbf{I} - \dfrac{2}{3}\begin{bmatrix} \frac{1}{3} & 0 & 0 \\ 0 & \frac{1}{6} & 0 \\ 0 & 0 & \frac{1}{3} \end{bmatrix}\begin{bmatrix} 3 & 0 & 0 \\ 0 & 6 & 0 \\ 0 & 0 & 3 \end{bmatrix}$

$\qquad = 3\mathbf{I} - \dfrac{2}{3}\begin{bmatrix} \frac{1}{3}\cdot 3+0+0 & 0+0+0 & 0+0+0 \\ 0+0+0 & \frac{1}{6}\cdot 6+0 & 0+0+0 \\ 0+0+0 & 0+0+0 & 0+0+\frac{1}{3}\cdot 3 \end{bmatrix}$

$\qquad = 3\mathbf{I} - \dfrac{2}{3}\begin{bmatrix} 1 & 0 & 0 \\ 0 & 1 & 0 \\ 0 & 0 & 1 \end{bmatrix} = \begin{bmatrix} 3 & 0 & 0 \\ 0 & 3 & 0 \\ 0 & 0 & 3 \end{bmatrix} - \begin{bmatrix} \frac{2}{3} & 0 & 0 \\ 0 & \frac{2}{3} & 0 \\ 0 & 0 & \frac{2}{3} \end{bmatrix} = \begin{bmatrix} \frac{7}{3} & 0 & 0 \\ 0 & \frac{7}{3} & 0 \\ 0 & 0 & \frac{7}{3} \end{bmatrix}$

43. $(\mathbf{DC})\mathbf{A} = \left\{ \begin{bmatrix} 1 & 0 & 0 \\ 0 & 1 & 1 \\ 1 & 2 & 1 \end{bmatrix}\begin{bmatrix} -1 & 1 \\ 0 & 3 \\ 2 & 4 \end{bmatrix} \right\}\mathbf{A} = \begin{bmatrix} -1+0+0 & 1+0+0 \\ 0+0+2 & 0+3+4 \\ -1+0+2 & 1+6+4 \end{bmatrix}\mathbf{A}$

$\qquad = \begin{bmatrix} -1 & 1 \\ 2 & 7 \\ 1 & 11 \end{bmatrix}\begin{bmatrix} 1 & -2 \\ 0 & 3 \end{bmatrix} = \begin{bmatrix} -1+0 & 2+3 \\ 2+0 & -4+21 \\ 1+0 & -2+33 \end{bmatrix} = \begin{bmatrix} -1 & 5 \\ 2 & 17 \\ 1 & 31 \end{bmatrix}$

45. Impossible: $\mathbf{A}$ is not a square matrix, so $\mathbf{A}^2$ is not defined.

47. $\mathbf{B}^3 = \left(\mathbf{B}^2\right)\mathbf{B} = \begin{bmatrix} 0 & 0 & -1 \\ 2 & -1 & 0 \\ 0 & 0 & 2 \end{bmatrix}^2 \mathbf{B} = \begin{bmatrix} 0 & 0 & -1 \\ 2 & -1 & 0 \\ 0 & 0 & 2 \end{bmatrix}\begin{bmatrix} 0 & 0 & -1 \\ 2 & -1 & 0 \\ 0 & 0 & 2 \end{bmatrix}\mathbf{B}$

$\qquad = \begin{bmatrix} 0 & 0 & -2 \\ -2 & 1 & -2 \\ 0 & 0 & 4 \end{bmatrix}\begin{bmatrix} 0 & 0 & -1 \\ 2 & -1 & 0 \\ 0 & 0 & 2 \end{bmatrix} = \begin{bmatrix} 0 & 0 & -4 \\ 2 & -1 & -2 \\ 0 & 0 & 8 \end{bmatrix}$

49. $(\mathbf{AIC})^{\text{T}} = \left(\begin{bmatrix} 1 & -1 & 0 \\ 0 & 1 & 1 \end{bmatrix} \begin{bmatrix} 1 & 0 & 0 \\ 0 & 1 & 0 \\ 0 & 0 & 1 \end{bmatrix} \begin{bmatrix} 1 & 0 \\ 2 & -1 \\ 0 & 1 \end{bmatrix} \right)^{\text{T}}$

$= \left(\begin{bmatrix} 1 & -1 & 0 \\ 0 & 1 & 1 \end{bmatrix} \begin{bmatrix} 1 & 0 \\ 2 & -1 \\ 0 & 1 \end{bmatrix} \right)^{\text{T}}$

$= \begin{bmatrix} -1 & 1 \\ 2 & 0 \end{bmatrix}^{\text{T}}$

$= \begin{bmatrix} -1 & 2 \\ 1 & 0 \end{bmatrix}$

51. $\left(\mathbf{BA}^{\text{T}} \right)^{\text{T}} = \left\{ \begin{bmatrix} 0 & 0 & -1 \\ 2 & -1 & 0 \\ 0 & 0 & 2 \end{bmatrix} \begin{bmatrix} 1 & 0 \\ -1 & 1 \\ 0 & 1 \end{bmatrix} \right\}^{\text{T}} = \begin{bmatrix} 0 & -1 \\ 3 & -1 \\ 0 & 2 \end{bmatrix}^{\text{T}} = \begin{bmatrix} 0 & 3 & 0 \\ -1 & -1 & 2 \end{bmatrix}$

53. $(2\mathbf{I})^2 - 2\mathbf{I}^2 = (2\mathbf{I})^2 - 2\mathbf{I} = \begin{bmatrix} 2 & 0 & 0 \\ 0 & 2 & 0 \\ 0 & 0 & 2 \end{bmatrix}^2 - \begin{bmatrix} 2 & 0 & 0 \\ 0 & 2 & 0 \\ 0 & 0 & 2 \end{bmatrix}$

$= \begin{bmatrix} 2 & 0 & 0 \\ 0 & 2 & 0 \\ 0 & 0 & 2 \end{bmatrix} \begin{bmatrix} 2 & 0 & 0 \\ 0 & 2 & 0 \\ 0 & 0 & 2 \end{bmatrix} - \begin{bmatrix} 2 & 0 & 0 \\ 0 & 2 & 0 \\ 0 & 0 & 2 \end{bmatrix} = \begin{bmatrix} 4 & 0 & 0 \\ 0 & 4 & 0 \\ 0 & 0 & 4 \end{bmatrix} - \begin{bmatrix} 2 & 0 & 0 \\ 0 & 2 & 0 \\ 0 & 0 & 2 \end{bmatrix} = \begin{bmatrix} 2 & 0 & 0 \\ 0 & 2 & 0 \\ 0 & 0 & 2 \end{bmatrix}$

55. $\mathbf{A}(\mathbf{I} - \mathbf{O}) = \mathbf{A}(\mathbf{I}) = \mathbf{AI}$. Since $\mathbf{I}$ is 3×3 and $\mathbf{A}$ has three columns, $\mathbf{AI} = \mathbf{A}$. Thus
$\mathbf{A}(\mathbf{I} - \mathbf{O}) = \mathbf{A} = \begin{bmatrix} 1 & -1 & 0 \\ 0 & 1 & 1 \end{bmatrix}$.

57. $(\mathbf{AB})(\mathbf{AB})^{\text{T}} = \begin{bmatrix} 1 & -1 & 0 \\ 0 & 1 & 1 \end{bmatrix} \begin{bmatrix} 0 & 0 & -1 \\ 2 & -1 & 0 \\ 0 & 0 & 2 \end{bmatrix} (\mathbf{AB})^{\text{T}} = \begin{bmatrix} -2 & 1 & -1 \\ 2 & -1 & 2 \end{bmatrix} (\mathbf{AB})^{\text{T}}$

$= \begin{bmatrix} -2 & 1 & -1 \\ 2 & -1 & 2 \end{bmatrix} \begin{bmatrix} -2 & 2 \\ 1 & -1 \\ -1 & 2 \end{bmatrix} = \begin{bmatrix} 6 & -7 \\ -7 & 9 \end{bmatrix}$

59. $\mathbf{AX} = \mathbf{B}$
$\mathbf{A} = \begin{bmatrix} 3 & 1 \\ 2 & -9 \end{bmatrix}$
$\mathbf{X} = \begin{bmatrix} x \\ y \end{bmatrix}$

$$\mathbf{B} = \begin{bmatrix} 6 \\ 5 \end{bmatrix}$$

The system is represented by $\begin{bmatrix} 3 & 1 \\ 2 & -9 \end{bmatrix}\begin{bmatrix} x \\ y \end{bmatrix} = \begin{bmatrix} 6 \\ 5 \end{bmatrix}$.

61. AX = B

$$\mathbf{A} = \begin{bmatrix} 2 & -1 & 3 \\ 5 & -1 & 2 \\ 3 & -2 & 2 \end{bmatrix}$$

$$\mathbf{X} = \begin{bmatrix} r \\ s \\ t \end{bmatrix}$$

$$\mathbf{B} = \begin{bmatrix} 9 \\ 5 \\ 11 \end{bmatrix}$$

The system is represented by $\begin{bmatrix} 2 & -1 & 3 \\ 5 & -1 & 2 \\ 3 & -2 & 2 \end{bmatrix}\begin{bmatrix} r \\ s \\ t \end{bmatrix} = \begin{bmatrix} 9 \\ 5 \\ 11 \end{bmatrix}$.

63. $\begin{bmatrix} 6 & 10 & 7 \end{bmatrix}\begin{bmatrix} 55 \\ 150 \\ 35 \end{bmatrix} = [6 \cdot 55 + 10 \cdot 150 + 7 \cdot 35]$

= [330 + 1500 + 245]
= [2075]
The value of the inventory is $2075.

65. Q = [5 2 4]

$$\mathbf{R} = \begin{bmatrix} 5 & 20 & 16 & 7 & 17 \\ 7 & 18 & 12 & 9 & 21 \\ 6 & 25 & 8 & 5 & 13 \end{bmatrix}$$

$$\mathbf{C} = \begin{bmatrix} 2500 \\ 1200 \\ 800 \\ 150 \\ 1500 \end{bmatrix}$$

$$\mathbf{QRC} = \mathbf{Q(RC)} = \mathbf{Q} \begin{bmatrix} 5 \cdot 2500 + 20 \cdot 1200 + 16 \cdot 800 + 7 \cdot 150 + 17 \cdot 1500 \\ 7 \cdot 2500 + 18 \cdot 1200 + 12 \cdot 800 + 9 \cdot 150 + 21 \cdot 1500 \\ 6 \cdot 2500 + 25 \cdot 1200 + 8 \cdot 800 + 5 \cdot 150 + 13 \cdot 1500 \end{bmatrix}$$

$$= \begin{bmatrix} 5 & 2 & 4 \end{bmatrix} \begin{bmatrix} 75,850 \\ 81,550 \\ 71,650 \end{bmatrix}$$

$$= [5(75,850) + 2(81,550) + 4(71,650)]$$

$$= [828,950]$$

The total cost of raw materials is $828,950.

67. a. Amount spent on goods:

coal industry: $\mathbf{D_C P} = \begin{bmatrix} 0 & 1 & 4 \end{bmatrix} \begin{bmatrix} 10,000 \\ 20,000 \\ 40,000 \end{bmatrix} = [180,000]$

elec. industry: $\mathbf{D_E P} = \begin{bmatrix} 20 & 0 & 8 \end{bmatrix} \begin{bmatrix} 10,000 \\ 20,000 \\ 40,000 \end{bmatrix} = [520,000]$

steel industry: $\mathbf{D_S P} = \begin{bmatrix} 30 & 5 & 0 \end{bmatrix} \begin{bmatrix} 10,000 \\ 20,000 \\ 40,000 \end{bmatrix} = [400,000]$

The coal industry spends $180,000, the electric industry spends $520,000, and the steel industry spends $400,000.

consumer 1: $\mathbf{D_1 P} = \begin{bmatrix} 3 & 2 & 5 \end{bmatrix} \begin{bmatrix} 10,000 \\ 20,000 \\ 40,000 \end{bmatrix} = [270,000]$

consumer 2: $\mathbf{D_2 P} = \begin{bmatrix} 0 & 17 & 1 \end{bmatrix} \begin{bmatrix} 10,000 \\ 20,000 \\ 40,000 \end{bmatrix} = [380,000]$

consumer 3: $\mathbf{D_3 P} = \begin{bmatrix} 4 & 6 & 12 \end{bmatrix} \begin{bmatrix} 10,000 \\ 20,000 \\ 40,000 \end{bmatrix} = [640,000]$

Consumer 1 pays $270,000, consumer 2 pays $380,000, and consumer 3 pays $640,000.

b. From Example 3 of Sec. 6.2, the number of units sold of coal, electricity, and steel are 57, 31, and 30, respectively. Thus the profit for coal is 10,000(57) – 180,000 = $390,000, the profit for elec. is 20,000(31) – 520,000 = $100,000, and the profit for steel is 40,000(30) – 400,000 = $800,000.

c. From (a), the total amount of money that is paid out by all the industries and consumers is 180,000 + 520,000 + 400,000 + 270,000 + 380,000 + 640,000 = $2,390,000.

d. The proportion of the total amount in (c) paid out by the industries is
$$\frac{180,000+520,000+400,000}{2,390,000}=\frac{110}{239}.$$

The proportion of the total amount in (c) paid by consumers is
$$\frac{270,000+380,000+640,000}{2,390,000}=\frac{129}{239}.$$

69.
$$\begin{bmatrix} 1 & 2 \\ 1 & 2 \end{bmatrix}\begin{bmatrix} 2 & -3 \\ -1 & \frac{3}{2} \end{bmatrix}$$

$$=\begin{bmatrix} 1(2)+(2)(-1) & 1(-3)+2\left(\frac{3}{2}\right) \\ 1(2)+2(-1) & 1(-3)+2\left(\frac{3}{2}\right) \end{bmatrix}$$

$$=\begin{bmatrix} 0 & 0 \\ 0 & 0 \end{bmatrix}$$

71.
$$\begin{bmatrix} 72.82 & -9.8 \\ 51.32 & -36.32 \end{bmatrix}$$

73.
$$\begin{bmatrix} 15.606 & 64.08 \\ -739.428 & 373.056 \end{bmatrix}$$

Principles in Practice 6.4

1. The corresponding system is
$$\begin{cases} 6A+B+3C=35 \\ 3A+2B+3C=22 \\ A+5B+3C=18 \end{cases}$$

Reduce the augmented coefficient matrix of the system.
$$\begin{bmatrix} 6 & 1 & 3 & | & 35 \\ 3 & 2 & 3 & | & 22 \\ 1 & 5 & 3 & | & 18 \end{bmatrix}$$

$$\xrightarrow{R_1 \leftrightarrow R_3}\begin{bmatrix} 1 & 5 & 3 & | & 18 \\ 3 & 2 & 3 & | & 22 \\ 6 & 1 & 3 & | & 35 \end{bmatrix}$$

$$\xrightarrow[-6R_1+R_3]{-3R_1+R_2}\begin{bmatrix} 1 & 5 & 3 & | & 18 \\ 0 & -13 & -6 & | & -32 \\ 0 & -29 & -15 & | & -73 \end{bmatrix}$$

$$\xrightarrow{-\frac{1}{13}R_2}\begin{bmatrix} 1 & 5 & 3 & | & 18 \\ 0 & 1 & \frac{6}{13} & | & \frac{32}{13} \\ 0 & -29 & -15 & | & -73 \end{bmatrix}$$

$$\xrightarrow[29R_2+R_3]{-5R_2+R_1}\begin{bmatrix} 1 & 0 & \frac{9}{13} & | & \frac{74}{13} \\ 0 & 1 & \frac{6}{13} & | & \frac{32}{13} \\ 0 & 0 & -\frac{21}{13} & | & -\frac{21}{13} \end{bmatrix}$$

$$\xrightarrow{-\frac{13}{21}R_3}\begin{bmatrix} 1 & 0 & \frac{9}{13} & | & \frac{74}{13} \\ 0 & 1 & \frac{6}{13} & | & \frac{32}{13} \\ 0 & 0 & 1 & | & 1 \end{bmatrix}$$

$$\xrightarrow[-\frac{6}{13}R_3+R_2]{-\frac{9}{13}R_3+R_1}\begin{bmatrix} 1 & 0 & 0 & | & 5 \\ 0 & 1 & 0 & | & 2 \\ 0 & 0 & 1 & | & 1 \end{bmatrix}$$

Thus there should be 5 blocks of A, 2 blocks of B, and 1 block of C suggested.

3. Let a, b, c, and d be the number of bags of foods A, B, C, and D, respectively. The corresponding system is
$$\begin{cases} 5a+5b+10c+5d=10,000 \\ 10a+5b+30c+10d=20,000 \\ 5a+15b+10c+25d=20,000 \end{cases}$$

Reduce the augmented coefficient matrix of the system.
$$\begin{bmatrix} 5 & 5 & 10 & 5 & | & 10,000 \\ 10 & 5 & 30 & 10 & | & 20,000 \\ 5 & 15 & 10 & 25 & | & 20,000 \end{bmatrix}$$

$$\xrightarrow{\frac{1}{5}R_1}\begin{bmatrix} 1 & 1 & 2 & 1 & | & 2000 \\ 10 & 5 & 30 & 10 & | & 20,000 \\ 5 & 15 & 10 & 25 & | & 20,000 \end{bmatrix}$$

$$\xrightarrow[-5R_1+R_3]{-10R_1+R_2}\begin{bmatrix} 1 & 1 & 2 & 1 & | & 2000 \\ 0 & -5 & 10 & 0 & | & 0 \\ 0 & 10 & 0 & 20 & | & 10,000 \end{bmatrix}$$

$$\xrightarrow{-\frac{1}{5}R_2}\begin{bmatrix} 1 & 1 & 2 & 1 & | & 2000 \\ 0 & 1 & -2 & 0 & | & 0 \\ 0 & 10 & 0 & 20 & | & 10,000 \end{bmatrix}$$

$$\begin{array}{c} -R_2 + R_1 \\ \xrightarrow{\hspace{1.5cm}} \\ -10R_2 + R_3 \end{array} \left[\begin{array}{cccc|c} 1 & 0 & 4 & 1 & 2000 \\ 0 & 1 & -2 & 0 & 0 \\ 0 & 0 & 20 & 20 & 10{,}000 \end{array}\right]$$

$$\xrightarrow{\frac{1}{20}R_3} \left[\begin{array}{cccc|c} 0 & 0 & 4 & 0 & 2000 \\ 0 & 1 & -2 & 0 & 0 \\ 0 & 0 & 1 & 1 & 500 \end{array}\right]$$

$$\begin{array}{c} -4R_3 + R_1 \\ \xrightarrow{\hspace{1.5cm}} \\ 2R_3 + R_2 \end{array} \left[\begin{array}{cccc|c} 1 & 0 & 0 & -3 & 0 \\ 0 & 1 & 0 & 2 & 1000 \\ 0 & 0 & 1 & 1 & 500 \end{array}\right]$$

This reduced matrix corresponds to the system $\begin{cases} a - 3d = 0 \\ b + 2d = 1000 \\ c + d = 500 \end{cases}$

Letting $d = r$, we get the general solution of the system:
$a = 3r$
$b = -2r + 1000$
$c = -r + 500$
$d = r$
Note that a, b, c, and d cannot be negative, given the context, hence $0 \le r \le 500$. One specific solution is when $r = 250$, then $a = 750$, $b = 500$, $c = 250$, and $d = 250$.

Problems 6.4

1. The first nonzero entry in row 2 is not to the right of the first nonzero entry in row 1, hence not reduced.

3. Reduced.

5. The first row consists entirely of zeros and is not below each row containing a nonzero entry, hence not reduced.

7. $\begin{bmatrix} 1 & 3 \\ 4 & 0 \end{bmatrix} \xrightarrow{-4R_1 + R_2} \begin{bmatrix} 1 & 3 \\ 0 & -12 \end{bmatrix}$

$\xrightarrow{-\frac{1}{12}R_2} \begin{bmatrix} 1 & 3 \\ 0 & 1 \end{bmatrix}$

$\xrightarrow{-3R_2 + R_1} \begin{bmatrix} 1 & 0 \\ 0 & 1 \end{bmatrix}$

9. $\begin{bmatrix} 2 & 4 & 6 \\ 1 & 2 & 3 \\ 1 & 2 & 3 \end{bmatrix} \xrightarrow{R_1 \leftrightarrow R_3} \begin{bmatrix} 1 & 2 & 3 \\ 1 & 2 & 3 \\ 2 & 4 & 6 \end{bmatrix}$

$\xrightarrow[-2R_1 + R_3]{-R_1 + R_2} \begin{bmatrix} 1 & 2 & 3 \\ 0 & 0 & 0 \\ 0 & 0 & 0 \end{bmatrix}$

11. $\begin{bmatrix} 2 & 0 & 3 & 1 \\ 1 & 4 & 2 & 2 \\ -1 & 3 & 1 & 4 \\ 0 & 2 & 1 & 0 \end{bmatrix} \xrightarrow{R_1 \leftrightarrow R_2} \begin{bmatrix} 1 & 4 & 2 & 2 \\ 2 & 0 & 3 & 1 \\ -1 & 3 & 1 & 4 \\ 0 & 2 & 1 & 0 \end{bmatrix}$

$\xrightarrow[R_1 + R_3]{-2R_1 + R_2} \begin{bmatrix} 1 & 4 & 2 & 2 \\ 0 & -8 & -1 & -3 \\ 0 & 7 & 3 & 6 \\ 0 & 2 & 1 & 0 \end{bmatrix} \xrightarrow{-\frac{1}{8}R_2} \begin{bmatrix} 1 & 4 & 2 & 2 \\ 0 & 1 & \frac{1}{8} & \frac{3}{8} \\ 0 & 7 & 3 & 6 \\ 0 & 2 & 1 & 0 \end{bmatrix}$

$\xrightarrow[\substack{-4R_2 + R_1 \\ -7R_2 + R_3 \\ -2R_2 + R_4}]{} \begin{bmatrix} 1 & 0 & \frac{3}{2} & \frac{1}{2} \\ 0 & 1 & \frac{1}{8} & \frac{3}{8} \\ 0 & 0 & \frac{17}{8} & \frac{27}{8} \\ 0 & 0 & \frac{3}{4} & -\frac{3}{4} \end{bmatrix} \xrightarrow{\frac{8}{17}R_3} \begin{bmatrix} 1 & 0 & \frac{3}{2} & \frac{1}{2} \\ 0 & 1 & \frac{1}{8} & \frac{3}{8} \\ 0 & 0 & 1 & \frac{27}{17} \\ 0 & 0 & \frac{3}{4} & -\frac{3}{4} \end{bmatrix}$

$\xrightarrow[\substack{-\frac{3}{2}R_3 + R_1 \\ -\frac{1}{8}R_3 + R_2 \\ -\frac{3}{4}R_3 + R_4}]{} \begin{bmatrix} 1 & 0 & 0 & -\frac{32}{17} \\ 0 & 1 & 0 & \frac{3}{17} \\ 0 & 0 & 1 & \frac{27}{17} \\ 0 & 0 & 0 & -\frac{33}{17} \end{bmatrix}$

$\xrightarrow{-\frac{17}{33}R_4} \begin{bmatrix} 1 & 0 & 0 & -\frac{32}{17} \\ 0 & 1 & 0 & \frac{3}{17} \\ 0 & 0 & 1 & \frac{27}{17} \\ 0 & 0 & 0 & 1 \end{bmatrix} \xrightarrow[\substack{-\frac{3}{17}R_4 + R_2 \\ -\frac{27}{17}R_4 + R_3}]{\frac{32}{17}R_4 + R_1} \begin{bmatrix} 1 & 0 & 0 & 0 \\ 0 & 1 & 0 & 0 \\ 0 & 0 & 1 & 0 \\ 0 & 0 & 0 & 1 \end{bmatrix}$

13. $\begin{bmatrix} 2 & -7 & | & 50 \\ 1 & 3 & | & 10 \end{bmatrix} \rightarrow \begin{bmatrix} 1 & 3 & | & 10 \\ 2 & -7 & | & 50 \end{bmatrix}$

$\rightarrow \begin{bmatrix} 1 & 3 & | & 10 \\ 0 & -13 & | & 30 \end{bmatrix} \rightarrow \begin{bmatrix} 1 & 3 & | & 10 \\ 0 & 1 & | & -\frac{30}{13} \end{bmatrix}$

$\rightarrow \begin{bmatrix} 1 & 0 & | & \frac{220}{13} \\ 0 & 1 & | & -\frac{30}{13} \end{bmatrix}$

Thus $x = \dfrac{220}{13}$ and $y = -\dfrac{30}{13}$.

15. $\begin{bmatrix} 3 & 1 & | & 4 \\ 12 & 4 & | & 2 \end{bmatrix} \rightarrow \begin{bmatrix} 3 & 1 & | & 4 \\ 0 & 0 & | & -14 \end{bmatrix} \rightarrow \begin{bmatrix} 1 & \frac{1}{3} & | & \frac{4}{3} \\ 0 & 0 & | & -14 \end{bmatrix} \rightarrow \begin{bmatrix} 1 & \frac{1}{3} & | & \frac{4}{3} \\ 0 & 0 & | & 1 \end{bmatrix} \rightarrow \begin{bmatrix} 1 & \frac{1}{3} & | & 0 \\ 0 & 0 & | & 1 \end{bmatrix}$

The last row indicates $0 = 1$, which is never true, so there is no solution.

17. $\begin{bmatrix} 1 & 2 & 1 & | & 4 \\ 3 & 0 & 2 & | & 5 \end{bmatrix} \rightarrow \begin{bmatrix} 1 & 2 & 1 & | & 4 \\ 0 & -6 & -1 & | & -7 \end{bmatrix} \rightarrow \begin{bmatrix} 1 & 2 & 1 & | & 4 \\ 0 & 1 & \frac{1}{6} & | & \frac{7}{6} \end{bmatrix} \rightarrow \begin{bmatrix} 1 & 0 & \frac{2}{3} & | & \frac{5}{3} \\ 0 & 1 & \frac{1}{6} & | & \frac{7}{6} \end{bmatrix},$

which gives $\begin{cases} x + \frac{2}{3}z = \frac{5}{3} \\ y + \frac{1}{6}z = \frac{7}{6} \end{cases}.$

Thus, $x = -\dfrac{2}{3}r + \dfrac{5}{3}$, $y = -\dfrac{1}{6}r + \dfrac{7}{6}$, $z = r$, where r is any real number.

19. $\begin{bmatrix} 1 & -3 & | & 0 \\ 2 & 2 & | & 3 \\ 5 & -1 & | & 1 \end{bmatrix} \rightarrow \begin{bmatrix} 1 & -3 & | & 0 \\ 0 & 8 & | & 3 \\ 0 & 14 & | & 1 \end{bmatrix} \rightarrow \begin{bmatrix} 1 & -3 & | & 0 \\ 0 & 1 & | & \frac{3}{8} \\ 0 & 14 & | & 1 \end{bmatrix} \rightarrow \begin{bmatrix} 1 & 0 & | & \frac{9}{8} \\ 0 & 1 & | & \frac{3}{8} \\ 0 & 0 & | & -\frac{17}{4} \end{bmatrix}$

From the third row, $0 = -\dfrac{17}{4}$, which is never true, so there is no solution.

21. $\begin{bmatrix} 1 & -1 & -3 & | & -5 \\ 2 & -1 & -4 & | & -8 \\ 1 & 1 & -1 & | & -1 \end{bmatrix} \rightarrow \begin{bmatrix} 1 & -1 & -3 & | & -5 \\ 0 & 1 & 2 & | & 2 \\ 0 & 2 & 2 & | & 4 \end{bmatrix} \rightarrow \begin{bmatrix} 1 & 0 & -1 & | & -3 \\ 0 & 1 & 2 & | & 2 \\ 0 & 0 & -2 & | & 0 \end{bmatrix}$

$\rightarrow \begin{bmatrix} 1 & 0 & -1 & | & -3 \\ 0 & 1 & 2 & | & 2 \\ 0 & 0 & 1 & | & 0 \end{bmatrix} \rightarrow \begin{bmatrix} 1 & 0 & 0 & | & -3 \\ 0 & 1 & 0 & | & 2 \\ 0 & 0 & 1 & | & 0 \end{bmatrix}$

Thus, $x = -3$, $y = 2$, $z = 0$.

23. $\begin{bmatrix} 2 & 0 & -4 & 8 \\ 1 & -2 & -2 & 14 \\ 1 & 1 & -2 & -1 \\ 3 & 1 & 1 & 0 \end{bmatrix} \rightarrow \begin{bmatrix} 1 & 0 & -2 & 4 \\ 1 & -2 & -2 & 14 \\ 1 & 1 & -2 & -1 \\ 3 & 1 & 1 & 0 \end{bmatrix} \rightarrow \begin{bmatrix} 1 & 0 & -2 & 4 \\ 0 & -2 & 0 & 10 \\ 0 & 1 & 0 & -5 \\ 0 & 1 & 7 & -12 \end{bmatrix}$

$\rightarrow \begin{bmatrix} 1 & 0 & -2 & 4 \\ 0 & 1 & 0 & -5 \\ 0 & -2 & 0 & 10 \\ 0 & 1 & 7 & -12 \end{bmatrix} \rightarrow \begin{bmatrix} 1 & 0 & -2 & 4 \\ 0 & 1 & 0 & -5 \\ 0 & 0 & 0 & 0 \\ 0 & 0 & 7 & -7 \end{bmatrix} \rightarrow \begin{bmatrix} 1 & 0 & -2 & 4 \\ 0 & 1 & 0 & -5 \\ 0 & 0 & 1 & -1 \\ 0 & 0 & 0 & 0 \end{bmatrix} \rightarrow \begin{bmatrix} 1 & 0 & 0 & 2 \\ 0 & 1 & 0 & -5 \\ 0 & 0 & 1 & -1 \\ 0 & 0 & 0 & 0 \end{bmatrix}$

Thus $x = 2$, $y = -5$, $z = -1$.

25. $\begin{bmatrix} 1 & -1 & -1 & -1 & -1 & 0 \\ 1 & 1 & -1 & -1 & -1 & 0 \\ 1 & 1 & 1 & -1 & -1 & 0 \\ 1 & 1 & 1 & 1 & -1 & 0 \end{bmatrix} \rightarrow \begin{bmatrix} 1 & -1 & -1 & -1 & -1 & 0 \\ 0 & 2 & 0 & 0 & 0 & 0 \\ 0 & 2 & 2 & 0 & 0 & 0 \\ 0 & 2 & 2 & 2 & 0 & 0 \end{bmatrix}$

$\rightarrow \begin{bmatrix} 1 & -1 & -1 & -1 & -1 & 0 \\ 0 & 1 & 0 & 0 & 0 & 0 \\ 0 & 2 & 2 & 0 & 0 & 0 \\ 0 & 2 & 2 & 2 & 0 & 0 \end{bmatrix} \rightarrow \begin{bmatrix} 1 & 0 & -1 & -1 & -1 & 0 \\ 0 & 1 & 0 & 0 & 0 & 0 \\ 0 & 0 & 2 & 0 & 0 & 0 \\ 0 & 0 & 2 & 2 & 0 & 0 \end{bmatrix}$

$\rightarrow \begin{bmatrix} 1 & 0 & -1 & -1 & -1 & 0 \\ 0 & 1 & 0 & 0 & 0 & 0 \\ 0 & 0 & 1 & 0 & 0 & 0 \\ 0 & 0 & 2 & 2 & 0 & 0 \end{bmatrix} \rightarrow \begin{bmatrix} 1 & 0 & 0 & -1 & -1 & 0 \\ 0 & 1 & 0 & 0 & 0 & 0 \\ 0 & 0 & 1 & 0 & 0 & 0 \\ 0 & 0 & 0 & 2 & 0 & 0 \end{bmatrix}$

$\rightarrow \begin{bmatrix} 1 & 0 & 0 & -1 & -1 & 0 \\ 0 & 1 & 0 & 0 & 0 & 0 \\ 0 & 0 & 1 & 0 & 0 & 0 \\ 0 & 0 & 0 & 1 & 0 & 0 \end{bmatrix} \rightarrow \begin{bmatrix} 1 & 0 & 0 & 0 & -1 & 0 \\ 0 & 1 & 0 & 0 & 0 & 0 \\ 0 & 0 & 1 & 0 & 0 & 0 \\ 0 & 0 & 0 & 1 & 0 & 0 \end{bmatrix}$

Thus, $x_1 = r$, $x_2 = 0$, $x_3 = 0$, $x_4 = 0$, and $x_5 = r$, where r is any number.

27. Let x = federal tax and y = state tax. Then $x = 0.25(312,000 - y)$ and $y = 0.10(312,000 - x)$. Equivalently,
$$\begin{cases} x + 0.25y = 78,000 \\ 0.10x + y = 31,200. \end{cases}$$

$\begin{bmatrix} 1 & 0.25 & 78,000 \\ 0.10 & 1 & 31,200 \end{bmatrix} \rightarrow \begin{bmatrix} 1 & 0.25 & 78,000 \\ 0 & 0.975 & 23,400 \end{bmatrix} \rightarrow \begin{bmatrix} 1 & 0.25 & 78,000 \\ 0 & 1 & 24,000 \end{bmatrix} \rightarrow \begin{bmatrix} 1 & 0 & 72,000 \\ 0 & 1 & 24,000 \end{bmatrix}$.

Thus $x = 72,000$ and $y = 24,000$, so the federal tax is \$72,000 and the state tax is \$24,000.

29. Let x = number of units of A produced, y = number of units of B produced, and z = number of units of C produced. Then
no. of units: $x + y + z = 11,000$
total cost: $4x + 5y + 7z + 17,000 = 80,000$
total profit: $x + 2y + 3z = 25,000$
Equivalently,

$$\begin{cases} x + y + z = 11,000 \\ 4x + 5y + 7z = 63,000 \\ x + 2y + 3z = 25,000 \end{cases}$$

$$\begin{bmatrix} 1 & 1 & 1 & | & 11,000 \\ 4 & 5 & 7 & | & 63,000 \\ 1 & 2 & 3 & | & 25,000 \end{bmatrix} \rightarrow \begin{bmatrix} 1 & 1 & 1 & | & 11,000 \\ 0 & 1 & 3 & | & 19,000 \\ 0 & 1 & 2 & | & 14,000 \end{bmatrix}$$

$$\rightarrow \begin{bmatrix} 1 & 0 & -2 & | & -8,000 \\ 0 & 1 & 3 & | & 19,000 \\ 0 & 0 & -1 & | & -5,000 \end{bmatrix} \rightarrow \begin{bmatrix} 1 & 0 & -2 & | & -8,000 \\ 0 & 1 & 3 & | & 19,000 \\ 0 & 0 & 1 & | & 5,000 \end{bmatrix} \rightarrow \begin{bmatrix} 1 & 0 & 0 & | & 2000 \\ 0 & 1 & 0 & | & 4000 \\ 0 & 0 & 1 & | & 5000 \end{bmatrix}$$

Thus $x = 2000$, $y = 4000$, and $z = 5000$, so 2000 units of A, 4000 units of B and 5000 units of C should be produced.

31. Let $x =$ number of brand X pills, $y =$ number of brand Y pills, and $z =$ number of brand Z pills. Considering the unit requirements gives the system

$$\begin{cases} 2x + 1y + 1z = 10 & \text{(vitamin A)} \\ 3x + 3y + 0z = 9 & \text{(vitamin D)} \\ 5x + 4y + 1z = 19 & \text{(vitamin E)} \end{cases}$$

$$\begin{bmatrix} 2 & 1 & 1 & | & 10 \\ 3 & 3 & 0 & | & 9 \\ 5 & 4 & 1 & | & 19 \end{bmatrix} \rightarrow \begin{bmatrix} 1 & \frac{1}{2} & \frac{1}{2} & | & 5 \\ 3 & 3 & 0 & | & 9 \\ 5 & 4 & 1 & | & 19 \end{bmatrix} \rightarrow \begin{bmatrix} 1 & \frac{1}{2} & \frac{1}{2} & | & 5 \\ 0 & \frac{3}{2} & -\frac{3}{2} & | & -6 \\ 0 & \frac{3}{2} & -\frac{3}{2} & | & -6 \end{bmatrix}$$

$$\rightarrow \begin{bmatrix} 1 & \frac{1}{2} & \frac{1}{2} & | & 5 \\ 0 & 1 & -1 & | & -4 \\ 0 & 0 & 0 & | & 0 \end{bmatrix} \rightarrow \begin{bmatrix} 1 & 0 & 1 & | & 7 \\ 0 & 1 & -1 & | & -4 \\ 0 & 0 & 0 & | & 0 \end{bmatrix}$$

Thus $\begin{cases} x = 7 - r \\ y = r - 4 \\ z = r \end{cases}$ where $r = 4, 5, 6, 7$.

The only solutions for the problem are $z = 4$, $x = 3$, and $y = 0$; $z = 5$, $x = 2$, and $y = 1$; $z = 6$, $x = 1$, and $y = 2$; $z = 7$, $x = 0$, and $y = 3$. Their respective costs (in cents) are 15, 23, 31, and 39.

a. The possible combinations are 3 of X, 4 of Z; 2 of X, 1 of Y, 5 of Z; 1 of X, 2 of Y, 6 of Z; 3 of Y, 7 of Z.

b. The combination 3 of X, 4 of Z costs 15 cents a day.

c. The least expensive combination is 3 of X, 4 of Z; the most expensive is 3 of Y, 7 of Z.

33. a. Let s, d, and g represent the number of units of S, D, and G, respectively. Then

$$\begin{cases} 12s + 20d + 32g = 220 & \text{(stock A)} \\ 16s + 12d + 28g = 176 & \text{(stock B)} \\ 8s + 28d + 36g = 264 & \text{(stock C)} \end{cases}$$

$$\begin{bmatrix} 12 & 20 & 32 & | & 220 \\ 16 & 12 & 28 & | & 176 \\ 8 & 28 & 36 & | & 264 \end{bmatrix} \xrightarrow[\left(\frac{1}{8}\right)R_3]{\begin{array}{c}\left(\frac{1}{4}\right)R_1 \\ \left(\frac{1}{4}\right)R_2\end{array}} \begin{bmatrix} 3 & 5 & 8 & | & 55 \\ 4 & 3 & 7 & | & 44 \\ 1 & \frac{7}{2} & \frac{9}{2} & | & 33 \end{bmatrix}$$

$$\xrightarrow{R_1 \leftrightarrow R_3} \begin{bmatrix} 1 & \frac{7}{2} & \frac{9}{2} & | & 33 \\ 4 & 3 & 7 & | & 44 \\ 3 & 5 & 8 & | & 55 \end{bmatrix}$$

$$\xrightarrow[-3R_1+R_3]{-4R_1+R_2} \begin{bmatrix} 1 & \frac{7}{2} & \frac{9}{2} & | & 33 \\ 0 & -11 & -11 & | & -88 \\ 0 & -\frac{11}{2} & -\frac{11}{2} & | & -44 \end{bmatrix}$$

$$\xrightarrow{-\frac{1}{11}R_2} \begin{bmatrix} 1 & \frac{7}{2} & \frac{9}{2} & | & 33 \\ 0 & 1 & 1 & | & 8 \\ 0 & -\frac{11}{2} & -\frac{11}{2} & | & -44 \end{bmatrix}$$

$$\xrightarrow[\frac{11}{2}R_2+R_3]{-\frac{7}{2}R_2+R_1} \begin{bmatrix} 1 & 0 & 1 & | & 5 \\ 0 & 1 & 1 & | & 8 \\ 0 & 0 & 0 & | & 0 \end{bmatrix}$$

Thus $s = 5 - r$, $d = 8 - r$, and $g = r$, where $r = 0, 1, 2, 3, 4, 5$.
The six possible combinations are given by

COMBINATION						
r	0	1	2	3	4	5
S	5	4	3	2	1	0
D	8	7	6	5	4	3
G	0	1	2	3	4	5

b. Computing the cost of each combination, we find that they are 4700, 4600, 4500, 4400, 4300, and 4200 dollars, respectively. Buying 3 units of Deluxe and 5 units of Gold Star ($s = 0$, $d = 3$, $g = 5$) minimizes the cost.

Principles in Practice 6.5

1. Write the coefficients matrix and reduce.

$$\begin{bmatrix} 5 & 3 & 4 \\ 6 & 8 & 7 \\ 3 & 1 & 2 \end{bmatrix} \xrightarrow{\frac{1}{5}R_1} \begin{bmatrix} 1 & \frac{3}{5} & \frac{4}{5} \\ 6 & 8 & 7 \\ 3 & 1 & 2 \end{bmatrix} \xrightarrow[-3R_1+R_3]{-6R_1+R_2} \begin{bmatrix} 1 & \frac{3}{5} & \frac{4}{5} \\ 0 & \frac{22}{5} & \frac{11}{5} \\ 0 & -\frac{4}{5} & -\frac{2}{5} \end{bmatrix}$$

$$\xrightarrow{\frac{5}{22}R_2} \begin{bmatrix} 1 & \frac{3}{5} & \frac{4}{5} \\ 0 & 1 & \frac{1}{2} \\ 0 & -\frac{4}{5} & -\frac{2}{5} \end{bmatrix} \xrightarrow[\frac{4}{5}R_2+R_3]{-\frac{3}{5}R_2+R_1} \begin{bmatrix} 1 & 0 & \frac{1}{2} \\ 0 & 1 & \frac{1}{2} \\ 0 & 0 & 0 \end{bmatrix}$$

The system has infinitely many solutions since there are two nonzero rows in the reduced coefficient matrix.

$$x + \frac{1}{2}z = 0$$

$$y + \frac{1}{2}z = 0$$

Let $z = r$, so $x = -\frac{1}{2}r$ and $y = -\frac{1}{2}r$, where r is any real number.

Problems 6.5

1. $\begin{bmatrix} 1 & 1 & -1 & -9 & -3 \\ 2 & 3 & 2 & 15 & 12 \\ 2 & 1 & 2 & 5 & 8 \end{bmatrix} \rightarrow \begin{bmatrix} 1 & 1 & -1 & -9 & 3 \\ 0 & 1 & 4 & 33 & 18 \\ 0 & -1 & 4 & 23 & 14 \end{bmatrix} \rightarrow \begin{bmatrix} 1 & 1 & -1 & -9 & -3 \\ 0 & 1 & 4 & 33 & 18 \\ 0 & 0 & 8 & 56 & 32 \end{bmatrix} \rightarrow \begin{bmatrix} 1 & 1 & -1 & -9 & -3 \\ 0 & 1 & 4 & 33 & 18 \\ 0 & 0 & 1 & 7 & 4 \end{bmatrix}$

$\rightarrow \begin{bmatrix} 1 & 1 & 0 & -2 & 1 \\ 0 & 1 & 0 & 5 & 2 \\ 0 & 0 & 1 & 7 & 4 \end{bmatrix} \rightarrow \begin{bmatrix} 1 & 0 & 0 & -7 & -1 \\ 0 & 1 & 0 & 5 & 2 \\ 0 & 0 & 1 & 7 & 4 \end{bmatrix}$

Thus $w = -1 + 7r$, $x = 2 - 5r$, $y = 4 - 7r$, $z = r$ (where r is any real number).

3. $\begin{bmatrix} 3 & -1 & -3 & -1 & -2 \\ 2 & -2 & -6 & -6 & -4 \\ 2 & -1 & -3 & -2 & -2 \\ 3 & 1 & 3 & 7 & 2 \end{bmatrix} \rightarrow \begin{bmatrix} 1 & -\frac{1}{3} & -1 & -\frac{1}{3} & -\frac{2}{3} \\ 2 & -2 & -6 & -6 & -4 \\ 2 & -1 & -3 & -2 & -2 \\ 3 & 1 & 3 & 7 & 2 \end{bmatrix}$

$\rightarrow \begin{bmatrix} 1 & -\frac{1}{3} & -1 & -\frac{1}{3} & -\frac{2}{3} \\ 0 & -\frac{4}{3} & -4 & -\frac{16}{3} & -\frac{8}{3} \\ 0 & -\frac{1}{3} & -1 & -\frac{4}{3} & -\frac{2}{3} \\ 0 & 2 & 6 & 8 & 4 \end{bmatrix} \rightarrow \begin{bmatrix} 1 & -\frac{1}{3} & -1 & -\frac{1}{3} & -\frac{2}{3} \\ 0 & 1 & 3 & 4 & 2 \\ 0 & -\frac{1}{3} & -1 & -\frac{4}{3} & -\frac{2}{3} \\ 0 & 2 & 6 & 8 & 4 \end{bmatrix} \rightarrow \begin{bmatrix} 1 & 0 & 0 & 1 & 0 \\ 0 & 1 & 3 & 4 & 2 \\ 0 & 0 & 0 & 0 & 0 \\ 0 & 0 & 0 & 0 & 0 \end{bmatrix}$

Thus, $w = -s$, $x = -3r - 4s + 2$, $y = r$, $z = s$ (where r and s are any real numbers).

5. $\begin{bmatrix} 1 & 1 & 3 & -1 & 2 \\ 2 & 1 & 5 & -2 & 0 \\ 2 & -1 & 3 & -2 & -8 \\ 3 & 2 & 8 & -3 & 2 \\ 1 & 0 & 2 & -1 & -2 \end{bmatrix} \rightarrow \begin{bmatrix} 1 & 1 & 3 & -1 & 2 \\ 0 & -1 & -1 & 0 & -4 \\ 0 & -3 & -3 & 0 & -12 \\ 0 & -1 & -1 & 0 & -4 \\ 0 & -1 & -1 & 0 & -4 \end{bmatrix} \rightarrow \begin{bmatrix} 1 & 1 & 3 & -1 & 2 \\ 0 & 1 & 1 & 0 & 4 \\ 0 & -3 & -3 & 0 & -12 \\ 0 & -1 & -1 & 0 & -4 \\ 0 & -1 & -1 & 0 & -4 \end{bmatrix} \rightarrow \begin{bmatrix} 1 & 0 & 2 & -1 & -2 \\ 0 & 1 & 1 & 0 & 4 \\ 0 & 0 & 0 & 0 & 0 \\ 0 & 0 & 0 & 0 & 0 \\ 0 & 0 & 0 & 0 & 0 \end{bmatrix}$

Thus, $w = -2r + s - 2$, $x = -r + 4$, $y = r$, $z = s$ (where r and s are any real numbers).

7. $\begin{bmatrix} 4 & -3 & 5 & -10 & 11 & -8 \\ 2 & 1 & 5 & 0 & 3 & 6 \end{bmatrix} \rightarrow \begin{bmatrix} 0 & -5 & -5 & -10 & 5 & -20 \\ 2 & 1 & 5 & 0 & 3 & 6 \end{bmatrix}$

$\rightarrow \begin{bmatrix} 2 & 1 & 5 & 0 & 3 & 6 \\ 0 & -5 & -5 & -10 & 5 & -20 \end{bmatrix} \rightarrow \begin{bmatrix} 2 & 1 & 5 & 0 & 3 & 6 \\ 0 & 1 & 1 & 2 & -1 & 4 \end{bmatrix}$

$\rightarrow \begin{bmatrix} 2 & 0 & 4 & -2 & 4 & 2 \\ 0 & 1 & 1 & 2 & -1 & 4 \end{bmatrix} \rightarrow \begin{bmatrix} 1 & 0 & 2 & -1 & 2 & 1 \\ 0 & 1 & 1 & 2 & -1 & 4 \end{bmatrix}$

Thus, $x_1 = -2r + s - 2t + 1$, $x_2 = -r - 2s + t + 4$, $x_3 = r$, $x_4 = s$, $x_5 = t$ (where r, s, and t are any real numbers).

9. The system is homogeneous with fewer equations than unknowns ($2 < 3$), so there are infinitely many solutions.

11. $\begin{bmatrix} 3 & -4 \\ 1 & 5 \\ 4 & -1 \end{bmatrix} \rightarrow \begin{bmatrix} 1 & 5 \\ 3 & -4 \\ 4 & -1 \end{bmatrix} \rightarrow \begin{bmatrix} 1 & 5 \\ 0 & -19 \\ 0 & -21 \end{bmatrix} \rightarrow \begin{bmatrix} 1 & 5 \\ 0 & 1 \\ 0 & -21 \end{bmatrix} \rightarrow \begin{bmatrix} 1 & 0 \\ 0 & 1 \\ 0 & 0 \end{bmatrix} = \mathbf{A}$

A has $k = 2$ nonzero rows. Number of unknowns is $n = 2$. Thus $k = n$, so the system has the trivial solution only.

13. $\begin{bmatrix} 1 & 1 & 1 \\ 1 & 0 & -1 \\ 1 & -2 & -5 \end{bmatrix} \rightarrow \begin{bmatrix} 1 & 1 & 1 \\ 0 & -1 & -2 \\ 0 & -3 & -6 \end{bmatrix} \rightarrow \begin{bmatrix} 1 & 1 & 1 \\ 0 & 1 & 2 \\ 0 & -3 & -6 \end{bmatrix} \rightarrow \begin{bmatrix} 1 & 0 & -1 \\ 0 & 1 & 2 \\ 0 & 0 & 0 \end{bmatrix} = \mathbf{A}$

A has $k = 2$ nonzero rows. Number of unknowns is $n = 3$. Thus $k < n$, so the system has infinitely many solutions.

15. $\begin{bmatrix} 1 & 1 \\ 3 & -4 \end{bmatrix} \rightarrow \begin{bmatrix} 1 & 1 \\ 0 & -7 \end{bmatrix} \rightarrow \begin{bmatrix} 1 & 1 \\ 0 & 1 \end{bmatrix} \rightarrow \begin{bmatrix} 1 & 0 \\ 0 & 1 \end{bmatrix}$

The solution is $x = 0$, $y = 0$.

17. $\begin{bmatrix} 1 & 6 & -2 \\ 2 & -3 & 4 \end{bmatrix} \rightarrow \begin{bmatrix} 1 & 6 & -2 \\ 0 & -15 & 8 \end{bmatrix} \rightarrow \begin{bmatrix} 1 & 6 & -2 \\ 0 & 1 & -\frac{8}{15} \end{bmatrix} \rightarrow \begin{bmatrix} 1 & 0 & \frac{6}{5} \\ 0 & 1 & -\frac{8}{15} \end{bmatrix}$

The solution is $x = -\frac{6}{5}r$, $y = \frac{8}{15}r$, $z = r$.

19. $\begin{bmatrix} 1 & 1 \\ 3 & -4 \\ 5 & -8 \end{bmatrix} \rightarrow \begin{bmatrix} 1 & 1 \\ 0 & -7 \\ 0 & -13 \end{bmatrix} \rightarrow \begin{bmatrix} 1 & 1 \\ 0 & 1 \\ 0 & -13 \end{bmatrix} \rightarrow \begin{bmatrix} 1 & 0 \\ 0 & 1 \\ 0 & 0 \end{bmatrix}$

The solution is $x = 0$, $y = 0$.

21. $\begin{bmatrix} 1 & 1 & 1 \\ 0 & -7 & -14 \\ 0 & -2 & -4 \\ 0 & -5 & -10 \end{bmatrix} \rightarrow \begin{bmatrix} 1 & 1 & 1 \\ 0 & 1 & 2 \\ 0 & -2 & -4 \\ 0 & -5 & -10 \end{bmatrix} \rightarrow \begin{bmatrix} 1 & 0 & -1 \\ 0 & 1 & 2 \\ 0 & 0 & 0 \\ 0 & 0 & 0 \end{bmatrix}$

The solution is $x = r$, $y = -2r$, $z = r$.

23. $\begin{bmatrix} 1 & 1 & 1 & 4 \\ 1 & 1 & 0 & 5 \\ 2 & 1 & 3 & 4 \\ 1 & -3 & 2 & -9 \end{bmatrix} \rightarrow \begin{bmatrix} 1 & 1 & 1 & 4 \\ 0 & 0 & -1 & 1 \\ 0 & -1 & 1 & -4 \\ 0 & -4 & 1 & -13 \end{bmatrix} \rightarrow \begin{bmatrix} 1 & 1 & 1 & 4 \\ 0 & -1 & 1 & -4 \\ 0 & 0 & -1 & 1 \\ 0 & -4 & 1 & -13 \end{bmatrix}$

$\rightarrow \begin{bmatrix} 1 & 1 & 1 & 4 \\ 0 & 1 & -1 & 4 \\ 0 & 0 & -1 & 1 \\ 0 & -4 & 1 & -13 \end{bmatrix} \rightarrow \begin{bmatrix} 1 & 0 & 2 & 0 \\ 0 & 1 & -1 & 4 \\ 0 & 0 & -1 & 1 \\ 0 & 0 & -3 & 3 \end{bmatrix} \rightarrow \begin{bmatrix} 1 & 0 & 2 & 0 \\ 0 & 1 & -1 & 4 \\ 0 & 0 & 1 & -1 \\ 0 & 0 & -3 & 3 \end{bmatrix} \rightarrow \begin{bmatrix} 1 & 0 & 0 & 2 \\ 0 & 1 & 0 & 3 \\ 0 & 0 & 1 & -1 \\ 0 & 0 & 0 & 0 \end{bmatrix}$

The solution is $w = -2r$, $x = -3r$, $y = r$, $z = r$.

Principles in Practice 6.6

1. $\begin{bmatrix} 1 & 3 \\ 2 & 4 \end{bmatrix}\begin{bmatrix} -2 & 1.5 \\ 1 & -0.5 \end{bmatrix} = \begin{bmatrix} 1 & 0 \\ 0 & 1 \end{bmatrix}$

Yes, they are inverses.

3. $\begin{bmatrix} \mathbf{E} | \mathbf{I} \end{bmatrix} = \begin{bmatrix} 3 & 1 & 2 & | & 1 & 0 & 0 \\ 2 & 2 & 2 & | & 0 & 1 & 0 \\ 2 & 1 & 3 & | & 0 & 0 & 1 \end{bmatrix}$

$\xrightarrow{R_1 \leftrightarrow R_2} \begin{bmatrix} 2 & 2 & 2 & | & 0 & 1 & 0 \\ 3 & 1 & 2 & | & 1 & 0 & 0 \\ 2 & 1 & 3 & | & 0 & 0 & 1 \end{bmatrix}$

$\xrightarrow{\frac{1}{2}R_1} \begin{bmatrix} 1 & 1 & 1 & | & 0 & \frac{1}{2} & 0 \\ 3 & 1 & 2 & | & 1 & 0 & 0 \\ 2 & 1 & 3 & | & 0 & 0 & 1 \end{bmatrix}$

$$\xrightarrow[-2R_1+R_3]{-3R_1+R_2} \left[\begin{array}{ccc|ccc} 1 & 1 & 1 & 0 & \frac{1}{2} & 0 \\ 0 & -2 & -1 & 1 & -\frac{3}{2} & 0 \\ 0 & -1 & 1 & 0 & -1 & 1 \end{array}\right]$$

$$\xrightarrow{R_2 \leftrightarrow R_3} \left[\begin{array}{ccc|ccc} 1 & 1 & 1 & 0 & \frac{1}{2} & 0 \\ 0 & -1 & 1 & 0 & -1 & 1 \\ 0 & -2 & -1 & 1 & -\frac{3}{2} & 0 \end{array}\right]$$

$$\xrightarrow{-R_2} \left[\begin{array}{ccc|ccc} 1 & 1 & 1 & 0 & \frac{1}{2} & 0 \\ 0 & 1 & -1 & 0 & 1 & -1 \\ 0 & -2 & -1 & 1 & -\frac{3}{2} & 0 \end{array}\right]$$

$$\xrightarrow[2R_2+R_3]{-R_2+R_1} \left[\begin{array}{ccc|ccc} 1 & 0 & 2 & 0 & -\frac{1}{2} & 1 \\ 0 & 1 & -1 & 0 & 1 & -1 \\ 0 & 0 & -3 & 1 & \frac{1}{2} & -2 \end{array}\right]$$

$$\xrightarrow{-\frac{1}{3}R_3} \left[\begin{array}{ccc|ccc} 1 & 0 & 2 & 0 & -\frac{1}{2} & 1 \\ 0 & 1 & -1 & 0 & 1 & -1 \\ 0 & 0 & 1 & -\frac{1}{3} & -\frac{1}{6} & \frac{2}{3} \end{array}\right]$$

$$\xrightarrow[R_3+R_2]{-2R_3+R_1} \left[\begin{array}{ccc|ccc} 1 & 0 & 0 & \frac{2}{3} & -\frac{1}{6} & -\frac{1}{3} \\ 0 & 1 & 0 & -\frac{1}{3} & \frac{5}{6} & -\frac{1}{3} \\ 0 & 0 & 1 & -\frac{1}{3} & -\frac{1}{6} & \frac{2}{3} \end{array}\right]$$

$$\mathbf{E}^{-1} = \left[\begin{array}{ccc} \frac{2}{3} & -\frac{1}{6} & -\frac{1}{3} \\ -\frac{1}{3} & \frac{5}{6} & -\frac{1}{3} \\ -\frac{1}{3} & -\frac{1}{6} & \frac{2}{3} \end{array}\right]$$

$$\left[\mathbf{F}\,\middle|\,\mathbf{I}\right] = \left[\begin{array}{ccc|ccc} 2 & 1 & 2 & 1 & 0 & 0 \\ 3 & 2 & 3 & 0 & 1 & 0 \\ 4 & 3 & 4 & 0 & 0 & 1 \end{array}\right]$$

$$\xrightarrow{\frac{1}{2}R_1} \left[\begin{array}{ccc|ccc} 1 & \frac{1}{2} & 1 & \frac{1}{2} & 0 & 0 \\ 3 & 2 & 3 & 0 & 1 & 0 \\ 4 & 3 & 4 & 0 & 0 & 1 \end{array}\right]$$

$$\xrightarrow[-4R_1+R_3]{-3R_1+R_2} \left[\begin{array}{ccc|ccc} 1 & \frac{1}{2} & 1 & \frac{1}{2} & 0 & 0 \\ 0 & \frac{1}{2} & 0 & -\frac{3}{2} & 1 & 0 \\ 0 & 1 & 0 & -2 & 0 & 1 \end{array}\right]$$

$$\xrightarrow{2R_2} \left[\begin{array}{ccc|ccc} 1 & \frac{1}{2} & 1 & \frac{1}{2} & 0 & 0 \\ 0 & 1 & 0 & -3 & 2 & 0 \\ 0 & 1 & 0 & -2 & 0 & 1 \end{array}\right]$$

$$\xrightarrow[-R_2+R_3]{-\frac{1}{2}R_2+R_1} \left[\begin{array}{ccc|ccc} 1 & 0 & 1 & 2 & -1 & 0 \\ 0 & 1 & 0 & -3 & 2 & 0 \\ 0 & 0 & 0 & 1 & -2 & 1 \end{array}\right]$$

F does not reduce to **I** so **F** is not invertible.

$$\xrightarrow{\frac{1}{6}R_2} \left[\begin{array}{ccc|ccc} 1 & 0 & -2 & 1 & 0 & 0 \\ 0 & 1 & -1 & -\frac{5}{6} & \frac{1}{6} & 0 \\ 0 & 2 & 18 & -5 & 0 & 1 \end{array}\right]$$

$$\xrightarrow{-2R_2+R_3} \left[\begin{array}{ccc|ccc} 1 & 0 & -2 & 1 & 0 & 0 \\ 0 & 1 & -1 & -\frac{5}{6} & \frac{1}{6} & 0 \\ 0 & 0 & 20 & -\frac{10}{3} & -\frac{1}{3} & 1 \end{array}\right]$$

$$\xrightarrow{\frac{1}{20}R_3} \left[\begin{array}{ccc|ccc} 1 & 0 & -2 & 1 & 0 & 0 \\ 0 & 1 & -1 & -\frac{5}{6} & \frac{1}{6} & 0 \\ 0 & 0 & 1 & -\frac{1}{6} & -\frac{1}{60} & \frac{1}{20} \end{array}\right]$$

$$\xrightarrow[R_3+R_2]{2R_3+R_1} \left[\begin{array}{ccc|ccc} 1 & 0 & 0 & \frac{2}{3} & -\frac{1}{30} & \frac{1}{10} \\ 0 & 1 & 0 & -1 & \frac{3}{20} & \frac{1}{20} \\ 0 & 0 & 1 & -\frac{1}{6} & -\frac{1}{60} & \frac{1}{20} \end{array}\right]$$

$$\mathbf{A}^{-1} = \left[\begin{array}{ccc} \frac{2}{3} & -\frac{1}{30} & \frac{1}{10} \\ -1 & \frac{3}{20} & \frac{1}{20} \\ -\frac{1}{6} & -\frac{1}{60} & \frac{1}{20} \end{array}\right]$$

Problems 6.6

1. $\begin{bmatrix} 6 & 1 & 1 & 0 \\ 7 & 1 & 0 & 1 \end{bmatrix} \rightarrow \begin{bmatrix} 1 & \frac{1}{6} & \frac{1}{6} & 0 \\ 7 & 1 & 0 & 1 \end{bmatrix} \rightarrow \begin{bmatrix} 1 & \frac{1}{6} & \frac{1}{6} & 0 \\ 0 & -\frac{1}{6} & -\frac{7}{6} & 1 \end{bmatrix}$

$\rightarrow \begin{bmatrix} 1 & 0 & -1 & 1 \\ 0 & -\frac{1}{6} & -\frac{7}{6} & 1 \end{bmatrix} \rightarrow \begin{bmatrix} 1 & 0 & -1 & 1 \\ 0 & 1 & 7 & -6 \end{bmatrix}$

The inverse is $\begin{bmatrix} -1 & 1 \\ 7 & -6 \end{bmatrix}$.

3. $\begin{bmatrix} 2 & 2 & 1 & 0 \\ 2 & 2 & 0 & 1 \end{bmatrix} \rightarrow \begin{bmatrix} 2 & 2 & 1 & 0 \\ 0 & 0 & -1 & 1 \end{bmatrix} \rightarrow \begin{bmatrix} 1 & 1 & \frac{1}{2} & 0 \\ 0 & 0 & -1 & 1 \end{bmatrix}$

The given matrix is not invertible.

5. $\begin{bmatrix} 1 & 0 & 0 & 1 & 0 & 0 \\ 0 & -3 & 0 & 0 & 1 & 0 \\ 0 & 0 & 4 & 0 & 0 & 1 \end{bmatrix} \rightarrow \begin{bmatrix} 1 & 0 & 0 & 1 & 0 & 0 \\ 0 & 1 & 0 & 0 & -\frac{1}{3} & 0 \\ 0 & 0 & 1 & 0 & 0 & \frac{1}{4} \end{bmatrix}$

The inverse is $\begin{bmatrix} 1 & 0 & 0 \\ 0 & -\frac{1}{3} & 0 \\ 0 & 0 & \frac{1}{4} \end{bmatrix}$.

7. $\begin{bmatrix} 1 & 2 & 3 & 1 & 0 & 0 \\ 0 & 0 & 4 & 0 & 1 & 0 \\ 0 & 0 & 5 & 0 & 0 & 1 \end{bmatrix} \rightarrow \begin{bmatrix} 1 & 2 & 3 & 1 & 0 & 0 \\ 0 & 0 & 1 & 0 & \frac{1}{4} & 0 \\ 0 & 0 & 5 & 0 & 0 & 1 \end{bmatrix}$

$\rightarrow \begin{bmatrix} 1 & 2 & 0 & 1 & -\frac{3}{4} & 0 \\ 0 & 0 & 1 & 0 & \frac{1}{4} & 0 \\ 0 & 0 & 0 & 0 & -\frac{5}{4} & 1 \end{bmatrix}$

The given matrix is not invertible.

9. The matrix is not square, so it is not invertible.

11.
$$\left[\begin{array}{ccc|ccc} 1 & 2 & 3 & 1 & 0 & 0 \\ 0 & 1 & 2 & 0 & 1 & 0 \\ 0 & 0 & 1 & 0 & 0 & 1 \end{array}\right] \rightarrow \left[\begin{array}{ccc|ccc} 1 & 0 & -1 & 1 & -2 & 0 \\ 0 & 1 & 2 & 0 & 1 & 0 \\ 0 & 0 & 1 & 0 & 0 & 1 \end{array}\right]$$

$$\rightarrow \left[\begin{array}{ccc|ccc} 1 & 0 & 0 & 1 & -2 & 1 \\ 0 & 1 & 0 & 0 & 1 & -2 \\ 0 & 0 & 1 & 0 & 0 & 1 \end{array}\right]$$

The inverse is $\left[\begin{array}{ccc} 1 & -2 & 1 \\ 0 & 1 & -2 \\ 0 & 0 & 1 \end{array}\right]$.

13.
$$\left[\begin{array}{ccc|ccc} 7 & 0 & -2 & 1 & 0 & 0 \\ 0 & 1 & 0 & 0 & 1 & 0 \\ -3 & 0 & 1 & 0 & 0 & 1 \end{array}\right] \rightarrow \left[\begin{array}{ccc|ccc} 1 & 0 & -\frac{2}{7} & \frac{1}{7} & 0 & 0 \\ 0 & 1 & 0 & 0 & 1 & 0 \\ -3 & 0 & 1 & 0 & 0 & 1 \end{array}\right]$$

$$\rightarrow \left[\begin{array}{ccc|ccc} 1 & 0 & -\frac{2}{7} & \frac{1}{7} & 0 & 0 \\ 0 & 1 & 0 & 0 & 1 & 0 \\ 0 & 0 & \frac{1}{7} & \frac{3}{7} & 0 & 1 \end{array}\right] \rightarrow \left[\begin{array}{ccc|ccc} 1 & 0 & 0 & 1 & 0 & 2 \\ 0 & 1 & 0 & 0 & 1 & 0 \\ 0 & 0 & \frac{1}{7} & \frac{3}{7} & 0 & 1 \end{array}\right] \rightarrow \left[\begin{array}{ccc|ccc} 1 & 0 & 0 & 1 & 0 & 2 \\ 0 & 1 & 0 & 0 & 1 & 0 \\ 0 & 0 & 1 & 3 & 0 & 7 \end{array}\right]$$

The inverse is $\left[\begin{array}{ccc} 1 & 0 & 2 \\ 0 & 1 & 0 \\ 3 & 0 & 7 \end{array}\right]$.

15.
$$\left[\begin{array}{ccc|ccc} 2 & 1 & 0 & 1 & 0 & 0 \\ 4 & -1 & 5 & 0 & 1 & 0 \\ 1 & -1 & 2 & 0 & 0 & 1 \end{array}\right] \rightarrow \left[\begin{array}{ccc|ccc} 1 & -1 & 2 & 0 & 0 & 1 \\ 4 & -1 & 5 & 0 & 1 & 0 \\ 2 & 1 & 0 & 1 & 0 & 0 \end{array}\right]$$

$$\rightarrow \left[\begin{array}{ccc|ccc} 1 & -1 & 2 & 0 & 0 & 1 \\ 0 & 3 & -3 & 0 & 1 & -4 \\ 0 & 3 & -4 & 1 & 0 & -2 \end{array}\right] \rightarrow \left[\begin{array}{ccc|ccc} 1 & -1 & 2 & 0 & 0 & 1 \\ 0 & 1 & -1 & 0 & \frac{1}{3} & -\frac{4}{3} \\ 0 & 3 & -4 & 1 & 0 & -2 \end{array}\right]$$

$$\rightarrow \left[\begin{array}{ccc|ccc} 1 & 0 & 1 & 0 & \frac{1}{3} & -\frac{1}{3} \\ 0 & 1 & -1 & 0 & \frac{1}{3} & -\frac{4}{3} \\ 0 & 0 & -1 & 1 & -1 & 2 \end{array}\right] \rightarrow \left[\begin{array}{ccc|ccc} 1 & 0 & 1 & 0 & \frac{1}{3} & -\frac{1}{3} \\ 0 & 1 & -1 & 0 & \frac{1}{3} & -\frac{4}{3} \\ 0 & 0 & 1 & -1 & 1 & -2 \end{array}\right] \rightarrow \left[\begin{array}{ccc|ccc} 1 & 0 & 0 & 1 & -\frac{2}{3} & \frac{5}{3} \\ 0 & 1 & 0 & -1 & \frac{4}{3} & -\frac{10}{3} \\ 0 & 0 & 1 & -1 & 1 & -2 \end{array}\right].$$

The inverse is $\left[\begin{array}{ccc} 1 & -\frac{2}{3} & \frac{5}{3} \\ -1 & \frac{4}{3} & -\frac{10}{3} \\ -1 & 1 & -2 \end{array}\right]$.

17. $\begin{bmatrix} 1 & 2 & 3 & | & 1 & 0 & 0 \\ 1 & 3 & 5 & | & 0 & 1 & 0 \\ 1 & 5 & 12 & | & 0 & 0 & 1 \end{bmatrix} \rightarrow \begin{bmatrix} 1 & 2 & 3 & | & 1 & 0 & 0 \\ 0 & 1 & 2 & | & -1 & 1 & 0 \\ 0 & 3 & 9 & | & -1 & 0 & 1 \end{bmatrix}$

$\rightarrow \begin{bmatrix} 1 & 0 & -1 & | & 3 & -2 & 0 \\ 0 & 1 & 2 & | & -1 & 1 & 0 \\ 0 & 0 & 3 & | & 2 & -3 & 1 \end{bmatrix} \rightarrow \begin{bmatrix} 1 & 0 & -1 & | & 3 & -2 & 0 \\ 0 & 1 & 2 & | & -1 & 1 & 0 \\ 0 & 0 & 1 & | & \frac{2}{3} & -1 & \frac{1}{3} \end{bmatrix} \rightarrow \begin{bmatrix} 1 & 0 & 0 & | & \frac{11}{3} & -3 & \frac{1}{3} \\ 0 & 1 & 0 & | & -\frac{7}{3} & 3 & -\frac{2}{3} \\ 0 & 0 & 1 & | & \frac{2}{3} & -1 & \frac{1}{3} \end{bmatrix}$

The inverse is $\begin{bmatrix} \frac{11}{3} & -3 & \frac{1}{3} \\ -\frac{7}{3} & 3 & -\frac{2}{3} \\ \frac{2}{3} & -1 & \frac{1}{3} \end{bmatrix}$.

19. $\mathbf{X} = \begin{bmatrix} x_1 \\ x_2 \end{bmatrix} = \mathbf{A}^{-1}\mathbf{B} = \begin{bmatrix} 1 & 2 \\ 8 & 1 \end{bmatrix}\begin{bmatrix} 2 \\ 4 \end{bmatrix} = \begin{bmatrix} 10 \\ 20 \end{bmatrix} \Rightarrow x_1 = 10,\ x_2 = 20$

21. $\begin{bmatrix} 6 & 5 & | & 1 & 0 \\ 1 & 1 & | & 0 & 1 \end{bmatrix} \rightarrow \begin{bmatrix} 1 & 1 & | & 0 & 1 \\ 6 & 5 & | & 1 & 0 \end{bmatrix} \rightarrow \begin{bmatrix} 1 & 1 & | & 0 & 1 \\ 0 & -1 & | & 1 & -6 \end{bmatrix} \rightarrow \begin{bmatrix} 1 & 1 & | & 0 & 1 \\ 0 & 1 & | & -1 & 6 \end{bmatrix} \rightarrow \begin{bmatrix} 1 & 0 & | & 1 & -5 \\ 0 & 1 & | & -1 & 6 \end{bmatrix}$

$\begin{bmatrix} x \\ y \end{bmatrix} = \mathbf{A}^{-1}\mathbf{B} = \begin{bmatrix} 1 & -5 \\ -1 & 6 \end{bmatrix}\begin{bmatrix} 2 \\ -3 \end{bmatrix} = \begin{bmatrix} 17 \\ -20 \end{bmatrix} \Rightarrow x = 17,\ y = -20$

23. $\begin{bmatrix} 3 & 1 & | & 1 & 0 \\ 3 & -1 & | & 0 & 1 \end{bmatrix} \rightarrow \begin{bmatrix} 1 & \frac{1}{3} & | & \frac{1}{3} & 0 \\ 3 & -1 & | & 0 & 1 \end{bmatrix} \rightarrow \begin{bmatrix} 1 & \frac{1}{3} & | & \frac{1}{3} & 0 \\ 0 & -2 & | & -1 & 1 \end{bmatrix} \rightarrow \begin{bmatrix} 1 & \frac{1}{3} & | & \frac{1}{3} & 0 \\ 0 & 1 & | & \frac{1}{2} & -\frac{1}{2} \end{bmatrix}$

$\rightarrow \begin{bmatrix} 1 & 0 & | & \frac{1}{6} & \frac{1}{6} \\ 0 & 1 & | & \frac{1}{2} & -\frac{1}{2} \end{bmatrix}$

$\begin{bmatrix} x \\ y \end{bmatrix} = \mathbf{A}^{-1}\mathbf{B} = \begin{bmatrix} \frac{1}{6} & \frac{1}{6} \\ \frac{1}{2} & -\frac{1}{2} \end{bmatrix}\begin{bmatrix} 5 \\ 7 \end{bmatrix} = \begin{bmatrix} 2 \\ -1 \end{bmatrix} \Rightarrow x = 2,\ y = -1$

25. The coefficient matrix is not invertible. The method of reduction yields
$\begin{bmatrix} 2 & 6 & | & 2 \\ 3 & 9 & | & 3 \end{bmatrix} \rightarrow \begin{bmatrix} 1 & 3 & | & 1 \\ 3 & 9 & | & 3 \end{bmatrix} \rightarrow \begin{bmatrix} 1 & 3 & | & 1 \\ 0 & 0 & | & 0 \end{bmatrix}$.
Thus $x = -3r + 1,\ y = r$.

27. $\begin{bmatrix} 1 & 2 & 1 & | & 1 & 0 & 0 \\ 3 & 0 & 1 & | & 0 & 1 & 0 \\ 1 & -1 & 1 & | & 0 & 0 & 1 \end{bmatrix} \to \begin{bmatrix} 1 & 2 & 1 & | & 1 & 0 & 0 \\ 0 & -6 & -2 & | & -3 & 1 & 0 \\ 0 & -3 & 0 & | & -1 & 0 & 1 \end{bmatrix}$

$\to \begin{bmatrix} 1 & 2 & 1 & | & 1 & 0 & 0 \\ 0 & 1 & \frac{1}{3} & | & \frac{1}{2} & -\frac{1}{6} & 0 \\ 0 & -3 & 0 & | & -1 & 0 & 1 \end{bmatrix} \to \begin{bmatrix} 1 & 0 & \frac{1}{3} & | & 0 & \frac{1}{3} & 0 \\ 0 & 1 & \frac{1}{3} & | & \frac{1}{2} & -\frac{1}{6} & 0 \\ 0 & 0 & 1 & | & \frac{1}{2} & -\frac{1}{2} & 1 \end{bmatrix}$

$\to \begin{bmatrix} 1 & 0 & 0 & | & -\frac{1}{6} & \frac{1}{2} & -\frac{1}{3} \\ 0 & 1 & 0 & | & \frac{1}{3} & 0 & -\frac{1}{3} \\ 0 & 0 & 1 & | & \frac{1}{2} & -\frac{1}{2} & 1 \end{bmatrix}$

$\begin{bmatrix} x \\ y \\ z \end{bmatrix} = \mathbf{A}^{-1}\mathbf{B} = \begin{bmatrix} -\frac{1}{6} & \frac{1}{2} & -\frac{1}{3} \\ \frac{1}{3} & 0 & -\frac{1}{3} \\ \frac{1}{2} & -\frac{1}{2} & 1 \end{bmatrix}\begin{bmatrix} 4 \\ 2 \\ 1 \end{bmatrix} = \begin{bmatrix} 0 \\ 1 \\ 2 \end{bmatrix}$

Thus, $x = 0$, $y = 1$, $z = 2$.

29. $\begin{bmatrix} 1 & 1 & 1 & | & 1 & 0 & 0 \\ 1 & -1 & 1 & | & 0 & 1 & 0 \\ 1 & -1 & -1 & | & 0 & 0 & 1 \end{bmatrix} \to \begin{bmatrix} 1 & 1 & 1 & | & 1 & 0 & 0 \\ 0 & -2 & 0 & | & -1 & 1 & 0 \\ 0 & -2 & -2 & | & -1 & 0 & 1 \end{bmatrix}$

$\to \begin{bmatrix} 1 & 1 & 1 & | & 1 & 0 & 0 \\ 0 & 1 & 0 & | & \frac{1}{2} & -\frac{1}{2} & 0 \\ 0 & -2 & -2 & | & -1 & 0 & 1 \end{bmatrix} \to \begin{bmatrix} 1 & 0 & 1 & | & \frac{1}{2} & \frac{1}{2} & 0 \\ 0 & 1 & 0 & | & \frac{1}{2} & -\frac{1}{2} & 0 \\ 0 & 0 & -2 & | & 0 & -1 & 1 \end{bmatrix}$

$\to \begin{bmatrix} 1 & 0 & 1 & | & \frac{1}{2} & \frac{1}{2} & 0 \\ 0 & 1 & 0 & | & \frac{1}{2} & -\frac{1}{2} & 0 \\ 0 & 0 & 1 & | & 0 & \frac{1}{2} & -\frac{1}{2} \end{bmatrix} \to \begin{bmatrix} 1 & 0 & 0 & | & \frac{1}{2} & 0 & \frac{1}{2} \\ 0 & 1 & 0 & | & \frac{1}{2} & -\frac{1}{2} & 0 \\ 0 & 0 & 1 & | & 0 & \frac{1}{2} & -\frac{1}{2} \end{bmatrix}$

$\begin{bmatrix} x \\ y \\ z \end{bmatrix} = \mathbf{A}^{-1}\mathbf{B} = \begin{bmatrix} \frac{1}{2} & 0 & \frac{1}{2} \\ \frac{1}{2} & -\frac{1}{2} & 0 \\ 0 & \frac{1}{2} & -\frac{1}{2} \end{bmatrix}\begin{bmatrix} 2 \\ 1 \\ 0 \end{bmatrix} = \begin{bmatrix} 1 \\ \frac{1}{2} \\ \frac{1}{2} \end{bmatrix}$

Thus, $x = 1$, $y = \frac{1}{2}$, $z = \frac{1}{2}$.

31. The coefficient matrix is not invertible. The method of reduction yields

$$\begin{bmatrix} 1 & 3 & 3 & | & 7 \\ 2 & 1 & 1 & | & 4 \\ 1 & 1 & 1 & | & 4 \end{bmatrix} \rightarrow \begin{bmatrix} 1 & 3 & 3 & | & 7 \\ 0 & -5 & -5 & | & -10 \\ 0 & -2 & -2 & | & -3 \end{bmatrix} \rightarrow \begin{bmatrix} 1 & 3 & 3 & | & 7 \\ 0 & 1 & 1 & | & 2 \\ 0 & -2 & -2 & | & -3 \end{bmatrix} \rightarrow \begin{bmatrix} 1 & 0 & 0 & | & 1 \\ 0 & 1 & 1 & | & 2 \\ 0 & 0 & 0 & | & 1 \end{bmatrix}.$$

The third row indicates that $0 = 1$, which is never true, so there is no solution.

33.

$$\begin{bmatrix} 1 & 0 & 2 & 1 & | & 1 & 0 & 0 & 0 \\ 1 & -1 & 0 & 2 & | & 0 & 1 & 0 & 0 \\ 2 & 1 & 0 & 1 & | & 0 & 0 & 1 & 0 \\ 1 & 2 & 1 & 1 & | & 0 & 0 & 0 & 1 \end{bmatrix} \rightarrow \begin{bmatrix} 1 & 0 & 2 & 1 & | & 1 & 0 & 0 & 0 \\ 0 & -1 & -2 & 1 & | & -1 & 1 & 0 & 0 \\ 0 & 1 & -4 & -1 & | & -2 & 0 & 1 & 0 \\ 0 & 2 & -1 & 0 & | & -1 & 0 & 0 & 1 \end{bmatrix}$$

$$\rightarrow \begin{bmatrix} 1 & 0 & 2 & 1 & | & 1 & 0 & 0 & 0 \\ 0 & 1 & 2 & -1 & | & 1 & -1 & 0 & 0 \\ 0 & 0 & -6 & 0 & | & -3 & 1 & 1 & 0 \\ 0 & 0 & -5 & 2 & | & -3 & 2 & 0 & 1 \end{bmatrix} \rightarrow \begin{bmatrix} 1 & 0 & 2 & 1 & | & 1 & 0 & 0 & 0 \\ 0 & 1 & 2 & -1 & | & 1 & -1 & 0 & 0 \\ 0 & 0 & 1 & 0 & | & \frac{1}{2} & -\frac{1}{6} & -\frac{1}{6} & 0 \\ 0 & 0 & -5 & 2 & | & -3 & 2 & 0 & 1 \end{bmatrix}$$

$$\rightarrow \begin{bmatrix} 1 & 0 & 0 & 1 & | & 0 & \frac{1}{3} & \frac{1}{3} & 0 \\ 0 & 1 & 0 & -1 & | & 0 & -\frac{2}{3} & \frac{1}{3} & 0 \\ 0 & 0 & 1 & 0 & | & \frac{1}{2} & -\frac{1}{6} & -\frac{1}{6} & 0 \\ 0 & 0 & 0 & 2 & | & -\frac{1}{2} & \frac{7}{6} & -\frac{5}{6} & 1 \end{bmatrix} \rightarrow \begin{bmatrix} 1 & 0 & 0 & 1 & | & 0 & \frac{1}{3} & \frac{1}{3} & 0 \\ 0 & 1 & 0 & -1 & | & 0 & -\frac{2}{3} & \frac{1}{3} & 0 \\ 0 & 0 & 1 & 0 & | & \frac{1}{2} & -\frac{1}{6} & -\frac{1}{6} & 0 \\ 0 & 0 & 0 & 1 & | & -\frac{1}{4} & \frac{7}{12} & -\frac{5}{12} & \frac{1}{2} \end{bmatrix}$$

$$\rightarrow \begin{bmatrix} 1 & 0 & 0 & 0 & | & \frac{1}{4} & -\frac{1}{4} & \frac{3}{4} & -\frac{1}{2} \\ 0 & 1 & 0 & 0 & | & -\frac{1}{4} & -\frac{1}{12} & -\frac{1}{12} & \frac{1}{2} \\ 0 & 0 & 1 & 0 & | & \frac{1}{2} & -\frac{1}{6} & -\frac{1}{6} & 0 \\ 0 & 0 & 0 & 1 & | & -\frac{1}{4} & \frac{7}{12} & -\frac{5}{12} & \frac{1}{2} \end{bmatrix}$$

$$\begin{bmatrix} w \\ x \\ y \\ z \end{bmatrix} = \mathbf{A}^{-1}\mathbf{B} = \begin{bmatrix} \frac{1}{4} & -\frac{1}{4} & \frac{3}{4} & -\frac{1}{2} \\ -\frac{1}{4} & -\frac{1}{12} & -\frac{1}{12} & \frac{1}{2} \\ \frac{1}{2} & -\frac{1}{6} & -\frac{1}{6} & 0 \\ -\frac{1}{4} & \frac{7}{12} & -\frac{5}{12} & \frac{1}{2} \end{bmatrix} \begin{bmatrix} 4 \\ 12 \\ 12 \\ 12 \end{bmatrix} = \begin{bmatrix} 1 \\ 3 \\ -2 \\ 7 \end{bmatrix}$$

Thus, $w = 1$, $x = 3$, $y = -2$, $z = 7$.

35. $\mathbf{I} - \mathbf{A} = \begin{bmatrix} 1 & 0 \\ 0 & 1 \end{bmatrix} - \begin{bmatrix} 5 & -2 \\ 1 & 2 \end{bmatrix} = \begin{bmatrix} -4 & 2 \\ -1 & -1 \end{bmatrix}$

$\begin{bmatrix} -4 & 2 & | & 1 & 0 \\ -1 & -1 & | & 0 & 1 \end{bmatrix} \rightarrow \begin{bmatrix} -1 & -1 & | & 0 & 1 \\ -4 & 2 & | & 1 & 0 \end{bmatrix} \rightarrow \begin{bmatrix} 1 & 1 & | & 0 & -1 \\ -4 & 2 & | & 1 & 0 \end{bmatrix} \rightarrow \begin{bmatrix} 1 & 1 & | & 0 & -1 \\ 0 & 6 & | & 1 & -4 \end{bmatrix}$

$\rightarrow \begin{bmatrix} 1 & 1 & | & 0 & -1 \\ 0 & 1 & | & \frac{1}{6} & -\frac{2}{3} \end{bmatrix} \rightarrow \begin{bmatrix} 1 & 0 & | & -\frac{1}{6} & -\frac{1}{3} \\ 0 & 1 & | & \frac{1}{6} & -\frac{2}{3} \end{bmatrix}$

Thus, $(\mathbf{I} - \mathbf{A})^{-1} = \begin{bmatrix} -\frac{1}{6} & -\frac{1}{3} \\ \frac{1}{6} & -\frac{2}{3} \end{bmatrix}$.

37. Let x = number of model A and y = number of model B.

   a. The system is

$$\begin{cases} x + y = 100 & \text{(painting)} \\ \frac{1}{2}x + y = 80 & \text{(polishing)} \end{cases}$$

Let $\mathbf{A} = \begin{bmatrix} 1 & 1 \\ \frac{1}{2} & 1 \end{bmatrix}$.

$\begin{bmatrix} 1 & 1 & | & 1 & 0 \\ \frac{1}{2} & 1 & | & 0 & 1 \end{bmatrix} \rightarrow \begin{bmatrix} 1 & 1 & | & 1 & 0 \\ 0 & \frac{1}{2} & | & -\frac{1}{2} & 1 \end{bmatrix} \rightarrow \begin{bmatrix} 1 & 1 & | & 1 & 0 \\ 0 & 1 & | & -1 & 2 \end{bmatrix} \rightarrow \begin{bmatrix} 1 & 0 & | & 2 & -2 \\ 0 & 1 & | & -1 & 2 \end{bmatrix}$

$\begin{bmatrix} x \\ y \end{bmatrix} = \mathbf{A}^{-1} \begin{bmatrix} 100 \\ 80 \end{bmatrix} = \begin{bmatrix} 2 & -2 \\ -1 & 2 \end{bmatrix} \begin{bmatrix} 100 \\ 80 \end{bmatrix} = \begin{bmatrix} 40 \\ 60 \end{bmatrix}$

Thus 40 of model A and 60 of model B can be produced.

   b. The system is

$$\begin{cases} 10x + 7y = 800 & \text{(widgets)} \\ 14x + 10y = 1130 & \text{(shims)} \end{cases}$$

Let $\mathbf{A} = \begin{bmatrix} 10 & 7 \\ 14 & 10 \end{bmatrix}$.

$\begin{bmatrix} 10 & 7 & | & 1 & 0 \\ 14 & 10 & | & 0 & 1 \end{bmatrix} \rightarrow \begin{bmatrix} 1 & \frac{7}{10} & | & \frac{1}{10} & 0 \\ 14 & 10 & | & 0 & 1 \end{bmatrix}$

$\rightarrow \begin{bmatrix} 1 & \frac{7}{10} & | & \frac{1}{10} & 0 \\ 0 & \frac{1}{5} & | & -\frac{7}{5} & 1 \end{bmatrix} \rightarrow \begin{bmatrix} 1 & \frac{7}{10} & | & \frac{1}{10} & 0 \\ 0 & 1 & | & -7 & 5 \end{bmatrix} \rightarrow \begin{bmatrix} 1 & 0 & | & 5 & -\frac{7}{2} \\ 0 & 1 & | & -7 & 5 \end{bmatrix}$

$\begin{bmatrix} x \\ y \end{bmatrix} = \mathbf{A}^{-1} \begin{bmatrix} 800 \\ 1130 \end{bmatrix} = \begin{bmatrix} 5 & -\frac{7}{2} \\ -7 & 5 \end{bmatrix} \begin{bmatrix} 800 \\ 1130 \end{bmatrix} = \begin{bmatrix} 45 \\ 50 \end{bmatrix}$

Thus 45 of model A and 50 of model B can be produced.

39. a. $\left(B^{-1}A^{-1}\right)(AB) = B^{-1}\left(A^{-1}A\right)B$

$= B^{-1}IB = B^{-1}B = I$

Since an invertible matrix has exactly one inverse, $B^{-1}A^{-1}$ is the inverse of **AB**.

b. From Part (a),

$(AB)^{-1} = B^{-1}A^{-1}$

$= \begin{bmatrix} 1 & 1 \\ 1 & 2 \end{bmatrix}\begin{bmatrix} 1 & 2 \\ 3 & 4 \end{bmatrix} = \begin{bmatrix} 4 & 6 \\ 7 & 10 \end{bmatrix}$

41. $P^TP = \begin{bmatrix} \frac{3}{5} & \frac{4}{5} \\ -\frac{4}{5} & \frac{3}{5} \end{bmatrix}\begin{bmatrix} \frac{3}{5} & -\frac{4}{5} \\ \frac{4}{5} & \frac{3}{5} \end{bmatrix} = \begin{bmatrix} 1 & 0 \\ 0 & 1 \end{bmatrix} = I$, so

$P^T = P^{-1}$. Yes, **P** is orthogonal.

43. Let x be the number of shares of D, y be the number of shares of E, and z be the number of shares of F. We get the following equations.

$60x + 80y + 30z = 500{,}000$

$0.16(60x) + 0.12(80y) + 0.09(30z)$
$= 0.1368(60x + 80y + 30z)$

$z = 4y$

Simplify the first equation.

$6x + 8y + 3z = 50{,}000$

Simplify the second equation.

$9.6x + 9.6y + 2.7z = 8.208x + 10.944y + 4.104z$

$1.392x - 1.344y - 1.404z = 0$

$1392x - 1344y - 1404z = 0$

$116x - 112y - 117z = 0$

Simplify the third equation.

$4y - z = 0$

Thus we solve the following system of equations.

$6x + 8y + 3z = 50{,}000$

$116x - 112y - 117z = 0$

$4y - z = 0$

The coefficient matrix is

$A = \begin{bmatrix} 6 & 8 & 3 \\ 116 & -112 & -117 \\ 0 & 4 & -1 \end{bmatrix}$.

$\left[A \mid I\right] = \begin{bmatrix} 6 & 8 & 3 & 1 & 0 & 0 \\ 116 & -112 & -117 & 0 & 1 & 0 \\ 0 & 4 & -1 & 0 & 0 & 1 \end{bmatrix}$

$\xrightarrow{\frac{1}{6}R_1} \begin{bmatrix} 1 & \frac{4}{3} & \frac{1}{2} & \frac{1}{6} & 0 & 0 \\ 116 & -112 & -117 & 0 & 1 & 0 \\ 0 & 4 & -1 & 0 & 0 & 1 \end{bmatrix}$

$\xrightarrow{-116R_1 + R_2} \begin{bmatrix} 1 & \frac{4}{3} & \frac{1}{2} & \frac{1}{6} & 0 & 0 \\ 0 & -\frac{800}{3} & -175 & -\frac{58}{3} & 1 & 0 \\ 0 & 4 & -1 & 0 & 0 & 1 \end{bmatrix}$

$\xrightarrow[-\frac{1}{4}R_3]{-\frac{3}{800}R_2} \begin{bmatrix} 1 & \frac{4}{3} & \frac{1}{2} & \frac{1}{6} & 0 & 0 \\ 0 & 1 & \frac{21}{32} & \frac{29}{400} & -\frac{3}{800} & 0 \\ 0 & -1 & \frac{1}{4} & 0 & 0 & -\frac{1}{4} \end{bmatrix}$

$\xrightarrow{R_2 + R_3} \begin{bmatrix} 1 & \frac{4}{3} & \frac{1}{2} & \frac{1}{6} & 0 & 0 \\ 0 & 1 & \frac{21}{32} & \frac{29}{400} & -\frac{3}{800} & 0 \\ 0 & 0 & \frac{29}{32} & \frac{29}{400} & -\frac{3}{800} & -\frac{1}{4} \end{bmatrix}$

$\xrightarrow{\frac{32}{29}R_3} \begin{bmatrix} 1 & \frac{4}{3} & \frac{1}{2} & \frac{1}{6} & 0 & 0 \\ 0 & 1 & \frac{21}{32} & \frac{29}{400} & -\frac{3}{800} & 0 \\ 0 & 0 & 1 & \frac{2}{25} & -\frac{3}{725} & -\frac{8}{29} \end{bmatrix}$

$\xrightarrow[-\frac{21}{32}R_3 + R_2]{-\frac{1}{2}R_3 + R_1} \begin{bmatrix} 1 & \frac{4}{3} & 0 & \frac{19}{150} & \frac{3}{1450} & \frac{4}{29} \\ 0 & 1 & 0 & \frac{1}{50} & -\frac{3}{2900} & \frac{21}{116} \\ 0 & 0 & 1 & \frac{2}{25} & -\frac{3}{725} & -\frac{8}{29} \end{bmatrix}$

$\xrightarrow{-\frac{4}{3}R_2 + R_1} \begin{bmatrix} 1 & 0 & 0 & \frac{1}{10} & \frac{1}{290} & -\frac{3}{29} \\ 0 & 1 & 0 & \frac{1}{50} & -\frac{3}{2900} & \frac{21}{116} \\ 0 & 0 & 1 & \frac{2}{25} & -\frac{3}{725} & -\frac{8}{29} \end{bmatrix}$

$\begin{bmatrix} x \\ y \\ z \end{bmatrix} = \begin{bmatrix} \frac{1}{10} & \frac{1}{290} & -\frac{3}{29} \\ \frac{1}{50} & -\frac{3}{2900} & \frac{21}{116} \\ \frac{2}{25} & -\frac{3}{725} & -\frac{8}{29} \end{bmatrix}\begin{bmatrix} 50{,}000 \\ 0 \\ 0 \end{bmatrix} = \begin{bmatrix} 5000 \\ 1000 \\ 4000 \end{bmatrix}$

They should buy 5000 shares of company D, 1000 shares of company E, and 4000 shares of company F.

45. a. $\begin{bmatrix} 2.05 & 1.28 \\ 0.73 & 1.71 \end{bmatrix}$

b. $\begin{bmatrix} \frac{84}{41} & \frac{105}{82} \\ \frac{30}{41} & \frac{70}{41} \end{bmatrix}$

47. $\begin{bmatrix} 2.75 & -1.59 & -1.11 \\ -0.48 & 1.43 & 0.00 \\ -1.22 & 0.32 & 2.22 \end{bmatrix}$

49. $\begin{bmatrix} w \\ x \\ y \\ z \end{bmatrix} = \begin{bmatrix} \frac{2}{5} & 4 & \frac{1}{2} & -\frac{3}{7} \\ \frac{5}{9} & -\frac{2}{3} & -4 & -1 \\ 0 & 1 & -\frac{4}{9} & \frac{5}{6} \\ \frac{1}{2} & 0 & 4 & -\frac{1}{3} \end{bmatrix}^{-1} \begin{bmatrix} \frac{14}{13} \\ \frac{7}{8} \\ 9 \\ \frac{4}{7} \end{bmatrix} = \begin{bmatrix} 14.44 \\ 0.03 \\ -0.80 \\ 10.33 \end{bmatrix}$

$w = 14.44,\ x = 0.03,\ y = -0.80,\ z = 10.33$

Problems 6.7

1. $A = \begin{bmatrix} \frac{200}{1200} & \frac{500}{1500} \\ \frac{400}{1200} & \frac{200}{1500} \end{bmatrix}$

$D = \begin{bmatrix} 600 \\ 805 \end{bmatrix}$

$X = (I - A)^{-1} D = \begin{bmatrix} 1290 \\ 1425 \end{bmatrix}$

The total value of other production costs is

$P_A + P_B = \dfrac{600}{1200}(1290) + \dfrac{800}{1500}(1425) = 1405$

3. $A = \begin{bmatrix} \frac{15}{100} & \frac{30}{120} & \frac{45}{180} \\ \frac{25}{100} & \frac{30}{120} & \frac{60}{180} \\ \frac{50}{100} & \frac{40}{120} & \frac{60}{180} \end{bmatrix}$

a. $D = \begin{bmatrix} 15 \\ 10 \\ 35 \end{bmatrix}$

$X = (I - A)^{-1} D = \begin{bmatrix} 134.29 \\ 162.25 \\ 234.35 \end{bmatrix}$

b. $D = \begin{bmatrix} 10 \\ 10 \\ 10 \end{bmatrix}$

$X = (I - A)^{-1} D = \begin{bmatrix} 68.59 \\ 84.50 \\ 108.69 \end{bmatrix}$

5. $A = \begin{bmatrix} \frac{400}{1000} & \frac{200}{1000} & \frac{200}{1000} \\ \frac{200}{1000} & \frac{400}{1000} & \frac{100}{1000} \\ \frac{200}{1000} & \frac{100}{1000} & \frac{300}{1000} \end{bmatrix}$

$D = \begin{bmatrix} 300 \\ 350 \\ 450 \end{bmatrix}$

$X = (I - A)^{-1} D = \begin{bmatrix} 1301 \\ 1215 \\ 1188 \end{bmatrix}$

7. $A = \begin{bmatrix} \frac{400}{1000} & \frac{200}{1000} & \frac{200}{1000} \\ \frac{200}{1000} & \frac{400}{1000} & \frac{100}{1000} \\ \frac{200}{1000} & \frac{100}{1000} & \frac{300}{1000} \end{bmatrix}$

$D = \begin{bmatrix} 300 \\ 400 \\ 500 \end{bmatrix}$

$X = (I - A)^{-1} D = \begin{bmatrix} 1382 \\ 1344 \\ 1301 \end{bmatrix}$

9. $A = \begin{bmatrix} \frac{1}{10} & \frac{1}{3} & \frac{1}{4} \\ \frac{1}{10} & \frac{1}{10} & \frac{1}{3} \\ \frac{1}{10} & \frac{1}{10} & \frac{1}{10} \end{bmatrix}$

$D = \begin{bmatrix} 300 \\ 200 \\ 500 \end{bmatrix}$

$(I - A)X = D$

Reducing $\left[\begin{array}{ccc|c} \frac{9}{10} & -\frac{1}{3} & -\frac{1}{4} & 300 \\ -\frac{1}{10} & \frac{9}{10} & -\frac{1}{3} & 200 \\ -\frac{1}{10} & -\frac{1}{10} & \frac{9}{10} & 500 \end{array} \right]$ with a calculator results in $\left[\begin{array}{ccc|c} 1 & 0 & 0 & 736.39 \\ 0 & 1 & 0 & 563.29 \\ 0 & 0 & 1 & 699.96 \end{array} \right]$.

Thus 736.39 units of coal, 563.29 units of steel, and 699.96 units of railroad services need to be produced.

Chapter 6 Review Problems

1. $2\begin{bmatrix} 3 & 4 \\ -5 & 1 \end{bmatrix} - 3\begin{bmatrix} 1 & 0 \\ 2 & 4 \end{bmatrix} = \begin{bmatrix} 6 & 8 \\ -10 & 2 \end{bmatrix} - \begin{bmatrix} 3 & 0 \\ 6 & 12 \end{bmatrix} = \begin{bmatrix} 3 & 8 \\ -16 & -10 \end{bmatrix}$

3. $\begin{bmatrix} 1 & 7 \\ 2 & -3 \\ 1 & 0 \end{bmatrix}\begin{bmatrix} 1 & 0 & -2 \\ 0 & 6 & 1 \end{bmatrix} = \begin{bmatrix} 1+0 & 0+42 & -2+7 \\ 2+0 & 0-18 & -4-3 \\ 1+0 & 0+0 & -2+0 \end{bmatrix} = \begin{bmatrix} 1 & 42 & 5 \\ 2 & -18 & -7 \\ 1 & 0 & -2 \end{bmatrix}$

5. $\begin{bmatrix} 2 & 3 \\ -1 & 3 \end{bmatrix}\left(\begin{bmatrix} 2 & 3 \\ 7 & 6 \end{bmatrix} - \begin{bmatrix} 1 & 8 \\ 4 & 4 \end{bmatrix} \right) = \begin{bmatrix} 2 & 3 \\ -1 & 3 \end{bmatrix}\begin{bmatrix} 1 & -5 \\ 3 & 2 \end{bmatrix} = \begin{bmatrix} 11 & -4 \\ 8 & 11 \end{bmatrix}$

7. $2\begin{bmatrix} 1 & -2 \\ 3 & 1 \end{bmatrix}^2 [1 \quad -2]^T = 2\begin{bmatrix} -5 & -4 \\ 6 & -5 \end{bmatrix}\begin{bmatrix} 1 \\ -2 \end{bmatrix} = 2\begin{bmatrix} 3 \\ 16 \end{bmatrix} = \begin{bmatrix} 6 \\ 32 \end{bmatrix}$

9. $(2A)^T - 3I^2 = 2A^T - 3I = 2\begin{bmatrix} 1 & -1 \\ 1 & 2 \end{bmatrix} - \begin{bmatrix} 3 & 0 \\ 0 & 3 \end{bmatrix}$

$= \begin{bmatrix} 2 & -2 \\ 2 & 4 \end{bmatrix} - \begin{bmatrix} 3 & 0 \\ 0 & 3 \end{bmatrix} = \begin{bmatrix} -1 & -2 \\ 2 & 1 \end{bmatrix}$

11. $B^3 + I^5 = \begin{bmatrix} 1 & 0 \\ 0 & 2 \end{bmatrix}^3 + \begin{bmatrix} 1 & 0 \\ 0 & 1 \end{bmatrix}^5 = \begin{bmatrix} 1 & 0 \\ 0 & 8 \end{bmatrix} + \begin{bmatrix} 1 & 0 \\ 0 & 1 \end{bmatrix} = \begin{bmatrix} 2 & 0 \\ 0 & 9 \end{bmatrix}$

13. $\begin{bmatrix} 5x \\ 7x \end{bmatrix} = \begin{bmatrix} 15 \\ y \end{bmatrix}$

$5x = 15$, or $x = 3$

$7x = y$, $7 \cdot 3 = y$, or $y = 21$

15. $\begin{bmatrix} 1 & 4 \\ 5 & 8 \end{bmatrix} \rightarrow \begin{bmatrix} 1 & 4 \\ 0 & -12 \end{bmatrix} \rightarrow \begin{bmatrix} 1 & 4 \\ 0 & 1 \end{bmatrix} \rightarrow \begin{bmatrix} 1 & 0 \\ 0 & 1 \end{bmatrix}$

17. $\begin{bmatrix} 2 & 4 & 7 \\ 1 & 2 & 4 \\ 5 & 8 & 2 \end{bmatrix} \rightarrow \begin{bmatrix} 1 & 2 & 4 \\ 2 & 4 & 7 \\ 5 & 8 & 2 \end{bmatrix} \rightarrow \begin{bmatrix} 1 & 2 & 4 \\ 0 & 0 & -1 \\ 0 & -2 & -18 \end{bmatrix} \rightarrow \begin{bmatrix} 1 & 2 & 4 \\ 0 & -2 & -18 \\ 0 & 0 & -1 \end{bmatrix}$

$\rightarrow \begin{bmatrix} 1 & 2 & 4 \\ 0 & 1 & 9 \\ 0 & 0 & -1 \end{bmatrix} \rightarrow \begin{bmatrix} 1 & 0 & -14 \\ 0 & 1 & 9 \\ 0 & 0 & -1 \end{bmatrix} \rightarrow \begin{bmatrix} 1 & 0 & -14 \\ 0 & 1 & 9 \\ 0 & 0 & 1 \end{bmatrix} \rightarrow \begin{bmatrix} 1 & 0 & 0 \\ 0 & 1 & 0 \\ 0 & 0 & 1 \end{bmatrix}$

19. $\left[\begin{array}{cc|c} 2 & -5 & 0 \\ 4 & 3 & 0 \end{array}\right] \rightarrow \left[\begin{array}{cc|c} 2 & -5 & 0 \\ 0 & 13 & 0 \end{array}\right] \rightarrow \left[\begin{array}{cc|c} 1 & -\frac{5}{2} & 0 \\ 0 & 1 & 0 \end{array}\right] \rightarrow \left[\begin{array}{cc|c} 1 & 0 & 0 \\ 0 & 1 & 0 \end{array}\right]$

Thus $x = 0$, $y = 0$.

21. $\left[\begin{array}{ccc|c} 1 & 1 & 2 & 1 \\ 3 & -2 & -4 & -7 \\ 2 & -1 & -2 & 2 \end{array}\right] \rightarrow \left[\begin{array}{ccc|c} 1 & 1 & 2 & 1 \\ 0 & -5 & -10 & -10 \\ 0 & -3 & -6 & 0 \end{array}\right] \rightarrow \left[\begin{array}{ccc|c} 1 & 1 & 2 & 1 \\ 0 & 1 & 2 & 2 \\ 0 & -3 & -6 & 0 \end{array}\right] \rightarrow \left[\begin{array}{ccc|c} 1 & 0 & 0 & -1 \\ 0 & 1 & 2 & 2 \\ 0 & 0 & 0 & 6 \end{array}\right]$

Row three indicates that $0 = 6$, which is never true, so there is no solution.

23. $\left[\begin{array}{cc|cc} 1 & 5 & 1 & 0 \\ 3 & 9 & 0 & 1 \end{array}\right] \rightarrow \left[\begin{array}{cc|cc} 1 & 5 & 1 & 0 \\ 0 & -6 & -3 & 1 \end{array}\right] \rightarrow \left[\begin{array}{cc|cc} 1 & 5 & 1 & 0 \\ 0 & 1 & \frac{1}{2} & -\frac{1}{6} \end{array}\right]$

$\rightarrow \left[\begin{array}{cc|cc} 1 & 0 & -\frac{3}{2} & \frac{5}{6} \\ 0 & 1 & \frac{1}{2} & -\frac{1}{6} \end{array}\right] \Rightarrow \mathbf{A}^{-1} = \begin{bmatrix} -\frac{3}{2} & \frac{5}{6} \\ \frac{1}{2} & -\frac{1}{6} \end{bmatrix}$

25. $\begin{bmatrix} 1 & 3 & -2 & | & 1 & 0 & 0 \\ 4 & 1 & 0 & | & 0 & 1 & 0 \\ 3 & -2 & 2 & | & 0 & 0 & 1 \end{bmatrix} \rightarrow \begin{bmatrix} 1 & 3 & -2 & | & 1 & 0 & 0 \\ 0 & -11 & 8 & | & -4 & 1 & 0 \\ 0 & -11 & 8 & | & -3 & 0 & 1 \end{bmatrix}$

$\begin{bmatrix} 1 & 3 & -2 & | & 1 & 0 & 0 \\ 0 & -11 & 8 & | & -4 & 1 & 0 \\ 0 & 0 & 0 & | & 1 & -1 & 1 \end{bmatrix} \rightarrow \begin{bmatrix} 1 & 3 & -2 & | & 1 & 0 & 0 \\ 0 & 1 & -\frac{8}{11} & | & \frac{4}{11} & -\frac{1}{11} & 0 \\ 0 & 0 & 0 & | & 1 & -1 & 1 \end{bmatrix}$

$\rightarrow \begin{bmatrix} 1 & 0 & \frac{2}{11} & | & -\frac{1}{11} & \frac{3}{11} & 0 \\ 0 & 1 & -\frac{8}{11} & | & \frac{4}{11} & -\frac{1}{11} & 0 \\ 0 & 0 & 0 & | & 1 & -1 & 1 \end{bmatrix} \Rightarrow$ no inverse exists

27. $\begin{bmatrix} 3 & 1 & 4 & | & 1 & 0 & 0 \\ 1 & 0 & 1 & | & 0 & 1 & 0 \\ 0 & 2 & 1 & | & 0 & 0 & 1 \end{bmatrix} \rightarrow \begin{bmatrix} 1 & 0 & 1 & | & 0 & 1 & 0 \\ 3 & 1 & 4 & | & 1 & 0 & 0 \\ 0 & 2 & 1 & | & 0 & 0 & 1 \end{bmatrix}$

$\rightarrow \begin{bmatrix} 1 & 0 & 1 & | & 0 & 1 & 0 \\ 0 & 1 & 1 & | & 1 & -3 & 0 \\ 0 & 2 & 1 & | & 0 & 0 & 1 \end{bmatrix} \rightarrow \begin{bmatrix} 1 & 0 & 1 & | & 0 & 1 & 0 \\ 0 & 1 & 1 & | & 1 & -3 & 0 \\ 0 & 0 & -1 & | & -2 & 6 & 1 \end{bmatrix}$

$\rightarrow \begin{bmatrix} 1 & 0 & 1 & | & 0 & 1 & 0 \\ 0 & 1 & 1 & | & 1 & -3 & 0 \\ 0 & 0 & 1 & | & 2 & -6 & -1 \end{bmatrix} \rightarrow \begin{bmatrix} 1 & 0 & 0 & | & -2 & 7 & 1 \\ 0 & 1 & 0 & | & -1 & 3 & 1 \\ 0 & 0 & 1 & | & 2 & -6 & -1 \end{bmatrix}$

$\begin{bmatrix} x \\ y \\ z \end{bmatrix} = \mathbf{A}^{-1}\mathbf{B} = \begin{bmatrix} -2 & 7 & 1 \\ -1 & 3 & 1 \\ 2 & -6 & -1 \end{bmatrix}\begin{bmatrix} 1 \\ 0 \\ 2 \end{bmatrix} = \begin{bmatrix} 0 \\ 1 \\ 0 \end{bmatrix}$

Thus $x = 0$, $y = 1$, $z = 0$.

29. $\mathbf{A}^2 = \mathbf{A}\mathbf{A} = \begin{bmatrix} 0 & 1 & 1 \\ 0 & 0 & 1 \\ 0 & 0 & 0 \end{bmatrix}\begin{bmatrix} 0 & 1 & 1 \\ 0 & 0 & 1 \\ 0 & 0 & 0 \end{bmatrix} = \begin{bmatrix} 0 & 0 & 1 \\ 0 & 0 & 0 \\ 0 & 0 & 0 \end{bmatrix}$

$\mathbf{A}^3 = \mathbf{A}^2\mathbf{A} = \begin{bmatrix} 0 & 0 & 1 \\ 0 & 0 & 0 \\ 0 & 0 & 0 \end{bmatrix}\begin{bmatrix} 0 & 1 & 1 \\ 0 & 0 & 1 \\ 0 & 0 & 0 \end{bmatrix} = \begin{bmatrix} 0 & 0 & 0 \\ 0 & 0 & 0 \\ 0 & 0 & 0 \end{bmatrix} = \mathbf{O}$

Since $\mathbf{A}^3 = \mathbf{O}$, every higher power of $\mathbf{A}$ is also $\mathbf{O}$, so $\mathbf{A}^{1000} = \mathbf{O}$.

Looking at $\begin{bmatrix} 0 & 1 & 1 & | & 1 & 0 & 0 \\ 0 & 0 & 1 & | & 0 & 1 & 0 \\ 0 & 0 & 0 & | & 0 & 0 & 1 \end{bmatrix}$, it is clear that there is no way of transforming the left side into $\mathbf{I}_3$,

since there is no way to get a nonzero entry in the first column. Thus $\mathbf{A}$ does not have an inverse.

31. a. Let x, y, and z represent the weekly doses of capsules of brand I, II, and III, respectively. Then

$$\begin{cases} x + y + 4z = 13 & \text{(vitamin A)} \\ x + 2y + 7z = 22 & \text{(vitamin B)} \\ x + 3y + 10z = 31 & \text{(vitamin C)} \end{cases}$$

$$\begin{bmatrix} 1 & 1 & 4 & | & 13 \\ 1 & 2 & 7 & | & 22 \\ 1 & 3 & 10 & | & 31 \end{bmatrix} \xrightarrow[-R_1 + R_3]{-R_1 + R_2} \begin{bmatrix} 1 & 1 & 4 & | & 13 \\ 0 & 1 & 3 & | & 9 \\ 0 & 2 & 6 & | & 18 \end{bmatrix}$$

$$\xrightarrow[-2R_2 + R_3]{-R_2 + R_1} \begin{bmatrix} 1 & 0 & 1 & | & 4 \\ 0 & 1 & 3 & | & 9 \\ 0 & 0 & 0 & | & 0 \end{bmatrix}$$

Thus $x = 4 - r$, $y = 9 - 3r$, and $z = r$, where $r = 0, 1, 2, 3$.
The four possible combinations are

Combination	x	y	z
1	4	9	0
2	3	6	1
3	2	3	2
4	1	0	3

    b. Computing the cost of each combination, we find that they are 83, 77, 71, and 65 cents, respectively. Thus combination 4, namely $x = 1$, $y = 0$, $z = 3$, minimizes weekly cost.

33. $\begin{bmatrix} 215 & 87 \\ 89 & 141 \end{bmatrix}$

35. $\mathbf{A} = \begin{bmatrix} \frac{10}{34} & \frac{20}{39} \\ \frac{15}{34} & \frac{14}{39} \end{bmatrix}$; $\mathbf{D} = \begin{bmatrix} 10 \\ 5 \end{bmatrix}$; $\mathbf{X} = (\mathbf{I} - \mathbf{A})^{-1}\mathbf{D} = \begin{bmatrix} 39.7 \\ 35.1 \end{bmatrix}$

Mathematical Snapshot Chapter 6

1. $\mathbf{A} = \begin{bmatrix} 20 & 40 & 30 & 10 \\ 30 & 0 & 10 & 10 \\ 10 & 0 & 30 & 50 \end{bmatrix}$

$$\mathbf{T} = \begin{bmatrix} 7 \\ 10 \\ 7 \\ 5 \end{bmatrix}$$

$$\mathbf{C} = \begin{bmatrix} 9 \\ 8 \\ 10 \end{bmatrix}$$

$$\mathbf{C}^{\mathrm{T}}(\mathbf{AT}) = \mathbf{C}^{\mathrm{T}} \left\{ \begin{bmatrix} 20 & 40 & 30 & 10 \\ 30 & 0 & 10 & 10 \\ 10 & 0 & 30 & 50 \end{bmatrix} \begin{bmatrix} 7 \\ 10 \\ 7 \\ 5 \end{bmatrix} \right\} = \mathbf{C}^{\mathrm{T}} \begin{bmatrix} 800 \\ 330 \\ 530 \end{bmatrix}$$

$$= \begin{bmatrix} 9 & 8 & 10 \end{bmatrix} \begin{bmatrix} 800 \\ 330 \\ 530 \end{bmatrix} = \begin{bmatrix} 15,140 \end{bmatrix}$$

The cost is $151.40.

3. It is not possible. Different combinations of lengths of stays can cost the same. For example, guest 1 staying for
20 days and guest 3 staying for 17 days costs the same as guest 1 staying for 15 days and guest 3 staying for 21 days (each costs $214.50).

Chapter 7

Principles in Practice 7.1

1. Let x = the number of type A magnets and
   y = the number of type B magnets.
   The cost for producing x type A magnets
   and y type B magnets is $50 + 0.90x + 0.70y$.
   The revenue for selling x type A magnets
   and y type B magnets is $2.00x + 1.50y$.
   Revenue is greater than cost when
   $2x + 1.5y > 50 + 0.9x + 0.7y$.
   $0.8y > -1.1x + 50$
   $y > -1.375x + 62.5$
   Sketch the dashed line $y = -1.375x + 62.5$
   and shade the half plane above the line. In
   order to make a profit, the number of
   magnets of types A and B must correspond
   to an ordered pair in the shaded region.
   Also, to take reality into account, both x and
   y must be positive (negative numbers of
   magnets are not feasible).

Problems 7.1

1.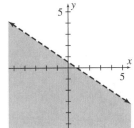

3.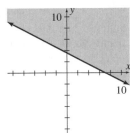

5.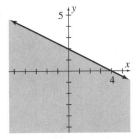

7.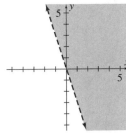

9.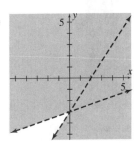

11.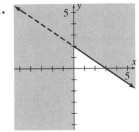

13.

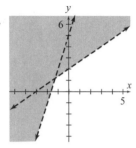

15.

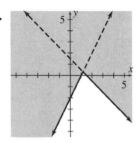

17.

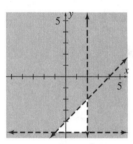

19.

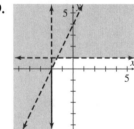

21.

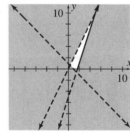

23.

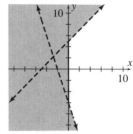

25. $6x + 4y \le 20$

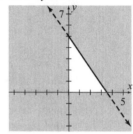

27. Let x be the amount purchased from supplier A, and y the amount purchased from B. The system of inequalities is
$x + y \le 100,$
$x \ge 0,$
$y \ge 0.$

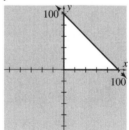

29. Since negative numbers of chairs cannot be produced, $x \ge 0$ and $y \ge 0$. The inequality for assembly time is $3x + 2y \le 240$. The inequality for painting time is $\dfrac{1}{2}x + y \le 80$.

The system of inequalities is
$$\begin{cases} 3x + 2y \le 240, \\ \dfrac{1}{2}x + y \le 80, \\ x \ge 0, \\ y \ge 0. \end{cases}$$

The region consists of points on or above

the *x*-axis and on or to the right of the *y*-axis. In addition, the points must be on or below the line $3x + 2y = 240$ and on or below the line $\frac{1}{2}x + y = 80$ (or, equivalently $x + 2y = 160$).

Problems 7.2

1. The feasible region appears below. The corner points are $(2, 0)$, $\left(\frac{47}{3}, \frac{41}{9}\right)$, $\left(\frac{45}{2}, 0\right)$. Evaluating P at each corner point, we find that P has a maximum value of $112\frac{1}{2}$ when $x = \frac{45}{2}$ and $y = 0$.

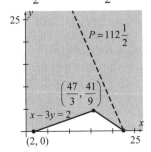

3. The feasible region appears below. The corner points are $(2, 3)$, $(0, 5)$, $(0, 7)$ and $\left(\frac{10}{3}, 7\right)$.

Evaluating Z at each point, we find that Z has a maximum value of -10 when $x = 2$ and $y = 3$.

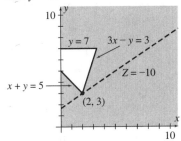

5. The feasible region is empty, so there is no optimum solution.

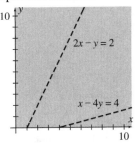

7. The feasible region is a line segment. The corner points are $(0, 1)$ and $(4, 5)$. Z has a minimum value of 3 when $x = 0$ and $y = 1$.

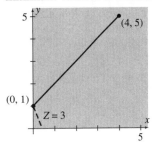

9. The feasible region is unbounded with 3 corner points. The member (see dashed line) of the family of lines $C = 3x + 2y$ which gives a minimum value of C, subject to the constraints, intersects the feasible region at corner point $\left(\frac{7}{3}, \frac{1}{3}\right)$ where $C = \frac{23}{3}$. Thus C has a minimum value of $\frac{23}{3}$ when $x = \frac{7}{3}$ and $y = \frac{1}{3}$. [Note: Here we chose the member of the family $y = \frac{1}{2}(-3x + C)$ whose *y*-intercept was *closest* to the origin and which had at least one point in common with the feasible region.]

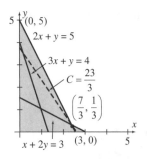

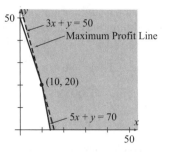

11. The feasible region is unbounded with 2 corner points. The family of lines given by $Z = 10x + 2y$ has members (see dashed lines for two sample members) that have arbitrarily large values of Z and that also intersect the feasible region. Thus no optimum solution exists.

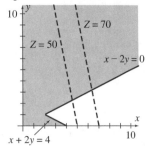

13. Let x and y be the number of trucks and spinning tops made per week, respectively. Then we are to maximize $P = 7x + 2y$ where

$$\begin{cases} x \geq 0 \\ y \geq 0 \\ 2x + y \leq 80 \text{ (for machine A)} \\ 3x + y \leq 50 \text{ (for machine B)} \\ 5x + y \leq 70 \text{ (for finishing)} \end{cases}$$

The feasible region is bounded. The corner points are $(0, 50)$, $(14, 0)$ and $(10, 20)$. Evaluating P at each corner point, we find that P is maximized at corner point $(10, 20)$, where its value is 110. Thus 10 trucks and 20 spinning tops should be made each week to give a maximum profit of $110.

15. Let x and y be the numbers of units of Food A and Food B, respectively, that are purchased. Then we are to minimize $C = 1.20x + 0.80y$, where

$$\begin{cases} x \geq 0, \\ y \geq 0, \\ 2x + 2y \geq 16 \text{ (for carbohydrates)}, \\ 4x + y \geq 20 \text{ (for protein)}. \end{cases}$$

The feasible region is unbounded. The corner points are $(8, 0)$, $(4, 4)$ and $(0, 20)$. C is minimized at corner point $(4, 4)$ where $C = 8$ (see the minimum cost line). Thus 4 units of Food A and 4 units of Food B gives a minimum cost of $8.

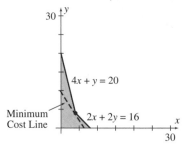

17. Let x and y be the numbers of tons of ores I and II, respectively, that are processed. Then we are to minimize $C = 50x + 60y$, where

$$\begin{cases} x \geq 0, \\ y \geq 0, \\ 100x + 200y \geq 3000 \text{ (for mineral A)}, \\ 200x + 50y \geq 2500 \text{ (for mineral B)}. \end{cases}$$

The feasible region is unbounded with 3 corner points. C is minimized at the corner point $(10, 10)$ where $C = 1100$ (see the minimum cost line). Thus 10 tons of ore I and 10 tons of ore II give a minimum cost of $1100.

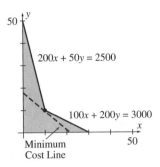

200x + 50y = 2500
100x + 200y = 3000
Minimum Cost Line

19. Let x and y be the number of chambers of type A and B, respectively. Then we are to minimize $C = 600{,}000x + 300{,}000y$, where

$$\begin{cases} x \geq 4, \\ y \geq 4, \\ 10x + 4y \geq 100 \ \text{(for polymer } P_1), \\ 20x + 30y \geq 420 \ \text{(for polymer } P_2). \end{cases}$$

The feasible region is unbounded with 3 corner points. Evaluating C at each corner point, we find C is minimized at corner point $(6, 10)$ where $C = 6{,}600{,}000$. Thus the solution is 6 chambers of type A and 10 chambers of type B.

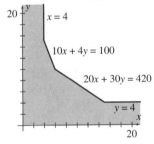

x = 4
10x + 4y = 100
20x + 30y = 420
y = 4

21. a. A builds x km of highway and y km of expressway, so B builds $(300 - x)$ km of highway and $(200 - y)$ km of expressway. Thus
$$D = 2x + 6y + 3(300 - x) + 5(200 - y)$$
$$= 1900 - x + y.$$

b. The first constraint is company A's construction limit.
The second constraint is company B's construction limit, which arises as follows:

$$(300 - x) + (200 - y) \leq 300,$$
$$500 - x - y \leq 300,$$
$$-x - y \leq -200,$$
$$x + y \geq 200.$$

The third constraint is the minimum contract for A.
The fourth constraint is the minimum contract for B, which arises as follows:

$$2(300 - x) + 8(200 - y) \geq 300,$$
$$2200 - 2x - 8y \geq 300,$$
$$-2x - 8y \geq -1900,$$
$$2x + 8y \leq 1900.$$

The fifth constraint reflects the fact that company A will not build more than 300 km of highway, since 300 km is the total being built; the sixth constraint is the corresponding constraint for the amount of expressway.

c. The feasible region (see below) is bounded. The corner points are

$(0, 200)$, $(150, 200)$, $\left(\dfrac{650}{3}, \dfrac{550}{3} \right)$,

$(300, 100)$, $(300, 0)$, and $(200, 0)$. Evaluating D at each corner point, we find that D is maximized at point $(0, 200)$, where $D = 2100$. That is, D is maximized when $x = 0$, $y = 200$.

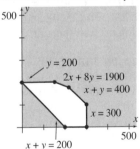

y = 200
2x + 8y = 1900
x + y = 400
x = 300
x + y = 200

23. $Z = 15.54$ when $x = 2.56$, $y = 6.74$

25. $Z = -75.98$ when $x = 9.48$, $y = 16.67$

Principles in Practice 7.3

1. Using the hint, the cost of shipping the TV sets is
   $Z = 18x + 24(25 - x) + 9y + 15(30 - y)$
   $= 1050 - 6x - 6y$.
   Since negative numbers of TV sets cannot be shipped, $x \geq 0$, $y \geq 0$, $25 - x \geq 0$, and $30 - y \geq 0$. Since warehouse C has only 45 TV sets, $x + y \leq 45$. Similarly, since warehouse D has only 40 TV sets, $25 - x + 30 - y \leq 45$ or $x + y \geq 10$.
   We need to minimize $Z = 1050 - 6x - 6y$ subject to the constraints
   $x + y \leq 45$,
   $x + y \geq 10$,
   $x \leq 25$,
   $y \leq 30$,
   $x \geq 0$, $y \geq 0$.

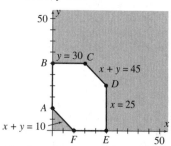

   The feasible region shown has corners
   $A = (0, 10)$, $B = (0, 30)$, $C = (15, 30)$,
   $D = (25, 20)$, $E = (25, 0)$, and $F = (10, 0)$.
   Evaluating the cost function at the corners gives
   $Z(A) = 1050 - 6(0) - 6(10) = 990$
   $Z(B) = 1050 - 6(0) - 6(30) = 870$
   $Z(C) = 1050 - 6(15) - 6(30) = 780$
   $Z(D) = 1050 - 6(25) - 6(20) = 780$
   $Z(E) = 1050 - 6(25) - 6(0) = 900$
   $Z(F) = 1050 - 6(10) - 6(0) = 990$
   The minimum value of Z is 780 which occurs at all points on the line segment joining C and D.
   This is $x = (1 - t)(15) + t(25) = 15 + 10t$ and $y = (1 - t)(30) + t(20) = 30 - 10t$ for $0 \leq t \leq 1$.
   Thus, ship $10t + 15$ TV sets from C to A,
   $-10t + 30$ TV sets from C to B,
   $25 - (10t + 15) = -10t + 10$ TV sets from D

to A, and $30 - (-10t + 30) = 10t$ TV sets from D to B, for $0 \leq t \leq 1$. The minimum cost is $780.

Problems 7.3

1. The feasible region is unbounded. Z is minimized at corner points $(2, 3)$ and $(5, 2)$, where its value is 33. Z is also minimized at all points on the line segment joining $(2, 3)$ and $(5, 2)$, so the solution is $Z = 33$ when
   $x = (1 - t)(2) + 5t = 2 + 3t$
   $y = (1 - t)(3) + 2t = 3 - t$ and $0 \leq t \leq 1$.

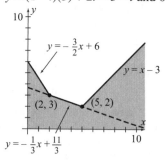

3. The feasible region appears below. The corner points are $(0, 0)$, $\left(0, \dfrac{8}{5}\right)$, $\left(\dfrac{36}{7}, \dfrac{4}{7}\right)$ and $(6, 0)$. Z is maximized at $\left(\dfrac{36}{7}, \dfrac{4}{7}\right)$ and $(6, 0)$, where its value is 84. Thus Z is also maximized at all points on the line segment joining $\left(\dfrac{36}{7}, \dfrac{4}{7}\right)$ and $(6, 0)$. The solution is
   $Z = 84$ when $x = (1 - t)\left(\dfrac{36}{7}\right) + 6t = \dfrac{6}{7}t + \dfrac{36}{7}$,
   $y = (1 - t)\left(\dfrac{4}{7}\right) + 0t = \dfrac{4}{7} - \dfrac{4}{7}t$ and $0 \leq t \leq 1$.

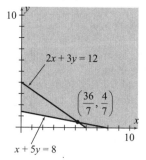

Principles in Practice 7.4

In these problems, the pivot entry is underlined.

1. Let x_1, x_2, and x_3 be the numbers of Type 1, Type 2, and Type 3 players, respectively, that the company produces. The situation is to maximize the profit
$P = 150x_1 + 250x_2 + 200x_3$, subject to the constraints
$$300x_1 + 300x_2 + 400x_3 \le 30,000$$
$$15x_1 + 15x_2 + 10x_3 \le 1200$$
$$2x_1 + 2x_2 + 3x_3 \le 180$$
$$x_1, x_2, x_3 \ge 0$$

The constraint inequalities can be simplified by dividing by the greatest common factor of the numbers involved. Thus, we will use
$$3x_1 + 3x_2 + 4x_3 \le 300$$
$$3x_1 + 3x_2 + 2x_3 \le 240$$
$$2x_1 + 2x_2 + 3x_3 \le 180$$
$$x_1, x_2, x_3 \ge 0$$

$$
\begin{array}{c|ccccccc|c}
 & x_1 & x_2 & x_3 & s_1 & s_2 & s_3 & P & b \\
\hline
s_1 & 3 & 3 & 4 & 1 & 0 & 0 & 0 & 300 \\
s_2 & 3 & \underline{3} & 2 & 0 & 1 & 0 & 0 & 240 \\
s_3 & 2 & 2 & 3 & 0 & 0 & 1 & 0 & 180 \\
\hline
P & -150 & -250 & -200 & 0 & 0 & 0 & 1 & 0
\end{array}
$$

$$
\begin{array}{c|ccccccc|c}
 & x_1 & x_2 & x_3 & s_1 & s_2 & s_3 & P & b \\
\hline
s_1 & 0 & 0 & 2 & 1 & -1 & 0 & 0 & 60 \\
x_2 & 1 & 1 & \frac{2}{3} & 0 & \frac{1}{3} & 0 & 0 & 80 \\
s_3 & 0 & 0 & \underline{\frac{5}{3}} & 0 & -\frac{2}{3} & 1 & 0 & 20 \\
\hline
P & 100 & 0 & -\frac{100}{3} & 0 & \frac{250}{3} & 0 & 1 & 20,000
\end{array}
$$

$$
\begin{array}{c|ccccccc|c}
 & x_1 & x_2 & x_3 & s_1 & s_2 & s_3 & P & b \\
\hline
s_1 & 0 & 0 & 0 & 1 & -\frac{1}{5} & -\frac{6}{5} & 0 & 36 \\
x_2 & 1 & 1 & 0 & 0 & \frac{3}{5} & -\frac{2}{5} & 0 & 72 \\
x_3 & 0 & 0 & 1 & 0 & -\frac{2}{5} & \frac{3}{5} & 0 & 12 \\
\hline
P & 100 & 0 & 0 & 0 & 70 & 20 & 1 & 20,400
\end{array}
$$

The maximum value of P is 20,400 when $x_1 = 0, x_2 = 72$, and $x_3 = 12$. The maximum profit is \$20,400 when 72 Type 2 players and 12 Type 3 players are produced and sold.

Problems 7.4

In these problems, the pivot entry is underlined.

1.
$$
\begin{array}{c|ccccc|cc}
 & x_1 & x_2 & s_1 & s_2 & Z & & \\
\hline
s_1 & 2 & 1 & 1 & 0 & 0 & 8 & 8 \\
s_2 & 2 & \underline{3} & 0 & 1 & 0 & 12 & 4 \\
\hline
Z & -1 & -2 & 0 & 0 & 1 & 0 &
\end{array}
$$

$$
\begin{array}{c|ccccc|c}
 & x_1 & x_2 & s_1 & s_2 & Z & \\
\hline
s_1 & \frac{4}{3} & 0 & 1 & -\frac{1}{3} & 0 & 4 \\
x_2 & \frac{2}{3} & 1 & 0 & \frac{1}{3} & 0 & 4 \\
\hline
Z & \frac{1}{3} & 0 & 0 & \frac{2}{3} & 1 & 8
\end{array}
$$

The solution is $Z = 8$ when $x_1 = 0, x_2 = 4$.

3.
$$
\begin{array}{c|ccccc|cc}
 & x_1 & x_2 & s_1 & s_2 & Z & & \\
\hline
s_1 & 3 & 2 & 1 & 0 & 0 & 5 & \frac{5}{2} \\
s_2 & -1 & \underline{3} & 0 & 1 & 0 & 3 & 1 \\
\hline
Z & 1 & -2 & 0 & 0 & 1 & 0 &
\end{array}
$$

$$
\begin{array}{c|ccccc|c}
 & x_1 & x_2 & s_1 & s_2 & Z & \\
\hline
s_1 & \frac{11}{3} & 0 & 1 & -\frac{2}{3} & 0 & 3 \\
x_2 & -\frac{1}{3} & 1 & 0 & \frac{1}{3} & 0 & 1 \\
\hline
Z & \frac{1}{3} & 0 & 0 & \frac{2}{3} & 1 & 2
\end{array}
$$

The solution is $Z = 2$ when $x_1 = 0, x_2 = 1$.

5.

	x_1	x_2	s_1	s_2	s_3	Z		
s_1	1	−1	1	0	0	0	1	1
s_2	1	2	0	1	0	0	8	8
s_3	1	1	0	0	1	0	5	5
Z	−8	−2	0	0	0	1	0	

	x_1	x_2	s_1	s_2	s_3	Z		
x_1	1	−1	1	0	0	0	1	
s_2	0	3	−1	1	0	0	7	$\frac{7}{3}$
s_3	0	2	−1	0	1	0	4	2
Z	0	−10	8	0	0	1	8	

	x_1	x_2	s_1	s_2	s_3	Z	
x_1	1	0	$\frac{1}{2}$	0	$\frac{1}{2}$	0	3
s_2	0	0	$\frac{1}{2}$	1	$-\frac{3}{2}$	0	1
x_2	0	1	$-\frac{1}{2}$	0	$\frac{1}{2}$	0	2
Z	0	0	3	0	5	1	28

The solution is $Z = 28$ when $x_1 = 3, x_2 = 2$.

7.

	x_1	x_2	x_3	s_1	s_2	Z		
s_1	1	2	0	1	0	0	10	5
s_2	2	2	1	0	1	0	10	5
Z	−3	−4	$-\frac{3}{2}$	0	0	1	0	

choosing s_2 as departing variable

	x_1	x_2	x_3	s_1	s_2	Z	
s_1	−1	0	−1	1	−1	0	0
x_2	1	1	$\frac{1}{2}$	0	$\frac{1}{2}$	0	5
Z	1	0	$\frac{1}{2}$	0	2	1	20

The solution is $Z = 20$ when
$x_1 = 0, x_2 = 5, x_3 = 0$

9. To obtain a standard linear programming problem, we write the second constraint as $-x_1 + 2x_2 + x_3 \le 2$.

	x_1	x_2	x_3	s_1	s_2	Z		
s_1	1	1	0	1	0	0	1	1
s_2	−1	2	1	0	1	0	2	
Z	−2	−1	1	0	0	1	0	

	x_1	x_2	x_3	s_1	s_2	Z	
x_1	1	1	0	1	0	0	1
s_2	0	3	1	1	1	0	3
Z	0	1	1	2	0	1	2

The solution is $Z = 2$ when
$x_1 = 1, x_2 = 0, x_3 = 0$.

11.

	x_1	x_2	s_1	s_2	s_3	s_4	Z		
s_1	2	−1	1	0	0	0	0	4	2
s_2	−1	2	0	1	0	0	0	6	
s_3	5	3	0	0	1	0	0	20	4
s_4	2	1	0	0	0	1	0	10	5
Z	−1	−1	0	0	0	0	1	0	

choosing x_1 as entering variable

	x_1	x_2	s_1	s_2	s_3	s_4	Z		
x_1	1	$-\frac{1}{2}$	$\frac{1}{2}$	0	0	0	0	2	
s_2	0	$\frac{3}{2}$	$\frac{1}{2}$	1	0	0	0	8	$\frac{16}{3}$
s_3	0	$\frac{11}{2}$	$-\frac{5}{2}$	0	1	0	0	10	$\frac{20}{11}$
s_4	0	2	−1	0	0	1	0	6	3
Z	0	$-\frac{3}{2}$	$\frac{1}{2}$	0	0	0	1	2	

	x_1	x_2	s_1	s_2	s_3	s_4	Z		
x_1	1	0	$\frac{3}{11}$	0	$\frac{1}{11}$	0	0	$\frac{32}{11}$	$\frac{32}{3}$
s_2	0	0	$\frac{13}{11}$	1	$-\frac{3}{11}$	0	0	$\frac{58}{11}$	$\frac{58}{13}$
x_2	0	1	$-\frac{5}{11}$	0	$\frac{2}{11}$	0	0	$\frac{20}{11}$	
s_4	0	0	$-\frac{1}{11}$	0	$-\frac{4}{11}$	1	0	$\frac{26}{11}$	
Z	0	0	$-\frac{2}{11}$	0	$\frac{3}{11}$	0	1	$\frac{52}{11}$	

$$\begin{array}{c} \\ x_1 \\ s_1 \\ x_2 \\ s_4 \\ Z \end{array} \begin{array}{cccccccc} x_1 & x_2 & s_1 & s_2 & s_3 & s_4 & Z & \\ \left[\begin{array}{ccccccc|c} 1 & 0 & 0 & -\frac{3}{13} & \frac{2}{13} & 0 & 0 & \frac{22}{13} \\ 0 & 0 & 1 & \frac{11}{13} & -\frac{3}{13} & 0 & 0 & \frac{58}{13} \\ 0 & 1 & 0 & \frac{5}{13} & \frac{1}{13} & 0 & 0 & \frac{50}{13} \\ 0 & 0 & 0 & \frac{1}{13} & -\frac{5}{13} & 1 & 0 & \frac{36}{13} \\ \hline 0 & 0 & 0 & \frac{2}{13} & \frac{3}{13} & 0 & 1 & \frac{72}{13} \end{array}\right] \end{array}$$

Thus the maximum value of Z is $\frac{72}{13}$, when

$x_1 = \frac{22}{13}$, $x_2 = \frac{50}{13}$. If we choose x_2 as the entering variable, then we have:

$$\begin{array}{c} \\ s_1 \\ s_2 \\ s_3 \\ s_4 \\ Z \end{array} \begin{array}{cccccccc} x_1 & x_2 & s_1 & s_2 & s_3 & s_4 & Z & \\ \left[\begin{array}{ccccccc|c} 2 & -1 & 1 & 0 & 0 & 0 & 0 & 4 \\ -1 & 2 & 0 & 1 & 0 & 0 & 0 & 6 \\ 5 & 3 & 0 & 0 & 1 & 0 & 0 & 20 \\ 2 & 1 & 0 & 0 & 0 & 1 & 0 & 10 \\ \hline -1 & -1 & 0 & 0 & 0 & 0 & 1 & 0 \end{array}\right] \end{array} \begin{array}{c} \\ 3 \\ \frac{20}{3} \\ 10 \\ \\ \end{array}$$

$$\begin{array}{c} \\ s_1 \\ x_2 \\ s_3 \\ s_4 \\ Z \end{array} \begin{array}{cccccccc} x_1 & x_2 & s_1 & s_2 & s_3 & s_4 & Z & \\ \left[\begin{array}{ccccccc|c} \frac{3}{2} & 0 & 1 & \frac{1}{2} & 0 & 0 & 0 & 7 \\ -\frac{1}{2} & 1 & 0 & \frac{1}{2} & 0 & 0 & 0 & 3 \\ \frac{13}{2} & 0 & 0 & -\frac{3}{2} & 1 & 0 & 0 & 11 \\ \frac{5}{2} & 0 & 0 & -\frac{1}{2} & 0 & 1 & 0 & 7 \\ \hline -\frac{3}{2} & 0 & 0 & \frac{1}{2} & 0 & 0 & 1 & 3 \end{array}\right] \end{array} \begin{array}{c} \frac{14}{3} \\ \\ \frac{22}{13} \\ \frac{14}{5} \\ \\ \end{array}$$

$$\begin{array}{c} \\ s_1 \\ x_2 \\ x_1 \\ s_4 \\ Z \end{array} \begin{array}{cccccccc} x_1 & x_2 & s_1 & s_2 & s_3 & s_4 & Z & \\ \left[\begin{array}{ccccccc|c} 0 & 0 & 1 & \frac{11}{13} & -\frac{3}{13} & 0 & 0 & \frac{58}{13} \\ 0 & 1 & 0 & \frac{5}{13} & \frac{1}{13} & 0 & 0 & \frac{50}{13} \\ 1 & 0 & 0 & -\frac{3}{13} & \frac{2}{13} & 0 & 0 & \frac{22}{13} \\ 0 & 0 & 0 & \frac{1}{13} & -\frac{5}{13} & 1 & 0 & \frac{36}{13} \\ \hline 0 & 0 & 0 & \frac{2}{13} & \frac{3}{13} & 0 & 1 & \frac{72}{13} \end{array}\right] \end{array}$$

The solution is $Z = \frac{72}{13}$ when $x_1 = \frac{22}{13}$,

$x_2 = \frac{50}{13}$.

13. To obtain a standard linear programming problem, we write the second constraint as $-x_1 - x_2 + x_3 \le 2$ and the third constraint as $x_1 - x_2 - x_3 \le 1$.

$$\begin{array}{c} \\ s_1 \\ s_2 \\ s_3 \\ W \end{array} \begin{array}{ccccccccc} x_1 & x_2 & x_3 & s_1 & s_2 & s_3 & W & \\ \left[\begin{array}{ccccccc|c} 4 & 3 & -1 & 1 & 0 & 0 & 0 & 1 \\ -1 & -1 & 1 & 0 & 1 & 0 & 0 & 2 \\ 1 & -1 & -1 & 0 & 0 & 1 & 0 & 1 \\ \hline -1 & 12 & -4 & 0 & 0 & 0 & 1 & 0 \end{array}\right] \end{array} \begin{array}{c} \\ 2 \\ \\ \\ \end{array}$$

$$\begin{array}{c} \\ s_1 \\ x_3 \\ s_3 \\ W \end{array} \begin{array}{ccccccccc} x_1 & x_2 & x_3 & s_1 & s_2 & s_3 & W & \\ \left[\begin{array}{ccccccc|c} 3 & 2 & 0 & 1 & 1 & 0 & 0 & 3 \\ -1 & -1 & 1 & 0 & 1 & 0 & 0 & 2 \\ 0 & -2 & 0 & 0 & 1 & 1 & 0 & 3 \\ \hline -5 & 8 & 0 & 0 & 4 & 0 & 1 & 8 \end{array}\right] \end{array} \begin{array}{c} 1 \\ \\ \\ \end{array}$$

$$\begin{array}{c} \\ x_1 \\ x_3 \\ s_3 \\ W \end{array} \begin{array}{ccccccccc} x_1 & x_2 & x_3 & s_1 & s_2 & s_3 & W & \\ \left[\begin{array}{ccccccc|c} 1 & \frac{2}{3} & 0 & \frac{1}{3} & \frac{1}{3} & 0 & 0 & 1 \\ 0 & -\frac{1}{3} & 1 & \frac{1}{3} & \frac{4}{3} & 0 & 0 & 3 \\ 0 & -2 & 0 & 0 & 1 & 1 & 0 & 3 \\ \hline 0 & \frac{34}{3} & 0 & \frac{5}{3} & \frac{17}{3} & 0 & 1 & 13 \end{array}\right] \end{array}$$

The solution is $W = 13$ when $x_1 = 1, x_2 = 0, x_3 = 3$.

15.

	x_1	x_2	x_3	x_4	s_1	s_2	s_3	s_4	Z		
s_1	1	−2	0	0	1	0	0	0	0	2	
s_2	1	1	0	0	0	1	0	0	0	5	
s_3	0	0	<u>1</u>	1	0	0	1	0	0	4	4
s_4	0	0	1	−2	0	0	0	1	0	7	7
Z	−60	0	−90	0	0	0	0	0	1	0	

	x_1	x_2	x_3	x_4	s_1	s_2	s_3	s_4	Z		
s_1	<u>1</u>	−2	0	0	1	0	0	0	0	2	2
s_2	1	1	0	0	0	1	0	0	0	5	5
x_3	0	0	1	1	0	0	1	0	0	4	
s_4	0	0	0	−3	0	0	−1	1	0	3	
Z	−60	0	0	90	0	0	90	0	1	360	

	x_1	x_2	x_3	x_4	s_1	s_2	s_3	s_4	Z		
x_1	1	−2	0	0	1	0	0	0	0	2	
s_2	0	<u>3</u>	0	0	−1	1	0	0	0	3	1
x_3	0	0	1	1	0	0	1	0	0	4	
s_4	0	0	0	−3	0	0	−1	1	0	3	
Z	0	−120	0	90	60	0	90	0	1	480	

	x_1	x_2	x_3	x_4	s_1	s_2	s_3	s_4	Z	
x_1	1	0	0	0	$\frac{1}{3}$	$\frac{2}{3}$	0	0	0	4
x_2	0	1	0	0	$-\frac{1}{3}$	$\frac{1}{3}$	0	0	0	1
x_3	0	0	1	1	0	0	1	0	0	4
s_4	0	0	0	−3	0	0	−1	1	0	3
Z	0	0	0	90	20	40	90	0	1	600

The solution is $Z = 600$ for $x_1 = 4, x_2 = 1, x_3 = 4, x_4 = 0$.

17. Let x_1 and x_2 denote the numbers of boxes transported from A and B, respectively. The revenue received is $R = 0.75x_1 + 0.50x_2$. We want to maximize R subject to

$$2x_1 + x_2 \le 2400 \quad \text{(volume)},$$
$$3x_1 + 5x_2 \le 36{,}800 \ \text{(weight)},$$
$$x_1, x_2 \ge 0.$$

	x_1	x_2	s_1	s_2	R		
s_1	<u>2</u>	1	1	0	0	2400	1200
s_2	3	5	0	1	0	36,800	$12,266\frac{2}{3}$
R	$-\frac{3}{4}$	$-\frac{1}{2}$	0	0	1	0	

$$\begin{array}{c} \begin{array}{ccccc} x_1 & x_2 & s_1 & s_2 & R \end{array} \\ \begin{array}{c} x_1 \\ s_2 \\ R \end{array} \left[\begin{array}{ccccc|c} 1 & \frac{1}{2} & \frac{1}{2} & 0 & 0 & 1200 \\ 0 & \frac{7}{2} & -\frac{3}{2} & 1 & 0 & 33,200 \\ \hline 0 & -\frac{1}{8} & \frac{3}{8} & 0 & 1 & 900 \end{array} \right] \begin{array}{c} 2400 \\ 9485\frac{5}{7} \\ \\ \end{array} \end{array}$$

$$\begin{array}{c} \begin{array}{ccccc} x_1 & x_2 & s_1 & s_2 & R \end{array} \\ \begin{array}{c} x_2 \\ s_2 \\ R \end{array} \left[\begin{array}{ccccc|c} 2 & 1 & 1 & 0 & 0 & 2400 \\ -7 & 0 & -5 & 1 & 0 & 24,800 \\ \hline \frac{1}{4} & 0 & \frac{1}{2} & 0 & 1 & 1200 \end{array} \right] \end{array}$$

Thus 0 boxes from A and 2400 from B give a maximum revenue of $1200.

19. Let x_1, x_2, and x_3 denote the numbers of chairs, rockers, and chaise lounges produced, respectively. We want to maximize

$R = 21x_1 + 24x_2 + 36x_3$ subject to

$$\begin{aligned} x_1 + x_2 + x_3 &\le 400, \\ x_1 + x_2 + 2x_3 &\le 500, \\ 2x_1 + 3x_2 + 5x_3 &\le 1450, \\ x_1, x_2, x_3 &\ge 0. \end{aligned}$$

$$\begin{array}{c} \begin{array}{ccccccc} x_1 & x_2 & x_3 & s_1 & s_2 & s_3 & R \end{array} \\ \begin{array}{c} s_1 \\ s_2 \\ s_3 \\ R \end{array} \left[\begin{array}{ccccccc|c} 1 & 1 & 1 & 1 & 0 & 0 & 0 & 400 \\ 1 & 1 & 2 & 0 & 1 & 0 & 0 & 500 \\ 2 & 3 & 5 & 0 & 0 & 1 & 0 & 1450 \\ \hline -21 & -24 & -36 & 0 & 0 & 0 & 1 & 0 \end{array} \right] \begin{array}{c} 400 \\ 250 \\ 290 \\ \\ \end{array} \end{array}$$

$$\begin{array}{c} \begin{array}{ccccccc} x_1 & x_2 & x_3 & s_1 & s_2 & s_3 & R \end{array} \\ \begin{array}{c} s_1 \\ x_3 \\ s_3 \\ R \end{array} \left[\begin{array}{ccccccc|c} \frac{1}{2} & \frac{1}{2} & 0 & 1 & -\frac{1}{2} & 0 & 0 & 150 \\ \frac{1}{2} & \frac{1}{2} & 1 & 0 & \frac{1}{2} & 0 & 0 & 250 \\ -\frac{1}{2} & \frac{1}{2} & 0 & 0 & -\frac{5}{2} & 1 & 0 & 200 \\ \hline -3 & -6 & 0 & 0 & 18 & 0 & 1 & 9000 \end{array} \right] \begin{array}{c} 300 \\ 500 \\ 400 \\ \\ \end{array} \end{array}$$

$$\begin{array}{c} \begin{array}{ccccccc} x_1 & x_2 & x_3 & s_1 & s_2 & s_3 & R \end{array} \\ \begin{array}{c} x_2 \\ x_3 \\ s_3 \\ R \end{array} \left[\begin{array}{ccccccc|c} 1 & 1 & 0 & 2 & -1 & 0 & 0 & 300 \\ 0 & 0 & 1 & -1 & 1 & 0 & 0 & 100 \\ -1 & 0 & 0 & -1 & -2 & 1 & 0 & 50 \\ \hline 3 & 0 & 0 & 12 & 12 & 0 & 1 & 10,800 \end{array} \right] \end{array}$$

The production of 0 chairs, 300 rockers, and 100 chaise lounges gives the maximum revenue of $10,800.

Principles in Practice 7.5

1. Let x_1, x_2, x_3 be the numbers of device 1, device 2, and device 3, respectively, that the company produces. The situation is to maximize the profit $P = 50x_1 + 50x_2 + 50x_3$ subject to the constraints

$$5.5x_1 + 5.5x_2 + 6.5x_3 \leq 190,$$
$$3.5x_1 + 6.5x_2 + 7.5x_3 \leq 180,$$
$$4.5x_1 + 6.0x_2 + 6.5x_3 \leq 165,$$

and $x_1, x_2, x_3 \geq 0$.

The matrices are shown rounded to 2 decimal places, although the exact values are used in the row operations.

Since the indicators are equal, we choose the first column as the pivot column.

	x_1	x_2	x_3	s_1	s_2	s_3	P	b
s_1	5.5	5.5	6.5	1	0	0	0	190
s_2	3.5	6.5	7.5	0	1	0	0	180
s_3	4.5	6.0	6.5	0	0	1	0	165
P	−50	−50	−50	0	0	0	1	0

	x_1	x_2	x_2	s_1	s_2	s_3	P	b
x_1	1	1	1.18	0.18	0	0	0	34.55
s_2	0	3	3.36	−0.64	1	0	0	59.09
s_3	0	1.50	1.18	−0.82	0	1	0	9.55
P	0	0	9.09	9.09	0	0	1	1727.27

An optimum solution is $x_1 = 35$, $x_2 = 0$, $x_3 = 0$, and $P = 1727$. However, x_2 is a nonbasic variable and its indicator is 0, so we check for multiple solutions. Treating x_2 as an entering variable, the following table is obtained:

	x_1	x_2	x_3	s_1	s_2	s_3	P	b
x_1	1	0	0.39	0.73	0	−0.67	0	28.18
s_2	0	0	1.00	1.00	1	−2.00	0	40.00
x_2	0	1	0.79	−0.55	0	0.67	0	6.36
P	0	0	9.09	9.09	0	0	1	1727.27

Another optimum solution is $x_1 = 28$, $x_2 = 6$, $x_3 = 0$, and $P = 1727$.

Thus, the optimum solution is for the company to produce $(1 - t)35 + 28t = 35 - 7t$ of device 1, $(1 - t)0 + 6t = 6t$ of device 2, and none of device 3, for $0 \leq t \leq 1$.

Problems 7.5

1. Yes; for the table, x_2 is the entering variable and the quotients $\dfrac{6}{2}$ and $\dfrac{3}{1}$ tie for being the smallest.

3.

	x_1	x_2	s_1	s_2	s_3	Z	
s_1	4	−3	1	0	0	0	4
s_2	3	−1	0	1	0	0	6
s_3	5	0	0	0	1	0	8
Z	−2	−7	0	0	0	1	0

The entering variable is x_2. Since no quotients exist, the problem has an unbounded solution. Thus, no optimum solution (unbounded).

5.

	x_1	x_2	s_1	s_2	s_3	Z		
s_1	2	−2	1	0	0	0	4	
s_2	−1	2	0	1	0	0	4	2
s_3	3	1	0	0	1	0	6	6
Z	4	−8	0	0	0	1	0	

	x_1	x_2	s_1	s_2	s_3	Z		
s_1	1	0	1	1	0	0	8	8
x_2	$-\frac{1}{2}$	1	0	$\frac{1}{2}$	0	0	2	
s_3	$\frac{7}{2}$	0	0	$-\frac{1}{2}$	1	0	4	$\frac{8}{7}$
Z	0	0	0	4	0	1	16	

Z has a maximum of 16 when $x_1 = 0$, $x_2 = 2$. Since x_1 is nonbasic for the last table and its indicator is 0, there may be multiple optimum solutions. Treating x_1 as an entering variable, we have

	x_1	x_2	s_1	s_2	s_3	Z	
s_1	0	0	1	$\frac{8}{7}$	$-\frac{2}{7}$	0	$\frac{48}{7}$
x_2	0	1	0	$\frac{3}{7}$	$\frac{1}{7}$	0	$\frac{18}{7}$
x_1	1	0	0	$-\frac{1}{7}$	$\frac{2}{7}$	0	$\frac{8}{7}$
Z	0	0	0	4	0	1	16

Here $Z = 16$ when $x_1 = \frac{8}{7}$, $x_2 = \frac{18}{7}$. Thus multiple optimum solutions exist. Hence Z is maximum when $x_1 = (1-t)(0) + \frac{8}{7}t = \frac{8}{7}t$,

$x_2 = (1-t)(2) + \frac{18}{7}t = 2 + \frac{4}{7}t$, and $0 \le t \le 1$.

For the last table s_3 is nonbasic and its indicator is 0. If we continue the process for determining other optimum solutions, we return to the second table.

7.

	x_1	x_2	x_3	s_1	s_2	s_3	Z		
s_1	9	3	−2	1	0	0	0	5	$\frac{5}{3}$
s_2	4	2	−1	0	1	0	0	2	1
s_3	1	−4	1	0	0	1	0	3	
Z	−5	−6	−1	0	0	0	1	0	

	x_1	x_2	x_3	s_1	s_2	s_3	Z	
s_1	3	0	$-\frac{1}{2}$	1	$-\frac{3}{2}$	0	0	2
x_2	2	1	$-\frac{1}{2}$	0	$\frac{1}{2}$	0	0	1
s_3	9	0	−1	0	2	1	0	7
Z	7	0	−4	0	3	0	1	6

For the last table, x_3 is the entering variable. Since no quotients exist, the problem has an unbounded solution. Thus, no optimum solution (unbounded).

9. To obtain a standard linear programming problem, we write the second constraint as $4x_1 + x_2 \le 6$.

	x_1	x_2	x_3	s_1	s_2	Z		
s_1	2	1	1	1	0	0	7	$\frac{7}{2}$
s_2	4	1	0	0	1	0	6	$\frac{3}{2}$
Z	−6	−2	−1	0	0	1	0	

	x_1	x_2	x_3	s_1	s_2	Z		
s_1	0	$\frac{1}{2}$	1	1	$-\frac{1}{2}$	0	4	4
x_1	1	$\frac{1}{4}$	0	0	$\frac{1}{4}$	0	$\frac{3}{2}$	
Z	0	$-\frac{1}{2}$	−1	0	$\frac{3}{2}$	1	9	

	x_1	x_2	x_3	s_1	s_2	Z		
x_3	0	$\frac{1}{2}$	1	1	$-\frac{1}{2}$	0	4	8
x_1	1	$\frac{1}{4}$	0	0	$\frac{1}{4}$	0	$\frac{3}{2}$	6
Z	0	0	0	1	1	1	13	

Z has a maximum value of 13 when

$x_1 = \dfrac{3}{2}, x_2 = 0, x_3 = 4$. Since x_2 is nonbasic for the last table and its indicator is 0, there may be multiple optimum solutions. Treating x_2 as an entering variable, we have

$$
\begin{array}{c}
\begin{array}{cccccc} x_1 & x_2\ x_3\ s_1 & & s_2 & Z \end{array} \\
\begin{array}{c} x_3 \\ x_2 \\ Z \end{array}
\left[
\begin{array}{cccccc|c}
-2 & 0 & 1 & 1 & -1 & 0 & 1 \\
4 & 1 & 0 & 0 & 1 & 0 & 6 \\
\hline
0 & 0 & 0 & 1 & 1 & 1 & 13
\end{array}
\right]
\end{array}
$$

Here $Z = 13$ when $x_1 = 0$, $x_2 = 6$, $x_3 = 1$. Thus multiple optimum solutions exist. Hence Z is maximum when

$$x_1 = (1-t)\left(\frac{3}{2}\right) + 0t = \frac{3}{2} - \frac{3}{2}t,$$

$$x_2 = (1-t)(0) + 6t = 6t,$$

$x_3 = (1-t)(4) + (1)t = 4 - 3t$, and $0 \le t \le 1$. For the last table, x_1 is nonbasic and its indicator is 0. If we continue the process for determining other optimum solutions, we return to the third table.

11. Let x_1, x_2, and x_3 denote the numbers of chairs, rockers, and chaise lounges produced, respectively. We want to maximize $R = 24x_1 + 32x_2 + 48x_3$ subject to

$x_1 + x_2 + x_3 \le 400,$

$x_1 + x_2 + 2x_3 \le 600,$

$2x_1 + 3x_2 + 5x_3 \le 1500,$

$x_1, x_2, x_3 \ge 0.$

$$
\begin{array}{c}
\begin{array}{ccccccc} x_1 & x_2 & x_3 & s_1 & s_2 & s_3 & R \end{array} \\
\begin{array}{c} s_1 \\ s_2 \\ s_3 \\ R \end{array}
\left[
\begin{array}{ccccccc|c}
1 & 1 & 1 & 1 & 0 & 0 & 0 & 400 \\
1 & 1 & 2 & 0 & 1 & 0 & 0 & 600 \\
2 & 3 & 5 & 0 & 0 & 1 & 0 & 1500 \\
\hline
-24 & -32 & -48 & 0 & 0 & 0 & 1 & 0
\end{array}
\right]
\begin{array}{c} 400 \\ 300 \\ 300 \\ \end{array}
\end{array}
$$

choosing s_3 as departing variable

$$
\begin{array}{c}
\begin{array}{ccccccc} x_1 & x_2 & x_3 & s_1 & s_2 & s_3 & R \end{array} \\
\begin{array}{c} s_1 \\ s_2 \\ x_3 \\ R \end{array}
\left[
\begin{array}{ccccccc|c}
\frac{3}{5} & \frac{2}{5} & 0 & 1 & 0 & -\frac{1}{5} & 0 & 100 \\
\frac{1}{5} & -\frac{1}{5} & 0 & 0 & 1 & -\frac{2}{5} & 0 & 0 \\
\frac{2}{5} & \frac{3}{5} & 1 & 0 & 0 & \frac{1}{5} & 0 & 300 \\
\hline
-\frac{24}{5} & -\frac{16}{5} & 0 & 0 & 0 & \frac{48}{5} & 1 & 14,400
\end{array}
\right]
\begin{array}{c} \frac{500}{3} \\ 0 \\ 750 \\ \end{array}
\end{array}
$$

$$\begin{array}{c} \quad\ x_1\ \ x_2\ \ x_3\ \ s_1\ \ s_2\ \ \ s_3\ \ \ R \\ \begin{array}{c} s_1 \\ x_1 \\ x_3 \\ R \end{array} \left[\begin{array}{ccccccc|cc} 0 & \underline{1} & 0 & 1 & -3 & 1 & 0 & 100 & 100 \\ 1 & -1 & 0 & 0 & 5 & -2 & 0 & 0 & 0 \\ 0 & 1 & 1 & 0 & -2 & 1 & 0 & 300 & 300 \\ \hdashline 0 & -8 & 0 & 0 & 24 & 0 & 1 & 14{,}400 \end{array} \right] \end{array}$$

$$\begin{array}{c} \quad\ x_1\ x_2\ x_3\ s_1\ \ s_2\ \ s_3\ \ \ R \\ \begin{array}{c} x_2 \\ x_1 \\ x_3 \\ R \end{array} \left[\begin{array}{ccccccc|cc} 0 & 1 & 0 & 1 & -3 & 1 & 0 & 100 & \\ 1 & 0 & 0 & 1 & \underline{2} & -1 & 0 & 100 & 50 \\ 0 & 0 & 1 & -1 & 1 & 0 & 0 & 200 & 200 \\ \hdashline 0 & 0 & 0 & 8 & 0 & 8 & 1 & 15{,}200 \end{array} \right] \end{array}$$

The maximum value of R is 15,200 when $x_1 = 100$, $x_2 = 100$, $x_3 = 200$. Since s_2 is nonbasic for the last table and its indicator is 0, there may be multiple optimum solutions. Treating s_2 as an entering variable, we have

$$\begin{array}{c} \quad\quad x_1\ \ x_2\ x_3\ \ s_1\ \ s_2\ \ \ s_3\ \ \ R \\ \begin{array}{c} x_2 \\ \\ s_2 \\ \\ x_3 \\ \\ R \end{array} \left[\begin{array}{ccccccc|c} \frac{3}{2} & 1 & 0 & \frac{5}{2} & 0 & -\frac{1}{2} & 0 & 250 \\ \frac{1}{2} & 0 & 0 & \frac{1}{2} & 1 & -\frac{1}{2} & 0 & 50 \\ -\frac{1}{2} & 0 & 1 & -\frac{3}{2} & 0 & \frac{1}{2} & 0 & 150 \\ \hdashline 0 & 0 & 0 & 8 & 0 & 8 & 1 & 15{,}200 \end{array} \right] \end{array}$$

Here $R = 15{,}200$ when $x_1 = 0$, $x_2 = 250$, $x_3 = 150$.

Thus multiple optimum solutions exist.

Hence R is maximum when

$$x_1 = (1-t)(100) + 0t = 100 - 100t,$$

$$x_2 = (1-t)(100) + 250t = 100 + 150t$$

$$x_3 = (1-t)(200) + 150t = 200 - 50t, \text{ and}$$

$0 \le t \le 1$. For the last table, x_1 is nonbasic and its indicator is 0. If we continue the process for determining other optimum solutions, we return to the fourth table. If we were to initially choose s_2 as the departing variable, then

$$\begin{array}{c} \quad\ x_1\ \ \ x_2\ \ \ x_3\ s_1\ s_2\ s_3\ \ R \\ \begin{array}{c} s_1 \\ s_2 \\ s_3 \\ R \end{array} \left[\begin{array}{ccccccc|cc} 1 & 1 & 1 & 1 & 0 & 0 & 0 & 400 & 400 \\ 1 & 1 & \underline{2} & 0 & 1 & 0 & 0 & 600 & 300 \\ 2 & 3 & 5 & 0 & 0 & 1 & 0 & 1500 & 300 \\ \hdashline -24 & -32 & -48 & 0 & 0 & 0 & 1 & 0 \end{array} \right] \end{array}$$

$$
\begin{array}{c}
\begin{array}{ccccccc} x_1 & x_2 & x_3 & s_1 & s_2 & s_3 & R \end{array}\\
\begin{array}{c} s_1 \\ x_3 \\ s_3 \\ R \end{array}
\left[\begin{array}{ccccccc|c|c}
\frac{1}{2} & \frac{1}{2} & 0 & 1 & -\frac{1}{2} & 0 & 0 & 100 & 200 \\
\frac{1}{2} & \frac{1}{2} & 1 & 0 & \frac{1}{2} & 0 & 0 & 300 & 600 \\
-\frac{1}{2} & \frac{1}{2} & 0 & 0 & -\frac{5}{2} & 1 & 0 & 0 & 0 \\
\hline
0 & -8 & 0 & 0 & 24 & 0 & 1 & 14,400 &
\end{array}\right]
\end{array}
$$

$$
\begin{array}{c}
\begin{array}{ccccccc} x_1 & x_2 & x_3 & s_1 & s_2 & s_3 & R \end{array}\\
\begin{array}{c} s_1 \\ x_3 \\ x_2 \\ R \end{array}
\left[\begin{array}{ccccccc|c|c}
1 & 0 & 0 & 1 & 2 & -1 & 0 & 100 & 50 \\
1 & 0 & 1 & 0 & 3 & -1 & 0 & 300 & 100 \\
-1 & 1 & 0 & 0 & -5 & 2 & 0 & 0 & \\
\hline
-8 & 0 & 0 & 0 & -16 & 16 & 1 & 14,400 &
\end{array}\right]
\end{array}
$$

$$
\begin{array}{c}
\begin{array}{ccccccc} x_1 & x_2 & x_3 & s_1 & s_2 & s_3 & R \end{array}\\
\begin{array}{c} s_2 \\ x_3 \\ x_2 \\ R \end{array}
\left[\begin{array}{ccccccc|c|c}
\frac{1}{2} & 0 & 0 & \frac{1}{2} & 1 & -\frac{1}{2} & 0 & 50 & 100 \\
-\frac{1}{2} & 0 & 1 & -\frac{3}{2} & 0 & \frac{1}{2} & 0 & 150 & \\
\frac{3}{2} & 1 & 0 & \frac{5}{2} & 0 & -\frac{1}{2} & 0 & 250 & \frac{500}{3} \\
\hline
0 & 0 & 0 & 8 & 0 & 8 & 1 & 15,200 &
\end{array}\right]
\end{array}
$$

the maximum value of R is 15,200 when $x_1 = 0, x_2 = 250, x_3 = 150$. For the last table, x_1 is nonbasic and its indicator is 0. Treating x_1 as an entering variable, we have

$$
\begin{array}{c}
\begin{array}{ccccccc} x_1 & x_2 & x_3 & s_1 & s_2 & s_3 & R \end{array}\\
\begin{array}{c} x_1 \\ x_3 \\ x_2 \\ R \end{array}
\left[\begin{array}{ccccccc|c}
1 & 0 & 0 & 1 & 2 & -1 & 0 & 100 \\
0 & 0 & 1 & -1 & 1 & 0 & 0 & 200 \\
0 & 1 & 0 & 1 & -3 & 1 & 0 & 100 \\
\hline
0 & 0 & 0 & 8 & 0 & 8 & 1 & 15,200
\end{array}\right]
\end{array}
$$

Here $R = 15,200$ when $x_1 = 100, x_2 = 100, x_3 = 200$. For the last table, s_2 is nonbasic and its indicator is 0. If we continue the process of determining other optimum solutions, we return to the table corresponding to the solution $x_1 = 0, x_2 = 250, x_3 = 150$.

Thus, the maximum revenue is $15,200 when $x_1 = 100 - 100t$, $x_2 = 100 + 150t$, $x_3 = 200 - 50t$, and $0 \le t \le 1$

Principles in Practice 7.6

1. Using the hint, $1000 - x_1$ standard and $800 - x_2$ deluxe snowboards must be manufactured at plant II. The constraints for plant I are $x_1 + x_2 \le 1200$ and $x_2 - x_1 \le 200$. The constraints for plant II are $(1000 - x_1) + (800 - x_2) \le 1000$ or $x_1 + x_2 \ge 800$. The quantity to be maximized is the profit
$P = 40x_1 + 60x_2 + 45(1000 - x_1) + 50(800 - x_2)$
$= -5x_1 + 10x_2 + 85,000$ subject to the constraints

$x_1 + x_2 \leq 1200,$

$-x_1 + x_2 \leq 200,$

$x_1 + x_2 \geq 800,$

and $x_1, x_2 \geq 0$.

Note that maximizing $Z = -5x_1 + 10x_2$ also maximizes the profit. The corresponding equations are:

$$x_1 + x_2 + s_1 = 1200,$$
$$-x_1 + x_2 + s_2 = 200,$$
$$x_1 + x_2 - s_3 + t = 800.$$

The artificial objective equation is $W = -5x_1 + 10x_2 - Mt$.

The augmented coefficient matrix is:

x_1	x_2	s_1	s_2	s_3	t	W	
1	1	1	0	0	0	0	1200
−1	1	0	1	0	0	0	200
1	1	0	0	−1	1	0	800
5	−10	0	0	0	M	1	0

The simplex tables follow.

	x_1	x_2	s_1	s_2	s_3	t	W	
s_1	1	1	1	0	0	0	0	1200
s_2	−1	1	0	1	0	0	0	200
t	1	1	0	0	−1	1	0	800
W	$5-M$	$-10-M$	0	0	M	0	1	$-800M$

	x_1	x_2	s_1	s_2	s_3	t	W	
s_1	2	0	1	−1	0	0	0	1000
x_2	−1	1	0	1	0	0	0	200
t	2	0	0	−1	−1	1	0	600
W	$-5-2M$	0	0	$10+M$	M	0	1	$2000-600M$

	x_1	x_2	s_1	s_2	s_3	t	W	
s_1	0	0	1	0	1	−1	0	400
x_2	0	1	0	$\frac{1}{2}$	$-\frac{1}{2}$	$\frac{1}{2}$	0	500
x_1	1	0	0	$-\frac{1}{2}$	$-\frac{1}{2}$	$\frac{1}{2}$	0	300
W	0	0	0	$\frac{15}{2}$	$-\frac{5}{2}$	$\frac{5}{2}+M$	1	3500

Delete the t-column since $t = 0$ and return to Z.

$$
\begin{array}{c}
\quad\;\; x_1\;\; x_2\;\; s_1\quad s_2\;\; s_3\;\; Z \\
\begin{array}{c}
s_3 \\[6pt]
x_2 \\[10pt]
x_1 \\[10pt]
Z
\end{array}
\left[
\begin{array}{cccccc|c}
0 & 0 & 1 & 0 & 1 & 0 & 400 \\[4pt]
0 & 1 & \frac{1}{2} & \frac{1}{2} & 0 & 0 & 700 \\[4pt]
1 & 0 & \frac{1}{2} & -\frac{1}{2} & 0 & 0 & 500 \\ \hline
0 & 0 & \frac{5}{2} & \frac{15}{2} & 0 & 0 & 4500
\end{array}
\right]
\end{array}
$$

Thus, $x_1 = 500$, $x_2 = 700$, and $Z = 4500$. Plant I should manufacture 500 standard and 700 deluxe snowboards. Plant II should manufacture $1000 - 500 = 500$ standard and $800 - 700 = 100$ deluxe snowboards. The maximum profit is $P = -5(500) + 10(700) + 85,000 = \$89,500$.

Problems 7.6

1.
$$
\begin{array}{c}
\quad x_1\;\; x_2\;\; s_1\;\; s_2\quad t_2\;\; W \\
\left[
\begin{array}{cccccc|c}
1 & 1 & 1 & 0 & 0 & 0 & 6 \\
-1 & 1 & 0 & -1 & 1 & 0 & 4 \\ \hline
-2 & -1 & 0 & 0 & M & 1 & 0
\end{array}
\right]
\end{array}
$$

$$
\begin{array}{c}
\qquad\quad x_1\qquad x_2\quad\; s_1\;\; s_2\; t_2\; W \\
\begin{array}{c}
s_1 \\[4pt]
t_2 \\[4pt]
W
\end{array}
\left[
\begin{array}{cccccc|c}
1 & 1 & 1 & 0 & 0 & 0 & 6 \\
-1 & \underline{1} & 0 & -1 & 1 & 0 & 4 \\ \hline
-2+M & -1-M & 0 & M & 0 & 1 & -4M
\end{array}
\right]
\begin{array}{c}
6 \\[4pt]
4 \\[4pt]
\;
\end{array}
\end{array}
$$

$$
\begin{array}{c}
\qquad x_1\;\; x_2\; s_1\; s_2\quad t_2\quad\; W \\
\begin{array}{c}
s_1 \\
x_2 \\
W
\end{array}
\left[
\begin{array}{cccccc|c}
\underline{2} & 0 & 1 & 1 & -1 & 0 & 2 \\
-1 & 1 & 0 & -1 & 1 & 0 & 4 \\ \hline
-3 & 0 & 0 & -1 & M+1 & 1 & 4
\end{array}
\right]
\begin{array}{c}
1 \\
\; \\
\;
\end{array}
\end{array}
$$

$$
\begin{array}{c}
\qquad x_1\; x_2\; s_1\quad s_2\;\; Z \\
\begin{array}{c}
x_1 \\[4pt]
x_2 \\[4pt]
Z
\end{array}
\left[
\begin{array}{ccccc|c}
1 & 0 & \frac{1}{2} & \frac{1}{2} & 0 & 1 \\[4pt]
0 & 1 & \frac{1}{2} & -\frac{1}{2} & 0 & 5 \\ \hline
0 & 0 & \frac{3}{2} & \frac{1}{2} & 1 & 7
\end{array}
\right]
\end{array}
$$

The maximum is $Z = 7$ when $x_1 = 1, x_2 = 5$.

3.

$$\begin{array}{c} \begin{array}{ccccccc} x_1 & x_2 & x_3 & s_1 & s_2 & t_2 & W \end{array} \\ \left[\begin{array}{ccccccc|c} 1 & 2 & 1 & 1 & 0 & 0 & 0 & 5 \\ -1 & 1 & 1 & 0 & -1 & 1 & 0 & 1 \\ \hline -2 & -1 & 1 & 0 & 0 & M & 1 & 0 \end{array}\right] \end{array}$$

$$\begin{array}{c} \begin{array}{ccccccc} x_1 & x_2 & x_3 & s_1 & s_2 & t_2 & W \end{array} \\ \begin{array}{c} s_1 \\ t_2 \\ W \end{array}\left[\begin{array}{ccccccc|cc} 1 & 2 & 1 & 1 & 0 & 0 & 0 & 5 & \frac{5}{2} \\ -1 & \underline{1} & 1 & 0 & -1 & 1 & 0 & 1 & 1 \\ \hline -2+M & -1-M & 1-M & 0 & M & 0 & 1 & -M & \end{array}\right] \end{array}$$

$$\begin{array}{c} \begin{array}{ccccccc} x_1 & x_2 & x_3 & s_1 & s_2 & t_2 & W \end{array} \\ \begin{array}{c} s_1 \\ x_2 \\ W \end{array}\left[\begin{array}{ccccccc|cc} \underline{3} & 0 & -1 & 1 & 2 & -2 & 0 & 3 & 1 \\ -1 & 1 & 1 & 0 & -1 & 1 & 0 & 1 & \\ \hline -3 & 0 & 2 & 0 & -1 & 1+M & 1 & 1 & \end{array}\right] \end{array}$$

$$\begin{array}{c} \begin{array}{cccccc} x_1 & x_2 & x_3 & s_1 & s_2 & Z \end{array} \\ \begin{array}{c} x_1 \\ x_2 \\ Z \end{array}\left[\begin{array}{cccccc|c} 1 & 0 & -\frac{1}{3} & \frac{1}{3} & \frac{2}{3} & 0 & 1 \\ 0 & 1 & \frac{2}{3} & \frac{1}{3} & -\frac{1}{3} & 0 & 2 \\ \hline 0 & 0 & 1 & 1 & 1 & 1 & 4 \end{array}\right] \end{array}$$

The maximum is $Z = 4$ when $x_1 = 1, x_2 = 2, x_3 = 0$.

5.

$$\begin{array}{c} \begin{array}{cccccc} x_1 & x_2 & x_3 & s_1 & t_2 & W \end{array} \\ \left[\begin{array}{cccccc|c} 1 & 1 & 1 & 1 & 0 & 0 & 10 \\ 1 & -1 & -1 & 0 & 1 & 0 & 6 \\ \hline -3 & -2 & -1 & 0 & M & 1 & 0 \end{array}\right] \end{array}$$

$$\begin{array}{c} \begin{array}{cccccc} x_1 & x_2 & x_3 & s_1 & t_2 & W \end{array} \\ \begin{array}{c} s_1 \\ t_2 \\ W \end{array}\left[\begin{array}{cccccc|cc} 1 & 1 & 1 & 1 & 0 & 0 & 10 & 10 \\ \underline{1} & -1 & -1 & 0 & 1 & 0 & 6 & 6 \\ \hline -3-M & -2+M & -1+M & 0 & 0 & 1 & -6M & \end{array}\right] \end{array}$$

$$\begin{array}{c} \begin{array}{cccccc} x_1 & x_2 & x_3 & s_1 & t_2 & W \end{array} \\ \begin{array}{c} s_1 \\ x_1 \\ W \end{array}\left[\begin{array}{cccccc|cc} 0 & \underline{2} & 2 & 1 & -1 & 0 & 4 & 2 \\ 1 & -1 & -1 & 0 & 1 & 0 & 6 & \\ \hline 0 & -5 & -4 & 0 & 3+M & 1 & 18 & \end{array}\right] \end{array}$$

$$
\begin{array}{c}
\begin{array}{ccccc} x_1 & x_2 & x_3 & s_1 & W \end{array}\\
\begin{array}{c} x_2 \\ x_1 \\ Z \end{array}
\left[\begin{array}{ccccc|c}
0 & 1 & 1 & \frac{1}{2} & 0 & 2 \\
1 & 0 & 0 & \frac{1}{2} & 0 & 8 \\
\hline
0 & 0 & 1 & \frac{5}{2} & 1 & 28
\end{array}\right]
\end{array}
$$

The maximum is $Z = 28$ when $x_1 = 8$, $x_2 = 2$, and $x_3 = 0$.

7.
$$
\begin{array}{c}
\begin{array}{ccccccc} x_1 & x_2 & s_1 & s_2 & s_3 & t_3 & W \end{array}\\
\left[\begin{array}{ccccccc|c}
1 & -1 & 1 & 0 & 0 & 0 & 0 & 1 \\
1 & 2 & 0 & 1 & 0 & 0 & 0 & 8 \\
1 & 1 & 0 & 0 & -1 & 1 & 0 & 5 \\
\hline
-1 & 10 & 0 & 0 & 0 & M & 1 & 0
\end{array}\right]
\end{array}
$$

$$
\begin{array}{c}
\begin{array}{ccccccc} x_1 & x_2 & s_1 & s_2 & s_3 & t_3 & W \end{array}\\
\begin{array}{c} s_1 \\ s_2 \\ t_3 \\ W \end{array}
\left[\begin{array}{ccccccc|c}
\underline{1} & -1 & 1 & 0 & 0 & 0 & 0 & 1 \\
1 & 2 & 0 & 1 & 0 & 0 & 0 & 8 \\
1 & 1 & 0 & 0 & -1 & 1 & 0 & 5 \\
\hline
-1-M & 10-M & 0 & 0 & M & 0 & 1 & -5M
\end{array}\right]
\begin{array}{c} 1 \\ 8 \\ 5 \end{array}
\end{array}
$$

$$
\begin{array}{c}
\begin{array}{ccccccc} x_1 & x_2 & s_1 & s_2 & s_3 & t_3 & W \end{array}\\
\begin{array}{c} x_1 \\ s_2 \\ t_3 \\ W \end{array}
\left[\begin{array}{ccccccc|c}
1 & -1 & 1 & 0 & 0 & 0 & 0 & 1 \\
0 & 3 & -1 & 1 & 0 & 0 & 0 & 7 \\
0 & \underline{2} & -1 & 0 & -1 & 1 & 0 & 4 \\
\hline
0 & 9-2M & 1+M & 0 & M & 0 & 1 & 1-4M
\end{array}\right]
\begin{array}{c} \\ \frac{7}{3} \\ 2 \end{array}
\end{array}
$$

$$
\begin{array}{c}
\begin{array}{ccccccc} x_1 & x_2 & s_1 & s_2 & s_3 & t_3 & W \end{array}\\
\begin{array}{c} x_1 \\ s_2 \\ x_2 \\ W \end{array}
\left[\begin{array}{ccccccc|c}
1 & 0 & \frac{1}{2} & 0 & -\frac{1}{2} & \frac{1}{2} & 0 & 3 \\
0 & 0 & \frac{1}{2} & 1 & \frac{3}{2} & -\frac{3}{2} & 0 & 1 \\
0 & 1 & -\frac{1}{2} & 0 & -\frac{1}{2} & \frac{1}{2} & 0 & 2 \\
\hline
0 & 0 & \frac{11}{2} & 0 & \frac{9}{2} & -\frac{9}{2}+M & 1 & -17
\end{array}\right]
\end{array}
$$

For the above table, $t_3 = 0$. Thus $W = Z$.

The maximum is $Z = -17$ when $x_1 = 3, x_2 = 2$.

9. We write the third constraint as $-x_1 + x_2 + x_3 \geq 6$.

x_1	x_2	x_3	s_1	s_2	s_3	t_2	t_3	W	
1	1	1	1	0	0	0	0	0	1
1	−1	1	0	−1	0	1	0	0	2
−1	1	1	0	0	−1	0	1	0	6
−3	2	−1	0	0	0	M	M	1	0

	x_1	x_2	x_3	s_1	s_2	s_3	t_2	t_3	W		
s_1	1	1	1	1	0	0	0	0	0	1	1
t_2	1	−1	1	0	−1	0	1	0	0	2	2
t_3	−1	1	1	0	0	−1	0	1	0	6	6
W	−3	2	$-1-2M$	0	M	M	0	0	1	$-8M$	

	x_1	x_2	x_3	s_1	s_2	s_3	t_2	t_3	W	
x_3	1	1	1	1	0	0	0	0	0	1
t_2	0	−2	0	−1	−1	0	1	0	0	1
t_3	−2	0	0	−1	0	−1	0	1	0	5
W	$-2+2M$	$3+2M$	0	$1+2M$	M	M	0	0	1	$1-6M$

There is no solution (empty feasible region).

11.

x_1	x_2	s_1	s_3	t_2	t_3	W	
1	−1	1	0	0	0	0	4
−1	1	0	0	1	0	0	4
1	0	0	−1	0	1	0	6
3	−2	0	0	M	M	1	0

	x_1	x_2	s_1	s_3	t_2	t_3	W		
s_1	1	−1	1	0	0	0	0	4	
t_2	−1	1	0	0	1	0	0	4	4
t_3	1	0	0	−1	0	1	0	6	
W	3	$-2-M$	0	M	0	0	1	$-10M$	

	x_1	x_2	s_1	s_3	t_2	t_3	W		
s_1	0	0	1	0	1	0	0	8	
x_2	−1	1	0	0	1	0	0	4	
t_3	1	0	0	−1	0	1	0	6	6
W	$1-M$	0	0	M	$2+M$	0	1	$8-6M$	

	x_1	x_2	s_1	s_3	t_2	t_3	W	
s_1	0	0	1	0	1	0	0	8
x_2	0	1	0	-1	1	1	0	10
x_1	1	0	0	-1	0	1	0	6
W	0	0	0	1	$2+M$	$-1+M$	1	2

For the above table, $t_2 = t_3 = 0$. Thus $W = Z$.

The maximum is $Z = 2$ when $x_1 = 6, x_2 = 10$.

13. Let x_1 and x_2 denote the numbers of Standard and Executive bookcases produced, respectively, each week. We want to maximize the profit function $P = 35x_1 + 40x_2$ subject to

$$2x_1 + 3x_2 \le 400,$$
$$3x_1 + 4x_2 \le 500,$$
$$3x_1 + 4x_2 \ge 250,$$
$$x_1, x_2 \ge 0.$$

The artificial objective function is $W = P - Mt_3$.

x_1	x_2	s_1	s_2	s_3	t_3	W	
2	3	1	0	0	0	0	400
3	4	0	1	0	0	0	500
3	4	0	0	-1	1	0	250
-35	-40	0	0	0	M	1	0

	x_1	x_2	s_1	s_2	s_3	t_3	W		
s_1	2	3	1	0	0	0	0	400	$\frac{400}{3}$
s_2	3	4	0	1	0	0	0	500	125
t_3	3	4	0	0	-1	1	0	250	$\frac{125}{2}$
W	$-35-3M$	$-40-4M$	0	0	M	0	1	$-250M$	

	x_1	x_2	s_1	s_2	s_3	t_3	W		
s_1	$-\frac{1}{4}$	0	1	0	$\frac{3}{4}$	$-\frac{3}{4}$	0	$\frac{425}{2}$	$\frac{850}{3}$
s_2	0	0	0	1	1	-1	0	250	250
x_2	$\frac{3}{4}$	1	0	0	$-\frac{1}{4}$	$\frac{1}{4}$	0	$\frac{125}{2}$	
W	-5	0	0	0	-10	$10+M$	1	2500	

$$\begin{array}{c}\\ s_1 \\ s_3 \\ x_2 \\ P\end{array}\begin{array}{cccccc}x_1 & x_2 & s_1 & s_2 & s_3 & P \\ \left[\begin{array}{cccccc}-\frac{1}{4} & 0 & 1 & -\frac{3}{4} & 0 & 0 \\ 0 & 0 & 0 & 1 & 1 & 0 \\ \frac{3}{4} & 1 & 0 & \frac{1}{4} & 0 & 0 \\ \hline -5 & 0 & 0 & 10 & 0 & 1\end{array}\right.\end{array}\begin{array}{c}25 \\ 250 \\ 125 \\ 5000\end{array}\begin{array}{c}\\ \\ \frac{500}{3}\end{array}$$

	x_1	x_2	s_1	s_2	s_3	P	
s_1	0	$\frac{1}{3}$	1	$-\frac{2}{3}$	0	0	$\frac{200}{3}$
s_3	0	0	0	1	1	0	250
x_1	1	$\frac{4}{3}$	0	$\frac{1}{3}$	0	0	$\frac{500}{3}$
P	0	$\frac{20}{3}$	0	$\frac{35}{3}$	0	1	$\frac{17{,}500}{3}$

This table indicates that, to maximize profit, the company should produce $\dfrac{500}{3} = 166\dfrac{2}{3}$ Standard and

0 Executive bookcases. Since an integer answer is preferable, note that $x_1 = 167$, $x_2 = 0$ does not satisfy the constraint $3x_1 + 4x_2 \le 500$, while $x_1 = 166$, $x_2 = 0$ satisfies all of the constraints. Thus the company should produce 166 Standard and 0 Executive bookcases each week.

15. Suppose I is the total investment. Let $x_1, x_2,$ and x_3 be the proportions invested in A, AA, and AAA bonds, respectively. If Z is the total annual yield expressed as a proportion of I, then $ZI = 0.08x_1I + 0.07x_2I + 0.06x_3I$, or equivalently, $Z = 0.08x_1 + 0.07x_2 + 0.06x_3$. We want to maximize Z subject to
$$x_1 + x_2 + x_3 = 1,$$
$$x_2 + x_3 \ge 0.50,$$
$$x_1 + x_2 \le 0.30,$$
$$x_1, x_2, x_3 \ge 0.$$
The artificial objective function is $W = Z - Mt_1 - Mt_2$.

x_1	x_2	x_3	s_2	s_3	t_1	t_2	W	
1	1	1	0	0	1	0	0	1
0	1	1	−1	0	0	1	0	0.5
1	1	0	0	1	0	0	0	0.3
−0.08	−0.07	−0.06	0	0	M	M	1	0

	x_1	x_2	x_3	s_2	s_3	t_1	t_2	W		
t_1	1	1	1	0	0	1	0	0	1	1
t_2	0	1	1	−1	0	0	1	0	0.5	0.5
s_3	1	1	0	0	1	0	0	0	0.3	0.3
W	$-0.08 - M$	$-0.07 - 2M$	$-0.06 - 2M$	M	0	0	0	1	$-1.5M$	

$$
\begin{array}{c}
\begin{array}{ccccccccc} & x_1 & x_2 & x_3 & s_2 & s_3 & t_1 & t_2 & W \end{array}\\
\begin{array}{c} t_1 \\ t_2 \\ x_2 \\ \hline W \end{array}
\left[\begin{array}{cccccccc|c}
0 & 0 & 1 & 0 & -1 & 1 & 0 & 0 & 0.7 \\
-1 & 0 & \underline{1} & -1 & -1 & 0 & 1 & 0 & 0.2 \\
1 & 1 & 0 & 0 & 1 & 0 & 0 & 0 & 0.3 \\
\hline
-0.01+M & 0 & -0.06-2M & M & 0.07+2M & 0 & 0 & 1 & 0.021-0.9M
\end{array}\right]\begin{array}{c}0.7\\0.2\\ \\ \\ \end{array}
\end{array}
$$

$$
\begin{array}{c}
\begin{array}{ccccccccc} & x_1 & x_2 & x_3 & s_2 & s_3 & t_1 & t_2 & W \end{array}\\
\begin{array}{c} t_1 \\ x_3 \\ x_2 \\ \hline W \end{array}
\left[\begin{array}{cccccccc|c}
1 & 0 & 0 & 1 & 0 & 1 & -1 & 0 & 0.5 \\
-1 & 0 & 1 & -1 & -1 & 0 & 1 & 0 & 0.2 \\
\underline{1} & 1 & 0 & 0 & 1 & 0 & 0 & 0 & 0.3 \\
\hline
-0.07-M & 0 & 0 & -0.06-M & 0.01 & 0 & 0.06+2M & 1 & 0.033-0.5M
\end{array}\right]\begin{array}{c}0.5\\ \\0.3\\ \\ \end{array}
\end{array}
$$

$$
\begin{array}{c}
\begin{array}{ccccccccc} & x_1 & x_2 & x_3 & s_2 & s_3 & t_1 & t_2 & W \end{array}\\
\begin{array}{c} t_1 \\ x_3 \\ x_1 \\ \hline W \end{array}
\left[\begin{array}{cccccccc|c}
0 & -1 & 0 & \underline{1} & -1 & 1 & -1 & 0 & 0.2 \\
0 & 1 & 1 & -1 & 0 & 0 & 1 & 0 & 0.5 \\
1 & 1 & 0 & 0 & 1 & 0 & 0 & 0 & 0.3 \\
\hline
0 & 0.07+M & 0 & -0.06-M & 0.08+M & 0 & 0.06+2M & 1 & 0.054-0.2M
\end{array}\right]\begin{array}{c}0.2\\ \\ \\ \end{array}
\end{array}
$$

$$
\begin{array}{c}
\begin{array}{ccccccccc} & x_1 & x_2 & x_3 & s_2 & s_3 & t_1 & t_2 & W \end{array}\\
\begin{array}{c} s_2 \\ x_3 \\ x_1 \\ \hline W \end{array}
\left[\begin{array}{cccccccc|c}
0 & -1 & 0 & 1 & -1 & 1 & -1 & 0 & 0.2 \\
0 & 0 & 1 & 0 & -1 & 1 & 0 & 0 & 0.7 \\
1 & 1 & 0 & 0 & 1 & 0 & 0 & 0 & 0.3 \\
\hline
0 & 0.01 & 0 & 0 & 0.02 & 0.06+M & M & 1 & 0.066
\end{array}\right]
\end{array}
$$

For the above table, $t_1 = t_2 = 0$. Thus $W = Z$.

The fund should put 30% in A bonds, 0% in AA, and 70% in AAA for a yield of 6.6%.

Problems 7.7

1.
$$
\begin{array}{c}
\begin{array}{ccccccc} x_1 & x_2 & s_1 & s_2 & t_1 & t_2 & W \end{array}\\
\left[\begin{array}{ccccccc|c}
1 & -1 & -1 & 0 & 1 & 0 & 0 & 7 \\
2 & 1 & 0 & -1 & 0 & 1 & 0 & 9 \\
\hline
2 & 5 & 0 & 0 & M & M & 1 & 0
\end{array}\right]
\end{array}
$$

$$
\begin{array}{c}
\begin{array}{ccccccc} & x_1 & x_2 & s_1 & s_2 & t_1 & t_2 & W \end{array}\\
\begin{array}{c} t_1 \\ t_2 \\ \hline W \end{array}
\left[\begin{array}{ccccccc|c}
1 & -1 & -1 & 0 & 1 & 0 & 0 & 7 \\
\underline{2} & 1 & 0 & -1 & 0 & 1 & 0 & 9 \\
\hline
2-3M & 5 & M & M & 0 & 0 & 1 & -16M
\end{array}\right]\begin{array}{c}7\\ \frac{9}{2}\\ \\ \end{array}
\end{array}
$$

$$
\begin{array}{c}
\begin{array}{ccccccc}
x_1 & x_2 & s_1 & s_2 & t_1 & t_2 & W
\end{array}\\
\begin{array}{c}
t_1\\ x_1\\ W
\end{array}
\left[\begin{array}{ccccccc|c}
0 & -\frac{3}{2} & -1 & \frac{1}{2} & 1 & -\frac{1}{2} & 0 & \frac{5}{2}\\
1 & \frac{1}{2} & 0 & -\frac{1}{2} & 0 & \frac{1}{2} & 0 & \frac{9}{2}\\
\hdashline
0 & 4+\frac{3}{2}M & M & 1-\frac{1}{2}M & 0 & -1+\frac{3}{2}M & 1 & -9-\frac{5}{2}M
\end{array}\right]\,5
\end{array}
$$

$$
\begin{array}{c}
\begin{array}{ccccccc}
x_1 & x_2 & s_1 & s_2 & t_1 & t_2 & W
\end{array}\\
\begin{array}{c}
s_2\\ x_1\\ W
\end{array}
\left[\begin{array}{ccccccc|c}
0 & -3 & -2 & 1 & 2 & -1 & 0 & 5\\
1 & -1 & -1 & 0 & 1 & 0 & 0 & 7\\
\hdashline
0 & 7 & 2 & 0 & -2+M & M & 1 & -14
\end{array}\right]
\end{array}
$$

The minimum is $Z = 14$ when $x_1 = 7$, $x_2 = 0$.

3.
$$
\begin{array}{c}
\begin{array}{cccccc}
x_1 & x_2 & x_3 & s & t & W
\end{array}\\
\left[\begin{array}{ccccc|c}
1 & -1 & -1 & -1 & 1 & 0 & 18\\
\hdashline
12 & 6 & 3 & 0 & M & 1 & 0
\end{array}\right]
\end{array}
$$

$$
\begin{array}{c}
\begin{array}{cccccc}
x_1 & x_2 & x_3 & s & t & W
\end{array}\\
\begin{array}{c}
t\\ W
\end{array}
\left[\begin{array}{cccccc|c}
1 & -1 & -1 & -1 & 1 & 0 & 18\\
\hdashline
12-M & 6+M & 3+M & M & 0 & 1 & -18M
\end{array}\right]\,18
\end{array}
$$

$$
\begin{array}{c}
\begin{array}{cccccc}
x_1 & x_2 & x_3 & s & t & W
\end{array}\\
\begin{array}{c}
x_1\\ W
\end{array}
\left[\begin{array}{cccccc|c}
1 & -1 & -1 & -1 & 1 & 0 & 18\\
\hdashline
0 & 18 & 15 & 12 & -12+M & 1 & -216
\end{array}\right]
\end{array}
$$

The minimum is $Z = 216$ when $x_1 = 18, x_2 = 0, x_3 = 0$.

5. We write the second constraint as $-x_1 + x_3 \ge 4$.

$$
\begin{array}{c}
\begin{array}{cccccccc}
x_1 & x_2 & x_3 & s_1 & s_2 & s_3 & t_2 & W
\end{array}\\
\left[\begin{array}{cccccccc|c}
1 & 1 & 1 & 1 & 0 & 0 & 0 & 0 & 6\\
-1 & 0 & 1 & 0 & -1 & 0 & 1 & 0 & 4\\
0 & 1 & 1 & 0 & 0 & 1 & 0 & 0 & 5\\
\hdashline
2 & 3 & 1 & 0 & 0 & 0 & M & 1 & 0
\end{array}\right]
\end{array}
$$

$$
\begin{array}{c}
\begin{array}{cccccccc}
x_1 & x_2 & x_3 & s_1 & s_2 & s_3 & t_2 & W
\end{array}\\
\begin{array}{c}
s_1\\ t_2\\ s_3\\ W
\end{array}
\left[\begin{array}{cccccccc|c}
1 & 1 & 1 & 1 & 0 & 0 & 0 & 0 & 6\\
-1 & 0 & 1 & 0 & -1 & 0 & 1 & 0 & 4\\
0 & 1 & 1 & 0 & 0 & 1 & 0 & 0 & 5\\
\hdashline
2+M & 3 & 1-M & 0 & M & 0 & 0 & 1 & -4M
\end{array}\right]
\begin{array}{c}
6\\ 4\\ 5\\
\end{array}
\end{array}
$$

	x_1	x_2	x_3	s_1	s_2	s_3		t_2	W	
s_1	2	1	0	1	1	0		-1	0	2
x_3	-1	0	1	0	-1	0		1	0	4
s_3	1	1	0	0	1	1		-1	0	1
W	3	3	0	0	1	0		$-1+M$	1	-4

The minimum is $Z = 4$ when $x_1 = 0, x_2 = 0, x_3 = 4$.

7.

	x_1	x_2	x_3	s_3	t_1	t_2	W	
	1	2	1	0	1	0	0	4
	0	1	1	0	0	1	0	1
	1	1	0	1	0	0	0	6
	1	-1	-3	0	M	M	1	0

	x_1	x_2	x_3	s_3	t_1	t_2	W		
t_1	1	2	1	0	1	0	0	4	2
t_2	0	$\underline{1}$	1	0	0	1	0	1	1
s_3	1	1	0	1	0	0	0	6	6
W	$1-M$	$-1-3M$	$-3-2M$	0	0	0	1	$-5M$	

	x_1	x_2	x_3	s_3	t_1	t_2	W		
t_1	$\underline{1}$	0	-1	0	1	-2	0	2	2
x_2	0	1	1	0	0	1	0	1	
s_3	1	0	-1	1	0	-1	0	5	5
W	$1-M$	0	$-2+M$	0	0	$1+3M$	1	$1-2M$	

	x_1	x_2	x_3	s_3	t_1	t_2	W		
x_1	1	0	-1	0	1	-2	0	2	
x_2	0	1	$\underline{1}$	0	0	1	0	1	1
s_3	0	0	0	1	-1	1	0	3	
W	0	0	-1	0	$-1+M$	$3+M$	1	-1	

	x_1	x_2	x_3	s_3	$-Z$	
x_1	1	1	0	0	0	3
x_3	0	1	1	0	0	1
s_3	0	0	0	1	0	3
$-Z$	0	1	0	0	1	0

The minimum is $Z = 0$ when $x_1 = 3, x_2 = 0, x_3 = 1$.

9.

$$
\begin{array}{c}
\begin{array}{ccccccccc} x_1 & x_2 & x_3 & s_1 & s_2 & t_1 & t_2 & W \end{array} \\
\left[\begin{array}{cccccccc|c}
1 & 1 & 1 & -1 & 0 & 1 & 0 & 0 & 8 \\
-1 & 2 & 1 & 0 & -1 & 0 & 1 & 0 & 2 \\
\hdashline
1 & 8 & 5 & 0 & 0 & M & M & 1 & 0
\end{array}\right]
\end{array}
$$

$$
\begin{array}{c}
\begin{array}{ccccccccc} & x_1 & x_2 & x_3 & s_1 & s_2 & t_1 & t_2 & W \end{array} \\
\begin{array}{c} t_1 \\ t_2 \\ W \end{array}
\left[\begin{array}{cccccccc|c}
1 & 1 & 1 & -1 & 0 & 1 & 0 & 0 & 8 \\
-1 & 2 & 1 & 0 & -1 & 0 & 1 & 0 & 2 \\
\hdashline
1 & 8-3M & 5-2M & M & M & 0 & 0 & 1 & -10M
\end{array}\right]
\begin{array}{c} 8 \\ 1 \end{array}
\end{array}
$$

$$
\begin{array}{c}
\begin{array}{ccccccccc} & x_1 & x_2 & x_3 & s_1 & s_2 & t_1 & t_2 & W \end{array} \\
\begin{array}{c} t_1 \\ \\ x_2 \\ \\ W \end{array}
\left[\begin{array}{cccccccc|c}
\frac{3}{2} & 0 & \frac{1}{2} & -1 & \frac{1}{2} & 1 & -\frac{1}{2} & 0 & 7 \\
-\frac{1}{2} & 1 & \frac{1}{2} & 0 & -\frac{1}{2} & 0 & \frac{1}{2} & 0 & 1 \\
\hdashline
5-\frac{3}{2}M & 0 & 1-\frac{1}{2}M & M & 4-\frac{1}{2}M & 0 & -4+\frac{3}{2}M & 1 & -8-7M
\end{array}\right]
\begin{array}{c} \frac{14}{3} \end{array}
\end{array}
$$

$$
\begin{array}{c}
\begin{array}{ccccccccc} & x_1 & x_2 & x_3 & s_1 & s_2 & t_1 & t_2 & W \end{array} \\
\begin{array}{c} x_1 \\ \\ x_2 \\ \\ W \end{array}
\left[\begin{array}{cccccccc|c}
1 & 0 & \frac{1}{3} & -\frac{2}{3} & \frac{1}{3} & \frac{2}{3} & -\frac{1}{3} & 0 & \frac{14}{3} \\
0 & 1 & \frac{2}{3} & -\frac{1}{3} & -\frac{1}{3} & \frac{1}{3} & \frac{1}{3} & 0 & \frac{10}{3} \\
\hdashline
0 & 0 & -\frac{2}{3} & \frac{10}{3} & \frac{7}{3} & -\frac{10}{3}+M & -\frac{7}{3}+M & 1 & -\frac{94}{3}
\end{array}\right]
\begin{array}{c} 14 \\ 5 \end{array}
\end{array}
$$

$$
\begin{array}{c}
\begin{array}{ccccccccc} x_1 & x_2 & x_3 & s_1 & s_2 & t_1 & t_2 & W \end{array} \\
\begin{array}{c} x_1 \\ \\ x_3 \\ \\ W \end{array}
\left[\begin{array}{cccccccc|c}
1 & -\frac{1}{2} & 0 & -\frac{1}{2} & \frac{1}{2} & \frac{1}{2} & -\frac{1}{2} & 0 & 3 \\
0 & \frac{3}{2} & 1 & -\frac{1}{2} & -\frac{1}{2} & \frac{1}{2} & \frac{1}{2} & 0 & 5 \\
\hdashline
0 & 1 & 0 & 3 & 2 & -3+M & -2+M & 1 & -28
\end{array}\right]
\end{array}
$$

The minimum is $Z = 28$ when $x_1 = 3, x_2 = 0, x_3 = 5$.

11. Let $x_1, x_2,$ and x_3 denote the annual numbers of barrels of cement produced in kilns that use device A, device B, and no device, respectively. We want to minimize the annual emission control cost C (C in dollars) where

$$C = \frac{1}{4}x_1 + \frac{2}{5}x_2 + 0x_3 \text{ subject to}$$

$$x_1 + x_2 + x_3 = 3,300,000,$$

$$\frac{1}{2}x_1 + \frac{1}{4}x_2 + 2x_3 \leq 1,000,000,$$

$$x_1, x_2, x_3 \geq 0.$$

$$
\begin{array}{cccccc}
x_1 & x_2 & x_3 & s_2 & t_1 & W \\
\end{array}
$$

$$
\left[
\begin{array}{cccccc|c}
1 & 1 & 1 & 0 & 1 & 0 & 3,300,000 \\
\frac{1}{2} & \frac{1}{4} & 2 & 1 & 0 & 0 & 1,000,000 \\
\hline
\frac{1}{4} & \frac{2}{5} & 0 & 0 & M & 1 & 0
\end{array}
\right]
$$

$$
\begin{array}{ccccccc}
 & x_1 & x_2 & x_3 & s_2 & t_1 & W
\end{array}
$$

$$
\begin{array}{c}
t_1 \\
s_2 \\
W
\end{array}
\left[
\begin{array}{cccccc|c}
1 & 1 & 1 & 0 & 1 & 0 & 3,300,000 \\
\frac{1}{2} & \frac{1}{4} & 2 & 1 & 0 & 0 & 1,000,000 \\
\hline
\frac{1}{4}-M & \frac{2}{5}-M & -M & 0 & 0 & 1 & -3,300,000M
\end{array}
\right]
\begin{array}{c}
3,300,000 \\
500,000 \\
\end{array}
$$

$$
\begin{array}{ccccccc}
 & x_1 & x_2 & x_3 & s_2 & t_1 & W
\end{array}
$$

$$
\begin{array}{c}
t_1 \\
x_3 \\
W
\end{array}
\left[
\begin{array}{cccccc|c}
\frac{3}{4} & \frac{7}{8} & 0 & -\frac{1}{2} & 1 & 0 & 2,800,000 \\
\frac{1}{4} & \frac{1}{8} & 1 & \frac{1}{2} & 0 & 0 & 500,000 \\
\hline
\frac{1}{4}-\frac{3}{4}M & \frac{2}{5}-\frac{7}{8}M & 0 & \frac{1}{2}M & 0 & 1 & -2,800,000M
\end{array}
\right]
\begin{array}{c}
3,200,000 \\
4,000,000 \\
\end{array}
$$

$$
\begin{array}{cccccc}
 & x_1 & x_2 & x_3 & s_2 & t_1 & W
\end{array}
$$

$$
\begin{array}{c}
x_2 \\
x_3 \\
W
\end{array}
\left[
\begin{array}{cccccc|c}
\frac{6}{7} & 1 & 0 & -\frac{4}{7} & \frac{8}{7} & 0 & 3,200,000 \\
\frac{1}{7} & 0 & 1 & \frac{4}{7} & -\frac{1}{7} & 0 & 100,000 \\
\hline
-\frac{13}{140} & 0 & 0 & \frac{8}{35} & -\frac{16}{35}+M & 1 & -1,280,000
\end{array}
\right]
\begin{array}{c}
\frac{11,200,000}{3} \\
700,000 \\
\end{array}
$$

$$
\begin{array}{cccccc}
 & x_1 & x_2 & x_3 & s_2 & -C
\end{array}
$$

$$
\begin{array}{c}
x_2 \\
x_1 \\
-C
\end{array}
\left[
\begin{array}{ccccc|c}
0 & 1 & -6 & -4 & 0 & 2,600,000 \\
1 & 0 & 7 & 4 & 0 & 700,000 \\
\hline
0 & 0 & \frac{13}{20} & \frac{3}{5} & 1 & -1,215,000
\end{array}
\right]
$$

Thus the minimum value of C is 1,215,000 when $x_1 = 700,000$, $x_2 = 2,600,000$, $x_3 = 0$.

The plant should install device A on kilns producing 700,000 barrels annually, and device B on kilns producing 2,600,000 barrels annually.

13. Let x_1 = number of DVD players shipped from Akron to Columbus,

x_2 = number of DVD players shipped from Springfield to Columbus,

x_3 = number of DVD players shipped from Akron to Dayton,

x_4 = number of DVD players shipped from Springfield to Dayton.

We want to minimize $C = 5x_1 + 3x_2 + 7x_3 + 2x_4$ subject to

$$x_1 + x_2 = 150,$$
$$x_3 + x_4 = 150,$$
$$x_1 + x_3 \le 200,$$
$$x_2 + x_4 \le 150,$$
$$x_1, x_2, x_3, x_4 \ge 0.$$

x_1	x_2	x_3	x_4	s_3	s_4	t_1	t_2	W	
1	1	0	0	0	0	1	0	0	150
0	0	1	1	0	0	0	1	0	150
1	0	1	0	1	0	0	0	0	200
0	1	0	1	0	1	0	0	0	150
5	3	7	2	0	0	M	M	1	0

	x_1	x_2	x_3	x_4	s_3	s_4	t_1	t_2	W		
t_1	1	1	0	0	0	0	1	0	0	150	
t_2	0	0	1	$\underline{1}$	0	0	0	1	0	150	150
s_3	1	0	1	0	1	0	0	0	0	200	
s_4	0	1	0	1	0	1	0	0	0	150	150
W	$5-M$	$3-M$	$7-M$	$2-M$	0	0	0	0	1	$-300M$	

	x_1	x_2	x_3	x_4	s_3	s_4	t_1	t_2	W		
t_1	1	1	0	0	0	0	1	0	0	150	150
x_4	0	0	1	1	0	0	0	1	0	150	
s_3	1	0	1	0	1	0	0	0	0	200	
s_4	0	$\underline{1}$	−1	0	0	1	0	−1	0	0	0
W	$5-M$	$3-M$	5	0	0	0	0	$-2+M$	1	$-300-150M$	

	x_1	x_2	x_3	x_4	s_3	s_4	t_1	t_2	W		
t_1	$\underline{1}$	0	1	0	0	−1	1	1	0	150	150
x_4	0	0	1	1	0	0	0	1	0	150	
s_3	1	0	1	0	1	0	0	0	0	200	200
x_2	0	1	−1	0	0	1	0	−1	0	0	
W	$5-M$	0	$8-M$	0	0	$-3+M$	0	1	1	$-300-150M$	

	x_1	x_2	x_3	x_4	s_3	s_4	t_1	t_2	W	
x_1	1	0	1	0	0	−1	1	1	0	150
x_4	0	0	1	1	0	0	0	1	0	150
s_3	0	0	0	0	1	1	−1	−1	0	50
x_2	0	1	−1	0	0	1	0	−1	0	0
W	0	0	3	0	0	2	$-5+M$	$-4+M$	1	−1050

The retailer should ship as follows: to Columbus, 150 from Akron and 0 from Springfield; to Dayton, 0 from Akron and 150 from Springfield. The transportation cost is $1050.

If s_4 is chosen as the departing variable in the second table, the result is the same, although the final table is different:

	x_1	x_2	x_3	x_4	s_3	s_4	t_1	t_2	W	
x_1	1	1	0	0	0	0	1	0	0	150
x_3	0	-1	1	0	0	-1	0	1	0	0
s_3	0	0	0	0	1	1	-1	-1	0	50
x_4	0	1	0	1	0	1	0	0	0	150
W	0	3	0	0	0	5	$-5+M$	$-7+M$	1	-1050

15. a. Roll width $\begin{cases} 15" & 3 \quad 2 \quad 1 \quad 0 \\ 10" & 0 \quad 1 \quad 3 \quad 4 \end{cases}$

Trim loss $\quad 3 \quad 8 \quad \underline{3} \quad 8$

b. We want to minimize $L = 3x_1 + 8x_2 + 3x_3 + 8x_4$ subject to

$$3x_1 + 2x_2 + x_3 \geq 50,$$
$$x_2 + 3x_3 + 4x_4 \geq 60,$$
$$x_1, x_2, x_3, x_4 \geq 0.$$

x_1	x_2	x_3	x_4	s_1	s_2	t_1	t_2	W	
3	2	1	0	-1	0	1	0	0	50
0	1	3	4	0	-1	0	1	0	60
3	8	3	8	0	0	M	M	1	0

	x_1	x_2	x_3	x_4	s_1	s_2	t_1	t_2	W		
t_1	3	2	1	0	-1	0	1	0	0	50	50
t_2	0	1	3	4	0	-1	0	1	0	60	20
W	$3-3M$	$8-3M$	$3-4M$	$8-4M$	M	M	0	0	1	$-110M$	

	x_1	x_2	x_3	x_4	s_1	s_2	t_1	t_2	W		
t_1	3	$\frac{5}{3}$	0	$-\frac{4}{3}$	-1	$\frac{1}{3}$	1	$-\frac{1}{3}$	0	30	10
x_3	0	$\frac{1}{3}$	1	$\frac{4}{3}$	0	$-\frac{1}{3}$	0	$\frac{1}{3}$	0	20	
W	$3-3M$	$7-\frac{5}{3}M$	0	$4+\frac{4}{3}M$	M	$1-\frac{1}{3}M$	0	$-1+\frac{4}{3}M$	1	$-60-30M$	

	x_1	x_2	x_3	x_4	s_1	s_2	t_1	t_2	W	
x_1	1	$\frac{5}{9}$	0	$-\frac{4}{9}$	$-\frac{1}{3}$	$\frac{1}{9}$	$\frac{1}{3}$	$-\frac{1}{9}$	0	10
x_3	0	$\frac{1}{3}$	1	$\frac{4}{3}$	0	$-\frac{1}{3}$	0	$\frac{1}{3}$	0	20
W	0	$\frac{16}{3}$	0	$\frac{16}{3}$	1	$\frac{2}{3}$	$-1+M$	$-\frac{2}{3}+M$	1	-90

$x_1 = 10, x_2 = 0, x_3 = 20, x_4 = 0.$

c. 90 in.

Principles in Practice 7.8

1. Let x_1, x_2, and x_3 be the numbers respectively, of Type 1, Type 2, and Type 3 gadgets produced. The original problem is to maximize
$P = 300x_1 + 200x_2 + 200x_3$, subject to

$300x_1 + 220x_2 + 180x_3 \leq 60,000,$
$20x_1 + 40x_2 + 20x_3 \leq 2000,$
$3x_1 + x_2 + 2x_3 \leq 120,$
and $x_1, x_2, x_3 \geq 0.$

The dual problem is to minimize
$W = 60,000y_1 + 2000y_2 + 120y_3,$

subject to
$300y_1 + 20y_2 + 3y_3 \geq 300,$
$220y_1 + 40y_2 + y_3 \geq 200,$
$180y_1 + 20y_2 + 2y_3 \geq 200,$
and $y_1, y_2, y_3 \geq 0.$

3. Let x_1, x_2, and x_3 be the numbers, respectively, of devices 1, 2, and 3 produced.
The original problem is to maximize $P = 30x_1 + 20x_2 + 20x_3$, subject to
$30x_1 + 15x_2 + 10x_3 \leq 300$, $20x_1 + 30x_2 + 20x_3 \leq 400$, $40x_1 + 30x_2 + 25x_3 \leq 600$, and $x_1, x_2, x_3 \geq 0.$
The dual problem is to minimize $W = 300y_1 + 400y_2 + 600y_3$, subject to
$30y_1 + 20y_2 + 40y_3 \geq 30$, $15y_1 + 30y_2 + 30y_3 \geq 20$,
$10y_1 + 20y_2 + 25y_3 \geq 20$,
and $y_1, y_2, y_3 \geq 0.$
The tablex to maximize $Z = -W = -300y_1 - 400y_2 - 600y_3$ follow.

y_1	y_2	y_3	s_1	s_2	s_3	t_1	t_2	t_3	Z	
30	20	40	-1	0	0	1	0	0	0	30
15	30	30	0	-1	0	0	1	0	0	20
10	20	25	0	0	-1	0	0	1	0	20
300	400	600	0	0	0	M	M	M	1	0

$$\begin{array}{c} \\ t_1 \\ t_2 \\ t_3 \\ Z \end{array}
\begin{array}{c} y_1 \quad\quad y_2 \quad\quad y_3 \quad\ s_1 \ s_2 \ s_3 \ t_1 \ t_2 \ t_3 \ Z \\ \left[\begin{array}{ccccccccc|c} 30 & 20 & 40 & -1 & 0 & 0 & 1 & 0 & 0 & 0 & 30 \\ 15 & 30 & 30 & 0 & -1 & 0 & 0 & 1 & 0 & 0 & 20 \\ 10 & 20 & 25 & 0 & 0 & -1 & 0 & 0 & 1 & 0 & 20 \\ \hline 300-55M & 400-70M & 600-95M & M & M & M & 0 & 0 & 0 & 1 & -70M \end{array}\right] \end{array}$$

$$\begin{array}{c} \\ t_1 \\ y_3 \\ t_3 \\ Z \end{array}
\begin{array}{c} y_1 \quad\quad y_2 \quad\ y_3 \ s_1 \quad\ s_2 \quad\ s_3 \ t_1 \quad\ t_2 \quad\ t_3 \ Z \\ \left[\begin{array}{ccccccccc|c} 10 & -20 & 0 & -1 & \frac{4}{3} & 0 & 1 & -\frac{4}{3} & 0 & 0 & \frac{10}{3} \\ \frac{1}{2} & 1 & 1 & 0 & -\frac{1}{30} & 0 & 0 & \frac{1}{30} & 0 & 0 & \frac{2}{3} \\ -\frac{5}{2} & -5 & 0 & 0 & \frac{5}{6} & -1 & 0 & -\frac{5}{6} & 1 & 0 & \frac{10}{3} \\ \hline -\frac{15}{2}M & -200+25M & 0 & M & 20-\frac{13}{6}M & M & 0 & -20+\frac{19}{6}M & 0 & 1 & -400-\frac{20}{3}M \end{array}\right] \end{array}$$

$$\begin{array}{c} \\ y_1 \\ y_3 \\ t_3 \\ Z \end{array}
\begin{array}{c} y_1 \quad\ y_2 \quad\ y_3 \ s_1 \quad\ s_2 \quad s_3 \ t_1 \quad\ t_2 \quad t_3 \ Z \\ \left[\begin{array}{ccccccccc|c} 1 & -2 & 0 & -\frac{1}{10} & \frac{2}{15} & 0 & \frac{1}{10} & -\frac{2}{15} & 0 & 0 & \frac{1}{3} \\ 0 & 2 & 1 & \frac{1}{20} & -\frac{1}{10} & 0 & -\frac{1}{20} & \frac{1}{10} & 0 & 0 & \frac{1}{2} \\ 0 & -10 & 0 & -\frac{1}{4} & \frac{7}{6} & -1 & \frac{1}{4} & -\frac{7}{6} & 1 & 0 & \frac{25}{6} \\ \hline 0 & -200+10M & 0 & \frac{1}{4}M & 20-\frac{7}{6}M & M & \frac{3}{4}M & -20+\frac{13}{6}M & 0 & 1 & -400-\frac{25}{6}M \end{array}\right] \end{array}$$

$$\begin{array}{c} \\ s_2 \\ y_3 \\ t_3 \\ Z \end{array}
\begin{array}{c} y_1 \quad\quad\ y_2 \quad\ y_3 \quad\ s_1 \quad s_2 \ s_3 \quad\ t_1 \quad\quad t_2 \ t_3 \ Z \\ \left[\begin{array}{ccccccccc|c} \frac{15}{2} & -15 & 0 & -\frac{3}{4} & 1 & 0 & \frac{3}{4} & -1 & 0 & 0 & \frac{5}{2} \\ \frac{3}{4} & \frac{1}{2} & 1 & -\frac{1}{40} & 0 & 0 & \frac{1}{40} & 0 & 0 & 0 & \frac{3}{4} \\ -\frac{35}{4} & \frac{15}{2} & 0 & \frac{5}{8} & 0 & -1 & -\frac{5}{8} & 0 & 1 & 0 & \frac{5}{4} \\ \hline -150+\frac{35}{4}M & 100-\frac{15}{2}M & 0 & 15-\frac{5}{8}M & 0 & M & -15+\frac{13}{8}M & M & 0 & 1 & -450-\frac{5}{4}M \end{array}\right] \end{array}$$

$$\begin{array}{c} \\ s_2 \\ y_3 \\ y_2 \\ Z \end{array}
\begin{array}{c} y_1 \quad y_2\ y_3 \quad s_1 \quad s_2 \ s_3 \quad\ t_1 \quad\ t_2 \quad\quad t_3 \quad Z \\ \left[\begin{array}{ccccccccc|c} -10 & 0 & 0 & \frac{1}{2} & 1 & -2 & -\frac{1}{2} & -1 & 2 & 0 & 5 \\ \frac{4}{3} & 0 & 1 & -\frac{1}{15} & 0 & \frac{1}{15} & \frac{1}{15} & 0 & -\frac{1}{15} & 0 & \frac{2}{3} \\ -\frac{7}{6} & 1 & 0 & \frac{1}{12} & 0 & -\frac{2}{15} & -\frac{1}{12} & 0 & \frac{2}{15} & 0 & \frac{1}{6} \\ \hline -\frac{100}{3} & 0 & 0 & \frac{20}{3} & 0 & \frac{40}{3} & -\frac{20}{3}+M & M & -\frac{40}{3}+M & 1 & -\frac{1400}{3} \end{array}\right] \end{array}$$

The $t_1, t_2,$ and t_3 columns are no longer needed.

	y_1	y_2	y_3	s_1	s_2	s_3	Z	
s_2	0	0	$\frac{15}{2}$	0	1	$-\frac{3}{2}$	0	10
y_1	1	0	$\frac{3}{4}$	$-\frac{1}{20}$	0	$\frac{1}{20}$	0	$\frac{1}{2}$
y_2	0	1	$\frac{7}{8}$	$\frac{1}{40}$	0	$-\frac{3}{40}$	0	$\frac{3}{4}$
Z	0	0	25	5	0	15	1	-450

From this table, the maximum profit of $450 corresponds to $x_1 = 5$, $x_2 = 0$, and $x_3 = 15$. The company should produce 5 of device 1 and 15 of device 3.

Problems 7.8

1. Minimize $W = 5y_1 + 3y_2$

   subject to $y_1 - y_2 \geq 1$
   $$y_1 + y_2 \geq 2$$
   $$y_1, y_2 \geq 0$$

3. Maximize $W = 8y_1 + 2y_2$ subject to

   $y_1 - y_2 \leq 1,$
   $y_1 + 2y_2 \leq 8,$
   $y_1 + y_2 \leq 5,$
   $y_1, y_2 \geq 0.$

5. The second and third constraints can be written as $x_1 - x_2 \leq -3$ and $-x_1 - x_2 \leq -11$.
   Minimize $W = 13y_1 - 3y_2 - 11y_3$ subject to

   $-y_1 + y_2 - y_3 \geq 1,$
   $2y_1 - y_2 - y_3 \geq -1,$
   $y_1, y_2, y_3 \geq 0.$

7. The first constraint can be written as $-x_1 + x_2 + x_3 \geq -3$. Maximize $W = -3y_1 + 3y_2$ subject to

   $-y_1 + y_2 \leq 4,$
   $y_1 - y_2 \leq 4,$
   $y_1 + y_2 \leq 6,$
   $y_1, y_2 \geq 0.$

9. The dual is: Maximize $W = 2y_1 + 3y_2$

   subject to
   $$y_1 - y_2 \leq 2,$$
   $$-y_1 + 2y_2 \leq 2,$$
   $$2y_1 + y_2 \leq 5,$$
   $$y_1, y_2 \geq 0.$$

	y_1	y_2	s_1	s_2	s_3	W		
s_1	1	-1	1	0	0	0	2	
s_2	-1	2	0	1	0	0	2	1
s_3	2	1	0	0	1	0	5	5
W	-2	-3	0	0	0	1	0	

	y_1	y_2	s_1	s_2	s_3	W		
s_1	$\frac{1}{2}$	0	1	$\frac{1}{2}$	0	0	3	6
y_2	$-\frac{1}{2}$	1	0	$\frac{1}{2}$	0	0	1	
s_3	$\frac{5}{2}$	0	0	$-\frac{1}{2}$	1	0	4	$\frac{8}{5}$
W	$-\frac{7}{2}$	0	0	$\frac{3}{2}$	0	1	3	

	y_1	y_2	s_1	s_2	s_3	W	
s_1	0	0	1	$\frac{3}{5}$	$-\frac{1}{5}$	0	$\frac{11}{5}$
y_2	0	1	0	$\frac{2}{5}$	$\frac{1}{5}$	0	$\frac{9}{5}$
y_1	1	0	0	$-\frac{1}{5}$	$\frac{2}{5}$	0	$\frac{8}{5}$
W	0	0	0	$\frac{4}{5}$	$\frac{7}{5}$	1	$\frac{43}{5}$

The minimum is $Z = \dfrac{43}{5}$ when $x_1 = 0$,

$$x_2 = \frac{4}{5}, \quad x_3 = \frac{7}{5}.$$

11. The dual is: Minimize $W = 8y_1 + 12y_2$ subject to

$$y_1 + y_2 \geq 3,$$
$$2y_1 + 6y_2 \geq 8,$$
$$y_1, y_2 \geq 0.$$

$$
\begin{array}{c}
\begin{array}{ccccccc} y_1 & y_2 & s_1 & s_2 & t_1 & t_2 & U \end{array} \\
\left[\begin{array}{ccccccc|c}
1 & 1 & -1 & 0 & 1 & 0 & 0 & 3 \\
2 & 6 & 0 & -1 & 0 & 1 & 0 & 8 \\
\hline
8 & 12 & 0 & 0 & M & M & 1 & 0
\end{array}\right]
\end{array}
$$

$$
\begin{array}{cc}
& \begin{array}{ccccccc} y_1 & y_2 & s_1 & s_2 & t_1 & t_2 & U \end{array} \\
\begin{array}{c} t_1 \\ t_2 \\ U \end{array} &
\left[\begin{array}{ccccccc|c}
1 & 1 & -1 & 0 & 1 & 0 & 0 & 3 \\
2 & 6 & 0 & -1 & 0 & 1 & 0 & 8 \\
\hline
8-3M & 12-7M & M & M & 0 & 0 & 1 & -11M
\end{array}\right]
\begin{array}{c} 3 \\ \frac{4}{3} \\ {} \end{array}
\end{array}
$$

$$
\begin{array}{cc}
& \begin{array}{ccccccc} y_1 & y_2 & s_1 & s_2 & t_1 & t_2 & U \end{array} \\
\begin{array}{c} t_1 \\ y_2 \\ U \end{array} &
\left[\begin{array}{ccccccc|c}
\frac{2}{3} & 0 & -1 & \frac{1}{6} & 1 & -\frac{1}{6} & 0 & \frac{5}{3} \\
\frac{1}{3} & 1 & 0 & -\frac{1}{6} & 0 & \frac{1}{6} & 0 & \frac{4}{3} \\
\hline
4-\frac{2}{3}M & 0 & M & 2-\frac{1}{6}M & 0 & -2+\frac{7}{6}M & 1 & -16-\frac{5}{3}M
\end{array}\right]
\begin{array}{c} \frac{5}{2} \\ 4 \\ {} \end{array}
\end{array}
$$

$$
\begin{array}{cc}
& \begin{array}{ccccccc} y_1 & y_2 & s_1 & s_2 & t_1 & t_2 & U \end{array} \\
\begin{array}{c} y_1 \\ y_2 \\ U \end{array} &
\left[\begin{array}{ccccccc|c}
1 & 0 & -\frac{3}{2} & \frac{1}{4} & \frac{3}{2} & -\frac{1}{4} & 0 & \frac{5}{2} \\
0 & 1 & \frac{1}{2} & -\frac{1}{4} & -\frac{1}{2} & \frac{1}{4} & 0 & \frac{1}{2} \\
\hline
0 & 0 & 6 & 1 & -6+M & -1+M & 1 & -26
\end{array}\right]
\end{array}
$$

The maximum is $Z = 26$ when $x_1 = 6, x_2 = 1$.

13. The first constraint can be written as $x_1 - x_2 \geq -1$. The dual is: Maximize $W = -y_1 + 3y_2$ subject to

$$y_1 + y_2 \leq 6,$$
$$-y_1 + y_2 \leq 4,$$
$$y_1, y_2 \geq 0.$$

$$
\begin{array}{cc}
& \begin{array}{ccccc} y_1 & y_2 & s_1 & s_2 & W \end{array} \\
\begin{array}{c} s_1 \\ s_2 \\ W \end{array} &
\left[\begin{array}{ccccc|c}
1 & 1 & 1 & 0 & 0 & 6 \\
-1 & 1 & 0 & 1 & 0 & 4 \\
\hline
1 & -3 & 0 & 0 & 1 & 0
\end{array}\right]
\begin{array}{c} 6 \\ 4 \\ {} \end{array}
\end{array}
$$

$$
\begin{array}{cc}
& \begin{array}{ccccc} y_1 & y_2 & s_1 & s_2 & W \end{array} \\
\begin{array}{c} s_1 \\ y_2 \\ W \end{array} &
\left[\begin{array}{ccccc|c}
2 & 0 & 1 & -1 & 0 & 2 \\
-1 & 1 & 0 & 1 & 0 & 4 \\
\hline
-2 & 0 & 0 & 3 & 1 & 12
\end{array}\right]
\begin{array}{c} 1 \\ {} \\ {} \end{array}
\end{array}
$$

$$
\begin{array}{c}
\begin{array}{ccccc} y_1 & y_2 & s_1 & s_2 & W \end{array} \\
\begin{array}{c} y_1 \\ y_2 \\ W \end{array}
\left[
\begin{array}{ccccc|c}
1 & 0 & \frac{1}{2} & -\frac{1}{2} & 0 & 1 \\
0 & 1 & \frac{1}{2} & \frac{1}{2} & 0 & 5 \\
\hdashline
0 & 0 & 1 & 2 & 1 & 14
\end{array}
\right]
\end{array}
$$

The minimum is $Z = 14$ when $x_1 = 1, x_2 = 2$.

15. Let x_1 = amount spent on newspaper advertising,

x_2 = amount spent on radio advertising.

We want to minimize $C = x_1 + x_2$ subject to

$$40x_1 + 50x_2 \ge 80,000,$$
$$100x_1 + 25x_2 \ge 60,000,$$
$$x_1, x_2 \ge 0.$$

The dual is: Maximize

$W = 80,000 y_1 + 60,000 y_2$ subject to

$$40 y_1 + 100 y_2 \le 1,$$
$$50 y_1 + 25 y_2 \le 1,$$
$$y_1, y_2 \ge 0.$$

$$
\begin{array}{c}
\begin{array}{ccccc} y_1 & \ \ y_2 & s_1 & s_2 & W \end{array} \\
\begin{array}{c} s_1 \\ s_2 \\ W \end{array}
\left[
\begin{array}{ccccc|c}
40 & 100 & 1 & 0 & 0 & 1 \\
50 & 25 & 0 & 1 & 0 & 1 \\
\hdashline
-80,000 & -60,000 & 0 & 0 & 1 & 0
\end{array}
\right]
\begin{array}{c} \frac{1}{40} \\ \frac{1}{50} \\ \ \end{array}
\end{array}
$$

$$
\begin{array}{c}
\begin{array}{ccccc} y_1 & y_2 & s_1 & s_2 & \ \ W \end{array} \\
\begin{array}{c} s_1 \\ y_1 \\ W \end{array}
\left[
\begin{array}{ccccc|c}
0 & \underline{80} & 1 & -\frac{4}{5} & 0 & \frac{1}{5} \\
1 & \frac{1}{2} & 0 & \frac{1}{50} & 0 & \frac{1}{50} \\
\hdashline
0 & -20,000 & 0 & 1600 & 1 & 1600
\end{array}
\right]
\begin{array}{c} \frac{1}{400} \\ \frac{1}{25} \\ \ \end{array}
\end{array}
$$

$$
\begin{array}{c}
\begin{array}{ccccc} y_1 & y_2 & s_1 & s_2 & \ W \end{array} \\
\begin{array}{c} y_2 \\ y_1 \\ W \end{array}
\left[
\begin{array}{ccccc|c}
0 & 1 & \frac{1}{80} & -\frac{1}{100} & 0 & \frac{1}{400} \\
1 & 0 & -\frac{1}{160} & \frac{1}{40} & 0 & \frac{3}{160} \\
\hdashline
0 & 0 & 250 & 1400 & 1 & 1650
\end{array}
\right]
\end{array}
$$

The firm should spend $250 on newspaper advertising and $1400 on radio advertising for a cost of $1650.

17. Let y_1 = number of shipping clerk apprentices,

y_2 = number of shipping clerks,

y_3 = number of semiskilled workers,

y_4 = number of skilled workers.

We want to minimize $W = 6y_1 + 9y_2 + 8y_3 + 14y_4$ subject to

$$y_1 + y_2 \geq 60,$$
$$-2y_1 + y_2 \geq 0,$$
$$y_3 + y_4 \geq 90,$$
$$y_3 - 2y_4 \geq 0,$$
$$y_1, y_2, y_3, y_4 \geq 0.$$

The dual is: Maximize $Z = 60x_1 + 0x_2 + 90x_3 + 0x_4$ subject to

$$x_1 - 2x_2 \leq 6,$$
$$x_1 + x_2 \leq 9,$$
$$x_3 + x_4 \leq 8,$$
$$x_3 - 2x_4 \leq 14,$$
$$x_1, x_2, x_3, x_4 \geq 0.$$

	x_1	x_2	x_3	x_4	s_1	s_2	s_3	s_4	Z		
s_1	1	−2	0	0	1	0	0	0	0	6	
s_2	1	1	0	0	0	1	0	0	0	9	
s_3	0	0	1	1	0	0	1	0	0	8	8
s_4	0	0	1	−2	0	0	0	1	0	14	14
Z	−60	0	−90	0	0	0	0	0	1	0	

	x_1	x_2	x_3	x_4	s_1	s_2	s_3	s_4	Z		
s_1	1	−2	0	0	1	0	0	0	0	6	6
s_2	1	1	0	0	0	1	0	0	0	9	9
x_3	0	0	1	1	0	0	1	0	0	8	
s_4	0	0	0	−3	0	0	−1	1	0	6	
Z	−60	0	0	90	0	0	90	0	1	720	

	x_1	x_2	x_3	x_4	s_1	s_2	s_3	s_4	Z		
x_1	1	−2	0	0	1	0	0	0	0	6	
s_2	0	3	0	0	−1	1	0	0	0	3	1
x_3	0	0	1	1	0	0	1	0	0	8	
s_4	0	0	0	−3	0	0	−1	1	0	6	
Z	0	−120	0	90	60	0	90	0	1	1080	

$$
\begin{array}{c}
\begin{array}{ccccccccc} x_1 & x_2 & x_3 & x_4 & s_1 & s_2 & & s_3 & s_4 & Z \end{array} \\
\begin{array}{c} x_1 \\ x_2 \\ x_3 \\ s_4 \\ Z \end{array}
\left[
\begin{array}{ccccccccc|c}
1 & 0 & 0 & 0 & \frac{1}{3} & \frac{2}{3} & 0 & 0 & 0 & 8 \\
0 & 1 & 0 & 0 & -\frac{1}{3} & \frac{1}{3} & 0 & 0 & 0 & 1 \\
0 & 0 & 1 & 1 & 0 & 0 & 1 & 0 & 0 & 8 \\
0 & 0 & 0 & -3 & 0 & 0 & -1 & 1 & 0 & 6 \\
\hdashline
0 & 0 & 0 & 90 & 20 & 40 & 90 & 0 & 1 & 1200
\end{array}
\right]
\end{array}
$$

The company should employ 20 shipping clerk apprentices, 40 shipping clerks, 90 semiskilled workers, and 0 skilled workers for a total hourly wage of $1200.

Chapter 7 Review Problems

1.

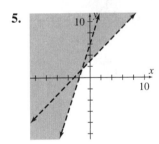

3.

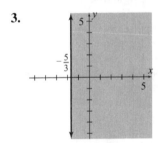

5.

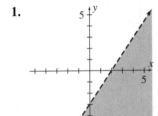

7.

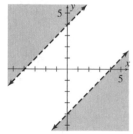

9.

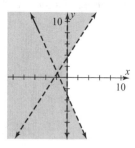

11. Feasible region follows. Corner points are (0, 0), (0, 2), (1, 3), (3, 1), (3, 0). Z is maximized at (3, 0) where its value is 3. Thus $Z = 3$ when $x = 3$ and $y = 0$.

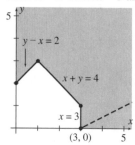

13. Feasible region is unbounded. Z is minimized at the corner point (0, 2) where its value is -2. Thus $Z = -2$ when $x = 0$ and $y = 2$.

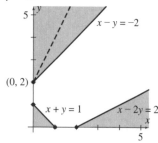

15. Feasible region follows. Corner points are $\left(\dfrac{20}{9}, \dfrac{10}{9}\right)$, (5, 0), and (4, 0). Z is minimized at $\left(\dfrac{20}{9}, \dfrac{10}{9}\right)$ where its value is $\dfrac{70}{9}$. Thus $Z = \dfrac{70}{9}$ when $x = \dfrac{20}{9}$ and $y = \dfrac{10}{9}$.

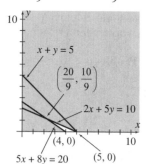

17. Feasible region follows. Corner points are (0, 0), (0, 4), (2, 3), and (4, 0). Z is maximized at (2, 3) and (4, 0) where its value is 36. Thus Z is maximized at all points on the line segment joining (2, 3) and (4, 0). The solution is $Z = 36$ when $x = (1 - t)(2) + 4t = 2 + 2t$, $y = (1 - t)(3) + 0t = 3 - 3t$, and $0 \le t \le 1$.

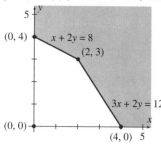

19.

$$
\begin{array}{c}
\begin{array}{ccccc} x_1 & x_2 & s_1 & s_2 & Z \end{array} \\
\begin{array}{c} s_1 \\ s_2 \\ Z \end{array}
\left[
\begin{array}{ccccc|c}
1 & \underline{6} & 1 & 0 & 0 & 12 \\
1 & 2 & 0 & 1 & 0 & 8 \\
\hline
-4 & -5 & 0 & 0 & 1 & 0
\end{array}
\right]
\begin{array}{c} 2 \\ 4 \\ {} \end{array}
\end{array}
$$

$$
\begin{array}{c}
\begin{array}{ccccc} x_1 & x_2 & s_1 & s_2 & Z \end{array} \\
\begin{array}{c} x_2 \\ {} \\ s_2 \\ {} \\ Z \end{array}
\left[
\begin{array}{ccccc|c}
\frac{1}{6} & 1 & \frac{1}{6} & 0 & 0 & 2 \\
\frac{2}{3} & 0 & -\frac{1}{3} & 1 & 0 & 4 \\
\hline
-\frac{19}{6} & 0 & \frac{5}{6} & 0 & 1 & 10
\end{array}
\right]
\begin{array}{c} 12 \\ {} \\ 6 \\ {} \\ {} \end{array}
\end{array}
$$

$$
\begin{array}{c}
\begin{array}{ccccc} x_1 & x_2 & s_1 & s_2 & Z \end{array} \\
\begin{array}{c} x_2 \\ {} \\ x_1 \\ {} \\ Z \end{array}
\left[
\begin{array}{ccccc|c}
0 & 1 & \frac{1}{4} & -\frac{1}{4} & 0 & 1 \\
1 & 0 & -\frac{1}{2} & \frac{3}{2} & 0 & 6 \\
\hline
0 & 0 & -\frac{3}{4} & \frac{19}{4} & 1 & 29
\end{array}
\right]
\begin{array}{c} 4 \\ {} \\ {} \\ {} \\ {} \end{array}
\end{array}
$$

$$
\begin{array}{c}
\begin{array}{ccccc} x_1 & x_2 & s_1 & s_2 & Z \end{array} \\
\begin{array}{c} s_1 \\ x_1 \\ Z \end{array}
\left[
\begin{array}{ccccc|c}
0 & 4 & 1 & -1 & 0 & 4 \\
1 & 2 & 0 & 1 & 0 & 8 \\
\hline
0 & 3 & 0 & 4 & 1 & 32
\end{array}
\right]
\end{array}
$$

Thus $Z = 32$ when $x_1 = 8$ and $x_2 = 0$.

21.

$$
\begin{array}{c}
\begin{array}{ccccccc} x_1 & x_2 & x_3 & s_1 & t_1 & W \end{array} \\
\left[
\begin{array}{cccccc|c}
1 & 2 & 3 & -1 & 1 & 0 & 5 \\
\hline
3 & 2 & 1 & 0 & M & 1 & 0
\end{array}
\right]
\end{array}
$$

$$
\begin{array}{c}
\begin{array}{cccccc} x_1 & x_2 & x_3 & s_1 & t_1 & W \end{array} \\
\begin{array}{c} t_1 \\ W \end{array}
\left[
\begin{array}{cccccc|c}
1 & 2 & \underline{3} & -1 & 1 & 0 & 5 \\
\hline
3-M & 2-2M & 1-3M & M & 0 & 1 & -5M
\end{array}
\right]
\begin{array}{c} \frac{5}{3} \\ {} \end{array}
\end{array}
$$

$$
\begin{array}{c}
\begin{array}{cccccc} x_1 & x_2 & x_3 & s_1 & t_1 & W \end{array} \\
\begin{array}{c} x_3 \\ {} \\ W \end{array}
\left[
\begin{array}{cccccc|c}
\frac{1}{3} & \frac{2}{3} & 1 & -\frac{1}{3} & \frac{1}{3} & 0 & \frac{5}{3} \\
\hline
\frac{8}{3} & \frac{4}{3} & 0 & \frac{1}{3} & -\frac{1}{3}+M & 1 & -\frac{5}{3}
\end{array}
\right]
\end{array}
$$

Thus $Z = \dfrac{5}{3}$ when $x_1 = 0$, $x_2 = 0$, and $x_3 = \dfrac{5}{3}$.

23.

	x_1	x_2	s_1	s_2	s_3	t_2	W	
	1	1	1	0	0	0	0	12
	1	1	0	−1	0	1	0	5
	1	0	0	0	1	0	0	10
	−1	−2	0	0	0	M	1	0

	x_1	x_2	s_1	s_2	s_3	t_2	W		
s_1	1	1	1	0	0	0	0	12	12
t_2	1	1	0	−1	0	1	0	5	5
s_3	1	0	0	0	1	0	0	10	
W	$-1-M$	$-2-M$	0	M	0	0	1	$-5M$	

	x_1	x_2	s_1	s_2	s_3	t_2	W		
s_1	0	0	1	1	0	−1	0	7	7
x_2	1	1	0	−1	0	1	0	5	
s_3	1	0	0	0	1	0	0	10	
W	1	0	0	−2	0	$2+M$	1	10	

	x_1	x_2	s_1	s_2	s_3	Z	
s_2	0	0	1	1	0	0	7
x_2	1	1	1	0	0	0	12
s_3	1	0	0	0	1	0	10
Z	1	0	2	0	0	1	24

Thus $Z = 24$ when $x_1 = 0$ and $x_2 = 12$.

25. We write the first constraint as $-x_1 + x_2 + x_3 \geq 1$.

	x_1	x_2	x_3	s_1	t_1	t_2	W	
	−1	1	1	−1	1	0	0	1
	6	3	2	0	0	1	0	12
	1	2	1	0	M	M	1	0

	x_1	x_2	x_3	s_1	t_1	t_2	W		
t_1	−1	1	1	−1	1	0	0	1	
t_2	6	3	2	0	0	1	0	12	2
W	$1-5M$	$2-4M$	$1-3M$	M	0	0	1	$-13M$	

	x_1	x_2	x_3	s_1	t_1	t_2	W		
t_1	0	$\frac{3}{2}$	$\frac{4}{3}$	−1	1	$\frac{1}{6}$	0	3	2
x_1	1	$\frac{1}{2}$	$\frac{1}{3}$	0	0	$\frac{1}{6}$	0	2	4
W	0	$\frac{3}{2}-\frac{3}{2}M$	$\frac{2}{3}-\frac{4}{3}M$	M	0	$-\frac{1}{6}+\frac{5}{6}M$	1	$-2-3M$	

$$\begin{array}{c} \\ x_2 \\ \\ x_1 \\ W \end{array} \begin{array}{cccccccc} x_1 & x_2 & x_3 & s_1 & t_1 & t_2 & W & \\ \left[\begin{array}{ccccccc|c} 0 & 1 & \frac{8}{9} & -\frac{2}{3} & \frac{2}{3} & \frac{1}{9} & 0 & 2 \\ 1 & 0 & -\frac{1}{9} & \frac{1}{3} & -\frac{1}{3} & \frac{1}{9} & 0 & 1 \\ \hdashline 0 & 0 & -\frac{2}{3} & 1 & -1+M & -\frac{1}{3}+M & 1 & -5 \end{array}\right] \end{array} \begin{array}{c} \frac{9}{4} \\ \\ \\ \end{array}$$

$$\begin{array}{c} \\ x_3 \\ \\ x_1 \\ -Z \end{array} \begin{array}{ccccc} x_1 & x_2 & x_3 & s_1 & -Z \\ \left[\begin{array}{ccccc|c} 0 & \frac{9}{8} & 1 & -\frac{3}{4} & 0 & \frac{9}{4} \\ 1 & \frac{1}{8} & 0 & \frac{1}{4} & 0 & \frac{5}{4} \\ \hdashline 0 & \frac{3}{4} & 0 & \frac{1}{2} & 1 & -\frac{7}{2} \end{array}\right] \end{array}$$

Thus $Z = \dfrac{7}{2}$ when $x_1 = \dfrac{5}{4}$, $x_2 = 0$, and $x_3 = \dfrac{9}{4}$.

27.

$$\begin{array}{c} \\ s_1 \\ s_2 \\ Z \end{array} \begin{array}{cccccc} x_1 & x_2 & x_3 & s_1 & s_2 & Z \\ \left[\begin{array}{cccccc|c} 4 & -1 & 0 & 1 & 0 & 0 & 2 \\ -8 & 2 & 5 & 0 & 1 & 0 & 2 \\ \hdashline -1 & -4 & -2 & 0 & 0 & 1 & 0 \end{array}\right] \end{array} \begin{array}{c} \\ 1 \\ \\ \end{array}$$

$$\begin{array}{c} \\ s_1 \\ \\ x_2 \\ Z \end{array} \begin{array}{cccccc} x_1 & x_2 & x_3 & s_1 & s_2 & Z \\ \left[\begin{array}{cccccc|c} 0 & 0 & \frac{5}{2} & 1 & \frac{1}{2} & 0 & 3 \\ -4 & 1 & \frac{5}{2} & 0 & \frac{1}{2} & 0 & 1 \\ \hdashline -17 & 0 & 8 & 0 & 2 & 1 & 4 \end{array}\right] \end{array}$$

For the last table, x_1 is the entering variable. Since no quotients exist, the problem has an unbounded solution. That is, no optimum solution (unbounded).

29. The dual is: Maximize $W = 35y_1 + 25y_2$ subject to

$$y_1 + y_2 \le 2,$$
$$2y_1 + y_2 \le 7,$$
$$3y_1 + y_2 \le 8,$$
$$y_1, y_2 \ge 0.$$

$$\begin{array}{c} \\ s_1 \\ s_2 \\ s_3 \\ W \end{array} \begin{array}{cccccc} y_1 & y_2 & s_1 & s_2 & s_3 & W \\ \left[\begin{array}{cccccc|c} 1 & 1 & 1 & 0 & 0 & 0 & 2 \\ 2 & 1 & 0 & 1 & 0 & 0 & 7 \\ 3 & 1 & 0 & 0 & 1 & 0 & 8 \\ \hdashline -35 & -25 & 0 & 0 & 0 & 1 & 0 \end{array}\right] \end{array} \begin{array}{c} 2 \\ \frac{7}{2} \\ \frac{8}{3} \\ \\ \end{array}$$

$$\begin{array}{c c} & \begin{array}{c c c c c c} y_1 & y_2 & s_1 & s_2 & s_3 & W \end{array} \\ \begin{array}{c} y_1 \\ s_2 \\ s_3 \\ W \end{array} & \left[\begin{array}{c c c c c c|c} 1 & 1 & 1 & 0 & 0 & 0 & 2 \\ 0 & -1 & -2 & 1 & 0 & 0 & 3 \\ 0 & -2 & -3 & 0 & 1 & 0 & 2 \\ \hline 0 & 10 & 35 & 0 & 0 & 1 & 70 \end{array}\right] \end{array}$$

Thus $Z = 70$ when $x_1 = 35, x_2 = 0$, and $x_3 = 0$.

31. Let x, y, and z denote the numbers of units of X, Y, and Z produced weekly, respectively. If P is the total profit obtained, we want to maximize
$P = 10x + 15y + 22z$ subject to
$x + 2y + 2z \le 40,$
$x + y + 2z \le 34,$
$x, y, z \ge 0.$

$$\begin{array}{c c} & \begin{array}{c c c c c c} x & y & z & s_1 & s_2 & P \end{array} \\ \begin{array}{c} s_1 \\ s_2 \\ P \end{array} & \left[\begin{array}{c c c c c c|c c} 1 & 2 & 2 & 1 & 0 & 0 & 40 & 20 \\ 1 & 1 & 2 & 0 & 1 & 0 & 34 & 17 \\ \hline -10 & -15 & -22 & 0 & 0 & 1 & 0 \end{array}\right] \end{array}$$

$$\begin{array}{c c} & \begin{array}{c c c c c c} x & y & z & s_1 & s_2 & P \end{array} \\ \begin{array}{c} s_1 \\ z \\ P \end{array} & \left[\begin{array}{c c c c c c|c c} 0 & 1 & 0 & 1 & -1 & 0 & 6 & 6 \\ \frac{1}{2} & \frac{1}{2} & 1 & 0 & \frac{1}{2} & 0 & 17 & 34 \\ \hline 1 & -4 & 0 & 0 & 11 & 1 & 374 \end{array}\right] \end{array}$$

$$\begin{array}{c} y \\ z \\ P \end{array} \left[\begin{array}{c c c c c c|c} 0 & 1 & 0 & 1 & -1 & 0 & 6 \\ \frac{1}{2} & 0 & 1 & -\frac{1}{2} & 1 & 0 & 14 \\ \hline 1 & 0 & 0 & 4 & 7 & 1 & 398 \end{array}\right]$$

Thus 0 units of X, 6 units of Y, and 14 units of Z give a maximum profit of $398.

33. Let x_{AC}, x_{AD}, x_{BC}, and x_{BD} denote the amounts (in hundreds of thousands of gallons) transported from A to C, A to D, B to C, and B to D, respectively. If C is the total transportation cost in thousands of dollars, we want to minimize $C = x_{AC} + 2x_{AD} + 2x_{BC} + 4x_{BD}$ subject to

$$x_{AC} + x_{AD} \le 6,$$
$$x_{BC} + x_{BD} \le 6,$$
$$x_{AC} + x_{BC} = 5,$$
$$x_{AD} + x_{BD} = 5,$$
$$x_{AC}, x_{AD}, x_{BC}, x_{BD} \ge 0.$$

	x_{AC}	x_{AD}	x_{BC}	x_{BD}	s_1	s_2	t_3	t_4	W	
	1	1	0	0	1	0	0	0	0	6
	0	0	1	1	0	1	0	0	0	6
	1	0	1	0	0	0	1	0	0	5
	0	1	0	1	0	0	0	1	0	5
	1	2	2	4	0	0	M	M	1	0

	x_{AC}	x_{AD}	x_{BC}	x_{BD}	s_1	s_2	t_3	t_4	W		
s_1	1	1	0	0	1	0	0	0	0	6	6
s_2	0	0	1	1	0	1	0	0	0	6	
t_3	1	0	1	0	0	0	1	0	0	5	5
t_4	0	1	0	1	0	0	0	1	0	5	
W	$1-M$	$2-M$	$2-M$	$4-M$	0	0	0	0	1	$-10M$	

	x_{AC}	x_{AD}	x_{BC}	x_{BD}	s_1	s_2	t_3	t_4	W		
s_1	0	1	-1	0	1	0	-1	0	0	1	1
s_2	0	0	1	1	0	1	0	0	0	6	
x_{AC}	1	0	1	0	0	0	1	0	0	5	
t_4	0	1	0	1	0	0	0	1	0	5	5
W	0	$2-M$	1	$4-M$	0	0	$-1+M$	0	1	$-5-5M$	

	x_{AC}	x_{AD}	x_{BC}	x_{BD}	s_1	s_2	t_3	t_4	W		
x_{AD}	0	1	-1	0	1	0	-1	0	0	1	
s_2	0	0	1	1	0	1	0	0	0	6	6
x_{AC}	1	0	1	0	0	0	1	0	0	5	5
t_4	0	0	1	1	-1	0	1	1	0	4	4
W	0	0	$3-M$	$4-M$	$-2+M$	0	1	0	1	$-7-4M$	

	x_{AC}	x_{AD}	x_{BC}	x_{BD}	s_1	s_2	t_3	t_4	W	
x_{AD}	0	1	0	1	0	0	0	1	0	5
s_2	0	0	0	0	1	1	-1	-1	0	2
x_{AC}	1	0	0	-1	1	0	0	-1	0	1
x_{BC}	0	0	1	1	-1	0	1	1	0	4
W	0	0	0	1	1	0	$-2+M$	$-3+M$	1	-19

The minimum value of C is 19, when $x_{AC} = 1, x_{AD} = 5, x_{BC} = 4$, and $x_{BD} = 0$. Thus 100,000 gal from A to C, 500,000 gal from A to D, and 400,000 gal from B to C give a minimum cost of \$19,000.

35. Let x and y represent daily consumption of foods A and B in 100-gram units. We want to minimize
$C = 8x + 22y$ subject to the constraints
$$8x + 4y \geq 176,$$
$$16x + 32y \geq 1024,$$
$$2x + 5y \geq 200,$$
$$x \geq 0,$$
$$y \geq 0.$$

The feasible region is unbounded with corner points $(100, 0)$, $\left(\dfrac{5}{2}, 39\right)$ and $(0, 44)$. C has a minimum

value at $(100, 0)$. Thus the animals should be fed $100 \times 100 = 10,000$ grams = 10 kilograms of food A
each day.

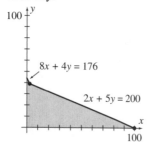

37.

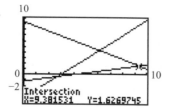

$Z = 129.83$ when $x = 9.38$, $y = 1.63$

Mathematical Snapshot Chapter 7

1.

	CURATIVE UNITS	TOXIC UNITS	RELATIVE DISCOMFORT
Drug (per ounce)	500	400	1
Radiation (per min)	1000	600	1
Requirement	≥ 2000	≤ 1400	

Let x_1 = number of ounces of drug and let x_2 = number of minutes of radiation. We want to minimize
the discomfort D, where $D = x_1 + x_2$, subject to
$$500x_1 + 1000x_2 \geq 2000,$$
$$400x_1 + 600x_2 \leq 1400,$$
where $x_1, x_2 \geq 0$.

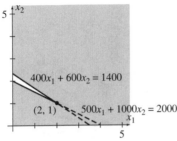

The corner points are $(0, 2)$, $\left(0, \dfrac{7}{3}\right)$, and $(2, 1)$.

At $(0, 2)$, $D = 0 + 2 = 2$;

at $\left(0, \dfrac{7}{3}\right)$, $D = 0 + \dfrac{7}{3} = \dfrac{7}{3}$;

at $(2, 1)$, $D = 2 + 1 = 3$.

Thus D is minimum at $(0, 2)$.

The patient should get 0 ounces of drug and 2 minutes of radiation.

3. Answers may vary.

Chapter 8

Problems 8.1

1.

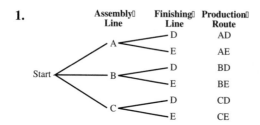

6 possible production routes

3.

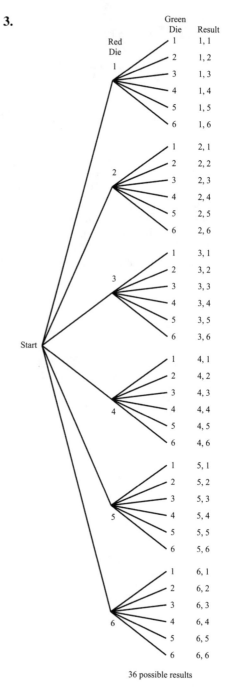

36 possible results

5. There are 5 science courses and 4 humanities courses. By the basic counting principle, the number of selections is $5 \cdot 4 = 20$.

7. There are 2 appetizers, 4 entrees, 4 desserts, and 3 beverages. By the basic counting principle, the number of possible complete dinners is $2 \cdot 4 \cdot 4 \cdot 3 = 96$.

9. For each of the 10 questions, there are 2 choices. By the basic counting principle, the number of ways to answer the examination is $2 \cdot 2 \cdots 2 = 2^{10} = 1024$.

11. $_6P_3 = \dfrac{6!}{(6-3)!} = \dfrac{6!}{3!} = 6 \cdot 5 \cdot 4 = 120$

13. $_6P_6 = \dfrac{6!}{(6-6)!} = \dfrac{6!}{0!} = \dfrac{6 \cdot 5 \cdot 4 \cdot 3 \cdot 2 \cdot 1}{1} = 720$

15. $_4P_2 \cdot {}_5P_3 = (4 \cdot 3)(5 \cdot 4 \cdot 3) = (12)(60) = 720$

17. $\dfrac{1000!}{999!} = \dfrac{1000 \cdot 999!}{999!} = 1000$

For most calculators, attempting to evaluate $\dfrac{1000!}{999!}$ results in an error message (because of the magnitude of the numbers involved).

19. A name for the firm is an ordered arrangement of the three last names. Thus the number of possible firm names is $_3P_3 = 3! = 3 \cdot 2 \cdot 1 = 6$.

21. The number of ways of selecting 3 of 8 contestants in an order is $_8P_3 = 8 \cdot 7 \cdot 6 = 336$.

23. On each roll of a die, there are 6 possible outcomes. By the basic counting principle, on 4 rolls the number of possible results is $6 \cdot 6 \cdot 6 \cdot 6 = 6^4 = 1296$.

25. The number of ways of selecting 3 of the 12 students in an order is $_{12}P_3 = 12 \cdot 11 \cdot 10 = 1320$.

27. The number of ways a student can choose 4 of the 6 items in an order is $_6P_4 = 6 \cdot 5 \cdot 4 \cdot 3 = 360$.

29. The number of ways to select six of the six different letters in the word MEADOW in an order is $_6P_6 = 6! = 6 \cdot 5 \cdot 4 \cdot 3 \cdot 2 \cdot 1 = 720$.

31. For an arrangement of books, order is important. The number of ways to arrange 5 of 7 books is $_7P_5 = 7 \cdot 6 \cdot 5 \cdot 4 \cdot 3 = 2520$.

All 7 books can be arranged in $_7P_7 = 7! = 5040$ ways.

33. After a "four of a kind" hand is dealt, the cards can be arranged so that the first four have the same face value, and order is not important, There are 13 possibilities for the first four cards (all 2's, all 3's, ..., all aces). The fifth card can be any one of the 48 cards that remain. By the basic counting principle, the number of "four of a kind" hands is $13 \cdot 48 = 624$.

35. The number of ways the waitress can place five of the five different sandwiches (and order is important) is $_5P_5 = 5! = 5 \cdot 4 \cdot 3 \cdot 2 \cdot 1 = 120$.

37. a. To fill the four offices by different people, 4 of 12 members must be selected, and order is important. This can be done in $_{12}P_4 = 12 \cdot 11 \cdot 10 \cdot 9 = 11{,}880$ ways.

    b. If the president and vice president must be different members, then there are 12 choices for president, 11 for vice president, 12 for secretary, and 12 for treasurer. By the basic counting principle, the offices can be filled in $12 \cdot 11 \cdot 12 \cdot 12 = 19{,}008$ ways.

39. There are 2 choices for the center position. After that choice is made, to fill the remaining four positions (and order is important), there are $_4P_4$ ways. By the basic counting principle, to assign positions to the five-member team there are $2 \cdot {_4P_4} = 2(4!) = 2(24) = 48$ ways.

41. There are $_3P_3$ ways to select the first three batters (order is important) and there are $_6P_6$ ways to select the remaining batters. By the basic counting principle, the number of possible batting orders is
$$_3P_3 \cdot {_6P_6} = 3! \cdot 6! = 6 \cdot 720 = 4320.$$

Problems 8.2

1. $_6C_4 = \dfrac{6!}{4!(6-4)!}$
$= \dfrac{6!}{4! \cdot 2!} = \dfrac{6 \cdot 5 \cdot 4!}{4!(2 \cdot 1)} = \dfrac{6 \cdot 5}{2 \cdot 1} = 15$

3. $_{100}C_{100} = \dfrac{100!}{100!(100-100)!} = \dfrac{1}{0!} = \dfrac{1}{1} = 1$

5. $_5P_3 \cdot {_4C_2} = 5 \cdot 4 \cdot 3 \dfrac{4!}{2!(4-2)!}$
$= 5 \cdot 4 \cdot 3 \dfrac{4 \cdot 3 \cdot 2!}{2!2!}$
$= 60 \cdot 6$
$= 360$

7. $_nC_r = \dfrac{n!}{r!(n-r)!}$

$_nC_{n-r} = \dfrac{n!}{(n-r)![n-(n-r)]!} = \dfrac{n!}{(n-r)!r!}$.

Thus $_nC_r = {_nC_{n-r}}$.

9. The number of ways of selecting 4 of 17 people so that order is not important is
$$_{17}C_4 = \dfrac{17!}{4!(17-4)!} = \dfrac{17!}{4! \cdot 13!}$$
$$= \dfrac{17 \cdot 16 \cdot 15 \cdot 14 \cdot 13!}{4 \cdot 3 \cdot 2 \cdot 1(13!)} = 2380$$

11. The number of ways of selecting 9 out of 13 questions (without regard to order) is
$$_{13}C_9 = \dfrac{13!}{9!(13-9)!} = \dfrac{13!}{9! \cdot 4!}$$
$$= \dfrac{13 \cdot 12 \cdot 11 \cdot 10 \cdot 9!}{9! \cdot 4 \cdot 3 \cdot 2 \cdot 1}$$
$$= 715.$$

13. The order of selecting 10 of the 74 dresses is of no concern. Thus the number of possible samples $_{74}C_{10} = \dfrac{74!}{10! \cdot (74-10)!} = \dfrac{74!}{10! \cdot 64!}$.

15. To score 80, 90, or 100, exactly 8, 9, or 10 questions must be correct, respectively. The number of ways in which 8 of 10 questions can be correct is
$$_{10}C_8 = \dfrac{10!}{8!(10-8)!} = \dfrac{10!}{8! \cdot 2!} = \dfrac{10 \cdot 9 \cdot 8!}{8! \cdot 2 \cdot 1} = 45.$$
For 9 of 10 questions, the number of ways is
$$_{10}C_9 = \dfrac{10!}{9!(10-9)!} = \dfrac{10!}{9! \cdot 1!} = \dfrac{10 \cdot 9!}{9! \cdot 1} = 10,$$
and for 10 of 10 questions, it is
$$_{10}C_{10} = \dfrac{10!}{10!(10-10)!} = \dfrac{10!}{10! \cdot 0!} = 1.$$
Thus the number of ways to score 80 or better is $45 + 10 + 1 = 56$.

17. The word MISSISSAUGA has 11 letters with repetition: one M, two I's, four S's, two A's, one U, and one G. Thus the number of distinguishable arrangements is
$$\dfrac{11!}{1! \cdot 2! \cdot 4! \cdot 2! \cdot 1! \cdot 1!} = \dfrac{11 \cdot 10 \cdot 9 \cdot 8 \cdot 7 \cdot 6 \cdot 5 \cdot 4!}{(2)4!(2)}$$
$$= 415,800.$$

19. The number of ways 4 heads and 3 tails can occur in 7 tosses of a coin is the same as the number of distinguishable permutations in the "word" HHHHTTT, which is

$$\frac{7!}{4! \cdot 3!} = \frac{7 \cdot 6 \cdot 5 \cdot 4!}{4!(6)} = 35.$$

21. Since the order in which the calls are made is important, the number of possible schedules for the 6 calls is $_6P_6 = 6! = 720$.

23. The number of ways to assign 9 scientists so 3 work on project A, 3 work on B, and 3 work on C is $\dfrac{9!}{3!3!3!} = 1680$.

25. A response to the true-false questions can be considered an ordered arrangement of 10 letters, 5 of which are T's and 5 of which are F's. The number of different responses is $\dfrac{10!}{5! \cdot 5!} = \dfrac{10 \cdot 9 \cdot 8 \cdot 7 \cdot 6 \cdot 5!}{5!(5 \cdot 4 \cdot 3 \cdot 2 \cdot 1)} = 252$.

27. The number of ways to assign 15 clients to 3 caseworkers (cells) with 5 clients to each caseworker is $\dfrac{15!}{5! \cdot 5! \cdot 5!} = 756,756$.

29. a. Seven flags must be arranged: two are red (type 1), three are green (type 2), and two are yellow (type 3). Thus the number of distinguishable arrangements (messages) is $\dfrac{7!}{2! \cdot 3! \cdot 2!} = 210$.

    b. If exactly two yellow flags are used, then seven flags are involved and the number of different messages is

$$\frac{7!}{2! \cdot 3! \cdot 2!} = 210.$$ If all three yellow

flags are used, then eight flags are involved and the number of different messages is $\dfrac{8!}{2! \cdot 3! \cdot 3!} = 560$. Thus if at least two yellow flags are used, the number of different messages is $210 + 560 = 770$.

31. The order in which the securities go into the portfolio is not important. The number of ways to select 8 of 12 stocks is $_{12}C_8$. The number of ways to select 4 of 7 bonds is $_7C_4$. By the basic counting principle, the number of ways to create the portfolio is

$$_{12}C_8 \cdot {}_7C_4 = \frac{12!}{8!(12-8)!} \cdot \frac{7!}{4!(7-4)!}$$

$$= \frac{12!}{8! \cdot 4!} \cdot \frac{7!}{4! \cdot 3!} = \frac{12 \cdot 11 \cdot 10 \cdot 9 \cdot 8!}{8! \cdot 4 \cdot 3 \cdot 2 \cdot 1} \cdot \frac{7 \cdot 6 \cdot 5 \cdot 4!}{4! \cdot 3 \cdot 2 \cdot 1}$$

$$= 495 \cdot 35 = 17,325.$$

33. a. Selecting 3 of the 3 males can be done in only 1 way.

    b. Selecting 4 of the 4 females can be done in only 1 way.

    c. Selecting 2 males and 2 females can be considered as a two-stage process. In the first stage, 2 of the 3 males are selected (and order is not important), which can be done in $_3C_2$ ways. In the second stage, 2 of the 4 females are selected, which can be done in $_4C_2$ ways. By the basic counting principle, the ways of selecting the subcommittee is $_3C_2 \cdot {}_4C_2 = \dfrac{3!}{2!(3-2)!} \cdot \dfrac{4!}{2!(4-2)!}$

$$= \frac{3!}{2! \cdot 1!} \cdot \frac{4!}{2! \cdot 2!} = 3 \cdot 6 = 18$$

35. There are 4 cards of a given denomination and the number of ways of selecting 3 cards of that denomination is $_4C_3$.
Since there are 13 denominations, the number of ways of selecting 3 cards of one denomination is $13 \cdot {}_4C_3$. After that selection is made, the 2 other cards must be of the same denomination (of which 12 denominations remain). Thus for the remaining 2 cards there are $12 \cdot {}_4C_2$ selections. By the basic counting principle, the number of possible full-house hands is

$$13 \cdot {}_4C_3 \cdot 12 \cdot {}_4C_2 = 13 \cdot \frac{4!}{3! \cdot 1!} \cdot 12 \cdot \frac{4!}{2! \cdot 2!}$$

$$= 13 \cdot 4 \cdot 12 \cdot 6 = 3744.$$

37. This situation can be considered as placing 18 tourists into 3 cells: 7 tourist go to the 7-passenger tram, 8 go to the 8-passenger tram, and 3 tourists remain at the bottom of the mountain. This can be done in

$$\frac{18!}{7! \cdot 8! \cdot 3!} = 5,250,960 \text{ ways.}$$

Principles in Practice 8.3

1. This is a combination problem because the order in which the videos are selected is not important. The number of possible choices is the number of ways 3 videos can be selected from 400 without regard to order.

$$_{400}C_3 = \frac{400!}{3!(400-3)!} = \frac{400!}{3!397!}$$

$$= \frac{400 \cdot 399 \cdot 398 \cdot 397!}{3!397!}$$

$$= \frac{400 \cdot 399 \cdot 398}{3 \cdot 2}$$

$$= 10,586,800$$

Problems 8.3

1. {9D, 9H, 9C, 9S]

3. {1H, 1T, 2H, 2T, 3H, 3T, 4H, 4T, 5H, 5T, 6H, 6T}

5. [64, 69, 60, 61, 46, 49, 40, 41, 96, 94, 90, 91, 06, 04, 09, 01, 16, 14, 19, 10]

7. a. {RR, RW, RB, WR, WW, WB, BR, BW, BB};

　　b. {RW, RB, WR, WB, BR, BW}

9. Sample space consists of ordered sets of six elements and each element is H or T. Since there are two possibilities for each toss (H or T), and there are six tosses, by he basic counting principle, the number of sample points is $2 \cdot 2 \cdot 2 \cdot 2 \cdot 2 \cdot 2 = 2^6 = 64$.

11. Sample space consists of ordered pairs where the first element indicates the card drawn (52 possibilities) and the second element indicates the number on the die (6 possibilities). By the basic counting principle, the number of sample points is $52 \cdot 6 = 312$.

13. Sample space consists of combinations of 52 cards taken 10 at a time. Thus the number of sample points is $_{52}C_{10}$.

15. The sample points that are either in E, or in F, or in both E and F are 1, 3, 5, 7, and 9. Thus $E \cup F = \{1, 3, 5, 7, 9\}$.

17. The sample points in S that are not in E are 2, 4, 6, 7, 8, 9, and 10. Thus $E' = \{2, 4, 6, 7, 8, 9, 10\}$.

The sample points common to both E' and F are 7 and 9. Thus $E' \cap F = \{7, 9\}$.

19. The sample points in S that are not in F are 1, 2, 4, 6, 8, and 10. Thus $F' = \{1, 2, 4, 6, 8, 10\}$.

21. $(F \cap G)' = \varnothing' = S$

23. $E_1 \cap E_2 \neq \varnothing$; $E_1 \cap E_3 \neq \varnothing$; $E_1 \cap E_4 = \varnothing$; $E_2 \cap E_3 = \varnothing$; $E_2 \cap E_4 \neq \varnothing$; $E_3 \cap E_4 = \varnothing$. Thus E_1 and E_4, E_2 and E_3, and E_3 and E_4 are mutually exclusive.

25. $E \cap F \neq \varnothing$, $E \cap G = \varnothing$, $E \cap H \neq \varnothing$, $E \cap I \neq \varnothing$, $F \cap G \neq \varnothing$, $F \cap H \neq \varnothing$ $F \cap I = \varnothing$, $G \cap H = \varnothing$, $G \cap I = \varnothing$, $H \cap I \neq \varnothing$. Thus E and G, F and I, G and H, and G and I are mutually exclusive.

27. a.　$S = \{\text{HHH, HHT, HTH, HTT, THH,}$
　　　　$\text{THT, TTH, TTT}\}$

b.　$E_1 = \{\text{HHH, HHT, HTH, HTT, THH,}$
　　　$\text{THT, TTH}\}$

c.　$E_2 = \{\text{HHT, HTH, HTT, THH, THT,}$
　　　$\text{TTH, TTT}\}$

d.　$E_1 \cup E_2 = \{\text{HHH, HHT, HTH, HTT,}$
　　　$\text{THH, THT, TTH, TTT}\} = S$

e.　$E_1 \cap E_2 = \{\text{HHT, HTH, HTT, THH,}$
　　　$\text{THT, TTH}\}$

f.　$(E_1 \cup E_2)' = S' = \varnothing$

g.　$(E_1 \cap E_2)' = \{\text{HHT, HTH, HTT, THH,}$
　　　$\text{THT, TTH}\}' = \{\text{HHH, TTT}\}$

29. a.　$\{\text{ABC, ACB, BAC, BCA, CAB, CBA}\}$

b.　$\{\text{ABC, ACB}\}$

c.　$\{\text{BAC, BCA, CAB, CBA}\}$

31. Using the properties in Table 8.1, we have
$(E \cap F) \cap (E \cap F')$
$= (E \cap F \cap E) \cap F'$　　[property 15]
$= (E \cap E \cap F) \cap F'$　　[property 11]
$= (E \cap E) \cap (F \cap F')$　[porperty 15]
$= E \cap \varnothing$　　　　　　[property 5]
$= \varnothing$　　　　　　　　　[property 9]
Thus
$(E \cap F) \cap (E \cap F') = \varnothing$, so $E \cap F$ and $E \cap F'$
are mutually exclusive.

Problems 8.4

1. $3000 P(E) = 3000(0.25) = 750$

3. a. $P(E') = 1 - P(E) = 1 - 0.2 = 0.8$

b. $P(E \cup F) = P(E) + P(F) - P(E \cap F)$
　　　$= 0.2 + 0.3 - 0.1 = 0.4$

5. If E and F are mutually exclusive, then
$E \cap F = \varnothing$.
Thus $P(E \cap F) = P(\varnothing) = 0$. Since it is
given that $P(E \cap F) = 0.831 \neq 0$, E and F
are not mutually exclusive.

7. a.　E_8
　　　$= \{(2, 6), (3, 5), (4, 4), (5, 3), (6, 2)\}$
　　　$P(E_8) = \dfrac{n(E_8)}{n(S)} = \dfrac{5}{36}$

b.　$E_{2 \text{ or } 3} = \{(1, 1), (1, 2), (2, 1)\}$
　　　$P(E_{2 \text{ or } 3}) = \dfrac{n(E_{2 \text{ or } 3})}{n(S)} = \dfrac{3}{36} = \dfrac{1}{12}$

c.　$E_{3, 4, \text{ or } 5} = \{(1,2),(2,1),(1,3),(2,2),(3,1),$
　　　　　　　　$(1,4),(2,3),(3,2),(4,1)\}$
　　　$P(E_{3, 4, \text{ or } 5}) = \dfrac{n(E_{3, 4, \text{ or } 5})}{n(S)} = \dfrac{9}{36} = \dfrac{1}{4}$

d.　$E_{12 \text{ or } 13} = E_{12}$, since E_{13} is an
　　　impossible event.
　　　$E_{12} = \{(6,6)\}$
　　　$P(E_{12 \text{ or } 13}) = \dfrac{n(E_{12 \text{ or } 13})}{n(S)} = \dfrac{1}{36}$

e.　$E_2 = \{(1,1)\}$
　　　$E_4 = \{(1,3),(2,2),(3,1)\}$
　　　$E_6 = \{(1,5),(2,4),(3,3,),(4,2),(5,1)\}$
　　　$E_8 = \{(2,6),(3,5),(4,4),(5,3),(6,2)\}$
　　　$E_{10} = \{(4,6),(5,5),(6,4)\}$
　　　$E_{12} = \{(6,6)\}$
　　　$P(E_{\text{even}}) = P(E_2) + P(E_4)$
　　　$+ P(E_6) + P(E_8) + P(E_{10}) + P(E_{12})$
　　　$= \dfrac{1}{36} + \dfrac{3}{36} + \dfrac{5}{36} + \dfrac{5}{36} + \dfrac{3}{36} + \dfrac{1}{36}$
　　　$= \dfrac{18}{36} = \dfrac{1}{2}$

f. $P(E_{\text{odd}}) = 1 - P(E_{\text{even}}) = 1 - \dfrac{1}{2} = \dfrac{1}{2}$

g. $E'_{\text{less than 10}} = E_{10} \cup E_{11} \cup E_{12}$

$= \{(4,6),(5,5),(6,4)\} \cup \{(5,6),(6,5)\}$

$\cup \{(6,6)\}$

$= \{(4,6),(5,5),(6,4),(5,6),(6,5),(6,6)\}$.

$P(E_{\text{less than 10}}) = 1 - P(E'_{\text{less than 10}})$

$= 1 - \dfrac{6}{36} = \dfrac{30}{36} = \dfrac{5}{6}$.

9. $n(S) = 52$.

a. $P(\text{king of hearts})$

$= \dfrac{n(E_{\text{king of hearts}})}{n(S)} = \dfrac{1}{52}$

b. $P(\text{diamond}) = \dfrac{n(E_{\text{diamond}})}{n(S)} = \dfrac{13}{52} = \dfrac{1}{4}$

c. $P(\text{jack}) = \dfrac{n(E_{\text{jack}})}{n(S)} = \dfrac{4}{52} = \dfrac{1}{13}$

d $P(\text{red}) = \dfrac{n(E_{\text{red}})}{n(S)} = \dfrac{26}{52} = \dfrac{1}{2}$

e. Because a heart is not a club,
$E_{\text{heart}} \cap E_{\text{club}} = \varnothing$.
Thus
$P(E_{\text{heart or club}}) = P(E_{\text{heart}} \cup E_{\text{club}})$
$= P(E_{\text{heart}}) + P(E_{\text{club}})$
$= \dfrac{n(E_{\text{heart}})}{n(S)} + \dfrac{n(E_{\text{club}})}{n(S)} = \dfrac{13}{52} + \dfrac{13}{52}$
$= \dfrac{26}{52} = \dfrac{1}{2}$

f. $E_{\text{club and 4}} = \{4\text{C}\}$

$P(E_{\text{club and 4}}) = \dfrac{n(E_{\text{club and 4}})}{n(S)} = \dfrac{1}{52}$

g. $P(\text{club or 4})$
$= P(\text{club}) + P(4) - P(\text{club and 4})$
$= \dfrac{13}{52} + \dfrac{4}{52} - \dfrac{1}{52} = \dfrac{16}{52} = \dfrac{4}{13}$

h. $E_{\text{red and king}} = \{\text{KH, KD}\}$

$P(\text{red and king}) = \dfrac{n(E_{\text{red and king}})}{n(S)}$

$= \dfrac{2}{52} = \dfrac{1}{26}$

i. $E_{\text{spade and heart}} = \varnothing$
Thus $P(\text{spade and heart}) = 0$

11. $n(S) = 2 \cdot 6 \cdot 52 = 624$

a. $P(\text{tail, 3, queen of hearts})$
$= \dfrac{n(E_{\text{T,3,QH}})}{n(S)} = \dfrac{1 \cdot 1 \cdot 1}{624} = \dfrac{1}{624}$

b. $P(\text{tail, 3, queen})$
$= \dfrac{n(E_{\text{T,3,Q}})}{n(S)} = \dfrac{1 \cdot 1 \cdot 4}{624} = \dfrac{1}{156}$

c. $P(\text{head, 2 or 3, queen})$
$= \dfrac{n(E_{\text{H,2 or 3,Q}})}{n(S)} = \dfrac{1 \cdot 2 \cdot 4}{624} = \dfrac{1}{78}$

d. $P(\text{head, even, diamond})$
$= \dfrac{n(E_{\text{H,E,D}})}{n(S)} = \dfrac{1 \cdot 3 \cdot 13}{624} = \dfrac{1}{16}$

13. $n(S) = 52 \cdot 51 \cdot 50 = 132{,}600$

a. $P(\text{all kings}) = \dfrac{4 \cdot 3 \cdot 2}{132{,}600} = \dfrac{1}{5525}$

b. $P(\text{all hearts}) = \dfrac{13 \cdot 12 \cdot 11}{132{,}600} = \dfrac{11}{850}$

15. $n(S) = 2 \cdot 2 \cdot 2 = 8$

    a. $E_{3 \text{ girls}} = \{GGG\}$

$$P(3 \text{ girls}) = \frac{n(E_{3 \text{ girls}})}{n(S)} = \frac{1}{8}$$

    b. $E_{1 \text{ boy}} = \{BGG, GBG, GGB\}$

$$P(1 \text{ boy}) = \frac{n(E_{1 \text{ boy}})}{n(S)} = \frac{3}{8}$$

    c. $E_{\text{no girl}} = \{BBB\}$

$$P(\text{no girl}) = \frac{n(E_{\text{no girl}})}{n(S)} = \frac{1}{8}$$

    d. $P(\text{at least 1 girl}) = 1 - P(\text{no girl})$

$$= 1 - \frac{1}{8} = \frac{7}{8}$$

17. The sample space consists of 60 stocks. Thus $n(S) = 60$.

    a. $P(6\% \text{ or more}) = \dfrac{n(E_{6\% \text{ or more}})}{n(S)}$

$$= \frac{48}{60} = \frac{4}{5}$$

    b. $P(\text{less than } 6\%) = 1 - P(6\% \text{ or more})$

$$= 1 - \frac{4}{5} = \frac{1}{5}$$

19. $n(S) = 40$
Of the 40 students, 4 received an A, 10 a B, 14 a C, 10 a D, and 2 an F.

    a. $P(A) = \dfrac{n(E_A)}{n(S)} = \dfrac{4}{40} = \dfrac{1}{10} = 0.1$

    b. $P(A \text{ or } B) = \dfrac{n(E_{A \text{ or } B})}{n(S)} = \dfrac{4 + 10}{40}$

$$= \frac{14}{40} = 0.35$$

    c. $P(\text{neither D nor F}) = P(A, B, \text{ or } C)$

$$= \frac{n(E_{A, B, \text{ or } C})}{n(S)}$$

$$= \frac{4 + 10 + 14}{40} = \frac{28}{40} = 0.7$$

    d. $P(\text{no F}) = 1 - P(F) = 1 - \dfrac{n(E_F)}{n(S)}$

$$= 1 - \frac{2}{40} = \frac{38}{40} = 0.95$$

    e. Let N = number of students. Then $n(S) = N$. Of the N students, $0.10N$ received an A, $0.25N$ a B, $0.35N$ a C, $0.25N$ a D, $0.05N$ an F.

$$P(A) = \frac{0.10N}{N} = 0.1$$

$$P(A \text{ or } B) = \frac{0.10N + 0.25N}{N}$$

$$= \frac{0.35N}{N} = 0.35$$

$$P(\text{neither D nor F}) = P(A, B, \text{ or } C)$$

$$= \frac{0.10N + 0.25N + 0.35N}{N}$$

$$= \frac{0.70N}{N} = 0.7$$

$$P(\text{no F}) = 1 - P(F)$$

$$= 1 - \frac{0.05N}{N} = 1 - 0.05 = 0.95$$

21. The sample space consists of combinations of 2 people selected from 5. Thus

$$n(S) = {}_5C_2 = \frac{5!}{2! \cdot 3!} = \frac{5 \cdot 4}{2} = 10. \text{ Because}$$

there are only 2 women in the group, the number of possible 2-woman committees is 1. Thus $P(2 \text{ women}) = \dfrac{n(E_{2 \text{ women}})}{n(S)} = \dfrac{1}{10}$.

23. Number of ways to answer exam is
$2^{10} = 1024 = n(S)$.

  a. There is only one way to achieve 100 points, namely to answer each question correctly. Thus

  $$P(100 \text{ points}) = \frac{n(E_{100 \text{ points}})}{n(S)} = \frac{1}{1024}.$$

  b. Number of ways to score 90 points = number of ways that exactly one question is answered incorrectly = 10.
  Thus
  $P(90 \text{ or more points})$
  $= P(90 \text{ points}) + P(100 \text{ points})$
  $$= \frac{10}{1024} + \frac{1}{1024} = \frac{11}{1024}.$$

25. A poker hand is a 5-card deal from 52 cards. Thus $n(S) = {}_{52}C_5$. In 52 cards, there are 4 cards of a particular denomination. Thus, for a four of a kind, the number of ways of selecting 4 of 4 cards of a particular denomination is ${}_4C_4$. Since there are 13 denominations, 4 cards of the same denomination can be dealt in $13 \cdot {}_4C_4$ ways. For the remaining card, there are 12 denominations that are possible, and for each denomination there are ${}_4C_1$ ways of dealing a card. Thus

$$P(\text{four of a kind}) = \frac{n(E_{\text{four of a kind}})}{n(S)}$$
$$= \frac{13 \cdot {}_4C_4 \cdot 12 \cdot {}_4C_1}{{}_{52}C_5}$$
$$= \frac{13 \cdot 12 \cdot 4}{{}_{52}C_5}$$

27. $n(S) = {}_{100}C_3 = \dfrac{100!}{3! \cdot 97!} = 161,700$

  a. $n(E_{3 \text{ females}}) = {}_{35}C_3 = \dfrac{35!}{3! \cdot 32!} = 6545$

  $$P(E_{3 \text{ females}}) = \frac{n(E_{3 \text{ females}})}{n(S)}$$
  $$= \frac{6545}{161,700} \approx 0.040$$

  b. The number of ways of selecting one professor is 15; the number of ways of selecting two associate professors is ${}_{24}C_2$.

  Thus $n(E_{1 \text{ professor \& 2 associate professors}})$
  $$= 15 \cdot \frac{24!}{2! \cdot 22!} = 15 \cdot 276 = 4140.$$
  Therefore,
  $P(E_{1 \text{ professor \& 2 associate professors}})$
  $$= \frac{4140}{161,700} \approx 0.026.$$

29. Shiloh needs to win 3 more rounds to win the game and Caitlin needs to win 5 more rounds. Shiloh's probability of winning is

$$\sum_{k=0}^{4} \frac{{}_7C_k}{2^7}$$
$$= \frac{1}{2^7} \sum_{k=0}^{4} {}_7C_k$$
$$= \frac{1}{2^7} ({}_7C_0 + {}_7C_1 + {}_7C_2 + {}_7C_3 + {}_7C_4)$$
$$= \frac{1}{2^7} (1 + 7 + 21 + 35 + 35)$$
$$= \frac{99}{128}$$

Shiloh's share of the pot is then
$$\frac{99}{128}(\$25) \approx \$19.34.$$

31. Let $p = P(1) = P(3) = P(5)$. Then
$2p = P(2) = P(4) = P(6)$. Since $P(S) = 1$,
then $3(p) + 3(2p) = 1$, $9p = 1$, $p = p(1) = \dfrac{1}{9}$.

33. a. Of the 100 voters, 51 favor the tax increase. Thus
$P(\text{favors tax increase}) = \dfrac{51}{100} = 0.51$.

   b. Of the 100 voters, 44 oppose the tax increase. Thus
$P(\text{opposes tax increase}) = \dfrac{44}{100} = 0.44$.

   c. Of the 100 voters, 3 are Republican with no opinion. Thus
$P(\text{is a Republican with no opinion})$
$= \dfrac{3}{100}$
$= 0.03$.

35. $\dfrac{P(E)}{P(E')} = \dfrac{P(E)}{1 - P(E)} = \dfrac{\frac{4}{5}}{1 - \left(\frac{4}{5}\right)} = \dfrac{\frac{4}{5}}{\frac{1}{5}} = \dfrac{4}{1}$

The odds are 4:1.

37. $\dfrac{P(E)}{P(E')} = \dfrac{P(E)}{1 - P(E)} = \dfrac{0.7}{1 - 0.7} = \dfrac{0.7}{0.3} = \dfrac{7}{3}$
The odds are 7:3.

39. $P(E) = \dfrac{7}{7 + 5} = \dfrac{7}{12}$

41. $P(E) = \dfrac{4}{4 + 10} = \dfrac{4}{14} = \dfrac{2}{7}$

43. Odds that it will rain tomorrow
$= \dfrac{P(\text{rain})}{P(\text{no rain})} = \dfrac{0.75}{1 - 0.75} = \dfrac{0.75}{0.25} = 3$.
The odds are 3:1.

Problems 8.5

1. a. $P(E|F) = \dfrac{n(E \cap F)}{n(F)} = \dfrac{1}{5}$

   b. Using the result of part (a),
$P(E'|F) = 1 - P(E|F) = 1 - \dfrac{1}{5} = \dfrac{4}{5}$.

   c. $F' = \{3, 7, 8, 9\}$ so
$P(E|F') = \dfrac{n(E \cap F')}{n(F')} = \dfrac{1}{4}$.

   d. $P(F|E) = \dfrac{n(F \cap E)}{n(E)} = \dfrac{1}{2}$

   e. $F \cap G = \{5, 6\}$ so
$P(E|F \cap G)$
$= \dfrac{n(E \cap (F \cap G))}{n(F \cap G)} = \dfrac{0}{2} = 0.$

3. $P(E|E) = \dfrac{P(E \cap E)}{P(E)} = \dfrac{P(E)}{P(E)} = 1$

5. $P(E'|F) = 1 - P(E|F) = 1 - 0.57 = 0.43$

7. a. $P(E|F) = \dfrac{P(E \cap F)}{P(F)} = \dfrac{1/6}{1/3} = \dfrac{1}{2}$

   b. $P(F|E) = \dfrac{P(F \cap E)}{P(E)} = \dfrac{1/6}{1/4} = \dfrac{2}{3}$

9. a. $P(F|E) = \dfrac{P(F \cap E)}{P(E)} = \dfrac{1/6}{1/4} = \dfrac{2}{3}$

   b. $P(E \cup F) = P(E) + P(F) - P(E \cap F)$
$\dfrac{7}{12} = \dfrac{1}{4} + P(F) - \dfrac{1}{6}$
Thus $P(F) = \dfrac{7}{12} - \dfrac{1}{4} + \dfrac{1}{6} = \dfrac{1}{2}.$

c. From part (b) $P(F) = \dfrac{1}{2}$.

Then $P(E \mid F) = \dfrac{P(E \cap F)}{P(F)} = \dfrac{1/6}{1/2} = \dfrac{1}{3}$.

d. $P(E) = P(E \cap F) + P(E \cap F')$

$\dfrac{1}{4} = \dfrac{1}{6} + P(E \cap F')$

so $P(E \cap F') = \dfrac{1}{4} - \dfrac{1}{6} = \dfrac{1}{12}$.

Then $P(E \mid F') = \dfrac{P(E \cap F')}{P(F')}$

$= \dfrac{1/12}{1 - 1/2} = \dfrac{1/12}{1/2} = \dfrac{1}{6}$.

11. a. $P(F) = \dfrac{125}{200} = \dfrac{5}{8}$

b. $P(F \mid \text{II}) = \dfrac{n(F \cap \text{II})}{n(\text{II})} = \dfrac{35}{58}$

c. $P(O \mid \text{I}) = \dfrac{n(O \cap \text{I})}{n(\text{I})} = \dfrac{22}{78} = \dfrac{11}{39}$

d. $P(\text{III}) = \dfrac{64}{200} = \dfrac{8}{25}$

e. $P(\text{III} \mid O) = \dfrac{n(\text{III} \cap O)}{n(O)} = \dfrac{10}{47}$

f. $P(\text{II} \mid N') = \dfrac{n(\text{II} \cap N')}{n(N')}$

$= \dfrac{35 + 15}{125 + 47} = \dfrac{50}{172} = \dfrac{25}{86}$

13. a. $P(A \mid B) = \dfrac{P(A \cap B)}{P(B)} = \dfrac{0.20}{0.40} = \dfrac{1}{2}$

b. $P(B \mid A) = \dfrac{P(B \cap A)}{P(A)} = \dfrac{0.20}{0.45} = \dfrac{4}{9}$

15. $S = \{BB, BG, GG, GB\}$
Let $E = \{\text{at least one girl}\} = \{BG, GG, GB\}$,
$F = \{\text{at least one boy}\} = \{BB, BG, GB\}$.
Thus $P(E \mid F) = \dfrac{n(E \cap F)}{n(F)} = \dfrac{2}{3}$.

17. $S = \{HHH, HHT, HTH, THH, THT, TTH,$
$TTT\}$.
Let $E = \{\text{exactly two tails}\}$
$= \{HTT, THT, TTH\}$,
$F = \{\text{second toss is a tail}\}$
$= \{HTH, HTT, TTH, TTT\}$,
$G = \{\text{second toss is a head}\}$
$= \{HHH, HHT, THH, THT\}$.

a. $P(E \mid F) = \dfrac{n(E \cap F)}{n(F)} = \dfrac{2}{4} = \dfrac{1}{2}$

b. $P(E \mid G) = \dfrac{n(E \cap G)}{n(G)} = \dfrac{1}{4}$

19. $P(< 4 \mid \text{odd}) = \dfrac{n(< 4 \cap \text{odd})}{n(\text{odd})}$

$= \dfrac{n(\{1, 3\})}{n(\{1, 3, 5\})} = \dfrac{2}{3}$

21. *Method 1.* The usual sample space has
36 outcomes, where the event "two 1's" is
$\{(1, 1)\}$. Note that
$\{\text{at least one 1}\}' = \{\text{no 1's}\}$, and the event
"no 1's" occurs in $5 \cdot 5 = 25$ ways. Thus
$P(\text{two 1's} \mid \text{at least one 1})$

$= \dfrac{n(\text{two 1's} \cap \text{at least one 1})}{n(\text{at least one 1})}$

$= \dfrac{n(\{(1, 1)\})}{36 - 25} = \dfrac{1}{11}$

Method 2. From the usual sample space, we
find that the reduced sample space for "at
least one 1" (which has 11 outcomes) is
$\{(1, 1), (1, 2), (1, 3), (1, 4), (1, 5), (1, 6),$
$(2, 1), (3, 1), (4, 1), (5, 1)\}$.

Thus $P(\text{two 1's} \mid \text{at least one 1}) = \dfrac{1}{11}$.

23. The usual sample space consists of ordered pairs (R, G), where R = no. on red die and G = no. on green die. Now, $n(\text{green is even}) = 6 \cdot 3 = 18$, because the red die can show any of six numbers and the green any of three: 2, 4, or 6. Also,
$n(\text{total of } 7 \cap \text{green even}) = n(\{(5, 2), (3, 4), (1, 6)\}) = 3$.

Thus $P(\text{total of } 7 | \text{green even}) = \dfrac{n(\text{total of } 7 \cap \text{green even})}{n(\text{green even})} = \dfrac{3}{18} = \dfrac{1}{6}$.

25. The usual sample space consists of 36 ordered pairs. Let $E = \{\text{total} > 7\}$ and $F = \{\text{first toss} > 3\}$. Then
$n(F) = 3 \cdot 6 = 18$ and $n(E \cap F)$
$= n(\{(4, 4), (4, 5), (4, 6), (5, 3), (5, 4), (5, 5), (5, 6), (6, 2), (6, 3), (6, 4), (6, 5), (6, 6)\})$
$= 12$

Thus $P\left(E|F\right) = \dfrac{n(E \cap F)}{n(F)} = \dfrac{12}{18} = \dfrac{2}{3}$.

27. $P(K \mid H) = \dfrac{n(K \cap H)}{n(H)} = \dfrac{1}{13}$

29. Let $E = \{\text{second card is not a face card}\}$ and $F = \{\text{first card is a face card}\}$.

$P(E \mid F) = \dfrac{n(E \cap F)}{n(F)} = \dfrac{12 \cdot \frac{51-11}{51}}{12} = \dfrac{40}{51}$

31. $P(K_1 \cap Q_2 \cap J_3) = P(K_1)P(Q_2 \mid K_1)P(J_3 \mid (K_1 \cap Q_2))$

$= \dfrac{4}{52} \cdot \dfrac{4}{51} \cdot \dfrac{4}{50} = \dfrac{8}{16,575}$

33. $P(J_1 \cap J_2 \cap J_3) = P(J_1)P(J_2 \mid J_1)P(J_3 \mid (J_1 \cap J_2))$

$= \dfrac{4}{52} \cdot \dfrac{3}{51} \cdot \dfrac{2}{50} = \dfrac{1}{5525}$

35. Let $D = \{\text{two diamonds}\}$ and
$R = \{\text{first card red}\}$. We have $D \cap R = \{\text{two diamonds}\} = D$ and

$P(D) = \dfrac{13}{52} \cdot \dfrac{12}{51}$.

Thus $P(D \mid R) = \dfrac{P(D \cap R)}{P(R)} = \dfrac{\frac{13}{52} \cdot \frac{12}{51}}{\frac{26}{52}} = \dfrac{2}{17}$.

37. a. $P(U) = P(F \cap U) + P(O \cap U) + P(N \cap U)$
$= P(F)P(U|F) + P(O)P(U|O) + P(N)P(U|N)$
$= (0.60)(0.45) + (0.30)(0.55) + (0.10)(0.35)$
$= 0.47 = \dfrac{47}{100}$

b. $P(F \mid U) = \dfrac{P(F \cap U)}{P(U)} = \dfrac{(0.60)(0.45)}{0.47} = \dfrac{27}{47}$

39. a. After the first draw, if the rabbit drawn is red, then 4 rabbits remain, 3 of which are yellow.

$P(\text{second is yellow} \mid \text{first is red}) = \dfrac{3}{4}$

b. After red rabbit is replaced, 5 rabbits remain, 3 of which are yellow.

$P(\text{second is yellow} \mid \text{first is red}) = \dfrac{3}{5}$

41. $P(W) = P(\text{Box } 1 \cap W) + P(\text{Box } 2 \cap W)$

$= P(\text{Box } 1)P(W \mid \text{Box } 1) + P(\text{Box } 2)P(W \mid \text{Box } 2) = \dfrac{1}{2} \cdot \dfrac{2}{5} + \dfrac{1}{2} \cdot \dfrac{2}{4} = \dfrac{9}{20}$

43. $P(W_2) = P(B1 \cap G_1 \cap W_2) + P(B1 \cap R_1 \cap W_2) + P(B2 \cap W_1 \cap W_2)$

$= P(B1)P\big(G_1 \big| B1\big)P\big(W_2 \big| (G_1 \cap B1)\big) + P(B1)P\big(R_1 \big| B1\big)P\big(W_2 \big| (R_1 \cap B1)\big)$
$\qquad + P(B2)P\big(W_1 \big| B2\big)P\big(W_2 \big| (W_1 \cap B2)\big)$

$= \dfrac{1}{2} \cdot \dfrac{1}{2} \cdot \dfrac{1}{3} + \dfrac{1}{2} \cdot \dfrac{1}{2} \cdot \dfrac{1}{3} + \dfrac{1}{2} \cdot \dfrac{1}{2} \cdot \dfrac{1}{3} = \dfrac{1}{4}$

45. $P(\text{Und.}) = P(MS \cap \text{Und.}) + P(DS \cap \text{Und.})$

$\qquad = P(MS)P(\text{Und.}|MS) + P(DS)P(\text{Und}|DS)$

$\qquad = \dfrac{20,000}{60,000} \cdot \dfrac{1}{100} + \dfrac{40,000}{60,000} \cdot \dfrac{3}{100}$

$\qquad = \dfrac{7}{300}$

47. $P(\text{Def}) = P(A \cap \text{Def}) + P(B \cap \text{Def}) + P(C \cap \text{Def})$

$\qquad = P(A)P(\text{Def} \mid A) + P(B)P(\text{Def} \mid B) + P(C)P(\text{Def} \mid C)$

$\qquad = (0.10)(0.06) + (0.20)(0.04) + (0.70)(0.05) = 0.049$

49. a. $P(D \cap V) = P(D)P(V \mid D) = (0.40)(0.15) = 0.06$

b. $P(V) = P(D \cap V) + P(R \cap V) + P(I \cap V)$

$\qquad = P(D)P(V \mid D) + P(R)P(V \mid R) + P(I)P(V \mid I)$
$\qquad = (0.40)(0.15) + (0.35)(0.20) + (0.25)(0.10)$
$\qquad = 0.155$

51. $P(3 \text{ Fem}|\text{at least one Fem})$

$= \dfrac{P(3 \text{ Fem} \cap \text{at least one Fem})}{P(\text{at least one Fem})}$

$= \dfrac{P(3 \text{ Fem})}{1 - P(\text{no Fem})} = \dfrac{\frac{_6C_3}{_{11}C_3}}{1 - \frac{_5C_3}{_{11}C_3}} = \dfrac{\frac{4}{33}}{1 - \frac{2}{33}} = \dfrac{4}{31}$

Problems 8.6

1. a. $P(E \cap F) = P(E)P(F) = \dfrac{1}{3} \cdot \dfrac{3}{4} = \dfrac{1}{4}$

b. $P(E \cup F) = P(E) + P(F) - P(E \cap F)$

$= \dfrac{1}{3} + \dfrac{3}{4} - \dfrac{1}{4} = \dfrac{5}{6}$

c. $P(E \mid F) = P(E) = \dfrac{1}{3}$

d. $P(E' \mid F) = 1 - P(E \mid F) = 1 - \dfrac{1}{3} = \dfrac{2}{3}$

e. $P(E \cap F') = P(E)P(F') = \dfrac{1}{3} \cdot \dfrac{1}{4} = \dfrac{1}{12}$

f. $P(E \cup F') = P(E) + P(F') - P(E \cap F')$

$= \dfrac{1}{3} + \dfrac{1}{4} - \dfrac{1}{12} = \dfrac{1}{2}$

g. $P(E \mid F') = \dfrac{P(E \cap F')}{P(F')} = \dfrac{1/12}{1/4} = \dfrac{1}{3}$

3. $P(E \cap F) = P(E)P(F)$,

$\dfrac{1}{9} = \dfrac{2}{7} \cdot P(F)$ so $P(F) = \dfrac{1}{9} \cdot \dfrac{7}{2} = \dfrac{7}{18}$

5. $P(E)P(F) = \dfrac{3}{4} \cdot \dfrac{8}{9} = \dfrac{2}{3} = P(E \cap F)$

Since $P(E)P(F) = P(E \cap F)$, events E and F are independent.

7. Let $F = \{\text{full service}\}$ and $I = \{\text{increase in value}\}$.

$P(F) = \dfrac{400}{600} = \dfrac{2}{3}$

and $P(F \mid I) = \dfrac{n(F \cap I)}{n(I)} = \dfrac{320}{480} = \dfrac{2}{3}$

Since $P(F \mid I) = P(F)$, events F and I are independent.

9. Let S be the usual sample space consisting of ordered pairs of the form (R, G), where the first component of each pair represents the number showing on the red die, and the second component represents the number on the green die. Then $n(S) = 6 \cdot 6 = 36$. For E, any number of four can occur on the red die, and any number on the green die. Thus $n(E) = 4 \cdot 6 = 24$. For F we have $F = \{(2, 6), (3, 5), (4, 4), (5, 3), (6, 2)\}$, so $n(F) = 5$.

Also, $E \cap F = \{(4, 4), (5, 3), (6, 2)\}$, so $n(E \cap F) = 3$. Thus

$P(E)P(F) = \dfrac{24}{36} \cdot \dfrac{5}{36} = \dfrac{5}{54}$ and

$P(E \cap F) = \dfrac{3}{36} = \dfrac{1}{12}$. Since

$P(E)P(F) \ne P(E \cap F)$, events E and F are dependent.

11. $S = \{HH, HT, TH, TT\}$,
$E = \{HT, TH, TT\}$,
$F = \{HT, TH\}$, and $E \cap F = \{HT, TH\}$.

Thus $P(E) = \dfrac{3}{4}$

$P(F) = \dfrac{2}{4} = \dfrac{1}{2}$, and

$P(E \cap F) = \dfrac{2}{4} = \dfrac{1}{2}$. We have

$P(E)P(F) = \dfrac{3}{4} \cdot \dfrac{1}{2} = \dfrac{3}{8} \ne P(E \cap F)$, so

events E and F are dependent.

13. Let S be the set of ordered pairs whose first (*second*) component represents the number on the first (*second*) chip. Then $n(S) = 7 \cdot 7 = 49$, $n(E) = 1 \cdot 7 = 7$, and $n(F) = 7 \cdot 1 = 7$. For G, if the first chip is 1, 3, 5 or 7, then the second chip must be 2, 4 or 6; if the first chip is 2, 4 or 6, the second must be 1, 3, 5 or 7. Thus $n(G) = 4 \cdot 3 + 3 \cdot 4 = 24$.

    a. $E \cap F = \{(3,3)\}$, so $P(E \cap F) = \dfrac{1}{49}$. Since

$$P(E)P(F) = \frac{7}{49} \cdot \frac{7}{49} = \frac{1}{49} = P(E \cap F),$$ events E and F are independent.

    b. $E \cap G = \{(3,2),(3,4),(3,6)\}$,

         so $P(E \cap G) = \dfrac{3}{49}$. Since $P(E)P(G) = \dfrac{7}{49} \cdot \dfrac{24}{49} = \dfrac{24}{343} \neq P(E \cap G)$, events E and G are dependent.

    c. $F \cap G = \{(2,3),(4,3),(6,3)\}$

         so $P(F \cap G) = \dfrac{3}{49}$.

         Since $P(F)P(G) = \dfrac{7}{49} \cdot \dfrac{24}{49} = \dfrac{24}{343} \neq P(F \cap G)$.

         Events F and G are dependent.

    d. $E \cap F \cap G = \varnothing$, so $P(E \cap F \cap G) = 0$.

         However, $P(E)P(F)P(G) \neq 0 = P(E \cap F \cap G)$,

         so events E, F and G are not independent.

15. $P(E \cap F) = P(E)P(F \mid E)$, thus

$$P(E) = \frac{P(E \cap F)}{P(F \mid E)} = \frac{0.3}{0.4} = 0.75$$

     Since $P(E) = 0.75 \neq 0.5 = P(E \mid F)$, E and F are dependent.

17. Let $E = \{\text{red } 4\}$ and $F = \{\text{green} > 4\}$. Assume E and F are independent.

$$P(E \cap F) = P(E)P(F) = \frac{1}{6} \cdot \frac{1}{3} = \frac{1}{18}$$

19. Let $F = \{\text{first person attends regularly}\}$ and $S = \{\text{second person attends regularly}\}$.

     Then $P(F \cap S) = P(F)P(S) = \dfrac{1}{5} \cdot \dfrac{1}{5} = \dfrac{1}{25}$.

21. Because of replacement, assume the cards selected on the draws are independent events.
P(ace, then face card, then spade) = P(ace) · P(face card) · P(spade)

$$= \frac{4}{52} \cdot \frac{12}{52} \cdot \frac{13}{52} = \frac{3}{676}$$

23. a. $P(\text{Bill gets A} \cap \text{Jim gets A} \cap \text{Linda gets A})$

$= P(\text{Bill gets A}) \cdot P(\text{Jim gets A}) \cdot P(\text{Linda gets A})$

$= \dfrac{3}{4} \cdot \dfrac{1}{2} \cdot \dfrac{4}{5} = \dfrac{3}{10}.$

b. $P(\text{Bill no A} \cap \text{Jim no A} \cap \text{Linda no A})$

$= P(\text{Bill no A}) \cdot P(\text{Jim no A}) \cdot P(\text{Linda no A})$

$= \dfrac{1}{4} \cdot \dfrac{1}{2} \cdot \dfrac{1}{5} = \dfrac{1}{40}$

c. $P(\text{Bill no A} \cap \text{Jim no A} \cap \text{Linda gets A})$

$= P(\text{Bill no A}) \cdot P(\text{Jim no A}) \cdot P(\text{Linda gets A})$

$= \dfrac{1}{4} \cdot \dfrac{1}{2} \cdot \dfrac{4}{5} = \dfrac{1}{10}$

25. Let $A = \{\text{A survives 15 more years}\}$, $B = \{\text{B survives 15 more years}\}$.

a. $P(A \cap B) = P(A)P(B) = \dfrac{2}{3} \cdot \dfrac{3}{5} = \dfrac{2}{5}$

b. $P(A' \cap B) = P(A')P(B) = \dfrac{1}{3} \cdot \dfrac{3}{5} = \dfrac{1}{5}$

c. $A \cap B'$ and $A' \cap B$ are mutually exclusive.

$P[(A \cap B') \cup (A' \cap B)] = P(A)P(B') + P(A')P(B) = \dfrac{2}{3} \cdot \dfrac{2}{5} + \dfrac{1}{3} \cdot \dfrac{3}{5} = \dfrac{7}{15}$

d. $P(\text{at least one survives}) = P(\text{exactly one survives}) + P(\text{both survive}) = \dfrac{7}{15} + \dfrac{2}{5} = \dfrac{13}{15}.$

e. $P(\text{neither survives}) = 1 - P(\text{at least one survives}) = 1 - \dfrac{13}{15} = \dfrac{2}{15}.$

27. Assume the colors selected on the draws are independent events.

a. $P(W_1 \cap G_2) = P(W_1)P(G_2) = \dfrac{7}{18} \cdot \dfrac{6}{18} = \dfrac{7}{54}$

b. $P[(R_1 \cap W_2) \cup (W_1 \cap R_2)] = P(R_1)P(W_2) + P(W_1)P(R_2) = \dfrac{5}{18} \cdot \dfrac{7}{18} + \dfrac{7}{18} \cdot \dfrac{5}{18} = \dfrac{35}{162}$

29. Assume that the selections are independent.

$P(\text{both red} \cup \text{both white} \cup \text{both green}) = \dfrac{3}{19} \cdot \dfrac{3}{19} + \dfrac{7}{19} \cdot \dfrac{7}{19} + \dfrac{9}{19} \cdot \dfrac{9}{19} = \dfrac{139}{361}$

31. Assume that the draws are independent.
P(particular 1st ticket $\cap$ particular 2nd ticket)

$$= \frac{1}{20} \cdot \frac{1}{20} = \frac{1}{400}$$

P(sum is 35) $= P\{(20, 15), (19, 16), (18, 17), (17, 18), (16, 19), (15, 20)\}$

$$= 6\left(\frac{1}{400}\right) = \frac{3}{200}$$

33. a. $\dfrac{1}{12} \cdot \dfrac{1}{12} \cdot \dfrac{1}{12} = \dfrac{1}{1728}$

b. To get exactly one even, there are $_3C_1 = 3$ ways.
P(one even and two odd) $= 3[P(\text{even 1st spin}) \cdot P(\text{odd 2nd spin}) \cdot P(\text{odd 3rd spin})]$

$$= 3\left(\frac{6}{12} \cdot \frac{6}{12} \cdot \frac{6}{12}\right) = \frac{3}{8}.$$

35. a. The number of ways of getting exactly four correct answers out of five is $_5C_4 = 5$. Each of these

ways has a probability of $\dfrac{1}{4} \cdot \dfrac{1}{4} \cdot \dfrac{1}{4} \cdot \dfrac{1}{4} \cdot \dfrac{3}{4} = \dfrac{3}{1024}$. Thus

$$P(\text{exactly 4 correct}) = 5 \cdot \frac{3}{1024} = \frac{15}{1024}.$$

b. P(at least 4 correct) $= P$(exactly 4) $+ P$(exactly 5)

$$= \frac{15}{1024} + \frac{1}{4} \cdot \frac{1}{4} \cdot \frac{1}{4} \cdot \frac{1}{4} \cdot \frac{1}{4} = \frac{1}{64}$$

c. The number of ways of getting exactly three correct answers out of five is

$_5C_3 = 10$. Each of these ways has a probability of $\dfrac{1}{4} \cdot \dfrac{1}{4} \cdot \dfrac{1}{4} \cdot \dfrac{3}{4} \cdot \dfrac{3}{4} = \dfrac{9}{1024}$, so

$$P(\text{exactly 3 correct}) = 10 \cdot \frac{9}{1024} = \frac{45}{512}. \text{ Thus}$$

P(3 or more correct) $= P$(exactly 3) $+ P$(at least 4)

$$= \frac{45}{512} + \frac{1}{64} = \frac{53}{512}.$$

37. A wrong majority decision can occur in one of two mutually exclusive ways: exactly two wrong recommendations, or three wrong recommendations. Exactly two wrong recommendations can occur in $_3C_2 = 3$ mutually exclusive ways. Thus
P(wrong majority decision)
$= [(0.04)(0.05)(0.9) + (0.04)(0.95)(0.1) + (0.96)(0.05)(0.1)] + (0.04)(0.05)(0.1)$
$= 0.0106.$

Problems 8.7

1. $P(E \mid D) = \dfrac{P(E)P(D \mid E)}{P(E)P(D \mid E) + P(F)P(D \mid F)} = \dfrac{\frac{2}{5} \cdot \frac{1}{10}}{\frac{2}{5} \cdot \frac{1}{10} + \frac{3}{5} \cdot \frac{1}{5}} = \dfrac{1}{4}$

For the second part, $P(D' \mid F) = 1 - P(D \mid F) = 1 - \dfrac{1}{5} = \dfrac{4}{5}$, and

$P(D' \mid E) = 1 - P(D \mid E) = 1 - \dfrac{1}{10} = \dfrac{9}{10}$. Then

$P(F \mid D') = \dfrac{P(F)P(D' \mid F)}{P(E)P(D' \mid E) + P(F)P(D' \mid F)} = \dfrac{\frac{3}{5} \cdot \frac{4}{5}}{\frac{2}{5} \cdot \frac{9}{10} + \frac{3}{5} \cdot \frac{4}{5}} = \dfrac{4}{7}$.

3. $D = \{\text{is Democrat}\}$,
$R = \{\text{is Republican}\}$,
$I = \{\text{is Independent}\}$,
$V = \{\text{voted}\}$.

$P(D \mid V) = \dfrac{P(D)P(V \mid D)}{P(D)P(V \mid D) + P(R)P(V \mid R) + P(I)P(V \mid I)}$

$= \dfrac{(0.42)(0.25)}{(0.42)(0.25) + (0.33)(0.27) + (0.25)(0.15)}$

$= \dfrac{175}{386} \approx 0.453$

5. $D = \{\text{has the disease}\}$
$D' = \{\text{does not have the disease}\}$
$R = \{\text{positive reaction}\}$
$N = \{\text{negative reaction}\} = R'$

a. $P(D \mid R) = \dfrac{P(D)P(R \mid D)}{P(D)P(R \mid D) + P(D')P(R \mid D')} = \dfrac{(0.03)(0.86)}{(0.03)(0.86) + (0.97)(0.07)} = \dfrac{258}{937} \approx 0.275$

b. $P(D \mid N) = \dfrac{P(D)P(N \mid D)}{P(D)P(N \mid D) + P(D')P(N \mid D')} = \dfrac{(0.03)(0.14)}{(0.03)(0.14) + (0.97)(0.93)} = \dfrac{14}{3021} \approx 0.005$

7. $B_1 = \{\text{first bag selected}\}$
$B_2 = \{\text{second bag selected}\}$
$R = \{\text{red jelly bean drawn}\}$
$P(B_1) = P(B_2) = \dfrac{1}{2}$.

$P(B_1 \mid R) = \dfrac{P(B_1)P(R \mid B_1)}{P(B_1)P(R \mid B_1) + P(B_2)P(R \mid B_2)} = \dfrac{\frac{1}{2} \cdot \frac{4}{6}}{\frac{1}{2} \cdot \frac{4}{6} + \frac{1}{2} \cdot \frac{2}{5}} = \dfrac{5}{8}$.

9. $A = \{$unit from line A$\}$
$B = \{$unit from line B$\}$
$D = \{$defective unit$\}$.

$$P(A) = \frac{300}{800} = \frac{3}{8}$$

$$P(B) = \frac{500}{800} = \frac{5}{8}$$

$$P(A \mid D) = \frac{P(A)P(D \mid A)}{P(A)P(D \mid A) + P(B)P(D \mid B)} = \frac{\frac{3}{8} \cdot \frac{2}{100}}{\frac{3}{8} \cdot \frac{2}{100} + \frac{5}{8} \cdot \frac{5}{100}} = \frac{6}{31}$$

11. $C = \{$call made$\}$
$T = \{$on time for meeting$\}$

$$P(C \mid T) = \frac{P(C)P(T \mid C)}{P(C)P(T \mid C) + P(C')P(T \mid C')}$$

$$= \frac{(0.95)(0.9)}{(0.95)(0.9) + (0.05)(0.75)} = \frac{114}{119} \approx 0.958$$

13. $W = \{$walking reported$\}$
$B = \{$bicycling reported$\}$
$R = \{$running reported$\}$
$C = \{$completed requirement$\}$

$$P(W \mid C) = \frac{P(W)P(C \mid W)}{P(W)P(C \mid W) + P(B)P(C \mid B) + P(R)P(C \mid R)} = \frac{\frac{1}{2} \cdot \frac{9}{10}}{\frac{1}{2} \cdot \frac{9}{10} + \frac{1}{4} \cdot \frac{4}{5} + \frac{1}{4} \cdot \frac{2}{3}} = \frac{27}{49} \approx 0.551$$

55.1% would be expected to report walking.

15. $J = \{$had Japanese-made car$\}$
$E = \{$had European-made car$\}$
$A = \{$had American-made car$\}$
$B = \{$buy same make again$\}$

$$P(J \mid B) = \frac{P(J)P(B \mid J)}{P(J)P(B \mid J) + P(E)P(B \mid E) + P(A)P(B \mid A)} = \frac{\frac{3}{5} \cdot \frac{85}{100}}{\frac{3}{5} \cdot \frac{85}{100} + \frac{1}{10} \cdot \frac{50}{100} + \frac{3}{10} \cdot \frac{40}{100}} = \frac{3}{4}$$

17. $P = \{$pass the exam$\}$
$A = \{$answer every question$\}$

$$P(A \mid P) = \frac{P(A)P(P \mid A)}{P(A)P(P \mid A) + P(A')P(P \mid A')} = \frac{(0.75)(0.8)}{(0.75)(0.8) + (0.25)(0.50)} = \frac{24}{29} \approx 0.828$$

19. $S = \{$signals sent$\}$
$D = \{$signals detected$\}$

$$P(S \mid D) = \frac{P(S)P(D \mid S)}{P(S)P(D \mid S) + P(S')P(D \mid S')} = \frac{\frac{2}{5} \cdot \frac{3}{5}}{\frac{2}{5} \cdot \frac{3}{5} + \frac{3}{5} \cdot \frac{1}{10}} = \frac{4}{5}$$

21. $S = \{\text{movie is a success}\}$
$U = \{\text{"Two Thumbs Up"}\}$

$$P(S \mid U) = \frac{P(S)P(U \mid S)}{P(S)P(U \mid S) + P(S')P(U \mid S')} = \frac{\frac{8}{10} \cdot \frac{70}{100}}{\frac{8}{10} \cdot \frac{70}{100} + \frac{2}{10} \cdot \frac{20}{100}} = \frac{14}{15} \approx 0.933$$

23. $S = \{\text{is substandard request}\}$
$C = \{\text{is considered substandard request by Blackwell}\}$

a. $P(C) = P(S)P(C \mid S) + P(S')P(C \mid S') = (0.20)(0.75) + (0.8)(0.15) = 0.27 = \dfrac{27}{100}$

b. $P(S \mid C) = \dfrac{P(S)P(C \mid S)}{P(S)P(C \mid S) + P(S')P(C \mid S')} = \dfrac{(0.20)(0.75)}{0.27} = \dfrac{0.15}{0.27} = \dfrac{15}{27} \approx 0.556$

c. $P(\text{Error}) = P(C' \cap S) + P(C \cap S')$
$= P(S)P(C' \mid S) + P(S')P(C \mid S')$
$= (0.20)(0.25) + (0.80)(0.15) = 0.17 = \dfrac{17}{100}$

25. a. $P(L \mid E) = \dfrac{P(L)P(E \mid L)}{P(L)P(E \mid L) + P(M)P(E \mid M) + P(H)P(E \mid H)}$

$= \dfrac{(0.25)(0.49)}{(0.25)(0.49) + (0.25)(0.64) + (0.5)(0.81)} \approx 0.18$

b. $P(M \mid E) = \dfrac{P(M)P(E \mid M)}{P(L)P(E \mid L) + P(M)P(E \mid M) + P(H)P(E \mid H)}$

$= \dfrac{(0.25)(0.64)}{(0.25)(0.49) + (0.25)(0.64) + (0.5)(0.81)} \approx 0.23$

c. $P(H \mid E) = \dfrac{P(H)P(E \mid H)}{P(L)P(E \mid L) + P(M)P(E \mid M) + P(H)P(E \mid H)}$

$= \dfrac{(0.5)(0.81)}{(0.25)(0.49) + (0.25)(0.64) + (0.5)(0.81)} \approx 0.59$

d. High quality

27. $F = \{\text{fair weather}\}$
$I = \{\text{inclement weather}\}$
$W = \{\text{predict fair weather}\}.$

$$P(F \mid W) = \frac{P(F)P(W \mid F)}{P(F)P(W \mid F) + P(I)P(W \mid I)} = \frac{(0.6)(0.7)}{(0.6)(0.7) + (0.4)(0.3)} = \frac{7}{9} \approx 0.78$$

Chapter 8 Review Problems

1. $_8P_3 = 8 \cdot 7 \cdot 6 = 336$

3. $_9C_7 = \dfrac{9!}{7!(9-7)!} = \dfrac{9!}{7! \cdot 2!}$
    $= \dfrac{9 \cdot 8 \cdot 7!}{7! \cdot 2 \cdot 1} = \dfrac{9 \cdot 8}{2} = 36$

5. For each of the first 3 characters there are 26 choices, while for each of the last 3 characters there are 10 choices. By the basic counting principle, the number of license plates that are possible is
    $26 \cdot 26 \cdot 26 \cdot 10 \cdot 10 \cdot 10 = 17{,}576{,}000.$

7. Each of the five switches has 2 possible positions. By the basic counting principle, the number of different codes is
    $2 \cdot 2 \cdot 2 \cdot 2 \cdot 2 = 2^5 = 32.$

9. A possibility for first, second, and third place is a selection of three of the seven teams so that order is important. Thus the number of ways the season can end is
    $_7P_3 = 7 \cdot 6 \cdot 5 = 210.$

11. The order of the group is not important. Thus the number of groups that can board is
    $_{11}C_6 = \dfrac{11!}{6! \cdot 5!} = \dfrac{11 \cdot 10 \cdot 9 \cdot 8 \cdot 7 \cdot 6!}{5 \cdot 4 \cdot 3 \cdot 2 \cdot 1 \cdot 6!} = 462.$

13. **a.** Three bulbs are selected from 24, and the order of selection is not important. Thus the number of possible selections is $_{24}C_3 = \dfrac{24!}{3!(24-3)!} = \dfrac{24!}{3! \cdot 21!}$
    $= \dfrac{24 \cdot 23 \cdot 22 \cdot 21!}{3 \cdot 2 \cdot 1 \cdot 21!} = \dfrac{24 \cdot 23 \cdot 22}{3 \cdot 2 \cdot 1} = 2024.$

    b. Only one bulb is defective and that bulb must be included in the selection. The other two bulbs must be selected from the 23 remaining bulbs and there are $_{23}C_2$ such selections possible. Thus the number of ways of selecting three bulbs

such that one is defective is
$1 \cdot {}_{23}C_2 = {}_{23}C_2 = \dfrac{23!}{2!(23-2)!} = \dfrac{23!}{2! \cdot 21!}$
$\dfrac{23 \cdot 22 \cdot 21!}{2 \cdot 1 \cdot 21!} = \dfrac{23 \cdot 22}{2 \cdot 1} = 253.$

15. In the word MISSISSIPPI, there are 11 letters with repetition: 1 M, 4 I's, 4 S's, and 2 P's. Thus the number of distinguishable permutations is
    $\dfrac{11!}{1! \cdot 4! \cdot 4! \cdot 2!} = 34{,}650.$

17. Of the nine professors, four go to Dalhousie University (Cell A), three go to St. Mary's (Cell B), and two are not assigned (Cell C). The number of possible assignments is
    $\dfrac{9!}{4! \cdot 3! \cdot 2!} = 1260.$

19. **a.** $E_1 \cup E_2 = \{1, 2, 3, 4, 5, 6, 7\}$

    b. $E_1 \cap E_2 = \{4, 5, 6\}$

    c. $E_1' \cup E_2 = \{7, 8\} \cup \{4, 5, 6, 7\}$
    $ = \{4, 5, 6, 7, 8\}$

    d. The intersection of any event and its complement is $\varnothing$.

    e. $(E_1 \cap E_2')'$
    $= (\{1, 2, 3, 4, 5, 6\} \cap \{1, 2, 3, 8\})'$
    $= \{1, 2, 3\}' = \{4, 5, 6, 7, 8\}$

    f. From (b), $E_1 \cap E_2 \neq \varnothing$, so E_1 and E_2 are not mutually exclusive.

21. **a.** $\{R_1R_2R_3, R_1R_2G_3, R_1G_2R_3,$
    $R_1G_2G_3, G_1R_2R_3, G_1R_2G_3,$
    $G_1G_2R_3, G_1G_2G_3\}$

    b. $\{R_1R_2G_3, R_1G_2R_3, G_1R_2R_3\}$

    c. $\{R_1R_2R_3, G_1G_2G_3\}$

23. $n(S) = {}_{10}C_2 = \dfrac{10!}{2! \cdot 8!} = \dfrac{10 \cdot 9 \cdot 8!}{2 \cdot 1 \cdot 8!} = \dfrac{10 \cdot 9}{2 \cdot 1} = 45$

Let E be the event that box is rejected. If box is rejected, the one defective chip must be in the two-chip sample and there are nine possibilities for the other chip. Thus

$n(E) = 9$ and $P(E) = \dfrac{n(E)}{n(S)} = \dfrac{9}{45} = \dfrac{1}{5} = 0.2$.

25. Number of ways to answer exam is $4^5 = 1024 = n(S)$. Let $E = \{$exactly two questions are incorrect$)$. The

number of ways of selecting two of the five questions that are incorrect is ${}_5C_2 = \dfrac{5!}{2! \cdot 3!} = 10$. However,

there are three ways to answer a question incorrectly. Since two questions are incorrect

$n(E) = 10 \cdot 3 \cdot 3 = 90$. Thus $P(E) = \dfrac{n(E)}{n(S)} = \dfrac{90}{1024} = \dfrac{45}{512}$.

27. a. There are 10 jelly beans in the bag.
$n(S) = 10 \cdot 10 = 100$
$n(E_{\text{both red}}) = 4 \cdot 4 = 16$

Thus $P(E_{\text{both red}}) = \dfrac{n(E_{\text{both red}})}{n(S)} = \dfrac{16}{100} = \dfrac{4}{25}$.

b. $n(S) = 10 \cdot 9 = 90$
$n(E_{\text{both red}}) = 4 \cdot 3 = 12$

Thus $P(E_{\text{both red}}) = \dfrac{12}{90} = \dfrac{2}{15}$.

29. $n(S) = 52 \cdot 52 \cdot 52$.

a. There are 26 black cards in a deck. Thus $n(E_{\text{all black}}) = 26 \cdot 26 \cdot 26$ and

$P(E_{\text{all black}}) = \dfrac{26 \cdot 26 \cdot 26}{52 \cdot 52 \cdot 52} = \dfrac{1}{8}$.

b. There are 13 diamonds in a deck, none of which are black. If $E =$ event that two cards are black and the other is a diamond, then E occurs if the diamond is the first, second, or third card. Thus

$n(E) = 13 \cdot 26 \cdot 26 + 26 \cdot 13 \cdot 26 + 26 \cdot 26 \cdot 13 = 3 \cdot 13 \cdot 26 \cdot 26$ and $P(E) = \dfrac{3 \cdot 13 \cdot 26 \cdot 26}{52 \cdot 52 \cdot 52} = \dfrac{3}{16}$.

31. $\dfrac{P(E)}{P(E')} = \dfrac{\frac{3}{8}}{1 - \left(\frac{3}{8}\right)} = \dfrac{\frac{3}{8}}{\frac{5}{8}} = \dfrac{3}{5}$ or 3:5

33. $P(E) = \dfrac{6}{6+1} = \dfrac{6}{7}$

35. $P\left(F'|H\right) = \dfrac{P(F' \cap H)}{P(H)} = \dfrac{\frac{10}{52}}{\frac{1}{4}} = \dfrac{10}{13}$

37. $P(S \cap M) = P(S)P(M \mid S)$
$= (0.6)(0.7) = 0.42$

39. a. The reduced sample space consists of
$\{(4, 1), (4, 2), (4, 3), (4, 4), (4, 5), (4, 6), (1, 4), (2, 4), (3, 4), (5, 4), (6, 4)\}.$
In two of these 11 points, the sum of the components is 7. Thus
$$P(\text{sum} = 7 \mid a\ 4\ \text{shows}) = \dfrac{2}{11}.$$

b. Out of 36 sample points, the event
$\{\text{getting a total of 7 and having a 4 show}\}$ is $\{(4, 3), (3, 4)\}$. Thus the probability of this event is
$$\dfrac{2}{36} = \dfrac{1}{18}.$$

41. The second number must be a 1 or 2, so the reduced sample space has $6 \cdot 2 = 12$ sample points. Of these, the event $\{\text{first number} \le \text{second number}\}$ consists of $(1, 1)$, $(1, 2)$, and $(2, 2)$. Thus the conditional probability is $\dfrac{3}{12} = \dfrac{1}{4}$.

43. a. $P(L' \mid F) = \dfrac{n(L' \cap F)}{n(F)} = \dfrac{160}{480} = \dfrac{1}{3}$

b. $P(L) = \dfrac{400}{600} = \dfrac{2}{3}$ and
$$P(L|M) = \dfrac{n(L \cap M)}{n(M)} = \dfrac{80}{120} = \dfrac{2}{3}.$$
Since $P(L|M) = P(L)$, events L and M are independent.

45. $P = \{\text{attend public college}\}$
$M = \{\text{from middle-class family}\}$
$$P(P) = \dfrac{125}{175} = \dfrac{5}{7}$$
$$P(P \mid M) = \dfrac{n(P \cap M)}{n(M)} = \dfrac{55}{80} = \dfrac{11}{16}$$
Since $P(P \mid M) \ne P(P)$, events P and M are dependent.

47. a. $P(\text{none take root}) = (0.3)(0.3)(0.3)(0.3) = 0.0081$

  b. The probability that a particular two shrubs take root and the remaining two do not is $(0.7)(0.7)(0.3)(0.3)$. The number of ways the two that take root can be chosen from the four shrubs is $_4C_2$. Thus $P(\text{exactly two take root}) = {}_4C_2(0.7)^2(0.3)^2 = 0.2646$.

  c. For at most two shrubs to take root, either none does, exactly one does, or exactly two do.
$P(\text{none}) + P(\text{exactly one}) + P(\text{exactly two})$

$$= 0.0081 + {}_4C_1(0.7)(0.3)^3 + 0.2646$$
$$= 0.0081 + 0.0756 + 0.2646$$
$$= 0.3483$$

49. $P(R_{\text{II}}) = P(G_{\text{I}})P(R_{\text{II}} \mid G_{\text{I}}) + P(R_{\text{I}})P(R_{\text{II}} \mid R_{\text{I}})$

$$= \frac{3}{5} \cdot \frac{4}{9} + \frac{2}{5} \cdot \frac{5}{9} = \frac{22}{45}.$$

51. $P(G \mid A) = \dfrac{P(G \cap A)}{P(A)} = \dfrac{0.1}{0.4} = \dfrac{1}{4}$

53. **a.** $F = \{\text{produced by first shift}\}$
$S = \{\text{produced by second shift}\}$
$D = \{\text{scratched}\}$
$P(D) = P(F)P(D|F) + P(S)P(D|S)$

$$= \frac{3000}{8000} \cdot (0.01) + \frac{5000}{8000} \cdot (0.02)$$
$$= 0.00375 + 0.0125 = 0.01625$$

  b. $P(F \mid D) = \dfrac{P(F)P(D \mid F)}{P(F)P(D \mid F) + P(S)P(D \mid S)}$

$$= \frac{0.00375}{0.01625} = \frac{3}{13} \approx 0.23$$

Mathematical Snapshot Chapter 8

  1. Trial and error should yield a critical value of around 0.645.

Chapter 9

Problems 9.1

1. $\mu = \sum_x x f(x) = 0(0.1) + 1(0.4) + 2(0.2) + 3(0.3) = 1.7$

$\text{Var}(X) = \sum_x x^2 f(x) - \mu^2 = [0^2(0.1) + 1^2(0.4) + 2^2(0.2) + 3^2(0.3)] - (1.7)^2 = 1.01$

$\sigma = \sqrt{\text{Var}(X)} = \sqrt{1.01} \approx 1.00$

3. $\mu = \sum_x x f(x) = 1\left(\dfrac{1}{4}\right) + 2\left(\dfrac{1}{4}\right) + 3\left(\dfrac{1}{2}\right) = \dfrac{9}{4} = 2.25$

$\text{Var}(X) = \sum_x x^2 f(x) - \mu^2 = \left[1^2\left(\dfrac{1}{4}\right) + 2^2\left(\dfrac{1}{4}\right) + 3^2\left(\dfrac{1}{2}\right)\right] - \left(\dfrac{9}{4}\right)^2 = \dfrac{11}{16} = 0.6875$

$\sigma = \sqrt{\dfrac{11}{16}} = \dfrac{\sqrt{11}}{4} \approx 0.83$

5. a. $P(X = 3) = 1 - [P(X = 5) + P(X = 6) + P(X = 7)] = 1 - [0.3 + 0.2 + 0.4] = 0.1$

b. $\mu = \sum_x x f(x) = 3(0.1) + 5(0.3) + 6(0.2) + 7(0.4) = 5.8$

c. $\sigma^2 = \sum_x x^2 f(x) - \mu^2 = [3^2(0.1) + 5^2(0.3) + 6^2(0.2) + 7^2(0.4)] - (5.8)^2 = 1.56$

7. Distribution of X:

$f(0) = \dfrac{1}{8}, f(1) = \dfrac{3}{8}, f(2) = \dfrac{3}{8}, f(3) = \dfrac{1}{8}$

$E(X) = \sum_x x f(x) = 0\left(\dfrac{1}{8}\right) + 1\left(\dfrac{3}{8}\right) + 2\left(\dfrac{3}{8}\right) + 3\left(\dfrac{1}{8}\right) = \dfrac{12}{8} = \dfrac{3}{2} = 1.5$

$$\sigma^2 = \text{Var}(X) = \sum_x x^2 f(x) - [E(x)]^2$$

$$= \left[0^2 \left(\frac{1}{8} \right) + 1^2 \left(\frac{3}{8} \right) + 2^2 \left(\frac{3}{8} \right) + 3^2 \left(\frac{1}{8} \right) \right] - \left(\frac{3}{2} \right)^2$$

$$= \frac{24}{8} - \frac{9}{4} = \frac{6}{8} = \frac{3}{4} = 0.75$$

$$\sigma = \sqrt{\frac{3}{4}} = \frac{\sqrt{3}}{2} \approx 0.87$$

9. The number of outcomes in the sample space is $_5C_2 = 10$.

Distribution of X:

$$f(0) = \frac{_2C_2}{10} = \frac{1}{10}, \, f(1) = \frac{_2C_1 \cdot _3C_1}{10} = \frac{3}{5},$$

$$f(2) = \frac{_3C_2}{10} = \frac{3}{10}$$

$$E(X) = \sum_x x f(x) = 0 \left(\frac{1}{10} \right) + 1 \left(\frac{3}{5} \right) + 2 \left(\frac{3}{10} \right) = \frac{6}{5} = 1.2$$

$$\sigma^2 = \sum_x x^2 f(x) - [E(x)]^2$$

$$= \left[0^2 \left(\frac{1}{10} \right) + 1^2 \left(\frac{3}{5} \right) + 2^2 \left(\frac{3}{10} \right) \right] - \left(\frac{6}{5} \right)^2$$

$$= \frac{9}{5} - \frac{36}{25} = \frac{9}{25} = 0.36$$

$$\sigma = \sqrt{\frac{9}{25}} = \frac{3}{5} = 0.6$$

11. $f(0) = P(X = 0) = \dfrac{_2C_2}{_5C_2} = \dfrac{1}{10}$

$f(1) = P(X = 1) = \dfrac{_3C_1 \cdot _2C_1}{_5C_2} = \dfrac{6}{10} = \dfrac{3}{5}$

$f(2) = P(X = 2) = \dfrac{_3C_2}{_5C_2} = \dfrac{3}{10}$

13. a. If X is the gain (in dollars), then $X = -2$ or 4998.
Distribution of X:

$$f(-2) = \frac{7999}{8000}, \; f(4998) = \frac{1}{8000}$$

$$E(x) = \sum_x xf(x)$$

$$= -2 \cdot \frac{7999}{8000} + 4998 \cdot \frac{1}{8000}$$

$$= -\frac{11,000}{8000} \approx -\$1.38 \text{ (a loss)}$$

b. Here $X = -4$ or 4996. Distribution of X:

$$f(-4) = \frac{7998}{8000}, \; f(4996) = \frac{2}{8000}$$

$$E(X) = \sum_x xf(x)$$

$$= -4 \cdot \frac{7998}{8000} + 4996 \cdot \frac{2}{8000}$$

$$= -\$2.75 \text{ (a loss)}$$

15. Let X = daily earnings (in dollars).
Distribution of X:

$$f(200) = \frac{4}{7}, f(-30) = \frac{3}{7}$$

$$E(X) = \sum_x x f(x)$$

$$= 200 \cdot \frac{4}{7} + (-30) \cdot \frac{3}{7}$$

$$= \frac{710}{7} \approx \$101.43$$

17. The probability that a person in the group is not hospitalized is
$1 - (0.001 + 0.002 + 0.003 + 0.004 + 0.008) = 0.982$.
Let X = gain (in dollars) to the company from a policy.
Distribution of X:

$f(10) = 0.982, f(-90) = 0.001, \; f(-190) = 0.002, \; f(-290) = 0.003,$

$f(-390) = 0.004, f(-490) = 0.008$

$E(X) = 10(0.982) + (-90)(0.001) + (-190)(0.002) + (-290)(0.003) \; + (-390)(0.004) + (-490)(0.008)$

$= \$3.00$

19. Let p = the annual premium (in dollars) per policy. If X = gain (in dollars) to the company from a policy, then either $X = p$ or $X = -(180,000 - p)$. We set $E(X) = 50$:

$$-(180,000 - p)(0.002) + p(0.998) = 50$$
$$-360 + 0.002p + 0.998p = 50$$
$$-360 + p = 50$$
$$p = \$410$$

21. Let X = gain (in dollars) on a play.

If 0 heads show, then $X = 0 - 1.25 = -\dfrac{5}{4}$.

If exactly 1 head shows, then $X = 1.00 - 1.25 = -\dfrac{1}{4}$.

If 2 heads show, then $X = 2.00 - 1.25 = \dfrac{3}{4}$.

Distribution of X:

$$f\left(-\frac{5}{4}\right) = \frac{1}{4}, f\left(-\frac{1}{4}\right) = \frac{1}{2}, f\left(\frac{3}{4}\right) = \frac{1}{4}$$

$$E(X) = \left(-\frac{5}{4}\right)\left(\frac{1}{4}\right) + \left(-\frac{1}{4}\right)\left(\frac{1}{2}\right) + \left(\frac{3}{4}\right)\left(\frac{1}{4}\right) = -\frac{1}{4} = -0.25$$

Thus there is an expected loss of \$0.25 on each play.
For a fair game, let p = amount (in dollars) paid to play.
Distribution of X:

$$f(-p) = \frac{1}{4}, f(1-p) = \frac{1}{2}, f(2-p) = \frac{1}{4}$$

We set $E(X) = 0$:

$$(-p)\frac{1}{4} + (1-p)\frac{1}{2} + (2-p)\frac{1}{4} = 0$$
$$-\frac{p}{4} + \frac{1}{2} - \frac{p}{2} + \frac{1}{2} - \frac{p}{4} = 0$$
$$1 - p = 0$$
$$p = 1$$

Thus you should pay \$1 for a fair game.

Principles in Practice 9.2

1. Here $p = 0.30$, $q = 1 - p = 0.70$, and $n = 4$.

$$P(X = x) = {}_nC_x p^x q^{n-x}, x = 0, 1, 2, 3, 4$$

$$P(X = 0) = {}_4C_0(0.3)^0(0.7)^4 = 0.2401 = \frac{2401}{10,000}$$

$$P(X = 1) = {}_4C_1(0.3)^1(0.7)^3 = 0.4116 = \frac{4116}{10,000}$$

$$P(X = 2) = {}_4C_2(0.3)^2(0.7)^2 = 0.2646 = \frac{2646}{10,000}$$

$P(X = 3) = {}_4C_3(0.3)^3(0.7)^1 = 0.0756$

$= \dfrac{756}{10,000}$

$P(X = 4) = {}_4C_4(0.3)^4(0.7)^0 = 0.0081$

$= \dfrac{81}{10,000}$

Problems 9.2

1. $f(0) = {}_2C_0\left(\dfrac{1}{5}\right)^0\left(\dfrac{4}{5}\right)^2 = \dfrac{2!}{0! \cdot 2!} \cdot 1 \cdot \dfrac{16}{25}$

$= 1 \cdot 1 \cdot \dfrac{16}{25} = \dfrac{16}{25}$

$f(1) = {}_2C_1\left(\dfrac{1}{5}\right)^1\left(\dfrac{4}{5}\right)^1 = \dfrac{2!}{1! \cdot 1!} \cdot \dfrac{1}{5} \cdot \dfrac{4}{5}$

$= 2 \cdot \dfrac{1}{5} \cdot \dfrac{4}{5} = \dfrac{8}{25}$

$f(2) = {}_2C_2\left(\dfrac{1}{5}\right)^2\left(\dfrac{4}{5}\right)^0 = \dfrac{2!}{2! \cdot 0!} \cdot \dfrac{1}{25} \cdot 1$

$= 1 \cdot \dfrac{1}{25} \cdot 1 = \dfrac{1}{25}.$

$\mu = np = 2 \cdot \dfrac{1}{5} = \dfrac{2}{5}$

$\sigma = \sqrt{npq} = \sqrt{2 \cdot \dfrac{1}{5} \cdot \dfrac{4}{5}}$

$= \sqrt{\dfrac{8}{25}} = \dfrac{2\sqrt{2}}{5}$

3. $f(0) = {}_3C_0\left(\dfrac{2}{3}\right)^0\left(\dfrac{1}{3}\right)^3 = 1 \cdot 1 \cdot \dfrac{1}{27} = \dfrac{1}{27}$

$f(1) = {}_3C_1\left(\dfrac{2}{3}\right)^1\left(\dfrac{1}{3}\right)^2 = \dfrac{3!}{1! \cdot 2!} \cdot \dfrac{2}{3} \cdot \dfrac{1}{9}$

$= 3 \cdot \dfrac{2}{3} \cdot \dfrac{1}{9} = \dfrac{2}{9}$

$f(2) = {}_3C_2\left(\dfrac{2}{3}\right)^2\left(\dfrac{1}{3}\right)^1 = \dfrac{3!}{2! \cdot 1!} \cdot \dfrac{4}{9} \cdot \dfrac{1}{3}$

$= 3 \cdot \dfrac{4}{9} \cdot \dfrac{1}{3} = \dfrac{4}{9}$

$f(3) = {}_3C_3\left(\dfrac{2}{3}\right)^3\left(\dfrac{1}{3}\right)^0 = \dfrac{3!}{3! \cdot 0!} \cdot \dfrac{8}{27} \cdot 1$

$= 1 \cdot \dfrac{8}{27} \cdot 1 = \dfrac{8}{27}$

$\mu = np = 3 \cdot \dfrac{2}{3} = 2; \ \sigma = \sqrt{npq} = \sqrt{3 \cdot \dfrac{2}{3} \cdot \dfrac{1}{3}}$

$= \sqrt{\dfrac{2}{3}} = \dfrac{\sqrt{6}}{3}$

5. $P(X = 5) = {}_6C_5(0.2)^5(0.8)^1$

$= 6(0.00032)(0.8) = 0.001536$

7. $P(X = 2) = {}_4C_2\left(\dfrac{4}{5}\right)^2\left(\dfrac{1}{5}\right)^2 = 6 \cdot \dfrac{16}{25} \cdot \dfrac{1}{25}$

$= \dfrac{96}{625} = 0.1536$

9. $P(X < 2) = P(X = 0) + P(X = 1)$

$= {}_5C_0\left(\dfrac{1}{2}\right)^0\left(\dfrac{1}{2}\right)^5 + {}_5C_1\left(\dfrac{1}{2}\right)^1\left(\dfrac{1}{2}\right)^4$

$= 1 \cdot 1 \cdot \dfrac{1}{32} + 5 \cdot \dfrac{1}{2} \cdot \dfrac{1}{16} = \dfrac{6}{32} = \dfrac{3}{16}$

11. Let $X =$ number of heads that occurs.

$p = \dfrac{1}{2}, n = 11$

$P(X = 8) = {}_{11}C_8\left(\dfrac{1}{2}\right)^8\left(\dfrac{1}{2}\right)^3$

$= 165 \cdot \dfrac{1}{256} \cdot \dfrac{1}{8}$

$= \dfrac{165}{2048} \approx 0.081$

13. Let X = number of green marbles drawn. The probability of selecting a green marble on any draw is $\dfrac{7}{12}$, $n = 4$.

$$P(X = 2) = {}_4C_2 \left(\frac{7}{12}\right)^2 \left(\frac{5}{12}\right)^2$$

$$= 6 \cdot \frac{49}{144} \cdot \frac{25}{144} = \frac{1225}{3456} \approx 0.3545$$

15. Let X = number of defective switches selected. The probability that a switch is defective is
$p = 0.02$, $n = 4$.

$$P(X = 2) = {}_4C_2 (0.02)^2 (0.98)^2$$
$$= 6(0.0004)(0.9604) \approx 0.002$$

17. Let X = number of heads that occurs.

$$p = \frac{1}{4}, n = 3$$

  a. $P(X = 2) = {}_3C_2 \left(\dfrac{1}{4}\right)^2 \left(\dfrac{3}{4}\right)^1 = \dfrac{9}{64}$

$$= 3 \cdot \frac{1}{16} \cdot \frac{3}{4}$$

  b. $P(X = 3) = {}_3C_3 \left(\dfrac{1}{4}\right)^3 \left(\dfrac{3}{4}\right)^0 = \dfrac{1}{64}$

$$= 1 \cdot \frac{1}{64} \cdot 1$$

  Thus

$$P(X = 2) + P(X = 3) = \frac{9}{64} + \frac{1}{64}$$

$$= \frac{10}{64} = \frac{5}{32}$$

19. Let X = number of defective in sample.

$$p = \frac{1}{5}, \ n = 6$$

$$P(X \le 1)$$
$$= P(X = 0) + P(X = 1)$$

$$= {}_6C_0 \left(\frac{1}{5}\right)^0 \left(\frac{4}{5}\right)^6 + {}_6C_1 \left(\frac{1}{5}\right)^1 \left(\frac{4}{5}\right)^5$$

$$= 1 \cdot 1 \cdot \frac{4096}{15,625} + 6 \cdot \frac{1}{5} \cdot \frac{1024}{3125}$$

$$= \frac{10,240}{15,625} = \frac{2048}{3125} \approx 0.655$$

21. Let X = number of hits in four at-bats.
$p = 0.300$, $n = 4$
$$P(X \ge 1) = 1 - P(X = 0)$$
$$= 1 - {}_4C_0 (0.300)^0 (0.700)^4$$
$$= 1 - 1 \cdot 1 \cdot (0.2401) = 0.7599$$

23. Let X = number of girls. The probability that a child is a girl is $p = \dfrac{1}{2}$. Here $n = 5$. We must find
$$P(X \ge 2) = 1 - P(X < 2)$$
$$= 1 - [P(X = 0) + P(X = 1)].$$

$$P(X = 0) = {}_5C_0 \left(\frac{1}{2}\right)^0 \left(\frac{1}{2}\right)^5 = 1 \cdot 1 \cdot \frac{1}{32} = \frac{1}{32}$$

$$P(X = 1) = {}_5C_1 \left(\frac{1}{2}\right)^1 \left(\frac{1}{2}\right)^4 = 5 \cdot \frac{1}{2} \cdot \frac{1}{16} = \frac{5}{32}$$

Thus,
$$P(X \ge 2) = 1 - [P(X = 0) + P(X = 1)]$$

$$= 1 - \left[\frac{1}{32} + \frac{5}{32}\right] = 1 - \frac{3}{16} = \frac{13}{16}$$

25. $\mu = 3, \sigma^2 = 2$

Since $\mu = np,$ then $np = 3.$ Since

$\sigma^2 = npq,$ then $(np)q = 2,$ or $3q = 2,$ so

$q = \dfrac{2}{3}.$ Thus, $p = 1 - q = 1 - \dfrac{2}{3} = \dfrac{1}{3}.$ Since

$np = 3,$ then $n \cdot \dfrac{1}{3} = 3,$ or $n = 9.$ Thus

$$P(X = 2) = {}_9C_2 \left(\frac{1}{3}\right)^2 \left(\frac{2}{3}\right)^7$$

$$= 36 \cdot \frac{1}{9} \cdot \frac{128}{2187} = \frac{512}{2187} \approx 0.234.$$

Problems 9.3

1. $\begin{bmatrix} \frac{1}{2} & \frac{2}{3} \\ -\frac{3}{2} & \frac{1}{3} \end{bmatrix}$

No, since the entry at row 2 column 1 is negative.

3. $\begin{bmatrix} \frac{1}{2} & \frac{1}{8} & \frac{1}{3} \\ -\frac{1}{4} & \frac{5}{8} & \frac{1}{3} \\ \frac{3}{4} & \frac{1}{4} & \frac{1}{3} \end{bmatrix}$

No, since there is a negative entry.

5. $\begin{bmatrix} 0.4 & 0 & 0.5 \\ 0.2 & 0.1 & 0.3 \\ 0.4 & 0.9 & 0.2 \end{bmatrix}$

Yes, since all entries are nonnegative and the sum of the entries in each column is 1.

7. $\begin{bmatrix} \frac{2}{3} & b \\ a & \frac{1}{4} \end{bmatrix}$

$\dfrac{2}{3} + a = 1,$ so $a = \dfrac{1}{3}.$

$b + \dfrac{1}{4} = 1,$ so $b = \dfrac{3}{4}.$

9. $\begin{bmatrix} 0.4 & a & a \\ a & 0.1 & b \\ 0.3 & b & c \end{bmatrix}$

$0.4 + a + 0.3 = 1,$ so $a = 0.3.$
$a + 0.1 + b = 1, 0.3 + 0.1 + b = 1,$ so $b = 0.6.$
$a + b + c = 1, 0.3 + 0.6 + c = 1,$ so $c = 0.1.$

11. $\begin{bmatrix} 0.4 \\ 0.6 \end{bmatrix}$

Yes, all entries are nonnegative and their sum is 1.

13. $\begin{bmatrix} 0.2 \\ 0.7 \\ 0.5 \end{bmatrix}$

No, the sum of the entries is not 1.

15. $\mathbf{X}_1 = \mathbf{TX}_0 = \begin{bmatrix} \frac{2}{3} & 1 \\ \frac{1}{3} & 0 \end{bmatrix} \begin{bmatrix} \frac{1}{4} \\ \frac{3}{4} \end{bmatrix} = \begin{bmatrix} \frac{11}{12} \\ \frac{1}{12} \end{bmatrix}$

$\mathbf{X}_2 = \mathbf{TX}_1 = \begin{bmatrix} \frac{2}{3} & 1 \\ \frac{1}{3} & 0 \end{bmatrix} \begin{bmatrix} \frac{11}{12} \\ \frac{1}{12} \end{bmatrix} = \begin{bmatrix} \frac{25}{36} \\ \frac{11}{36} \end{bmatrix}$

$\mathbf{X}_3 = \mathbf{TX}_2 = \begin{bmatrix} \frac{2}{3} & 1 \\ \frac{1}{3} & 0 \end{bmatrix} \begin{bmatrix} \frac{25}{36} \\ \frac{11}{36} \end{bmatrix} = \begin{bmatrix} \frac{83}{108} \\ \frac{25}{108} \end{bmatrix}$

17. $\mathbf{X}_1 = \mathbf{TX}_0 = \begin{bmatrix} 0.3 & 0.5 \\ 0.7 & 0.5 \end{bmatrix} \begin{bmatrix} 0.4 \\ 0.6 \end{bmatrix} = \begin{bmatrix} 0.42 \\ 0.58 \end{bmatrix}$

$\mathbf{X}_2 = \mathbf{TX}_1 = \begin{bmatrix} 0.3 & 0.5 \\ 0.7 & 0.5 \end{bmatrix} \begin{bmatrix} 0.42 \\ 0.58 \end{bmatrix} = \begin{bmatrix} 0.416 \\ 0.584 \end{bmatrix}$

$\mathbf{X}_3 = \mathbf{TX}_2 = \begin{bmatrix} 0.3 & 0.5 \\ 0.7 & 0.5 \end{bmatrix} \begin{bmatrix} 0.416 \\ 0.584 \end{bmatrix} = \begin{bmatrix} 0.4168 \\ 0.5832 \end{bmatrix}$

19. $X_1 = TX_0 = \begin{bmatrix} 0.1 & 0 & 0.3 \\ 0.2 & 0.4 & 0.3 \\ 0.7 & 0.6 & 0.4 \end{bmatrix} \begin{bmatrix} 0.2 \\ 0 \\ 0.8 \end{bmatrix} = \begin{bmatrix} 0.26 \\ 0.28 \\ 0.46 \end{bmatrix}$

$X_2 = TX_1 = \begin{bmatrix} 0.1 & 0 & 0.3 \\ 0.2 & 0.4 & 0.3 \\ 0.7 & 0.6 & 0.4 \end{bmatrix} \begin{bmatrix} 0.26 \\ 0.28 \\ 0.46 \end{bmatrix} = \begin{bmatrix} 0.164 \\ 0.302 \\ 0.534 \end{bmatrix}$

$X_3 = TX_2 = \begin{bmatrix} 0.1 & 0 & 0.3 \\ 0.2 & 0.4 & 0.3 \\ 0.7 & 0.6 & 0.4 \end{bmatrix} \begin{bmatrix} 0.164 \\ 0.302 \\ 0.534 \end{bmatrix} = \begin{bmatrix} 0.1766 \\ 0.3138 \\ 0.5096 \end{bmatrix}$

21. a. $T^2 = \begin{bmatrix} \frac{1}{4} & \frac{3}{4} \\ \frac{3}{4} & \frac{1}{4} \end{bmatrix} \begin{bmatrix} \frac{1}{4} & \frac{3}{4} \\ \frac{3}{4} & \frac{1}{4} \end{bmatrix} = \begin{bmatrix} \frac{5}{8} & \frac{3}{8} \\ \frac{3}{8} & \frac{5}{8} \end{bmatrix}$

$T^3 = T^2 T = \begin{bmatrix} \frac{5}{8} & \frac{3}{8} \\ \frac{3}{8} & \frac{5}{8} \end{bmatrix} \begin{bmatrix} \frac{1}{4} & \frac{3}{4} \\ \frac{3}{4} & \frac{1}{4} \end{bmatrix} = \begin{bmatrix} \frac{7}{16} & \frac{9}{16} \\ \frac{9}{16} & \frac{7}{16} \end{bmatrix}.$

b. Entry in row 2, column 1, of T^2 is $\frac{3}{8}$.

c. Entry in row 1, column 2 of T^3 is $\frac{9}{16}$.

23. a. $T^2 = \begin{bmatrix} 0 & 0.5 & 0.3 \\ 1 & 0.4 & 0.3 \\ 0 & 0.1 & 0.4 \end{bmatrix} \begin{bmatrix} 0 & 0.5 & 0.3 \\ 1 & 0.4 & 0.3 \\ 0 & 0.1 & 0.4 \end{bmatrix} = \begin{bmatrix} 0.50 & 0.23 & 0.27 \\ 0.40 & 0.69 & 0.54 \\ 0.10 & 0.08 & 0.19 \end{bmatrix}$

$T^3 = T^2 T = \begin{bmatrix} 0.50 & 0.23 & 0.27 \\ 0.40 & 0.69 & 0.54 \\ 0.10 & 0.08 & 0.19 \end{bmatrix} \begin{bmatrix} 0 & 0.5 & 0.3 \\ 1 & 0.4 & 0.3 \\ 0 & 0.1 & 0.4 \end{bmatrix} = \begin{bmatrix} 0.230 & 0.369 & 0.327 \\ 0.690 & 0.530 & 0.543 \\ 0.080 & 0.101 & 0.130 \end{bmatrix}$

b. Entry in row 2, column 1, of T^2 is 0.40.

c. Entry in row 1, column 2 of T^3 is 0.369.

25. $\mathbf{T} - \mathbf{I} = \begin{bmatrix} \frac{1}{2} & \frac{2}{3} \\ \frac{1}{2} & \frac{1}{3} \end{bmatrix} - \begin{bmatrix} 1 & 0 \\ 0 & 1 \end{bmatrix} = \begin{bmatrix} -\frac{1}{2} & \frac{2}{3} \\ \frac{1}{2} & -\frac{2}{3} \end{bmatrix}$

$\left[\begin{array}{cc|c} 1 & 1 & 1 \\ -\frac{1}{2} & \frac{2}{3} & 0 \\ \frac{1}{2} & -\frac{2}{3} & 0 \end{array} \right] \rightarrow \cdots \rightarrow \left[\begin{array}{cc|c} 1 & 0 & \frac{4}{7} \\ 0 & 1 & \frac{3}{7} \\ 0 & 0 & 0 \end{array} \right]$

$\mathbf{Q} = \begin{bmatrix} \frac{4}{7} \\ \frac{3}{7} \end{bmatrix}$

27. $\mathbf{T} - \mathbf{I} = \begin{bmatrix} \frac{1}{5} & \frac{3}{5} \\ \frac{4}{5} & \frac{2}{5} \end{bmatrix} - \begin{bmatrix} 1 & 0 \\ 0 & 1 \end{bmatrix} = \begin{bmatrix} -\frac{4}{5} & \frac{3}{5} \\ \frac{4}{5} & -\frac{3}{5} \end{bmatrix}$

$\left[\begin{array}{cc|c} 1 & 1 & 1 \\ -\frac{4}{5} & \frac{3}{5} & 0 \\ \frac{4}{5} & -\frac{3}{5} & 0 \end{array} \right] \rightarrow \cdots \rightarrow \left[\begin{array}{cc|c} 1 & 0 & \frac{3}{7} \\ 0 & 1 & \frac{4}{7} \\ 0 & 0 & 0 \end{array} \right]$

$\mathbf{Q} = \begin{bmatrix} \frac{3}{7} \\ \frac{4}{7} \end{bmatrix}$

29. $\mathbf{T} - \mathbf{I} = \begin{bmatrix} 0.4 & 0.6 & 0.6 \\ 0.3 & 0.3 & 0.1 \\ 0.3 & 0.1 & 0.3 \end{bmatrix} - \begin{bmatrix} 1 & 0 & 0 \\ 0 & 1 & 0 \\ 0 & 0 & 1 \end{bmatrix} = \begin{bmatrix} -0.6 & 0.6 & 0.6 \\ 0.3 & -0.7 & 0.1 \\ 0.3 & 0.1 & -0.7 \end{bmatrix}$

$\left[\begin{array}{ccc|c} 1 & 1 & 1 & 1 \\ -0.6 & 0.6 & 0.6 & 0 \\ 0.3 & -0.7 & 0.1 & 0 \\ 0.3 & 0.1 & -0.7 & 0 \end{array} \right] \rightarrow \cdots \rightarrow \left[\begin{array}{ccc|c} 1 & 0 & 0 & 0.5 \\ 0 & 1 & 0 & 0.25 \\ 0 & 0 & 1 & 0.25 \\ 0 & 0 & 0 & 0 \end{array} \right]$

$\mathbf{Q} = \begin{bmatrix} 0.5 \\ 0.25 \\ 0.25 \end{bmatrix}$

31. a. $\mathbf{T} = \begin{matrix} \\ \text{Flu} \\ \text{No flu} \end{matrix} \begin{matrix} \text{Flu} \quad \text{No flu} \\ \begin{bmatrix} 0.1 & 0.2 \\ 0.9 & 0.8 \end{bmatrix} \end{matrix}$

b. $X_0 = \begin{bmatrix} \frac{120}{200} \\ \frac{80}{200} \end{bmatrix} = \begin{bmatrix} 0.6 \\ 0.4 \end{bmatrix}.$

If a period is 4 days, then 8 days corresponds to 2 periods, and 12 days corresponds to 3 periods. The state vector corresponding to 8 days from now is

$X_2 = T^2 X_0 = \begin{bmatrix} 0.19 & 0.18 \\ 0.81 & 0.82 \end{bmatrix} \begin{bmatrix} 0.6 \\ 0.4 \end{bmatrix}$

$= \begin{bmatrix} 0.186 \\ 0.814 \end{bmatrix}.$

Thus $0.186(200) \approx 37$ students can be expected to have the flu 8 days from now.

The state vector corresponding to 12 days from now is

$X_3 = T^3 X_0 = \begin{bmatrix} 0.181 & 0.182 \\ 0.819 & 0.818 \end{bmatrix} \begin{bmatrix} 0.6 \\ 0.4 \end{bmatrix}$

$= \begin{bmatrix} 0.1814 \\ 0.8186 \end{bmatrix}.$

Thus $0.1814(200) \approx 36$ students can be expected to have the flu 12 days from now.

33. a. $\begin{array}{cc} & \text{A} \quad \text{B} \end{array}$
$T = \begin{array}{c} \text{A} \\ \text{B} \end{array} \begin{bmatrix} 0.7 & 0.4 \\ 0.3 & 0.6 \end{bmatrix}$

b. Wednesday corresponds to step 2.
$T^2 = \begin{bmatrix} 0.61 & 0.52 \\ 0.39 & 0.48 \end{bmatrix}.$
The probability is 0.61.

35. a. $\begin{array}{ccc} & \text{D} \quad \text{R} \quad \text{O} \end{array}$
$T = \begin{array}{c} \text{D} \\ \text{R} \\ \text{O} \end{array} \begin{bmatrix} 0.8 & 0.1 & 0.3 \\ 0.1 & 0.8 & 0.2 \\ 0.1 & 0.1 & 0.5 \end{bmatrix}$

b. $T^2 = \begin{bmatrix} 0.68 & 0.19 & 0.41 \\ 0.18 & 0.67 & 0.29 \\ 0.14 & 0.14 & 0.30 \end{bmatrix}$
The probability is 0.19.

c. $X_1 = TX_0$

$= \begin{bmatrix} 0.8 & 0.1 & 0.3 \\ 0.1 & 0.8 & 0.2 \\ 0.1 & 0.1 & 0.5 \end{bmatrix} \begin{bmatrix} 0.40 \\ 0.40 \\ 0.20 \end{bmatrix}$

$= \begin{bmatrix} 0.42 \\ 0.40 \\ 0.18 \end{bmatrix}$

40% are expected to be Republican.

37. a. $\begin{array}{cc} & \text{A} \quad \text{Compet.} \end{array}$
$T = \begin{array}{c} \text{A} \\ \text{Compet.} \end{array} \begin{bmatrix} 0.8 & 0.3 \\ 0.2 & 0.7 \end{bmatrix}$

b. $X_1 = TX_0 = \begin{bmatrix} 0.8 & 0.3 \\ 0.2 & 0.7 \end{bmatrix} \begin{bmatrix} 0.70 \\ 0.30 \end{bmatrix}$

$= \begin{bmatrix} 0.65 \\ 0.35 \end{bmatrix}$

A is expected to control 65% of the market.

c. $T - I = \begin{bmatrix} 0.8 & 0.3 \\ 0.2 & 0.7 \end{bmatrix} - \begin{bmatrix} 1 & 0 \\ 0 & 1 \end{bmatrix}$

$= \begin{bmatrix} -0.2 & 0.3 \\ 0.2 & -0.3 \end{bmatrix}$

$\begin{bmatrix} 1 & 1 & | & 1 \\ -0.2 & 0.3 & | & 0 \\ 0.2 & -0.3 & | & 0 \end{bmatrix} \to \dots \to \begin{bmatrix} 1 & 0 & | & 0.6 \\ 0 & 1 & | & 0.4 \\ 0 & 0 & | & 0 \end{bmatrix}$

$Q = \begin{bmatrix} 0.6 \\ 0.4 \end{bmatrix}$

In the long run, A can expect to control 60% of the market.

39. a.
$$\mathbf{T} = \begin{array}{c} \\ 1 \\ 2 \end{array}\begin{array}{cc} 1 & 2 \\ \begin{bmatrix} \dfrac{5}{7} & \dfrac{3}{7} \\ \dfrac{2}{7} & \dfrac{4}{7} \end{bmatrix} \end{array}$$

b. $\mathbf{X_2} = \mathbf{T^2 X_0} = \begin{bmatrix} \dfrac{31}{49} & \dfrac{27}{49} \\ \dfrac{18}{49} & \dfrac{22}{49} \end{bmatrix}\begin{bmatrix} \dfrac{1}{2} \\ \dfrac{1}{2} \end{bmatrix} = \begin{bmatrix} \dfrac{29}{49} \\ \dfrac{20}{49} \end{bmatrix}$

$\approx \begin{bmatrix} 0.5918 \\ 0.4082 \end{bmatrix}$

About 59.18% in compartment 1 and 40.82% in compartment 2.

c. $\mathbf{T - I} = \begin{bmatrix} \dfrac{5}{7} & \dfrac{3}{7} \\ \dfrac{2}{7} & \dfrac{4}{7} \end{bmatrix} - \begin{bmatrix} 1 & 0 \\ 0 & 1 \end{bmatrix} = \begin{bmatrix} -\dfrac{2}{7} & \dfrac{3}{7} \\ \dfrac{2}{7} & -\dfrac{3}{7} \end{bmatrix}$

$$\left[\begin{array}{cc|c} 1 & 1 & 1 \\ -\dfrac{2}{7} & \dfrac{3}{7} & 0 \\ \dfrac{2}{7} & -\dfrac{3}{7} & 0 \end{array}\right] \rightarrow \dots \rightarrow \left[\begin{array}{cc|c} 1 & 0 & \dfrac{3}{5} \\ 0 & 1 & \dfrac{2}{5} \\ 0 & 0 & 0 \end{array}\right]$$

$$\mathbf{Q} = \begin{bmatrix} \dfrac{3}{5} \\ \dfrac{2}{5} \end{bmatrix} = \begin{bmatrix} 0.6 \\ 0.4 \end{bmatrix}$$

In the long run, there will be 60% in compartment 1 and 40% in compartment 2.

41. a. $\mathbf{T - I} = \begin{bmatrix} \dfrac{3}{4} & \dfrac{1}{2} \\ \dfrac{1}{4} & \dfrac{1}{2} \end{bmatrix} - \begin{bmatrix} 1 & 0 \\ 0 & 1 \end{bmatrix} = \begin{bmatrix} -\dfrac{1}{4} & \dfrac{1}{2} \\ \dfrac{1}{4} & -\dfrac{1}{2} \end{bmatrix}$

$$\left[\begin{array}{cc|c} 1 & 1 & 1 \\ -\dfrac{1}{4} & \dfrac{1}{2} & 0 \\ \dfrac{1}{4} & -\dfrac{1}{2} & 0 \end{array}\right] \rightarrow \dots \rightarrow \left[\begin{array}{cc|c} 1 & 0 & \dfrac{2}{3} \\ 0 & 1 & \dfrac{1}{3} \\ 0 & 0 & 0 \end{array}\right]$$

$$\mathbf{Q} = \begin{bmatrix} \dfrac{2}{3} \\ \dfrac{1}{3} \end{bmatrix}$$

b. Presently, A accounts for 50% of sales and in long run A will account for $\dfrac{2}{3}$, or $66\dfrac{2}{3}\%$, of sales. Thus the percentage increase in sales above the present level is

$\dfrac{66\frac{2}{3} - 50}{50} \cdot 100\% = \dfrac{16\frac{2}{3}}{50} \cdot 100\%$

$= 33\dfrac{1}{3}\%$.

43. $\mathbf{T^2} = \mathbf{TT} = \begin{bmatrix} \dfrac{1}{2} & 1 \\ \dfrac{1}{2} & 0 \end{bmatrix}\begin{bmatrix} \dfrac{1}{2} & 1 \\ \dfrac{1}{2} & 0 \end{bmatrix} = \begin{bmatrix} \dfrac{3}{4} & \dfrac{1}{2} \\ \dfrac{1}{4} & \dfrac{1}{2} \end{bmatrix}$

Since all entries of $\mathbf{T^2}$ are positive, $\mathbf{T}$ is regular.

Chapter 9 Review Problems

1. $\mu = \sum_x x f(x) = 1 \cdot f(1) + 2 \cdot f(2) + 3 \cdot f(3)$

$= 1(0.7) + 2(0.1) + 3(0.2) = 1.5$

$\text{Var}(X) = \sum_x x^2 f(x) - \mu^2$

$= \left[1^2(0.7) + 2^2(0.1) + 3^2(0.2)\right] - (1.5)^2$

$= 0.65$

$\sigma = \sqrt{\text{Var}(X)} = \sqrt{0.65} \approx 0.81$

3. a. $n(S) = 2 \cdot 6 = 12$

$E_0 = \{H1\}, \quad E_1 = \{T1, H2\},$

$E_2 = \{T2, H3\}, \quad E_3 = \{T3, H4\},$

$E_4 = \{T4, H5\}, \quad E_5 = \{T5, H6\},$

$E_6 = \{T6\}$

$f(0) = P(E_0) = \dfrac{n(E_0)}{n(S)} = \dfrac{1}{12}$

$f(1) = P(E_1) = \dfrac{n(E_1)}{n(S)} = \dfrac{2}{12} = \dfrac{1}{6}$

Similarly, $f(2), f(3), f(4),$ and $f(5)$ equal $\dfrac{1}{6}$.

$f(6) = P(E_6) = \dfrac{n(E_6)}{n(S)} = \dfrac{1}{12}$

b. $E(X) = \displaystyle\sum_x x f(x)$

$= 0 \cdot \dfrac{1}{12} + \dfrac{1+2+3+4+5}{6} + 6 \cdot \dfrac{1}{12}$

$= 0 + \dfrac{15}{6} + \dfrac{6}{12} = \dfrac{36}{12} = 3$

5. Let X = gain (in dollars) on a play. If no 10 appears, then $X = 0 - \dfrac{1}{4} = -\dfrac{1}{4}$; if exactly one 10 appears, then $X = 1 - \dfrac{1}{4} = \dfrac{3}{4}$; if two 10's appear, then $X = 2 - \dfrac{1}{4} = \dfrac{7}{4}$.

$n(S) = 52 \cdot 52$. In a deck, there are 4 10's and 48 non 10's. Thus $n(E_{\text{no }10}) = 48 \cdot 48$.

The event $E_{\text{one }10}$ occurs if the first card is a 10 and the second is a non-10, or vice versa. Thus

$n(E_{\text{one }10}) = 4 \cdot 48 + 48 \cdot 4 = 2 \cdot 4 \cdot 48$.

$n(E_{\text{two }10's}) = 4 \cdot 4$.

Dist. of X:

$f\left(-\dfrac{1}{4}\right) = \dfrac{48 \cdot 48}{52 \cdot 52} = \dfrac{144}{169},$

$f\left(\dfrac{3}{4}\right) = \dfrac{2 \cdot 4 \cdot 48}{52 \cdot 52} = \dfrac{24}{169},$

$f\left(\dfrac{7}{4}\right) = \dfrac{4 \cdot 4}{52 \cdot 52} = \dfrac{1}{169}.$

$E(X) = -\dfrac{1}{4} \cdot \dfrac{144}{169} + \dfrac{3}{4} \cdot \dfrac{24}{169} + \dfrac{7}{4} \cdot \dfrac{1}{169}$

$= \dfrac{-144 + 72 + 7}{4 \cdot 169} = -\dfrac{65}{676} = -\dfrac{5}{52} \approx -0.10$

There is a loss of \$0.10 per play.

7. a. Let X = gain (in dollars) on each unit shipped. Then $P(X = -100) = 0.08$ and $P(X = 200) = 1 - 0.08 = 0.92$.

$E(X) = -100f(-100) + 200f(200)$

$= -100(0.08) + 200(0.92)$

$= \$176$ per unit

b. Since the expected gain per unit is \$176 and 4000 units are shipped per year, then expected annual profit is $4000(176) = \$704,000$.

9. $f(0) = {_4}C_0 (0.15)^0 (0.85)^4 \approx \dfrac{4!}{0!4!} \cdot 1(0.522)$

$= 0.522$

$f(1) = {_4}C_1 (0.15)^1 (0.85)^3$

$\approx \dfrac{4!}{1!3!} \cdot (0.15)(0.614) = 0.368$

$f(2) = {_4}C_2 (0.15)^2 (0.85)^2$

$= \dfrac{4!}{2!2!} \cdot (0.0225)(0.7225) \approx 0.098$

$f(3) = {_4}C_3 (0.15)^3 (0.85)^1$

$= \dfrac{4!}{3!1!} \cdot (0.003375)(0.85) \approx 0.011$

$f(4) = {_4}C_4 (0.15)^4 (0.85)^0$

$\approx \dfrac{4!}{4!0!} \cdot (0.000506)1 = 0.0005$

$\mu = np = 4(0.15) = 0.6$

$\sigma = \sqrt{npq} = \sqrt{4(0.15)(0.85)} \approx 0.71$

11. $P(X \leq 1) = P(X = 0) + P(X = 1)$

$$= {}_5C_0 \left(\frac{3}{4}\right)^0 \left(\frac{1}{4}\right)^5 + {}_5C_1 \left(\frac{3}{4}\right)^1 \left(\frac{1}{4}\right)^4$$

$$= 1 \cdot 1 \cdot \frac{1}{1024} + 5 \cdot \frac{3}{4} \cdot \frac{1}{256} = \frac{16}{1024} = \frac{1}{64}$$

13. The probability that a 2 or 3 results on one roll is $\frac{2}{6} = \frac{1}{3}$. Let X = number of 2's or 3's that appear on 4 rolls. Then X is binomial with $p = \frac{1}{3}$ and $n = 4$.

$$P(X = 3) = {}_4C_3 \left(\frac{1}{3}\right)^3 \left(\frac{2}{3}\right)^1 = 4 \cdot \frac{1}{27} \cdot \frac{2}{3} = \frac{8}{81}$$

15. Let X = number of heads that occur. Then X is binomial.

$$P(X = 0) = {}_5C_0 \left(\frac{2}{5}\right)^0 \left(\frac{3}{5}\right)^5$$

$$= 1 \cdot 1 \cdot \frac{243}{3125} = \frac{243}{3125}$$

$$P(X = 1) = {}_5C_1 \left(\frac{2}{5}\right)^1 \left(\frac{3}{5}\right)^4$$

$$= 5 \cdot \frac{2}{5} \cdot \frac{81}{625} = \frac{810}{3125}$$

$$P(X \geq 2) = 1 - [P(X = 0) + P(X = 1)]$$

$$= 1 - \left[\frac{243}{3125} + \frac{810}{3125}\right] = 1 - \frac{1053}{3125} = \frac{2072}{3125}$$

17. From column 1, $0.1 + a + 0.6 = 1$, so $a = 0.3$.
From column 2, $2a + b + b = 1$, so

$$2b = 1 - 2a, \text{ or } b = \frac{1 - 2a}{2} = \frac{1 - 2(0.3)}{2} = 0.2.$$

From column 3, $a + b + c = 1$, so
$c = 1 - a - b$, or $c = 1 - 0.3 - 0.2 = 0.5$.

19. $\mathbf{X}_1 = \mathbf{TX}_0 = \begin{bmatrix} 0.1 & 0.3 & 0.1 \\ 0.2 & 0.4 & 0.1 \\ 0.7 & 0.3 & 0.8 \end{bmatrix} \begin{bmatrix} 0.5 \\ 0 \\ 0.5 \end{bmatrix} = \begin{bmatrix} 0.10 \\ 0.15 \\ 0.75 \end{bmatrix}$

$\mathbf{X}_2 = \mathbf{TX}_1 = \begin{bmatrix} 0.1 & 0.3 & 0.1 \\ 0.2 & 0.4 & 0.1 \\ 0.7 & 0.3 & 0.8 \end{bmatrix} \begin{bmatrix} 0.10 \\ 0.15 \\ 0.75 \end{bmatrix}$

$= \begin{bmatrix} 0.130 \\ 0.155 \\ 0.715 \end{bmatrix}$

$\mathbf{X}_3 = \mathbf{TX}_2 = \begin{bmatrix} 0.1 & 0.3 & 0.1 \\ 0.2 & 0.4 & 0.1 \\ 0.7 & 0.3 & 0.8 \end{bmatrix} \begin{bmatrix} 0.130 \\ 0.155 \\ 0.715 \end{bmatrix}$

$= \begin{bmatrix} 0.1310 \\ 0.1595 \\ 0.7095 \end{bmatrix}$

21. a. $\mathbf{T}^2 = \mathbf{TT} = \begin{bmatrix} \frac{1}{7} & \frac{3}{7} \\ \frac{6}{7} & \frac{4}{7} \end{bmatrix} \begin{bmatrix} \frac{1}{7} & \frac{3}{7} \\ \frac{6}{7} & \frac{4}{7} \end{bmatrix} = \begin{bmatrix} \frac{19}{49} & \frac{15}{49} \\ \frac{30}{49} & \frac{34}{49} \end{bmatrix}$

$\mathbf{T}^3 = \mathbf{T}^2\mathbf{T} = \begin{bmatrix} \frac{19}{49} & \frac{15}{49} \\ \frac{30}{49} & \frac{34}{49} \end{bmatrix} \begin{bmatrix} \frac{1}{7} & \frac{3}{7} \\ \frac{6}{7} & \frac{4}{7} \end{bmatrix}$

$= \begin{bmatrix} \frac{109}{343} & \frac{117}{343} \\ \frac{234}{343} & \frac{226}{343} \end{bmatrix}$

b. From $\mathbf{T}^2$, entry in row 1, column 2, is $\frac{15}{49}$.

c. From $\mathbf{T}^3$, entry in row 2, column 1, is $\frac{234}{343}$.

23. $\mathbf{T} - \mathbf{I} = \begin{bmatrix} \frac{1}{3} & \frac{2}{3} \\ \frac{2}{3} & \frac{1}{3} \end{bmatrix} - \begin{bmatrix} 1 & 0 \\ 0 & 1 \end{bmatrix} = \begin{bmatrix} -\frac{2}{3} & \frac{2}{3} \\ \frac{2}{3} & -\frac{2}{3} \end{bmatrix}$

$\begin{bmatrix} 1 & 1 & | & 1 \\ -\frac{2}{3} & \frac{2}{3} & | & 0 \\ \frac{2}{3} & -\frac{2}{3} & | & 0 \end{bmatrix} \rightarrow \ldots \rightarrow \begin{bmatrix} 1 & 0 & | & \frac{1}{2} \\ 0 & 1 & | & \frac{1}{2} \\ 0 & 0 & | & 0 \end{bmatrix}$

$\mathbf{Q} = \begin{bmatrix} \frac{1}{2} \\ \frac{1}{2} \end{bmatrix}$

25. $\mathbf{T} = \begin{array}{c} \\ \text{Japanese} \\ \text{Non-Japanese} \end{array} \overset{\text{Japanese Non-Japanese}}{\begin{bmatrix} 0.8 & 0.6 \\ 0.2 & 0.4 \end{bmatrix}}$

a. $\mathbf{T}^2 = \begin{bmatrix} 0.8 & 0.6 \\ 0.2 & 0.4 \end{bmatrix}\begin{bmatrix} 0.8 & 0.6 \\ 0.2 & 0.4 \end{bmatrix} = \begin{bmatrix} 0.76 & 0.72 \\ 0.24 & 0.28 \end{bmatrix}$

From row 1, column 1, the probability that a person who currently owns a Japanese car will buy a Japanese car two cars later is 0.76. Thus 76% of people who currently own Japanese cars will own Japanese cars two cars later.

b. $\mathbf{X}_2 = \mathbf{T}^2 \mathbf{X}_0 = \begin{bmatrix} 0.76 & 0.72 \\ 0.24 & 0.28 \end{bmatrix}\begin{bmatrix} 0.6 \\ 0.4 \end{bmatrix} = \begin{bmatrix} 0.744 \\ 0.256 \end{bmatrix}$

Two cars from now, we expect 74.4% Japanese, 25.6% non-Japanese.

c. $\mathbf{T} - \mathbf{I} = \begin{bmatrix} 0.8 & 0.6 \\ 0.2 & 0.4 \end{bmatrix} - \begin{bmatrix} 1 & 0 \\ 0 & 1 \end{bmatrix} = \begin{bmatrix} -0.2 & 0.6 \\ 0.2 & -0.6 \end{bmatrix}$

$\begin{bmatrix} 1 & 1 & | & 1 \\ -0.2 & 0.6 & | & 0 \\ 0.2 & -0.6 & | & 0 \end{bmatrix} \rightarrow \ldots \rightarrow \begin{bmatrix} 1 & 0 & | & 0.75 \\ 0 & 1 & | & 0.25 \\ 0 & 0 & | & 0 \end{bmatrix}$

$\mathbf{Q} = \begin{bmatrix} 0.75 \\ 0.25 \end{bmatrix}$

In the long run, 75% Japanese cars, 25% non-Japanese cars.

Mathematical Snapshot Chapter 9

1. For $\mathbf{X}_0 = \begin{bmatrix} 0 \\ 1 \\ 0 \\ 0 \end{bmatrix}$ or $\begin{bmatrix} 0 \\ 0 \\ 1 \\ 0 \end{bmatrix}$, the first entry of the state vector is greater than 0.5 for $n = 7$ or greater. If

$$\mathbf{X}_0 = \begin{bmatrix} 0 \\ 1 \\ 0 \\ 0 \end{bmatrix}, \text{ then } \mathbf{T}^7 \mathbf{X}_0 \approx \begin{bmatrix} 0.5217 \\ 0.0000 \\ 0.4783 \\ 0.0000 \end{bmatrix}.$$

3. Against Always Defect,

$$\mathbf{T} = \begin{array}{c} \\ 1 \\ 2 \\ 3 \\ 4 \end{array} \begin{array}{cccc} 1 & 2 & 3 & 4 \\ \begin{bmatrix} 0 & 0 & 0 & 0 \\ 1 & 0.1 & 1 & 0.1 \\ 0 & 0 & 0 & 0 \\ 0 & 0.9 & 0 & 0.9 \end{bmatrix} \end{array}.$$

Against Always Cooperate,

$$\mathbf{T} = \begin{array}{c} \\ 1 \\ 2 \\ 3 \\ 4 \end{array} \begin{array}{cccc} 1 & 2 & 3 & 4 \\ \begin{bmatrix} 1 & 0.1 & 1 & 0.1 \\ 0 & 0 & 0 & 0 \\ 0 & 0.9 & 0 & 0.9 \\ 0 & 0 & 0 & 0 \end{bmatrix} \end{array}.$$

Against regular Tit-for-tat,

$$\mathbf{T} = \begin{array}{c} \\ 1 \\ 2 \\ 3 \\ 4 \end{array} \begin{array}{cccc} 1 & 2 & 3 & 4 \\ \begin{bmatrix} 1 & 0.1 & 0 & 0 \\ 0 & 0 & 1 & 0.1 \\ 0 & 0.9 & 0 & 0 \\ 0 & 0 & 0 & 0.9 \end{bmatrix} \end{array}.$$

Chapter 10

Principles in Practice 10.1

1. The graph of the greatest integer function is shown.

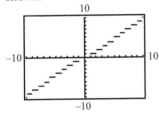

$\lim_{x \to a} f(x)$ does not exist when a is an integer since the limits are different depending on the side from which you approach the integer. $\lim_{x \to a} f(x)$ exists for all numbers which are not integers.

3. $\lim_{x \to 8} R(x) = \lim_{x \to 8} \left(500x - 6x^2 \right)$

   $= \lim_{x \to 8} 500x - \lim_{x \to 8} 6x^2$

   $= 500 \lim_{x \to 8} x - 6 \lim_{x \to 8} x^2 = 500(8) - 6(8)^2$

   $= 4000 - 384 = 3616$

5. As $h \to 0$, both the numerator and denominator approach 0. For $h \neq 0$,

   $\lim_{h \to 0} \dfrac{125 + 2(x+h) - (125 + 2x)}{h}$

   $= \lim_{h \to 0} \dfrac{125 + 2x + 2h - 125 - 2x}{h} = \lim_{h \to 0} \dfrac{2h}{h}$

   $= \lim_{h \to 0} 2 = 2$.

Problems 10.1

1. **a.** 1

   b. 0

   c. 1

3. **a.** 1

   b. does not exist

   c. 3

5. $f(-0.9) = -3.7 \qquad f(-1.1) = -4.3$
   $f(-0.99) = -3.97 \qquad f(-1.01) = -4.03$
   $f(-0.999) = -3.997 \qquad f(-1.001) = -4.003$
   estimate of limit: -4

7. $f(-0.1) \approx 0.9516 \qquad f(0.1) \approx 1.0517$
   $f(-0.01) \approx 0.9950 \qquad f(0.01) \approx 1.0050$
   $f(-0.001) \approx 0.9995 \qquad f(0.001) \approx 1.0005$
   estimate of limit: 1

9. $\lim_{x \to 2} 16 = 16$

11. $\lim_{t \to -5} \left(t^2 - 5 \right) = (-5)^2 - 5 = 25 - 5 = 20$

13. $\lim_{x \to -2} (3x^3 - 4x^2 + 2x - 3)$
    $= 3(-2)^3 - 4(-2)^2 + 2(-2) - 3$
    $= -24 - 16 - 4 - 3$
    $= -47$

15. $\lim_{t \to -3} \dfrac{t-2}{t+5} = \dfrac{\lim_{t \to -3} (t-2)}{\lim_{t \to -3} (t+5)} = \dfrac{-3-2}{-3+5}$
    $= \dfrac{-5}{2} = -\dfrac{5}{2}$

17. $\lim_{h \to 0} \dfrac{h}{h^2 - 7h + 1} = \dfrac{\lim_{h \to 0} h}{\lim_{h \to 0} \left(h^2 - 7h + 1 \right)}$
    $= \dfrac{0}{0^2 - 7(0) + 1} = 0$

19. $\lim_{p \to 4} \sqrt{p^2 + p + 5} = \sqrt{\lim_{p \to 4} \left(p^2 + p + 5 \right)}$
    $= \sqrt{4^2 + 4 + 5} = \sqrt{25} = 5$

21. $\displaystyle\lim_{x\to-2}\frac{x^2+2x}{x+2}=\lim_{x\to-2}\frac{x(x+2)}{x+2}=\lim_{x\to-2}x=-2$

23. $\displaystyle\lim_{x\to2}\frac{x^2-x-2}{x-2}=\lim_{x\to2}\frac{(x-2)(x+1)}{x-2}$
$\displaystyle=\lim_{x\to2}(x+1)=3$

25. $\displaystyle\lim_{x\to3}\frac{x^2-x-6}{x-3}=\lim_{x\to3}\frac{(x-3)(x+2)}{x-3}$
$\displaystyle=\lim_{x\to3}(x+2)$
$=5$

27. $\displaystyle\lim_{x\to3}\frac{x-3}{x^2-9}=\lim_{x\to3}\frac{x-3}{(x+3)(x-3)}=\lim_{x\to3}\frac{1}{x+3}=\frac{1}{6}$

29. $\displaystyle\lim_{x\to4}\frac{x^2-9x+20}{x^2-3x-4}=\lim_{x\to4}\frac{(x-4)(x-5)}{(x-4)(x+1)}$
$\displaystyle=\lim_{x\to4}\frac{x-5}{x+1}=-\frac{1}{5}$

31. $\displaystyle\lim_{x\to2}\frac{3x^2-x-10}{x^2+5x-14}=\lim_{x\to2}\frac{(3x+5)(x-2)}{(x+7)(x-2)}$
$\displaystyle=\lim_{x\to2}\frac{3x+5}{x+7}=\frac{11}{9}$

33. $\displaystyle\lim_{h\to0}\frac{(2+h)^2-2^2}{h}=\lim_{h\to0}\frac{\left[4+4h+h^2\right]-4}{h}=\lim_{h\to0}\frac{4h+h^2}{h}=\lim_{h\to0}\frac{h(4+h)}{h}=\lim_{h\to0}(4+h)=4$

35. $\displaystyle\lim_{h\to0}\frac{(x+h)^2-x^2}{h}=\lim_{h\to0}\frac{2xh+h^2}{h}=\lim_{h\to0}(2x+h)=2x$

37. $\displaystyle\lim_{h\to0}\frac{f(x+h)-f(x)}{h}=\lim_{h\to0}\frac{[7-3(x+h)]-(7-3x)}{h}=\lim_{h\to0}\frac{-3h}{h}=\lim_{h\to0}-3=-3$

39. $\displaystyle\lim_{h\to0}\frac{f(x+h)-f(x)}{h}=\lim_{h\to0}\frac{\left[(x+h)^2-3\right]-\left(x^2-3\right)}{h}$
$\displaystyle=\lim_{h\to0}\frac{x^2+2xh+h^2-3-\left(x^2-3\right)}{h}=\lim_{h\to0}\frac{2xh+h^2}{h}=\lim_{h\to0}(2x+h)=2x$

41. $\displaystyle\lim_{h\to 0}\frac{f(x+h)-f(x)}{h}=\lim_{h\to 0}\frac{[(x+h)^3-4(x+h)^2]-[x^3-4x^2]}{h}$

$\displaystyle\qquad\qquad = \lim_{h\to 0}\frac{x^3+3x^2h+3xh^2+h^3-4x^2-8xh-4h^2-x^3+4x^2}{h}$

$\displaystyle\qquad\qquad = \lim_{h\to 0}\frac{3x^2h+3xh^2+h^3-8xh-4h^2}{h}$

$\displaystyle\qquad\qquad = \lim_{h\to 0}\frac{h(3x^2+3xh+h^2-8x-4h)}{h}$

$\displaystyle\qquad\qquad = \lim_{h\to 0}(3x^2+3xh+h^2-8x-4h)=3x^2-8x$

43. $\displaystyle\lim_{x\to 6}\frac{\sqrt{x-2}-2}{x-6}=\lim_{x\to 6}\frac{\left(\sqrt{x-2}-2\right)\left(\sqrt{x-2}+2\right)}{(x-6)\left(\sqrt{x-2}+2\right)}$

$\displaystyle\quad = \lim_{x\to 6}\frac{(x-2)-4}{(x-6)\left(\sqrt{x-2}+2\right)}=\lim_{x\to 6}\frac{x-6}{(x-6)\left(\sqrt{x-2}+2\right)}$

$\displaystyle\quad = \lim_{x\to 6}\frac{1}{\sqrt{x-2}+2}=\frac{1}{4}$

45. a. $\displaystyle\lim_{T_c\to 0}\frac{T_h-T_c}{T_h}=\frac{T_h-0}{T_h}=\frac{T_h}{T_h}=1$

b. $\displaystyle\lim_{T_c\to T_h}\frac{T_h-T_c}{T_h}=\frac{T_h-T_h}{T_h}=\frac{0}{T_h}=0$

47.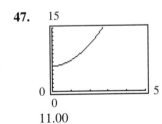

11.00

49.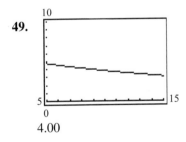

4.00

51. The graph of $C(p)$ is shown. (Negative amounts of impurities and money are not reasonable, so only the first quadrant is shown.)

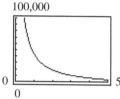

100,000

0 ⌞ 5

0

As p gets closer and closer to 0, the values of $C(p)$ increase without bound, so $\lim\limits_{p \to 0} C(p)$ does not exist.

Principles in Practice 10.2

1. The graph of $p(x)$ is shown.

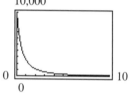

10,000

0 ⌞ 10

0

From the graph, it is apparent that $\lim\limits_{x \to \infty} p(x) = 0$. The graph starts out high and quickly drops down toward zero. According to this function, a low price corresponds to a high demand and a high price corresponds to a low demand.

3. The graph of $C(x)$ is shown.

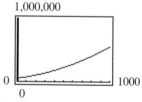

1,000,000

0 ⌞ 1000

0

From the graph it is apparent that $\lim\limits_{x \to \infty} C(x) = \infty$. This indicates that the cost increases without bound the more units that you make.

Problems 10.2

1. a. 2

b. 3

c. does not exist

d. $-\infty$

e. ∞

f. ∞

g. ∞

h. 0

i. 1

j. 1

k. 1

3. $\lim\limits_{x \to 3^+} (x - 2)$

As $x \to 3^+$, then $x - 2 \to 1$.

5. $\lim\limits_{x \to -\infty} 5x$

As x becomes very negative, so does $5x$. Thus $\lim\limits_{x \to -\infty} 5x = -\infty$.

7. $\lim\limits_{x \to 0^-} \dfrac{6x}{x^4} = \lim\limits_{x \to 0^-} \dfrac{6}{x^3} = -\infty$ since x^3 is

negative and close to 0 for $x \to 0^-$.

9. $\lim\limits_{x \to -\infty} x^2 = \infty$ since x^2 is positive for $x \to -\infty$.

11. $\lim\limits_{h \to 0^+} \sqrt{h} = 0$ since $\sqrt{h}$ is close to 0 when h is positive and close to 0.

13. $\lim\limits_{x \to -2^-} \dfrac{-3}{x + 2} = \infty$

15. $\lim\limits_{x \to 1^+} \left(4\sqrt{x-1}\right)$. As $x \to 1^+$, then $x - 1$ approaches 0 through positive values. So $\sqrt{x-1} \to 0$. Thus

$$\lim\limits_{x \to 1^+} \left(4\sqrt{x-1}\right) = 4 \cdot \lim\limits_{x \to 1^+} \sqrt{x-1} = 4 \cdot 0 = 0.$$

17. $\lim\limits_{x \to \infty} \sqrt{x+10}$

As x becomes very large, so does $x + 10$. Because square roots of very large numbers are very large, $\lim\limits_{x \to \infty} \sqrt{x+10} = \infty$.

19. $\lim\limits_{x \to \infty} \dfrac{3}{\sqrt{x}} = 3 \lim\limits_{x \to \infty} \dfrac{1}{x^{\frac{1}{2}}} = 3 \cdot 0 = 0$

21. $\lim\limits_{x \to \infty} \dfrac{x+8}{x-3} = \lim\limits_{x \to \infty} \dfrac{x}{x} = \lim\limits_{x \to \infty} 1 = 1$

23. $\lim\limits_{x \to -\infty} \dfrac{x^2-1}{x^3+4x-3} = \lim\limits_{x \to -\infty} \dfrac{x^2}{x^3} = \lim\limits_{x \to -\infty} \dfrac{1}{x} = 0$

25. $\lim\limits_{t \to \infty} \dfrac{3t^3+2t^2+9t-1}{5t^2-5} = \lim\limits_{t \to \infty} \dfrac{3t^3}{5t^2}$

$$= \lim\limits_{t \to \infty} \dfrac{3t}{5}$$
$$= \dfrac{3}{5} \lim\limits_{t \to \infty} t$$
$$= \infty$$

27. $\lim\limits_{x \to \infty} \dfrac{7}{2x+1} = \lim\limits_{x \to \infty} \dfrac{7}{2x}$

$$= \dfrac{7}{2} \cdot \lim\limits_{x \to \infty} \dfrac{1}{x} = \dfrac{7}{2} \cdot 0 = 0$$

29. $\lim\limits_{x \to \infty} \dfrac{3-4x-2x^3}{5x^3-8x+1} = \lim\limits_{x \to \infty} \dfrac{-2x^3}{5x^3}$

$$= \lim\limits_{x \to \infty} \dfrac{-2}{5} = -\dfrac{2}{5}$$

31. $\lim\limits_{x \to 3^-} \dfrac{x+3}{x^2-9} = \lim\limits_{x \to 3^-} \dfrac{x+3}{(x+3)(x-3)}$

$$= \lim\limits_{x \to 3^-} \dfrac{1}{x-3} = -\infty$$

33. $\lim\limits_{w \to \infty} \dfrac{2w^2-3w+4}{5w^2+7w-1} = \lim\limits_{w \to \infty} \dfrac{2w^2}{5w^2}$

$$= \lim\limits_{w \to \infty} \dfrac{2}{5} = \dfrac{2}{5}$$

35. $\lim\limits_{x \to \infty} \dfrac{6-4x^2+x^3}{4+5x-7x^2} = \lim\limits_{x \to \infty} \dfrac{x^3}{-7x^2}$

$$= \lim\limits_{x \to \infty} \dfrac{x}{-7} = -\infty$$

37. $\lim\limits_{x \to -3^-} \dfrac{5x^2+14x-3}{x^2+3x} = \lim\limits_{x \to -3^-} \dfrac{(5x-1)(x+3)}{x(x+3)}$

$$= \lim\limits_{x \to -3^-} \dfrac{5x-1}{x}$$
$$= \dfrac{-16}{-3}$$
$$= \dfrac{16}{3}$$

39. $\lim\limits_{x \to 1} \dfrac{x^2-3x+1}{x^2+1} = \dfrac{\lim\limits_{x \to 1}\left(x^2-3x+1\right)}{\lim\limits_{x \to 1}\left(x^2+1\right)}$

$$= \dfrac{-1}{2} = -\dfrac{1}{2}$$

41. As $x \to 1^+$, then $\dfrac{1}{x-1} \to \infty$. Thus

$$\lim\limits_{x \to 1^+} \left[1+\dfrac{1}{x-1}\right] = \infty$$

43. $\lim\limits_{x \to -7^-} \dfrac{x^2+1}{\sqrt{x^2-49}}$. As $x \to -7^-$, then

$x^2+1 \to 50$ and $\sqrt{x^2-49}$

approaches 0 through positive values. Thus

$\dfrac{x^2+1}{\sqrt{x^2-49}} \to \infty$.

45. As $x \to 0^+$, $x + x^2$ approaches 0 through

positive values. Thus $\dfrac{5}{x+x^2} \to \infty$.

47. $\lim\limits_{x \to 1^-} \dfrac{x}{x-1} = -\infty$

$\lim\limits_{x \to 1^+} \dfrac{x}{x-1} = \infty$

Answer: does not exist

49. As $x \to 1^+$, then $1 - x \to 0$ through

negative values. Thus, $\lim\limits_{x \to 1^+} \dfrac{-5}{1-x} = \infty$.

51. $\lim\limits_{x \to 0^+} |x| = \lim\limits_{x \to 0^+} x = 0$

$\lim\limits_{x \to 0^-} |x| = \lim\limits_{x \to 0^-} (-x) = 0$

Thus, $\lim\limits_{x \to 0} |x| = 0$.

53. $\lim\limits_{x \to -\infty} \dfrac{x+1}{x} = \lim\limits_{x \to -\infty} \dfrac{x}{x} = \lim\limits_{x \to -\infty} 1 = 1$

55. $f(x) = \begin{cases} 2 & \text{if } x \le 2 \\ 1 & \text{if } x > 2 \end{cases}$

   a. $\lim\limits_{x \to 2^+} f(x) = \lim\limits_{x \to 2^+} 1 = 1$

   b. $\lim\limits_{x \to 2^-} f(x) = \lim\limits_{x \to 2^-} 2 = 2$

   c. $\lim\limits_{x \to 2^+} f(x) \ne \lim\limits_{x \to 2^-} f(x)$, so $\lim\limits_{x \to 2} f(x)$

     does not exist.

   d. $\lim\limits_{x \to \infty} f(x) = \lim\limits_{x \to \infty} 1 = 1$

   e. $\lim\limits_{x \to -\infty} f(x) = \lim\limits_{x \to -\infty} 2 = 2$

57. $g(x) = \begin{cases} x & \text{if } x < 0 \\ -x & \text{if } x > 0 \end{cases}$

   a. $\lim\limits_{x \to 0^+} g(x) = \lim\limits_{x \to 0^+} (-x) = 0$

   b. $\lim\limits_{x \to 0^-} g(x) = \lim\limits_{x \to 0^-} x = 0$

   c. $\lim\limits_{x \to 0^+} g(x) = \lim\limits_{x \to 0^-} g(x) = 0$, so

     $\lim\limits_{x \to 0} g(x) = 0$

   d. $\lim\limits_{x \to \infty} g(x) = \lim\limits_{x \to \infty} (-x) = -\infty$

   e. $\lim\limits_{x \to -\infty} g(x) = \lim\limits_{x \to -\infty} x = -\infty$

59. $\lim\limits_{q \to -\infty} \bar{c} = \lim\limits_{q \to \infty} \left(\dfrac{5000}{q} + 6 \right) = 0 + 6 = 6$

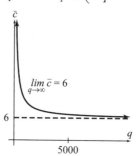

61. $\lim\limits_{t \to \infty} \left(50{,}000 - \dfrac{2000}{t+1} \right) = 50{,}000 - 0$

                             $= 50{,}000$

63. $\lim\limits_{x\to\infty} y = \lim\limits_{x\to\infty} \dfrac{900x}{10+45x} = \lim\limits_{x\to\infty} \dfrac{900x}{45x}$

$= \lim\limits_{x\to\infty} 20 = 20$

65. $1, 0.5, 0.525, 0.631, 0.912, 0.986, 0.998$;
conclude limit is 1.

67.

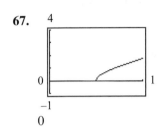

69.

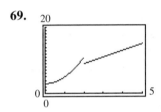

    a. 11

    b. 9

    c. does not exist

Problems 10.3

1. $f(x) = x^3 - 5x;\ x = 2$

    (i) f is defined at $x = 2$: $f(2) = -2$

    (ii) $\lim\limits_{x\to 2} f(x) = \lim\limits_{x\to 2} (x^3 - 5x)$
        $= 2^3 - 5(2) = -2$,
        which exists.

    (iii) $\lim\limits_{x\to 2} f(x) = -2 = f(2)$
        Thus f is continuous at $x = 2$.

3. $g(x) = \sqrt{2-3x}\ ; x = 0$

    (i) g is defined at $x = 0$; $g(0) = \sqrt{2}$.

    (ii) $\lim\limits_{x\to 0} g(x) = \lim\limits_{x\to 0} \sqrt{2-3x} = \sqrt{2}$, which
        exists,

    (iii) $\lim\limits_{x\to 0} g(x) = \sqrt{2} = g(0)$
        Thus g is continuous at $x = 0$.

5. $h(x) = \dfrac{x-4}{x+4}\ ; x = 4$

    (i) h is defined at $x = 4$, $h(4) = 0$.

    (ii) $\lim\limits_{x\to 4} h(x) = \lim\limits_{x\to 4} \dfrac{x-4}{x+4} = \dfrac{0}{8} = 0$, which
        exists.

    (iii) $\lim\limits_{x\to 4} h(x) = 0 = h(4)$
        Thus h is continuous at $x = 4$.

7. Continuous at -2 and 0 because f is a
rational function and at neither point is the
denominator zero.

9. Discontinuous at 3 and -3 because at both
points the denominator of this rational
function is 0.

11. $f(x) = \begin{cases} x+2 & \text{if } x \geq 2 \\ x^2 & \text{if } x < 2 \end{cases}$
f is defined at $x = 2$ and $x = 0$; $f(2) = 4$,
$f(0) = 0$.
Because $\lim\limits_{x\to 2^+} f(x) = \lim\limits_{x\to 2^+} (x+2) = 4$ and
$\lim\limits_{x\to 2^-} f(x) = \lim\limits_{x\to 2^-} x^2 = 4$, we have
$\lim\limits_{x\to 2} f(x) = 4$. In addition,
$\lim\limits_{x\to 0} f(x) = \lim\limits_{x\to 0} x^2 = 0$. Since
$\lim\limits_{x\to 2} f(x) = 4 = f(2)$ and

$\lim_{x \to 0} f(x) = 0 = f(0)$,

f is continuous at both 2 and 0.

Answer: Continuous at 2 and 0.

13. f is a polynomial function.

15. f is a rational function and the denominator is never zero.

17. None, because f is a polynomial function.

19. The denominator of this rational function is zero only when $x = -4$. Thus f is discontinuous only at $x = -4$.

21. None, because g is a polynomial function.
$$\left[g(x) = \frac{8}{15}x^6 - \frac{12}{5}x^4 + \frac{18}{5}x^2 - \frac{9}{5} \right]$$

23. $x^2 + 2x - 15 = 0$, $(x + 5)(x - 3) = 0$, $x = -5$ or 3. Discontinuous at -5 and 3.

25. $x^3 - x = 0$, $x\left(x^2 - 1\right) = 0$,

$x(x + 1)(x - 1) = 0$, $x = 0, \pm 1$. Discontinuous at $0, \pm 1$.

27. $x^2 + 1 = 0$ has no real roots, so no discontinuity exists.

29. $f(x) = \begin{cases} 1 & \text{if } x \geq 0 \\ -1 & \text{if } x < 0 \end{cases}$

For $x < 0$, $f(x) = -1$, which is a polynomial and hence continuous. For $x > 0$, $f(x) = 1$, which is a polynomial and hence continuous. Because

$\lim_{x \to 0^-} f(x) = \lim_{x \to 0^-} (-1) = -1$ and

$\lim_{x \to 0^+} f(x) = \lim_{x \to 0^+} 1 = 1$, $\lim_{x \to 0} f(x)$ does not

exist.

Thus f is discontinuous at $x = 0$.

31. $f(x) = \begin{cases} 0 & \text{if } x \leq 1 \\ x - 1 & \text{if } x > 1 \end{cases}$

For $x < 1$, $f(x) = 0$, which is a polynomial and hence continuous. For $x > 1$, $f(x) = x - 1$, which is a polynomial and hence continuous. For $x = 1$, f is defined $[f(1) = 0]$. Because $\lim_{x \to 1^-} f(x) = \lim_{x \to 1^-} 0 = 0$

and $\lim_{x \to 1^+} f(x) = \lim_{x \to 1^+} (x - 1) = 0$, then

$\lim_{x \to 1} f(x) = 0$. Since $\lim_{x \to 1} f(x) = 0 = f(0)$, f

is continuous at $x = 1$.

f has no discontinuities.

33. $f(x) = \begin{cases} x^2 + 1 & \text{if } x > 2 \\ 8x & \text{if } x < 2 \end{cases}$

For $x < 2$, $f(x) = 8x$, which is a polynomial and hence continuous. For $x > 2$,

$f(x) = x^2 + 1$, which is a polynomial and hence continuous.

Because f is not defined at $x = 2$, it is discontinuous there.

35.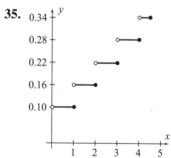

Discontinuous at 1, 2, 3, 4.

37.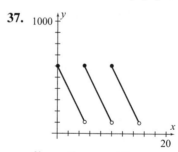

f is continuous at 2.
f is discontinuous at 5.
f is discontinuous at 10.

Principles in Practice 10.4

1. We need to solve $V(x) > 0$. The zeros of $V(x)$ occur when $x = 0$, $8 - 2x = 0$, and $10 - 2x = 0$, or $x = 0$, 4, and 5. These zeros determine the intervals $(-\infty, 0)$ $(0, 4)$, $(4, 5)$, and $(5, \infty)$. Using $x = -1$, 1, 4.5, and 6 for test points, we find the sign of $V(x)$:
$V(-1) = (-)(+)(+) = -$, so $V(x) < 0$ on $(-\infty, 0)$;
$V(1) = (+)(+)(+) = +$, so $V(x) > 0$ on $(0, 4)$;
$V(4.5) = (+)(-)(+) = -$, so $V(x) < 0$ on $(4, 5)$;
$V(6) = (+)(-)(-) = +$, so $V(x) > 0$ on $(5, \infty)$.
The volume is positive when $0 < x < 4$ or $5 < x$. However, $x > 5$ is unrealistic (as is $x < 0$) since the longest side of the piece of metal has length $2(5) = 10$ inches. Thus, the volume is positive when $0 < x < 4$.

Problems 10.4

1. $x^2 - 3x - 4 > 0$
$f(x) = x^2 - 3x - 4 = (x+1)(x-4)$ has zeros -1 and 4. By considering the intervals $(-\infty, -1)$, $(-1, 4)$, and $(4, \infty)$, we find $f(x) > 0$ on $(-\infty, -1)$ and $(4, \infty)$.
Answer: $(-\infty, -1)$, $(4, \infty)$

3. $x^2 - 3x - 10 \le 0$
$f(x) = (x+2)(x-5)$ has zeros -2 and 5. By considering the intervals $(-\infty, -2)$, $(-2, 5)$, and $(5, \infty)$, we find $f(x) < 0$ on $(-2, 5)$.
Answer: $[-2, 5]$

5. $2x^2 + 11x + 14 < 0$
$f(x) = 2x^2 + 11x + 14 = (2x+7)(x+2)$ has zeros $-\dfrac{7}{2}$ and -2. By considering the intervals $\left(-\infty, -\dfrac{7}{2}\right)$, $\left(-\dfrac{7}{2}, -2\right)$, and $(-2, \infty)$, we find

$f(x) < 0$ on $\left(-\dfrac{7}{2}, -2\right)$.

Answer: $\left(-\dfrac{7}{2}, -2\right)$

7. $x^2 + 4 < 0$. Since $x^2 + 4$ is always positive, the inequality $x^2 + 4 < 0$ has no solution.
Answer: no solution

9. $(x+2)(x-3)(x+6) \le 0$
$f(x) = (x+2)(x-3)(x+6)$ has zeros -2, 3, and -6. By considering the intervals $(-\infty, -6)$, $(-6, -2)$, $(-2, 3)$, and $(3, \infty)$, we find $f(x) < 0$ on $(-\infty, -6)$ and $(-2, 3)$.
Answer: $(-\infty, -6]$, $[-2, 3]$

11. $-x(x-5)(x+4) > 0$, or equivalently, $x(x-5)(x+4) < 0$.
$f(x) = x(x-5)(x+4)$ has zeros, 0, 5, and -4. By considering the intervals $(-\infty, -4)$, $(-4, 0)$, $(0, 5)$, and $(5, \infty)$, we find $f(x) < 0$ on $(-\infty, -4)$ and $(0, 5)$.
Answer: $(-\infty, -4)$, $(0, 5)$

13. $x^3 + 4x \ge 0$
$f(x) = x\left(x^2 + 4\right)$ has 0 as the only (real) zero. By considering the intervals $(-\infty, 0)$ and $(0, \infty)$, we find $f(x) > 0$ on $(0, \infty)$.
Answer: $[0, \infty)$

15. $x^3 + 8x^2 + 15x \le 0$
$f(x) = x(x+3)(x+5)$ has zeros 0, -3, and -5. By considering the intervals $(-\infty, -5)$, $(-5, -3)$, $(-3, 0)$, and $(0, \infty)$, we find $f(x) < 0$ on $(-\infty, -5)$ and $(-3, 0)$.
Answer: $(-\infty, -5]$, $[-3, 0]$

17. $\dfrac{x}{x^2 - 9} < 0$
$f(x) = \dfrac{x}{x^2 - 9}$ is discontinuous when $x = \pm 3$;
f has 0 as a zero. By considering the intervals $(-\infty, -3)$, $(-3, 0)$, $(0, 3)$, and $(3, \infty)$, we find $f(x) < 0$ on $(-\infty, -3)$ and $(0, 3)$.
Answer: $(-\infty, -3)$, $(0, 3)$

19. $\dfrac{4}{x-1} \geq 0$

$f(x) = \dfrac{4}{x-1}$ is discontinuous when $x = 1$,

and $f(x) = 0$ has no root. By considering the intervals $(-\infty, 1)$ and $(1, \infty)$, we find $f(x) > 0$ on $(1, \infty)$. Note also that $f(x) \neq 0$ for any x.
Answer: $(1, \infty)$

21. $\dfrac{x^2 - x - 6}{x^2 + 4x - 5} \geq 0$

$f(x) = \dfrac{x^2 - x - 6}{x^2 + 4x - 5} = \dfrac{(x-3)(x+2)}{(x+5)(x-1)}$ is

discontinuous at $x = -5$ and $x = 1$; f has zeros 3 and -2. By considering the intervals $(-\infty, -5)$, $(-5, -2)$, $(-2, 1)$, $(1, 3)$, and $(3, \infty)$, we find $f(x) > 0$ on $(-\infty, -5)$, $(-2, 1)$, and $(3, \infty)$.
Answer: $(-\infty, -5)$, $[-2, 1)$, $[3, \infty)$

23. $\dfrac{3}{x^2 + 6x + 5} \leq 0$

$f(x) = \dfrac{3}{x^2 + 6x + 5} = \dfrac{3}{(x+5)(x+1)}$ is never

zero, but is discontinuous at $x = -5$ and $x = -1$. By considering the intervals $(-\infty, -5)$, $(-5, -1)$, and $(-1, \infty)$, we find that $f(x) < 0$ on $(-5, -1)$.
Answer: $(-5, -1)$

25. $x^2 + 2x \geq 2$, or equivalently,

$x^2 + 2x - 2 \geq 0$. $f(x) = x^2 + 2x - 2$ has

zeros $-1 \pm \sqrt{3}$. By considering the intervals

$\left(-\infty, -1-\sqrt{3}\right)$ $\left(-1-\sqrt{3}, -1+\sqrt{3}\right)$, and

$\left(-1+\sqrt{3}, \infty\right)$, we find $f(x) > 0$ on

$\left(-\infty, -1-\sqrt{3}\right)$ and $\left(-1+\sqrt{3}, \infty\right)$.

Answer: $\left(-\infty, -1-\sqrt{3}\right], \left[-1+\sqrt{3}, \infty\right)$

27. Revenue = (no. of units)(price per unit). We
want $\quad q(28 - 0.2q) \geq 750$

$$0.2q^2 - 28q + 750 \leq 0$$
$$q^2 - 140q + 3750 \leq 0$$

Using the quadratic formula,

$q^2 - 140q + 3750 = 0$ when $q \approx 36.09$,

103.91. Thus $q^2 - 140q + 3750 \leq 0$ when

$36.09 \leq q \leq 103.91$, so sales revenue will be at least \$750 when between 37 and 103 units, inclusive, are sold.

29.

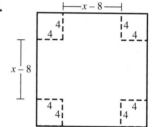

If x is the length of a side of the piece of aluminum, then the box will be 4 by $x - 8$ by $x - 8$.

$$4(x-8)^2 \geq 324$$
$$(x-8)^2 \geq 81$$
$$x^2 - 16x - 17 \geq 0$$
$$(x - 17)(x + 1) \geq 0$$

Solving gives $x \leq -1$ or $x \geq 17$. Since x must be positive, we have $x \geq 17$.
Answer: 17 in. by 17 in.

31.

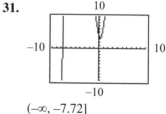

$(-\infty, -7.72]$

33.

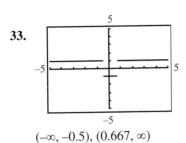

$(-\infty, -0.5), (0.667, \infty)$

Chapter 10 Review Problems

1. $\lim_{x \to -1} \left(2x^2 + 6x - 1\right) = 2(-1)^2 + 6(-1) - 1$
    $= -5$

3. $\lim_{x \to 3} \dfrac{x^2 - 9}{x^2 - 3x} = \lim_{x \to 3} \dfrac{(x+3)(x-3)}{x(x-3)}$
    $= \lim_{x \to 3} \dfrac{x+3}{x} = \dfrac{6}{3} = 2$

5. $\lim_{h \to 0} (x + h) = x + 0 = x$

7. $\lim_{x \to -4} \dfrac{x^3 + 4x^2}{x^2 + 2x - 8} = \lim_{x \to -4} \dfrac{x^2(x+4)}{(x+4)(x-2)}$
    $\lim_{x \to -4} \dfrac{x^2}{x-2} = \dfrac{16}{-6} = -\dfrac{8}{3}$

9. As $x \to \infty$, $x + 1 \to \infty$. Thus $\lim_{x \to \infty} \dfrac{2}{x+1} = 0$.

11. $\lim_{x \to \infty} \dfrac{2x+5}{7x-4} = \lim_{x \to \infty} \dfrac{2x}{7x} = \lim_{x \to \infty} \dfrac{2}{7} = \dfrac{2}{7}$

13. $\lim_{t \to 3^-} \dfrac{2t-3}{t-3} = -\infty$ and $\lim_{t \to 3^+} \dfrac{2t-3}{t-3} = \infty$. Thus
    $\lim_{t \to 3} \dfrac{2t-3}{t-3}$ does not exist.

15. $\lim_{x \to -\infty} \dfrac{x+3}{1-x} = \lim_{x \to -\infty} \dfrac{x}{-x} = \lim_{x \to -\infty} (-1) = -1$

17. $\lim_{x \to \infty} \dfrac{x^2 - 1}{(3x+2)^2} = \lim_{x \to \infty} \dfrac{x^2 - 1}{9x^2 + 12x + 4}$
    $= \lim_{x \to \infty} \dfrac{x^2}{9x^2} = \lim_{x \to \infty} \dfrac{1}{9} = \dfrac{1}{9}$

19. $\lim_{x \to 3^-} \dfrac{x+3}{x^2 - 9} = \lim_{x \to 3^-} \dfrac{x+3}{(x+3)(x-3)}$
    $= \lim_{x \to 3^-} \dfrac{1}{x-3} = -\infty$

21. As x becomes large, so does $3x$. Because the square roots of large numbers are also large, $\lim_{x \to \infty} \sqrt{3x} = \infty$.

23. $\lim_{x \to \infty} \dfrac{x^{100} + \dfrac{1}{x^3}}{\pi - x^{97}} = \lim_{x \to \infty} \dfrac{x^3\left(x^{100} + \dfrac{1}{x^3}\right)}{x^3\left(\pi - x^{97}\right)}$
    $= \lim_{x \to \infty} \dfrac{x^{103} + 1}{\pi x^3 - x^{100}} = \lim_{x \to \infty} \dfrac{x^{103}}{-x^{100}}$
    $= \lim_{x \to \infty} \left(-x^3\right) = -\infty$

25. $\lim_{x \to 1^-} f(x) = \lim_{x \to 1^-} x^2 = 1$
    $\lim_{x \to 1^+} f(x) = \lim_{x \to 1^+} x = 1$
    Thus $\lim_{x \to 1} f(x) = 1$.

27. $\lim_{x \to 4^+} \dfrac{\sqrt{x^2 - 16}}{4 - x} = \lim_{x \to 4^+} \dfrac{\sqrt{x-4}\sqrt{x+4}}{-(x-4)}$
    $= \lim_{x \to 4^+} -\dfrac{\sqrt{x+4}}{\sqrt{x-4}}$
    As $x \to 4^+$, $\sqrt{x-4}$ approaches 0 through positive values and $\sqrt{x+4} \to \sqrt{8}$. Thus
    $-\dfrac{\sqrt{x+4}}{\sqrt{x-4}} \to -\infty$.
    Answer: $-\infty$

29. $\lim\limits_{h\to 0} \dfrac{f(x+h)-f(x)}{h}$

$= \lim\limits_{h\to 0} \dfrac{[8(x+h)-2]-[8x-2]}{h}$

$= \lim\limits_{h\to 0} \dfrac{8h}{h} = \lim\limits_{h\to 0} 8 = 8$

31. $y = 23\left(1 - \dfrac{1}{1+2x}\right)$

Considering $\dfrac{1}{1+2x}$, we have

$\lim\limits_{x\to\infty} \dfrac{1}{1+2x} = \dfrac{1}{2}\cdot \lim\limits_{x\to\infty}\dfrac{1}{x} = \dfrac{1}{2}\cdot 0 = 0$. Thus

$\lim\limits_{x\to\infty} y = \lim\limits_{x\to\infty}\left[23\left(1 - \dfrac{1}{1+2x}\right)\right]$

$= 23(1-0) = 23$

Answer: 23

33. $f(x) = x + 5;\ x = 7$

(i) f is defined at $x = 7$; $f(7) = 12$

(ii) $\lim\limits_{x\to 7} f(x) = \lim\limits_{x\to 7}(x+5) = 7+5 = 12$,

which exists

(iii) $\lim\limits_{x\to 7} f(x) = 12 = f(7)$

Thus f is continuous at $x = 7$.

35. Since $f(x) = \dfrac{1}{5}x^2$ is polynomial function, it

is continuous everywhere.

37. $f(x) = \dfrac{x^2}{x+3}$ is a rational function and the

denominator is zero at $x = -3$. Thus f is
discontinuous at $x = -3$.

39. Since $f(x) = \dfrac{x-1}{2x^2+3}$ is a rational function

whose denominator is never zero, f is
continuous everywhere.

41. $f(x) = \dfrac{4-x^2}{x^2+3x-4} = \dfrac{4-x^2}{(x+4)(x-1)}$ is a

rational function and the denominator is
zero only when $x = -4$ or $x = 1$, so f is
discontinuous at $x = -4$, 1.

43. $f(x) = \begin{cases} x+4 & \text{if } x > -2 \\ 3x+6 & \text{if } x \le -2 \end{cases}$

For $x < -2$, $f(x) = 3x + 6$, which is a
polynomial and hence continuous. For
$x > -2$, $f(x) = x + 4$, which is a polynomial
and hence continuous. Because

$\lim\limits_{x\to -2^-} f(x) = \lim\limits_{x\to -2^-}(3x+6) = 0$ and

$\lim\limits_{x\to -2^+} f(x) = \lim\limits_{x\to -2^+}(x+4) = 2$, $\lim\limits_{x\to -2} f(x)$

does not exist. Thus f is discontinuous at
$x = -2$.

45. $x^2 + 4x - 12 > 0$

$f(x) = x^2 + 4x - 12 = (x+6)(x-2)$ has

zeros -6 and 2. By considering the intervals
$(-\infty, -6)$, $(-6, 2)$, and $(2, \infty)$, we find $f(x) > 0$
on $(-\infty, -6)$ and $(2, \infty)$.
Answer: $(-\infty, -6)$, $(2, \infty)$

47. $x^5 \le 7x^4$, $x^5 - 7x^4 \le 0$

$f(x) = x^5 - 7x^4 = x^4(x-7)$ has zeros 0 and

7.

By considering the intervals $(-\infty, 0)$, $(0, 7)$,
and $(7, \infty)$, we find $f(x) < 0$ on $(-\infty, 0)$ and
$(0, 7)$.
Answer: $(-\infty, 7]$

49. $\dfrac{x+5}{x^2-1} < 0$

$f(x) = \dfrac{x+5}{(x+1)(x-1)}$ is discontinuous when

$x = \pm 1$, and f has -5 as a zero. By
considering the intervals $(-\infty, -5)$, $(-5, -1)$,
$(-1, 1)$, and $(1, \infty)$, we find $f(x) < 0$ on
$(-\infty, -5)$ and $(-1, 1)$.
Answer: $(-\infty, -5)$, $(-1, 1)$

51. $\dfrac{x^2+3x}{x^2+2x-8} \geq 0$

$f(x) = \dfrac{x^2+3x}{x^2+2x-8} = \dfrac{x(x+3)}{(x+4)(x-2)}$ is

discontinuous when $x = -4, 2$ and has zeros
$x = -3, 0$. By considering the intervals
$(-\infty, -4), (-4, -3), (-3, 0), (0, 2),$ and $(2, \infty)$
we find $f(x) > 0$ on $(-\infty, -4), (-3, 0),$ and
$(2, \infty)$.
Answer: $(-\infty, -4), [-3, 0], (2, \infty)$

53.

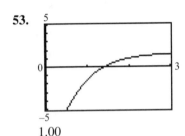

1.00

55.

0

57.

$[2.00, \infty)$

Mathematical Snapshot Chapter 10

1. $D = 8432e^{-rt}$
   A year from now, $t = 1$ and $D = 8000$. Thus
   $8000 = 8432e^{-r}$

   $e^{-r} = \dfrac{8000}{8432}$

   $-r = \ln\dfrac{8000}{8432}$

   $r = -\ln\dfrac{8000}{8432} \approx 0.053$

   The rate is 5.3%.

3. An exponential model assumes a fixed
   repayment rate. In reality, the repayment
   rate is constantly changing as a result of
   changing fiscal policy and other factors.

Chapter 11

Principles in Practice 11.1

1. $\dfrac{dH}{dt} = \dfrac{d}{dt}\left(6 + 40t - 16t^2\right)$

$= \displaystyle\lim_{h \to 0} \dfrac{H(t+h) - H(t)}{h}$

$= \displaystyle\lim_{h \to 0} \dfrac{\left[6 + 40(t+h) - 16(t+h)^2\right] - \left(6 + 40t - 16t^2\right)}{h}$

$= \displaystyle\lim_{h \to 0} \dfrac{6 + 40t + 40h - 16t^2 - 32th - 16h^2 - 6 - 40t + 16t^2}{h}$

$= \displaystyle\lim_{h \to 0} \dfrac{40h - 32th - 16h^2}{h} = \lim_{h \to 0} (40 - 32t - 16h)$

$= 40 - 32t$

$\dfrac{dH}{dt} = 40 - 32t$

Problems 11.1

1. a. $f(x) = x^3 + 3$, $P = (-2, -5)$

To begin, if $x = -3$, then $m_{PQ} = \dfrac{[(-3)^3 + 3] - (-5)}{-3 - (-2)} = 19$. If $x = -2.5$, then

$m_{PQ} = \dfrac{[(-2.5)^3 + 3] - (-5)}{-2.5 - (-2)} = 15.25.$

Continuing in this manner, we complete the table:

x-value of Q	-3	-2.5	-2.2	-2.1	-2.01	-2.001
m_{PQ}	19	15.25	13.24	12.61	12.0601	12.0060

b. We estimate that m_{tan} at P is 12.

3. $f(x) = x$

$f'(x) = \displaystyle\lim_{h \to 0} \dfrac{f(x+h) - f(x)}{h} = \lim_{h \to 0} \dfrac{(x+h) - x}{h} = \lim_{h \to 0} \dfrac{h}{h} = \lim_{h \to 0} 1 = 1$

5. $y = 3x + 5$. Let $y = f(x)$.

$$\frac{dy}{dx} = \lim_{h \to 0} \frac{f(x+h) - f(x)}{h}$$

$$= \lim_{h \to 0} \frac{[3(x+h) + 5] - [3x + 5]}{h}$$

$$= \lim_{h \to 0} \frac{3h}{h} = \lim_{h \to 0} 3 = 3$$

7. Let $f(x) = 5 - 4x$.

$$\frac{d}{dx}(5 - 4x) = \lim_{h \to 0} \frac{f(x+h) - f(x)}{h}$$

$$= \lim_{h \to 0} \frac{[5 - 4(x+h)] - [5 - 4x]}{h}$$

$$= \lim_{h \to 0} \frac{-4h}{h} = \lim_{h \to 0} (-4) = -4$$

9. $f(x) = 3$

$$f'(x) = \lim_{h \to 0} \frac{f(x+h) - f(x)}{h}$$

$$= \lim_{h \to 0} \frac{3 - 3}{h} = \lim_{h \to 0} \frac{0}{h} = \lim_{h \to 0} 0 = 0$$

11. Let $f(x) = x^2 + 4x - 8$.

$$\frac{d}{dx}\left(x^2 + 4x - 8\right)$$

$$= \lim_{h \to 0} \frac{f(x+h) - f(x)}{h}$$

$$= \lim_{h \to 0} \frac{\left[(x+h)^2 + 4(x+h) - 8\right] - \left[x^2 + 4x - 8\right]}{h}$$

$$= \lim_{h \to 0} \frac{x^2 + 2xh + h^2 + 4x + 4h - 8 - x^2 - 4x + 8}{h}$$

$$= \lim_{h \to 0} \frac{2xh + h^2 + 4h}{h}$$

$$= \lim_{h \to 0} (2x + h + 4) = 2x + 0 + 4 = 2x + 4$$

13. $p = f(q) = 3q^2 + 2q + 1$

$$\frac{dp}{dq} = \lim_{h \to 0} \frac{f(q+h) - f(q)}{h}$$

$$= \lim_{h \to 0} \frac{\left[3(q+h)^2 + 2(q+h) + 1\right] - \left[3q^2 + 2q + 1\right]}{h}$$

$$= \lim_{h \to 0} \frac{6qh + 3h^2 + 2h}{h}$$

$$= \lim_{h \to 0} (6q + 3h + 2) = 6q + 0 + 2 = 6q + 2$$

15. $y = f(x) = \dfrac{6}{x}$

$$y' = \lim_{h \to 0} \frac{f(x+h) - f(x)}{h} = \lim_{h \to 0} \frac{\frac{6}{x+h} - \frac{6}{x}}{h}$$

Multiplying the numerator and denominator by $x(x + h)$ gives

$$y' = \lim_{h \to 0} \frac{6x - 6(x+h)}{hx(x+h)} = \lim_{h \to 0} \frac{-6h}{hx(x+h)}$$

$$= \lim_{h \to 0} \left[-\frac{6}{x(x+h)} \right] = -\frac{6}{x(x+0)} = -\frac{6}{x^2}$$

17. $f(x) = \sqrt{x+2}$

$$f'(x) = \lim_{h \to 0} \frac{f(x+h) - f(x)}{h}$$

$$= \lim_{h \to 0} \frac{\sqrt{x+h+2} - \sqrt{x+2}}{h}$$

Rationalizing the numerator gives

$$\frac{\sqrt{x+h+2} - \sqrt{x+2}}{h}$$

$$= \frac{\sqrt{x+h+2} - \sqrt{x+2}}{h} \cdot \frac{\sqrt{x+h+2} + \sqrt{x+2}}{\sqrt{x+h+2} + \sqrt{x+2}}$$

$$= \frac{(x+h+2) - (x+2)}{h\left(\sqrt{x+h+2} + \sqrt{x+2}\right)} = \frac{1}{\sqrt{x+h+2} + \sqrt{x+2}}$$

Thus $f'(x) = \lim_{h \to 0} \dfrac{1}{\sqrt{x+h+2} + \sqrt{x+2}} = \dfrac{1}{2\sqrt{x+2}}$

19. $y = f(x) = x^2 + 4$

$$y' = \lim_{h \to 0} \frac{f(x+h) - f(x)}{h}$$

$$= \lim_{h \to 0} \frac{\left[(x+h)^2 + 4\right] - \left[x^2 + 4\right]}{h}$$

$$= \lim_{h \to 0} \frac{2xh + h^2}{h} = \lim_{h \to 0} (2x + h) = 2x + 0 = 2x$$

The slope at $(-2, 8)$ is $y'(-2) = 2(-2) = -4$.

21. $y = f(x) = 4x^2 - 5$

$$y' = \lim_{h \to 0} \frac{f(x+h) - f(x)}{h}$$

$$= \lim_{h \to 0} \frac{\left[4(x+h)^2 - 5\right] - \left[4x^2 - 5\right]}{h}$$

$$= \lim_{h \to 0} \frac{8xh + 4h^2}{h} = \lim_{h \to 0} (8x + 4h) = 8x$$

The slope when $x = 0$ is $y'(0) = 8(0) = 0$.

23. $y = x + 4$

$$y' = \lim_{h \to 0} \frac{[(x+h) + 4] - [x + 4]}{h} = \lim_{h \to 0} \frac{h}{h} = 1$$

If $x = 3$, then $y' = 1$. The tangent line at the point $(3, 7)$ is $y - 7 = 1(x - 3)$, or $y = x + 4$.

25. $y = x^2 + 2x + 3$

$$y' = \lim_{h \to 0} \frac{\left[(x+h)^2 + 2(x+h) + 3\right] - \left[x^2 + 2x + 3\right]}{h}$$

$$= \lim_{h \to 0} \frac{2xh + h^2 + 2h}{h}$$

$$= \lim_{h \to 0} (2x + h + 2) = 2x + 2$$

If $x = 1$, then $y' = 2(1) + 2 = 4$. The tangent line at the point $(1, 6)$ is $y - 6 = 4(x - 1)$, or $y = 4x + 2$.

27. $y = \dfrac{3}{x-1}$

$$y' = \lim_{h \to 0} \frac{\frac{3}{(x+h)-1} - \frac{3}{x-1}}{h}$$

$$= \lim_{h \to 0} \frac{\frac{3(x-1)-3(x+h-1)}{(x+h-1)(x-1)}}{h}$$

$$= \lim_{h \to 0} \frac{-3h}{h(x+h-1)(x-1)} = \lim_{h \to 0} \frac{-3}{(x+h-1)(x-1)}$$

$$= -\frac{3}{(x-1)^2}$$

If $x = 2$, then $y' = -\dfrac{3}{1} = -3$. The tangent line at

$(2, 3)$ is $y - 3 = -3(x-2)$, or $y = -3x + 9$.

29. $r = \left(\dfrac{\eta}{1+\eta}\right)\left(r_L - \dfrac{dC}{dD}\right)$

$$(1+\eta)r = \eta\left(r_L - \frac{dC}{dD}\right)$$

$$r + \eta r = \eta\left(r_L - \frac{dC}{dD}\right)$$

$$r = \eta\left(r_L - \frac{dC}{dD}\right) - \eta r$$

$$r = \eta\left(r_L - \frac{dC}{dD} - r\right)$$

$$\eta = \frac{r}{r_L - r - \frac{dC}{dD}}$$

31. -3.000, 13.445

33. -5.120, 0.038

35.

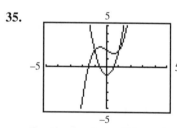

For the x-values of the points where the tangent to the graph of f is horizontal, the corresponding values of $f'(x)$ are 0. This is expected because the slope of a horizontal line is zero and the derivative gives the slope of the tangent line.

37. $n = 5$: $(z-x)\sum_{i=0}^{4} x^i z^{4-i} = (z-x)(z^4 + xz^3 + x^2 z^2 + x^3 z + x^4)$

$$= z^5 - xz^4 + xz^4 - x^2 z^3 + x^2 z^3 - x^3 z^2 + x^3 z^2 - x^4 z + x^4 z - x^5$$

$$= z^5 - x^5$$

$n = 3$: $(z-x)\sum_{i=0}^{2} x^i z^{2-i} = (z-x)(z^2 + xz + x^2)$

$$= z^3 - xz^2 + xz^2 - x^2 z + x^2 z - x^3$$

$$= z^3 - x^3$$

$f(x) = 4x^5 - 3x^3$

$f'(x) = \lim_{z \to x} \dfrac{f(z) - f(x)}{z - x}$

$\qquad = \lim_{z \to x} \dfrac{4z^5 - 3z^3 - (4x^5 - 3x^3)}{z - x}$

$\qquad = \lim_{z \to x} \dfrac{4(z^5 - x^5) - 3(z^3 - x^3)}{z - x}$

$\qquad = \lim_{z \to x} \dfrac{4(z-x)(z^4 + xz^3 + x^2 z^2 + x^3 z + x^4) - 3(z-x)(z^2 + xz + x^2)}{z - x}$

$\qquad = \lim_{z \to x} [4(z^4 + xz^3 + x^2 z^2 + x^3 z + x^4) - 3(z^2 + xz + x^2)]$

$\qquad = 4(5x^4) - 3(3x^2)$

$\qquad = 20x^4 - 9x^2$

Principles in Practice 11.2

1. $r'(q) = \dfrac{d}{dq}(50q - 0.3q^2)$

$\qquad = \dfrac{d}{dq}(50q) - \dfrac{d}{dq}\left(0.3q^2\right)$

$\qquad = 50\dfrac{d}{dq}(q) - 0.3\dfrac{d}{dq}\left(q^2\right)$

$\qquad = 50(1) - 0.3(2q) = 50 - 0.6q$

The marginal revenue is $r'(q) = 50 - 0.6q$.

Problems 11.2

1. $f(x) = 5$ is a constant function, so $f'(x) = 0$

3. $y = x^6$, $y' = 6x^{6-1} = 6x^5$

5. $y = x^{80}, \dfrac{dy}{dx} = 80x^{80-1} = 80x^{79}$

7. $f(x) = 9x^2, f'(x) = 9\left(2x^{2-1}\right) = 18x$

9. $g(w) = 8w^7, g'(w) = 8(7w^{7-1}) = 56w^6$

11. $y = \dfrac{2}{3}x^4, y' = \dfrac{2}{3}\left(4x^{4-1}\right) = \dfrac{8}{3}x^3$

13. $f(t) = \dfrac{t^7}{25}, f'(t) = \dfrac{1}{25}(7t^{7-1}) = \dfrac{7}{25}t^6$

15. $f(x) = x+3, f'(x) = 1+0 = 1$

17. $f'(x) = 4(2x) - 2(1) + 0 = 8x - 2$

19. $g'(p) = 4p^{4-1} - 3\left(3p^{3-1}\right) - 0 = 4p^3 - 9p^2$

21. $y' = 3x^{3-1} - \left(\dfrac{1}{2}x^{\frac{1}{2}-1}\right) = 3x^2 - \dfrac{1}{2\sqrt{x}}$

23. $y' = -13\left(3x^{3-1}\right) + 14(2x) - 2(1) + 0$

$= -39x^2 + 28x - 2$

25. $f'(x) = 2\left(0 - 4x^{4-1}\right) = -8x^3$

27. $g(x) = \dfrac{1}{3}\left(13 - x^4\right),$

$g'(x) = \dfrac{1}{3}\left(0 - 4x^{4-1}\right) = -\dfrac{4}{3}x^3$

29. $h(x) = 4x^4 + x^3 - \dfrac{9}{2}x^2 + 8x$

$h'(x) = 4\left(4x^{4-1}\right) + 3x^{3-1} - \dfrac{9}{2}(2x) + 8(1)$

$= 16x^3 + 3x^2 - 9x + 8$

31. $f(x) = \dfrac{3}{10}x^4 + \dfrac{7}{3}x^3$

$f'(x) = \dfrac{3}{10}\left(4x^3\right) + \dfrac{7}{3}\left(3x^2\right) = \dfrac{6}{5}x^3 + 7x^2$

33. $f'(x) = \dfrac{3}{5}x^{\frac{3}{5}-1} = \dfrac{3}{5}x^{-2/5}$

35. $y' = \dfrac{3}{4}x^{\left(\frac{3}{4}\right)-1} + 2\left(\dfrac{5}{3}x^{\left(\frac{5}{3}\right)-1}\right) = \dfrac{3}{4}x^{-\frac{1}{4}} + \dfrac{10}{3}x^{\frac{2}{3}}$

37. $f(x) = 11\sqrt{x} = 11x^{\frac{1}{2}},$

$f'(x) = 11\left(\dfrac{1}{2}\right)x^{\left(\frac{1}{2}\right)-1} = \dfrac{11}{2}x^{-\frac{1}{2}} = \dfrac{11}{2\sqrt{x}}$

39. $f(r) = 6r^{\frac{1}{3}}, f'(r) = 6\left(\dfrac{1}{3}r^{-\frac{2}{3}}\right) = 2r^{-\frac{2}{3}}$

41. $f(x) = x^{-4}, f'(x) = -4x^{-4-1} = -4x^{-5}$

43. $f(x) = x^{-3} + x^{-5} - 2x^{-6},$

$f'(x) = -3x^{-3-1} + \left(-5x^{-5-1}\right) - 2\left(-6x^{-6-1}\right)$

$= -3x^{-4} - 5x^{-6} + 12x^{-7}$

45. $y = \dfrac{1}{x} = x^{-1}$

$\dfrac{dy}{dx} = -1 \cdot x^{-1-1} = -x^{-2} = -\dfrac{1}{x^2}$

47. $y = \dfrac{8}{x^5} = 8x^{-5}$

$y' = 8\left(-5x^{-6}\right) = -40x^{-6}$

49. $g(x) = \dfrac{4}{3x^3} = \dfrac{4}{3}x^{-3}$

$g'(x) = \dfrac{4}{3}\left(-3x^{-4}\right) = -4x^{-4}$

51. $f(t) = \dfrac{1}{2}\left(\dfrac{1}{t}\right) = \dfrac{1}{2}t^{-1}$

$f'(t) = \dfrac{1}{2}\left(-1 \cdot t^{-2}\right) = -\dfrac{1}{2}t^{-2}$

53. $f(x) = \dfrac{1}{7}x + 7x^{-1}$

$f'(x) = \dfrac{1}{7}(1) + 7\left(-1x^{-2}\right) = \dfrac{1}{7} - 7x^{-2}$

55. $f(x) = -9x^{1/3} + 5x^{-2/5}$,

$f'(x) = -9\left(\dfrac{1}{3}x^{-\frac{2}{3}}\right) + 5\left(-\dfrac{2}{5}x^{-\frac{7}{5}}\right)$

$= -3x^{-\frac{2}{3}} - 2x^{-\frac{7}{5}}$

57. $q(x) = \dfrac{1}{\sqrt[3]{8}\sqrt[3]{x^2}} = \dfrac{1}{2x^{2/3}} = \dfrac{1}{2}x^{-2/3}$

$q'(x) = \dfrac{1}{2}\left(-\dfrac{2}{3}x^{-5/3}\right) = -\dfrac{1}{3}x^{-5/3}$

59. $y = \dfrac{2}{x^{\frac{1}{2}}} = 2x^{-\frac{1}{2}}$

$y' = 2\left(-\dfrac{1}{2}x^{-\frac{3}{2}}\right) = -x^{-\frac{3}{2}}$

61. $y = x^2\sqrt{x} = x^2\left(x^{\frac{1}{2}}\right) = x^{2+\left(\frac{1}{2}\right)} = x^{\frac{5}{2}}$

$y' = \dfrac{5}{2}x^{\frac{3}{2}}$

63. $f(x) = x\left(3x^2 - 10x + 7\right) = 3x^3 - 10x^2 + 7x$

$f'(x) = 9x^2 - 20x + 7$

65. $f(x) = x^3(3x)^2 = x^3\left(9x^2\right) = 9x^5$

$f'(x) = 45x^4$

67. $v(x) = x^{-\frac{2}{3}}(x+5) = x^{\frac{1}{3}} + 5x^{-\frac{2}{3}}$

$v'(x) = \dfrac{1}{3}x^{-\frac{2}{3}} - \dfrac{10}{3}x^{-\frac{5}{3}} = \dfrac{1}{3}x^{-\frac{5}{3}}(x-10)$

69. $f(q) = \dfrac{3q^2 + 4q - 2}{q} = \dfrac{3q^2}{q} + \dfrac{4q}{q} - \dfrac{2}{q^2}$

$= 3q + 4 - 2q^{-1}$

$f'(q) = 3(1) + 0 - 2(-q^{-2})$

$= 3 + 2q^{-2} = 3 + \dfrac{2}{q^2}$

71. $f(x) = (x+1)(x+3) = x^2 + 4x + 3$

$f'(x) = 2x + 4 = 2(x+2)$

73. $w(x) = \dfrac{x^2 + x^3}{x^2} = \dfrac{x^2}{x^2} + \dfrac{x^3}{x^2} = 1 + x$

$w'(x) = 0 + 1 = 1$

75. $y' = 6x + 4$

$y'\big|_{x=0} = 4$

$y'\big|_{x=2} = 16$

$y'\big|_{x=-3} = -14$

77. y is a constant, so $y' = 0$ for all x.

79. $y = 4x^2 + 5x + 6$

$y' = 8x + 5$

$y'\big|_{x=1} = 13$

An equation of the tangent line is
$y - 15 = 13(x - 1)$, or $y = 13x + 2$.

81. $y = \dfrac{1}{x^3} = x^{-3}$

$y' = -3x^{-4} = -\dfrac{3}{x^4}$

$y'\big|_{x=2} = -\dfrac{3}{16}$

An equation of the tangent line is

$y - \dfrac{1}{8} = -\dfrac{3}{16}(x-2)$, or $y = -\dfrac{3}{16}x + \dfrac{1}{2}$.

83. $y = 3 + x - 5x^2 + x^4$

$y' = 1 - 10x + 4x^3$.

When $x = 0$, then $y = 3$ and $y' = 1$. Thus an equation of the tangent line is
$y - 3 = 1(x - 0)$, or $y = x + 3$.

85. $y = \dfrac{5}{2}x^2 - x^3$

$y' = 5x - 3x^2$

A horizontal tangent line has slope 0, so we set $5x - 3x^2 = 0$. Then $x(5 - 3x) = 0$, $x = 0$

or $x = \dfrac{5}{3}$.

If $x = 0$, then $y = 0$. If $x = \dfrac{5}{3}$, $y = \dfrac{125}{54}$. This

gives the points $(0, 0)$ and $\left(\dfrac{5}{3}, \dfrac{125}{54}\right)$.

87. $y = x^2 - 5x + 3$

$y' = 2x - 5$

Setting $2x - 5 = 1$ gives $2x = 6$, $x = 3$. When $x = 3$, then $y = -3$. This gives the point $(3, -3)$.

89. $f(x) = \sqrt{x} + \dfrac{1}{\sqrt{x}} = x^{\frac{1}{2}} + x^{-\frac{1}{2}}$

$f'(x) = \dfrac{1}{2}x^{-\frac{1}{2}} - \dfrac{1}{2}x^{-\frac{3}{2}}$

$= \dfrac{1}{2\sqrt{x}} - \dfrac{1}{2x\sqrt{x}} = \dfrac{x-1}{2x\sqrt{x}}$

Thus $\dfrac{x-1}{2x\sqrt{x}} - f'(x) = \dfrac{x-1}{2x\sqrt{x}} - \dfrac{x-1}{2x\sqrt{x}} = 0$.

91. $y = x^3 - 3x$

$y'(x) = 3x^2 - 3$

$y'\big|_{x=2} = 3\left(2^2\right) - 3 = 9$

The tangent line at $(2, 2)$ is given by
$y - 2 = 9(x - 2)$, or $y = 9x - 16$.

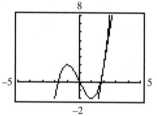

Principles in Practice 11.3

1. Here $\dfrac{dP}{dp} = 5$ and $\Delta p = 25.5 - 25 = 0.5$.

$\Delta P \approx \dfrac{dP}{dp}\Delta p = 5(0.5) = 2.5$

The profit increases by 2.5 units when the price is changed from 25 to 25.5 per unit.

3. $V'(r) = \dfrac{4}{3}\pi\left(3r^2\right) + 4\pi(2r) = 4\pi r^2 + 8\pi r$

When $r = 2$, $V'(r) = 4\pi(2)^2 + 8\pi(2) = 32\pi$
and

$V(r) = \dfrac{4}{3}\pi(2)^3 + 4\pi(2)^2$

$= \dfrac{32\pi}{3} + 16\pi = \dfrac{80}{3}\pi$.

The relative rate of change of the volume when $r = 2$ is $\dfrac{V'(2)}{V(2)} = \dfrac{32\pi}{\frac{80}{3}\pi} = \dfrac{6}{5} = 1.2$. Multiplying 1.2 by 100

gives the percentage rate of change: $(1.2)(100) = 120\%$.

Problems 11.3

1. $s = f(t) = 2t^2 + 3t$

If $\Delta t = 1$, then over [1, 2] we have

$\dfrac{\Delta s}{\Delta t} = \dfrac{f(2) - f(1)}{2 - 1} = \dfrac{14 - 5}{1} = 9.$

If $\Delta t = 0.5$, then over [1, 1.5] we have $\dfrac{\Delta s}{\Delta t} = \dfrac{f(1.5) - f(1)}{1.5 - 1} = \dfrac{9 - 5}{0.5} = 8.$

Continuing this way, we obtain the following table:

Δt	1	0.5	0.2	0.1	0.01	0.001
$\dfrac{\Delta s}{\Delta t}$	9	8	7.4	7.2	7.02	7.002

We estimate the velocity when $t = 1$ to be 7 m/s. With differentiation we get $v = \dfrac{ds}{dt} = 4t + 3,$

$\left.\dfrac{ds}{dt}\right|_{t=1} = 4(1) + 3 = 7$ m/s.

3. $s = f(t) = 2t^2 - 4t$

   a. When $t = 7$, then $s = 2(7^2) - 4(7) = 70$ m.

   b. $\dfrac{\Delta s}{\Delta t} = \dfrac{f(7.5) - f(7)}{0.5} = \dfrac{[2(7.5)^2 - 4(7.5)] - 70}{0.5} = 25$ m/s

   c. $v = \dfrac{ds}{dt} = 4t - 4.$ If $t = 7$, then $v = 4(7) - 4 = 24$ m/s

5. $s = f(t) = 2t^3 + 6$

   a. When $t = 1$, $s = 2(1)^3 + 6 = 8$ m.

b. $\dfrac{\Delta s}{\Delta t} = \dfrac{f(1.02) - f(1)}{0.02}$

$\qquad = \dfrac{\left[2(1.02)^3 + 6 \right] - 8}{0.02}$

$\qquad = 6.1208$ m/s

c. $v = \dfrac{ds}{dt} = 6t^2$. If $t = 1$, then

$\qquad v = 6(1)^2 = 6$ m/s

7. $s = f(t) = t^4 - 2t^3 + t$

a. When $t = 2$, $s = 2^4 - 2\left(2^3 \right) + 2 = 2$ m.

b. $\dfrac{\Delta s}{\Delta t} = \dfrac{f(2.1) - f(2)}{0.1}$

$\qquad = \dfrac{\left[(2.1)^4 - 2(2.1)^3 + 2.1 \right] - 2}{0.1}$

$\qquad = 10.261$ m/s

c. $v = \dfrac{ds}{dt} = 4t^3 - 6t^2 + 1$. If $t = 2$, then

$\qquad v = 4\left(2^3 \right) - 6\left(2^2 \right) + 1 = 9$ m/s

9. $\dfrac{dy}{dx} = \dfrac{25}{2} x^{\frac{3}{2}}$. If $x = 9$,

$\quad \dfrac{dy}{dx} = \dfrac{25}{2}(27) = 337.50$.

11. $\dfrac{dT}{dT_e} = 0 + 0.27(1 - 0) = 0.27$

13. $c = 500 + 10q$, $\dfrac{dc}{dq} = 10$. When $q = 100$,

$\quad \dfrac{dc}{dq} = 10$.

15. $\dfrac{dc}{dq} = 0.1(2q) + 3 = 0.2q + 3$. When $q = 5$,

$\quad \dfrac{dc}{dq} = 0.2(5) + 3 = 4$.

17. $\dfrac{dc}{dq} = 2q + 50$. Evaluating when $q = 15$, 16 and 17 gives 80, 82 and 84, respectively.

19. $\overline{c} = 0.01q + 5 + \dfrac{500}{q}$

$\quad c = \overline{c}q = 0.01q^2 + 5q + 500$

$\quad \dfrac{dc}{dq} = 0.02q + 5$

$\quad \left. \dfrac{dc}{dq} \right|_{q=50} = 6$

$\quad \left. \dfrac{dc}{dq} \right|_{q=100} = 7$

21. $c = \overline{c}q = 0.00002q^3 - 0.01q^2 + 6q + 20,000$

$\quad \dfrac{dc}{dq} = 0.00006q^2 - 0.02q + 6$

$\quad$ If $q = 100$, then $\dfrac{dc}{dq} = 4.6$. If $q = 500$, then

$\quad \dfrac{dc}{dq} = 11$.

23. $r = 0.8q$

$\quad \dfrac{dr}{dq} = 0.8$ for all q.

25. $r = 250q + 45q^2 - q^3$

$\quad \dfrac{dr}{dq} = 250 + 90q - 3q^2$. Evaluating when

$\quad q = 5$, 10 and 25 gives 625, 850 and 625, respectively.

27. $\dfrac{dc}{dq} = 6.750 - 0.000328(2q) = 6.750 - 0.000656q$

$\left.\dfrac{dc}{dq}\right|_{q=2000} = 6.750 - 0.000656(2000) = 5.438$

$\bar{c} = \dfrac{c}{q} = \dfrac{-10,484.69}{q} + 6.750 - 0.000328q$

$\bar{c}(2000) = \dfrac{-10,484.69}{2000} + 6.750 - 0.000328(2000)$
$\qquad\qquad = 0.851655$

29. $PR^{0.93} = 5,000,000$

$P = 5,000,000R^{-0.93}$

$\dfrac{dP}{dR} = -4,650,000R^{-1.93}$

31. a. $\dfrac{dy}{dx} = -1.5 - x$

$\left.\dfrac{dy}{dx}\right|_{x=6} = -1.5 - 6 = -7.5$

  b. Setting $-1.5 - x = -6$ gives $x = 4.5$.

33. a. $y' = 1$

  b. $\dfrac{y'}{y} = \dfrac{1}{x+4}$

  c. $y'(5) = 1$

  d. $\dfrac{1}{5+4} = \dfrac{1}{9} \approx 0.111$

  e. 11.1%

35. a. $y' = 6x$

  b. $\dfrac{y'}{y} = \dfrac{6x}{3x^2 + 7}$

  c. $y'(2) = 6(2) = 12$

d.　$\dfrac{12}{12+7} = \dfrac{12}{19} \approx 0.632$

e.　63.2%

37. a.　$y' = -3x^2$

b.　$\dfrac{y'}{y} = \dfrac{-3x^2}{8-x^3}$

c.　$y'(1) = -3$

d.　$\dfrac{-3}{8-1} = -\dfrac{3}{7} \approx -0.429$

e.　−42.9%

39.　$c = 0.3q^2 + 3.5q + 9$

$\dfrac{dc}{dq} = 0.6q + 3.5$

If $q = 10$, then $\dfrac{dc}{dq} = 0.6(10) + 3.5 = 9.5$. If

$q = 10$, then $c = 74$ and $\dfrac{\frac{dc}{dq}}{c}(100) = \dfrac{9.5}{74}(100) \approx 12.8\%$.

41. a.　$\dfrac{dr}{dq} = 30 - 0.6q$

b.　If $q = 10$, $\dfrac{r'}{r} = \dfrac{30-6}{300-30} = \dfrac{24}{270} = \dfrac{4}{45} \approx 0.09$.

c.　9%

43.　$\dfrac{W'}{W} = \dfrac{0.864t^{-0.568}}{2t^{0.432}} = \dfrac{0.432}{t}$

45.　The cost of $q = 20$ bikes is $q\bar{c} = 20(150) = \$3000$. The marginal cost, \$125, is the approximate cost of one additional bike. Thus the approximate cost of producing 21 bikes is $\$3000 + \$125 = \$3125$.

47.　\$5.07 per unit

Principles in Practice 11.4

1. $\dfrac{dR}{dx} = (2 - 0.15x)\dfrac{d}{dx}(225 + 20x) + (225 + 20x)\dfrac{d}{dx}(2 - 0.15x)$

   $= (2 - 0.15x)(20) + (225 + 20x)(-0.15)$

   $= 40 - 3x - 33.75 - 3x = 6.25 - 6x$

   $\dfrac{dR}{dx} = 6.25 - 6x$

Problems 11.4

1. $f'(x) = (4x + 1)(6) + (6x + 3)(4) = 24x + 6 + 24x + 12 = 48x + 18 = 6(8x + 3)$

3. $s'(t) = (5 - 3t)(3t^2 - 4t) + (t^3 - 2t^2)(-3) = 15t^2 - 20t - 9t^3 + 12t^2 - 3t^3 + 6t^2 = -12t^3 + 33t^2 - 20t$

5. $f'(r) = \left(3r^2 - 4\right)(2r - 5) + \left(r^2 - 5r + 1\right)(6r)$

   $= 6r^3 - 15r^2 - 8r + 20 + 6r^3 - 30r^2 + 6r = 12r^3 - 45r^2 - 2r + 20$

7. Without the product rule we have

   $f(x) = x^2\left(2x^2 - 5\right) = 2x^4 - 5x^2$

   $f'(x) = 8x^3 - 10x$

9. $y' = \left(x^2 + 3x - 2\right)(4x - 1) + \left(2x^2 - x - 3\right)(2x + 3)$

   $= \left(4x^3 + 12x^2 - 8x - x^2 - 3x + 2\right) + \left(4x^3 - 2x^2 - 6x + 6x^2 - 3x - 9\right)$

   $= 8x^3 + 15x^2 - 20x - 7$

11. $f'(w) = (w^2 + 3w - 7)(6w^2) + (2w^3 - 4)(2w + 3)$

    $= 6w^4 + 18w^3 - 42w^2 + 4w^4 + 6w^3 - 8w - 12$

    $= 10w^4 + 24w^3 - 42w^2 - 8w - 12$

13. $y' = \left(x^2 - 1\right)\left(9x^2 - 6\right) + \left(3x^3 - 6x + 5\right)(2x) - 4(8x + 2)$

    $= 9x^4 - 15x^2 + 6 + 6x^4 - 12x^2 + 10x - 32x - 8$

    $= 15x^4 - 27x^2 - 22x - 2$

15. $F'(p) = \dfrac{3}{2}\left[(5p^{1/2} - 2)(3) + (3p-1)\left(5 \cdot \dfrac{1}{2}p^{-1/2}\right)\right]$

$\qquad = \dfrac{3}{2}\left[15p^{1/2} - 6 + \dfrac{15}{2}p^{1/2} - \dfrac{5}{2}p^{-1/2}\right]$

$\qquad = \dfrac{3}{4}[45p^{1/2} - 12 - 5p^{-1/2}]$

17. $y = 7 \cdot \dfrac{2}{3}$ is a constant function, so $y' = 0$.

19. $y = 6x^3 + 47x^2 + 31x - 28$

$\qquad y' = 18x^2 + 94x + 31$

21. $f'(x) = \dfrac{(x-1)(5) - (5x)(1)}{(x-1)^2} = \dfrac{5x - 5 - 5x}{(x-1)^2}$

$\qquad = -\dfrac{5}{(x-1)^2}$

23. $f(x) = \dfrac{-13}{3x^5} = -\dfrac{13}{3}x^{-5}$

$\qquad f'(x) = -\dfrac{13}{3}(-5x^{-6}) = \dfrac{65}{3x^6}$

25. $y' = \dfrac{(x-1)(1) - (x+2)(1)}{(x-1)^2}$

$\qquad = \dfrac{x-1-x-2}{(x-1)^2}$

$\qquad = -\dfrac{3}{(x-1)^2}$

27. $h'(z) = \dfrac{\left(z^2 - 4\right)(-2) - (6 - 2z)(2z)}{\left(z^2 - 4\right)^2}$

$\qquad = \dfrac{-2z^2 + 8 - 12z + 4z^2}{\left(z^2 - 4\right)^2} = \dfrac{2z^2 - 12z + 8}{\left(z^2 - 4\right)^2}$

$\qquad = \dfrac{2\left(z^2 - 6z + 4\right)}{\left(z^2 - 4\right)^2}$

29. $y' = \dfrac{\left(x^2 - 5x\right)(16x - 2) - \left(8x^2 - 2x + 1\right)(2x - 5)}{\left(x^2 - 5x\right)^2}$

$= \dfrac{16x^3 - 82x^2 + 10x - \left(16x^3 - 44x^2 + 12x - 5\right)}{\left(x^2 - 5x\right)^2} = \dfrac{-38x^2 - 2x + 5}{\left(x^2 - 5x\right)^2}$

31. $y' = \dfrac{\left(2x^2 - 3x + 2\right)(2x - 4) - \left(x^2 - 4x + 3\right)(4x - 3)}{\left(2x^2 - 3x + 2\right)^2}$

$= \dfrac{4x^3 - 14x^2 + 16x - 8 - \left(4x^3 - 19x^2 + 24x - 9\right)}{\left(2x^2 - 3x + 2\right)^2}$

$= \dfrac{5x^2 - 8x + 1}{\left(2x^2 - 3x + 2\right)^2}$

33. $g'(x) = \dfrac{\left(x^{100} + 7\right)(0) - (1)\left(100x^{99}\right)}{\left(x^{100} + 7\right)^2} = -\dfrac{100x^{99}}{\left(x^{100} + 7\right)^2}$

35. $u(v) = \dfrac{v^3 - 8}{v} = \dfrac{v^3}{v} - \dfrac{8}{v} = v^2 - 8v^{-1}$

$u'(v) = 2v + 8v^{-2} = 2\left(v + \dfrac{4}{v^2}\right) = \dfrac{2(v^3 + 4)}{v^2}$

37. $y = \dfrac{3x^2 - x - 1}{\sqrt[3]{x}} = \dfrac{3x^2 - x - 1}{x^{\frac{1}{3}}} = 3x^{\frac{5}{3}} - x^{\frac{2}{3}} - x^{-\frac{1}{3}}$

$y' = 5x^{\frac{2}{3}} - \dfrac{2}{3}x^{-\frac{1}{3}} + \dfrac{1}{3}x^{-\frac{4}{3}} = 5x^{\frac{2}{3}} - \dfrac{2}{3x^{\frac{1}{3}}} + \dfrac{1}{3x^{\frac{4}{3}}}$

$= \dfrac{15x^2 - 2x + 1}{3x^{\frac{4}{3}}}$

39. $y' = -\dfrac{(x - 8)(0) - (4)(1)}{(x - 8)^2} + \dfrac{(3x + 1)(2) - (2x)(3)}{(3x + 1)^2}$

$= \dfrac{4}{(x - 8)^2} + \dfrac{2}{(3x + 1)^2}$

41. $y' = \dfrac{[(x+2)(x-4)](1)-(x-5)(2x-2)}{[(x+2)(x-4)]^2}$

$= \dfrac{x^2-2x-8-\left(2x^2-12x+10\right)}{[(x+2)(x-4)]^2}$

$= \dfrac{-\left(x^2-10x+18\right)}{[(x+2)(x-4)]^2}$

43. $s'(t) = \dfrac{\left[\left(t^2-1\right)\left(t^3+7\right)\right](2t+3)-\left(t^2+3t\right)\left(5t^4-3t^2+14t\right)}{\left[\left(t^2-1\right)\left(t^3+7\right)\right]^2}$

$= \dfrac{-3t^6-12t^5+t^4+6t^3-21t^2-14t-21}{\left[\left(t^2-1\right)\left(t^3+7\right)\right]^2}$

45. $y = 3x - \dfrac{\frac{2}{x}-\frac{3}{x-1}}{x-2} = 3x - \dfrac{\frac{2(x-1)-3x}{x(x-1)}}{x-2}$

$= 3x + \dfrac{x+2}{x(x-1)(x-2)} = 3x + \dfrac{x+2}{x^3-3x^2+2x}$

$y' = 3 + \dfrac{(x^3-3x^2+2x)(1)-(x+2)(3x^2-6x+2)}{[x(x-1)(x-2)]^2}$

$= 3 - \dfrac{2x^3+3x^2-12x+4}{[x(x-1)(x-2)]^2}$

47. $f'(x) = \dfrac{(a-x)(1)-(a+x)(-1)}{(a-x)^2} = \dfrac{2a}{(a-x)^2}$

49. $y = \left(4x^2+2x-5\right)\left(x^3+7x+4\right)$

$y' = \left(4x^2+2x-5\right)\left(3x^2+7\right)+\left(x^3+7x+4\right)(8x+2)$

$y'(-1) = (-3)(10)+(-4)(-6) = -6$

51. $y = \dfrac{6}{x-1}$

$y' = \dfrac{(x-1)(0)-(6)(1)}{(x-1)^2} = -\dfrac{6}{(x-1)^2}$

$y'(3) = -\dfrac{6}{2^2} = -\dfrac{3}{2}$

The tangent line is $y-3 = -\dfrac{3}{2}(x-3)$, or $y = -\dfrac{3}{2}x + \dfrac{15}{2}$.

53. $y = (2x+3)\left[2\left(x^4 - 5x^2 + 4\right)\right]$

$y' = (2x+3)\left[2\left(4x^3 - 10x\right)\right] + \left[2\left(x^4 - 5x^2 + 4\right)\right](2)$

$y'(0) = (3)(0) + [2(4)](2) = 16$

The tangent line is $y - 24 = 16(x - 0)$, or $y = 16x + 24$.

55. $y = \dfrac{x}{2x-6}$

$y' = \dfrac{(2x-6)(1)-x(2)}{(2x-6)^2} = \dfrac{-6}{(2x-6)^2}$

If $x = 1$, then $y = \dfrac{1}{2-6} = -\dfrac{1}{4}$ and $y' = \dfrac{-6}{(-4)^2} = \dfrac{-6}{16} = -\dfrac{3}{8}$.

Thus $\dfrac{y'}{y} = \dfrac{-\frac{3}{8}}{-\frac{1}{4}} = \dfrac{3}{2} = 1.5$.

57. $s = \dfrac{2}{t^3+1}$. When $t = 1$, then $s = 1$ m.

$v = \dfrac{ds}{dt} = \dfrac{(t^3+1)(0)-2(3t^2)}{(t^3+1)^2} = -\dfrac{6t^2}{(t^3+1)^2}$

If $t = 1$, then $v = -\dfrac{6}{4} = -1.5$ m/s.

59. $p = 50 - 0.01q$

$r = pq = 50q - 0.01q^2$

$\dfrac{dr}{dq} = 50 - 0.02q$

61. $p = \dfrac{108}{q+2} - 3$

$r = pq = \dfrac{108q}{q+2} - 3q$

$\dfrac{dr}{dq} = \dfrac{(q+2)(108) - (108q)(1)}{(q+2)^2} - 3$

$= \dfrac{216}{(q+2)^2} - 3$

63. $\dfrac{dC}{dI} = 0.672$

65. $C = 3 + I^{1/2} + 2I^{1/3}$

$\dfrac{dC}{dI} = 0 + \dfrac{1}{2}I^{-1/2} + \dfrac{2}{3}I^{-2/3} = \dfrac{1}{2\sqrt{I}} + \dfrac{2}{3\sqrt[3]{I^2}}$

When $I = 1$, then $\dfrac{dC}{dI} = \dfrac{1}{2} + \dfrac{2}{3} = \dfrac{7}{6}$.

$\dfrac{dS}{dI} = 1 - \dfrac{dC}{dI} = 1 - \dfrac{1}{2\sqrt{I}} - \dfrac{2}{3\sqrt[3]{I^2}}$

When $I = 1$, then $1 - \dfrac{dC}{dI} = 1 - \dfrac{7}{6} = -\dfrac{1}{6}$.

67. $\dfrac{dC}{dI} = \dfrac{\left(\sqrt{I}+4\right)\left(\frac{8}{\sqrt{I}} + 1.2\sqrt{I} - 0.2\right) - \left(16\sqrt{I} + 0.8\sqrt{I^3} - 0.2I\right)\left(\frac{1}{2\sqrt{I}}\right)}{\left(\sqrt{I}+4\right)^2}$

$\dfrac{dC}{dI}\bigg|_{I=36} \approx 0.615$, so $\dfrac{dS}{dI} \approx 1 - 0.615 = 0.385$ when $I = 36$.

69. Simplifying gives $C = 10 + 0.7I - 0.2I^{\frac{1}{2}}$

a. $\dfrac{dC}{dI} = 0.7 - 0.1I^{-\frac{1}{2}} = 0.7 - \dfrac{0.1}{\sqrt{I}}$

$\dfrac{dS}{dI} = 1 - \dfrac{dC}{dI} = 0.3 + \dfrac{0.1}{\sqrt{I}}$

$\dfrac{dS}{dI}\bigg|_{I=25} = 0.3 + \dfrac{0.1}{5} = 0.32$

b. $\dfrac{\frac{dC}{dI}}{C}$ when $I = 25$ is $\dfrac{0.7 - \frac{0.1}{5}}{10 + 0.7(25) - 0.2(5)} \approx 0.026$

71. $\dfrac{dc}{dq} = 6 \cdot \dfrac{(q+2)(2q) - q^2(1)}{(q+2)^2} = 6 \cdot \dfrac{q^2 + 4q}{(q+2)^2} = \dfrac{6q(q+4)}{(q+2)^2}$

73. $y = \dfrac{900x}{10 + 45x}$

$\dfrac{dy}{dx} = \dfrac{(10 + 45x)(900) - (900x)(45)}{(10 + 45x)^2}$

$\dfrac{dy}{dx}\bigg|_{x=2} = \dfrac{(100)(900) - (1800)(45)}{(100)^2} = \dfrac{9}{10}$

75. $y = \dfrac{0.7355x}{1 + 0.02744x}$

$\dfrac{dy}{dx} = \dfrac{(1 + 0.02744x)(0.7355) - (0.7355x)(0.02744)}{(1 + 0.02744x)^2}$

$= \dfrac{0.7355}{(1 + 0.02744x)^2}$

77. $\dfrac{d\bar{c}}{dq} = \dfrac{d}{dq}\left(\dfrac{c}{q}\right) = \dfrac{q \cdot \frac{dc}{dq} - c(1)}{q^2}$. When $q = 20$ we have $\dfrac{\frac{d\bar{c}}{dq}}{\bar{c}} = \dfrac{\frac{q \cdot \frac{dc}{dq} - c}{q^2}}{\bar{c}} = \dfrac{\frac{20(125) - 20(150)}{(20)^2}}{150} = -\dfrac{1}{120}$

Principles in Practice 11.5

1. By the chain rule,

$\dfrac{dy}{dt} = \dfrac{dy}{dx} \cdot \dfrac{dx}{dt} = \dfrac{d}{dx}\left(4x^2\right) \cdot \dfrac{d}{dt}(6t) = (8x)(6) = 48x$.

Since $x = 6t$, $\dfrac{dy}{dt} = 48(6t) = 288t$.

Problems 11.5

1. $\dfrac{dy}{dx} = \dfrac{dy}{du} \cdot \dfrac{du}{dx} = (2u - 2)(2x - 1) = \left[2\left(x^2 - x\right) - 2\right](2x - 1) = \left(2x^2 - 2x - 2\right)(2x - 1) = 4x^3 - 6x^2 - 2x + 2$

3. $\dfrac{dy}{dx} = \dfrac{dy}{dw} \cdot \dfrac{dw}{dx} = \left(-\dfrac{2}{w^3}\right)(-1) = \dfrac{2}{w^3} = \dfrac{2}{(2 - x)^3}$

5. $\dfrac{dw}{dt} = \dfrac{dw}{du} \cdot \dfrac{du}{dt} = (3u^2)\left[\dfrac{(t+1)-(t-1)}{(t+1)^2}\right] = 3u^2\left[\dfrac{2}{(t+1)^2}\right].$

If $t = 1$, then $u = \dfrac{1-1}{1+1} = 0$, so $\left.\dfrac{dw}{dt}\right|_{t=1} = 3(0)^2\left[\dfrac{2}{4}\right] = 0$.

7. $\dfrac{dy}{dx} = \dfrac{dy}{dw} \cdot \dfrac{dw}{dx} = (6w-8)(4x)$. If $x = 0$, then $\dfrac{dy}{dx} = 0$.

9. $y' = 6(3x+2)^5 \cdot \dfrac{d}{dx}(3x+2)$

$= 6(3x+2)^5(3) = 18(3x+2)^5$

11. $y' = 5(3+2x^3)^4 \cdot \dfrac{d}{dx}(3+2x^3)$

$= 5(3+2x^3)^4(6x^2)$

$= 30x^2(3+2x^3)$

13. $y' = 2 \cdot 100\left(x^3 - 8x^2 + x\right)^{99} \cdot \dfrac{d}{dx}\left(x^3 - 8x^2 + x\right)$

$= 200\left(x^3 - 8x^2 + x\right)^{99}\left(3x^2 - 16x + 1\right)$

$= 200\left(3x^2 - 16x + 1\right)\left(x^3 - 8x^2 + x\right)^{99}$

15. $y' = -3\left(x^2 - 2\right)^{-4} \cdot \dfrac{d}{dx}\left(x^2 - 2\right)$

$= -3\left(x^2 - 2\right)^{-4}(2x) = -6x\left(x^2 - 2\right)^{-4}$

17. $y' = 2\left(-\dfrac{5}{7}\right)(x^2 + 5x - 2)^{-12/7} \cdot \dfrac{d}{dx}(x^2 + 5x - 2)$

$= -\dfrac{10}{7}(2x+5)(x^2 + 5x - 2)^{-12/7}$

19. $y = \sqrt{5x^2 - x} = \left(5x^2 - x\right)^{\frac{1}{2}}$

$y' = \dfrac{1}{2}\left(5x^2 - x\right)^{-\frac{1}{2}}(10x - 1)$

$= \dfrac{1}{2}(10x - 1)\left(5x^2 - x\right)^{-\frac{1}{2}}$

21. $y = \sqrt[4]{2x-1} = (2x-1)^{\frac{1}{4}}$

$y' = \frac{1}{4}(2x-1)^{-\frac{3}{4}}(2) = \frac{1}{2}(2x-1)^{-\frac{3}{4}}$

23. $y = 2\sqrt[5]{\left(x^3+1\right)^2} = 2\left(x^3+1\right)^{\frac{2}{5}}$

$y' = 2\left(\frac{2}{5}\right)\left(x^3+1\right)^{-\frac{3}{5}}\left(3x^2\right) = \frac{12}{5}x^2\left(x^3+1\right)^{-\frac{3}{5}}$

25. $y = \frac{6}{2x^2-x+1} = 6\left(2x^2-x+1\right)^{-1}$

$y' = 6(-1)\left(2x^2-x+1\right)^{-2}(4x-1)$

$= -6(4x-1)\left(2x^2-x+1\right)^{-2}$

27. $y = \frac{1}{\left(x^2-3x\right)^2} = \left(x^2-3x\right)^{-2}$

$y' = -2\left(x^2-3x\right)^{-3}(2x-3)$

$= -2(2x-3)\left(x^2-3x\right)^{-3}$

29. $y = \frac{4}{\sqrt{9x^2+1}} = 4(9x^2+1)^{-1/2}$

$y' = 4\left(-\frac{1}{2}\right)(9x^2+1)^{-3/2}(18x)$

$= -36x(9x^2+1)^{-3/2}$

31. $y = \sqrt[3]{7x} + \sqrt[3]{7}x = (7x)^{\frac{1}{3}} + \sqrt[3]{7}x$

$y' = \frac{1}{3}(7x)^{-\frac{2}{3}}(7) + \sqrt[3]{7}(1) = \frac{7}{3}(7x)^{-\frac{2}{3}} + \sqrt[3]{7}$

33. $y' = x^2\left[5(x-4)^4(1)\right] + (x-4)^5(2x)$

$= x(x-4)^4[5x+2(x-4)]$

$= x(x-4)^4(7x-8)$

35. $y = 4x^2\sqrt{5x+1} = 4x^2(5x+1)^{\frac{1}{2}}$

$$y' = 4x^2\left(\frac{1}{2}(5x+1)^{-\frac{1}{2}}(5)\right) + \sqrt{5x+1}(8x)$$

$$= 10x^2(5x+1)^{-\frac{1}{2}} + 8x\sqrt{5x+1}$$

37. $y' = \left(x^2 + 2x - 1\right)^3(5) + (5x)\left[3\left(x^2 + 2x - 1\right)^2(2x+2)\right]$

$$= 5\left(x^2 + 2x - 1\right)^2\left[\left(x^2 + 2x - 1\right) + 3x(2x+2)\right]$$

$$= 5\left(x^2 + 2x - 1\right)^2\left(7x^2 + 8x - 1\right)$$

39. $y' = (8x-1)^3\left[4(2x+1)^3(2)\right] + (2x+1)^4\left[3(8x-1)^2(8)\right]$

$$= 8(8x-1)^2(2x+1)^3[(8x-1) + 3(2x+1)]$$

$$= 8(8x-1)^2(2x+1)^3(14x+2)$$

$$= 16(8x-1)^2(2x+1)^3(7x+1)$$

41. $y' = 12\left(\dfrac{x-3}{x+2}\right)^{11}\left[\dfrac{(x+2)(1)-(x-3)(1)}{(x+2)^2}\right]$

$$= 12\left(\frac{x-3}{x+2}\right)^{11}\left[\frac{5}{(x+2)^2}\right]$$

$$= \frac{60(x-3)^{11}}{(x+2)^{13}}$$

43. $y' = \dfrac{1}{2}\left(\dfrac{x-2}{x+3}\right)^{-\frac{1}{2}}\left[\dfrac{(x+3)(1)-(x-2)(1)}{(x+3)^2}\right]$

$$= \frac{5}{2(x+3)^2}\left(\frac{x-2}{x+3}\right)^{-\frac{1}{2}} = \frac{5}{2(x+3)^2}\sqrt{\frac{x+3}{x-2}}$$

45. $y' = \dfrac{\left(x^2+4\right)^3(2)-(2x-5)\left[3\left(x^2+4\right)^2(2x)\right]}{\left(x^2+4\right)^6}$

$= \dfrac{\left(x^2+4\right)^2\left\{\left(x^2+4\right)(2)-(2x-5)[3(2x)]\right\}}{\left(x^2+4\right)^6}$

$= \dfrac{2x^2+8-12x^2+30x}{\left(x^2+4\right)^4} = \dfrac{-10x^2+30x+8}{\left(x^2+4\right)^4}$

$= \dfrac{-2\left(5x^2-15x-4\right)}{\left(x^2+4\right)^4}$

47. $y' = \dfrac{(3x-1)^3\left[5(8x-1)^4(8)\right]-(8x-1)^5\left[3(3x-1)^2(3)\right]}{(3x-1)^6}$

$= \dfrac{(3x-1)^2(8x-1)^4[(3x-1)(40)-(8x-1)(9)]}{(3x-1)^6}$

$= \dfrac{(8x-1)^4(48x-31)}{(3x-1)^4}$

49. $y = 6\left(5x^2+2\right)\sqrt{x^4+5} = 6\left[\left(5x^2+2\right)\left(x^4+5\right)^{\frac{1}{2}}\right]$

$y' = 6\left[\left(5x^2+2\right)\cdot\dfrac{1}{2}\left(x^4+5\right)^{-\frac{1}{2}}\left(4x^3\right)+\left(x^4+5\right)^{\frac{1}{2}}(10x)\right]$

$= 6\left[\left(5x^2+2\right)\left(x^4+5\right)^{-\frac{1}{2}}\left(2x^3\right)+\left(x^4+5\right)^{\frac{1}{2}}(10x)\right]$

$= 12x\left[\left(5x^2+2\right)\left(x^4+5\right)^{-\frac{1}{2}}\left(x^2\right)+\left(x^4+5\right)^{\frac{1}{2}}(5)\right]$

Factoring out $\left(x^4+5\right)^{-\frac{1}{2}}$ gives

$y' = 12x\left(x^4+5\right)^{-\frac{1}{2}}\left[\left(5x^2+2\right)\left(x^2\right)+\left(x^4+5\right)(5)\right]$

$= 12x\left(x^4+5\right)^{-\frac{1}{2}}\left(10x^4+2x^2+25\right)$

51. $y' = 8 + \dfrac{(t+4)(1) - (t-1)(1)}{(t+4)^2} - 2\left(\dfrac{8t-7}{4}\right)\left(\dfrac{1}{4} \cdot 8\right)$

$= 8 + \dfrac{5}{(t+4)^2} - (8t-7) = 15 - 8t + \dfrac{5}{(t+4)^2}$

53. $y' = \dfrac{(x^3-5)^5[(2x+1)^3(2)(x+3)(1) + (x+3)^2(3)(2x+1)^2(2)] - (2x+1)^3(x+3)^2[5(x^3-5)^4(3x^2)]}{(x^3-5)^{10}}$

55. $\dfrac{dy}{dx} = \dfrac{dy}{du} \cdot \dfrac{du}{dx} = \left[3(5u+6)^2(5)\right]\left[4\left(x^2+1\right)^3(2x)\right]$

When $x = 0$, then $\dfrac{dy}{dx} = 0$.

57. $y' = 3\left(x^2 - 7x - 8\right)^2(2x-7)$

If $x = 8$, then slope $= y' = 3(64 - 56 - 8)^2(16 - 7) = 0$.

59. $y = \left(x^2 - 8\right)^{\frac{2}{3}}$

$y' = \dfrac{2}{3}\left(x^2-8\right)^{-\frac{1}{3}}(2x) = \dfrac{4x}{3\left(x^2-8\right)^{\frac{1}{3}}}$

If $x = 3$, then $y' = \dfrac{12}{3(1)} = 4$. Thus the tangent line is $y - 1 = 4(x-3)$, or $y = 4x - 11$.

61. $y' = \dfrac{(x+1)\left(\frac{1}{2}\right)(7x+2)^{-\frac{1}{2}}(7) - \sqrt{7x+2}\,(1)}{(x+1)^2}$

$= \dfrac{(x+1)\left(\frac{7}{2}\right)\frac{1}{\sqrt{7x+2}} - \sqrt{7x+2}}{(x+1)^2}$

If $x = 1$, then $y' = \dfrac{2\left(\frac{7}{2}\right)\left(\frac{1}{3}\right) - 3}{4} = -\dfrac{1}{6}$. The tangent line is $y - \dfrac{3}{2} = -\dfrac{1}{6}(x-1)$, or $y = -\dfrac{1}{6}x + \dfrac{5}{3}$.

63. $y = \left(x^2 + 9\right)^3$ and $y' = 6x\left(x^2+9\right)^2$. When

$x = 4$, then $y = (25)^3$ and $y' = 6(4)(25)^2$, so $\dfrac{y'}{y}(100) = \dfrac{6(4)(25)^2}{(25)^3}(100) = \dfrac{24}{25}(100) = 96\%$

65. $q = 5m, \; p = -0.4q + 50; \; m = 6$

$$\frac{dr}{dm} = \frac{dr}{dq} \cdot \frac{dq}{dm}$$

$r = pq = -0.4q^2 + 50q, \; \dfrac{dr}{dq} = -0.8q + 50, \;$. For

$m = 6$, then $q = 30$, so $\left. \dfrac{dr}{dq}\right|_{m=6} = -24 + 50 = 26$. Also, $\dfrac{dq}{dm} = 5$. Thus $\left. \dfrac{dr}{dm}\right|_{m=6} = (26)(5) = 130$.

67. $q = \dfrac{10m^2}{\sqrt{m^2 + 9}}$

$p = \dfrac{525}{q + 3}; \; m = 4$

$$\frac{dr}{dm} = \frac{dr}{dq} \cdot \frac{dq}{dm}$$

$r = pq = \dfrac{525q}{q + 3}$, so $\dfrac{dr}{dq} = 525 \cdot \dfrac{(q+3)(1) - q(1)}{(q+3)^2} = \dfrac{1575}{(q+3)^2} \cdot$

If $m = 4$, then $q = 32$, so $\left. \dfrac{dr}{dq}\right|_{m=4} = \dfrac{1575}{1225} = \dfrac{9}{7}$.

$$\frac{dq}{dm} = \frac{\left(m^2 + 9\right)^{\frac{1}{2}}(20m) - 10m^2 \cdot \frac{1}{2}\left(m^2 + 9\right)^{-\frac{1}{2}}(2m)}{m^2 + 9}$$

$$= \frac{\left(m^2 + 9\right)^{-\frac{1}{2}}\left[20m\left(m^2 + 9\right) - 10m^3\right]}{m^2 + 9}$$

$$= \frac{10m^3 + 180m}{\left(m^2 + 9\right)^{\frac{3}{2}}}$$

When $m = 4$, then $\dfrac{dq}{dm} = \dfrac{10(64) + 180(4)}{(25)^{\frac{3}{2}}} = \dfrac{1360}{125} = \dfrac{272}{25}$. Thus $\left. \dfrac{dr}{dm}\right|_{m=4} = \dfrac{9}{7} \cdot \dfrac{272}{25} \approx 13.99$.

69. a. $\dfrac{dp}{dq} = 0 - \dfrac{1}{2}\left(q^2 + 20\right)^{-\frac{1}{2}}(2q) = \dfrac{-q}{\sqrt{q^2 + 20}}$

b. $\dfrac{\frac{dp}{dq}}{p} = \dfrac{\frac{-q}{\sqrt{q^2+20}}}{100-\sqrt{q^2+20}}$

$$= -\dfrac{q}{\sqrt{q^2+20}\left(100-\sqrt{q^2+20}\right)}$$

$$= -\dfrac{q}{100\sqrt{q^2+20}-q^2-20}$$

c. $r = pq = 100q - q\sqrt{q^2+20}$

$\dfrac{dr}{dq}$

$$= 100 - \left[q \cdot \frac{1}{2}\left(q^2+20\right)^{-\frac{1}{2}}(2q) + \sqrt{q^2+20}(1) \right]$$

$$= 100 - \dfrac{q^2}{\sqrt{q^2+20}} - \sqrt{q^2+20}$$

71. $\dfrac{dc}{dp} = \dfrac{dc}{dq} \cdot \dfrac{dq}{dp} = (12+0.4q)(-1.5)$

When $p = 85$, then $q = 772.5$, so $\left.\dfrac{dc}{dp}\right|_{p=85} = -481.5$.

73. $\dfrac{dc}{dq} = \dfrac{\left(q^2+3\right)^{\frac{1}{2}}(10q) - \left(5q^2\right)\left[\frac{1}{2}\left(q^2+3\right)^{-\frac{1}{2}}(2q)\right]}{q^2+3}$

Multiplying numerator and denominator by $\left(q^2+3\right)^{\frac{1}{2}}$ gives

$$\dfrac{dc}{dq} = \dfrac{\left(q^2+3\right)(10q) - 5q^2(q)}{\left(q^2+3\right)^{\frac{3}{2}}} = \dfrac{5q^3+30q}{\left(q^2+3\right)^{\frac{3}{2}}} = \dfrac{5q\left(q^2+6\right)}{\left(q^2+3\right)^{\frac{3}{2}}}.$$

75. $\dfrac{dV}{dt} = \dfrac{dV}{dr} \cdot \dfrac{dr}{dt} = \left(4\pi r^2\right)\left[10^{-8}(2t)+10^{-7}\right]$. When $t = 10$, then $r = 10^{-8}\left(10^2\right)+10^{-7}(10)$

$= 10^{-6}+10^{-6} = 2(10)^{-6}$. Thus

$$\left.\dfrac{dV}{dt}\right|_{t=10} = 4\pi\left[2(10)^{-6}\right]^2\left[10^{-8}(2)(10)+10^{-7}\right] = 4\pi\left[4(10)^{-12}\right]\left[3\left(10^{-7}\right)\right] = 48\pi(10)^{-19}$$

77. a. $\dfrac{d}{dx}(I_x) = -0.001416x^3 + 0.01356x^3 + 1.696x - 34.9$

     If $x = 65$, $\dfrac{d}{dx}(I_x) = -256.238$.

b. If $x = 65$, $\dfrac{\frac{d}{dx}(I_x)}{I_x} \approx \dfrac{-256.238}{16,236.484} \approx -0.01578$

     If $x = 65$, the percentage rate of change is $\dfrac{\frac{d}{dx}(I_x)}{I_x} \cdot = \dfrac{-25,623.8}{16,236.484} = -1.578\%$.

79. By the chain rule, $\dfrac{dc}{dp} = \dfrac{dc}{dq} \cdot \dfrac{dq}{dp}$. We are given that $q = \dfrac{100}{p} = 100p^{-1}$, so $\dfrac{dq}{dp} = -100p^{-2} = \dfrac{-100}{p^2}$. Thus

     $\dfrac{dc}{dp} = \dfrac{dc}{dq}\left[\dfrac{-100}{p^2}\right]$. When $q = 200$, then $p = \dfrac{100}{200} = \dfrac{1}{2}$ and we are given that $\dfrac{dc}{dq} = 0.01$. Therefore

     $\dfrac{dc}{dp} = 0.01\left[\dfrac{-100}{\left(\frac{1}{2}\right)^2}\right] = -4$.

81. $\dfrac{dy}{dt} = \dfrac{dy}{dx} \cdot \dfrac{dx}{dt} = f'(x)g'(t)$. We are given that $g(2) = 3$, so $x = 3$ when $t = 2$. Thus

     $\left.\dfrac{dy}{dt}\right|_{t=2} = \left.\dfrac{dy}{dx}\right|_{x=g(2)} \cdot \left.\dfrac{dx}{dt}\right|_{t=2} = f'(3)g'(2) = 10(4) = 40$.

83. 86,111.37

Chapter 11 Review Problems

1. $f(x) = 2 - x^2$

     $f'(x) = \displaystyle\lim_{h\to 0}\dfrac{f(x+h)-f(x)}{h} = \lim_{h\to 0}\dfrac{\left[2-(x+h)^2\right]-\left(2-x^2\right)}{h}$

     $= \displaystyle\lim_{h\to 0}\dfrac{\left[2-x^2-2hx-h^2\right]-\left(2-x^2\right)}{h} = \lim_{h\to 0}\dfrac{-2hx-h^2}{h}$

     $= \displaystyle\lim_{h\to 0}\dfrac{-h(2x+h)}{h} = \lim_{h\to 0}-(2x+h) = -2x$

3. $f(x) = \sqrt{3x}$

$$f'(x) = \lim_{h \to 0} \frac{f(x+h) - f(x)}{h} = \lim_{h \to 0} \frac{\sqrt{3(x+h)} - \sqrt{3x}}{h}$$

$$= \lim_{h \to 0} \frac{\sqrt{3(x+h)} - \sqrt{3x}}{h} \cdot \frac{\sqrt{3(x+h)} + \sqrt{3x}}{\sqrt{3(x+h)} + \sqrt{3x}}$$

$$= \lim_{h \to 0} \frac{3(x+h) - 3x}{h\left(\sqrt{3(x+h)} + \sqrt{3x}\right)} = \lim_{h \to 0} \frac{3h}{h\left(\sqrt{3(x+h)} + \sqrt{3x}\right)}$$

$$= \lim_{h \to 0} \frac{3}{\sqrt{3(x+h)} + \sqrt{3x}}$$

$$= \frac{3}{\sqrt{3x} + \sqrt{3x}} = \frac{3}{2\sqrt{3x}} = \frac{\sqrt{3}}{2\sqrt{x}}$$

5. y is a constant function, so $y' = 0$.

7. $y' = 7\left(4x^3\right) - 6\left(3x^2\right) + 5(2x) + 0$

$$= 28x^3 - 18x^2 + 10x = 2x\left(14x^2 - 9x + 5\right)$$

9. $f(s) = s^2\left(s^2 + 2\right) = s^4 + 2s^2$

$$f'(s) = 4s^3 + 2(2s) = 4s^3 + 4s = 4s\left(s^2 + 1\right)$$

11. $y = \frac{1}{5}\left(x^2 + 1\right)$

$$y' = \frac{1}{5}(2x) = \frac{2x}{5}$$

13. $y' = (x^3 + 7x^2)(3x^2 - 2x) + (x^3 - x^2 + 5)(3x^2 + 14x)$

$$= 3x^5 + 19x^4 - 14x^3 + 3x^5 + 11x^4 - 14x^3 + 15x^2 + 70x$$

$$= 6x^5 + 30x^4 - 28x^3 + 15x^2 + 70x$$

15. $f'(x) = 100\left(2x^2 + 4x\right)^{99}(4x+4) = 400(x+1)[(2x)(x+2)]^{99}$

17. $y = 3(2x+1)^{-1}$

$$y' = 3(-1)(2x+1)^{-2}(2) = -\frac{6}{(2x+1)^2}$$

19. $y' = (8+2x)\left[(4)\left(x^2+1\right)^3(2x)\right]+\left(x^2+1\right)^4(2)$

$= 2\left(x^2+1\right)^3\left[4x(8+2x)+\left(x^2+1\right)\right]$

$= 2\left(x^2+1\right)^3\left(32x+8x^2+x^2+1\right)$

$= 2\left(x^2+1\right)^3\left(9x^2+32x+1\right)$

21. $f'(z) = \dfrac{\left(z^2+4\right)(2z)-\left(z^2-1\right)(2z)}{\left(z^2+4\right)^2} = \dfrac{10z}{\left(z^2+4\right)^2}$

23. $y = (4x-1)^{\frac{1}{3}}$

$y' = \dfrac{1}{3}(4x-1)^{-\frac{2}{3}}(4) = \dfrac{4}{3}(4x-1)^{-\frac{2}{3}}$

25. $y = (1-x^2)^{-\frac{1}{2}}$

$y' = \left(-\dfrac{1}{2}\right)(1-x^2)^{-\frac{3}{2}}(-2x) = x(1-x^2)^{-\frac{3}{2}}$

27. $h'(x) = (x-6)^4\left[3(x+5)^2\right]+(x+5)^3\left[4(x-6)^3\right]$

$= (x-6)^3(x+5)^2[3(x-6)+4(x+5)]$

$= (x-6)^3(x+5)^2(7x+2)$

29. $y' = \dfrac{(x+6)(5)-(5x-4)(1)}{(x+6)^2} = \dfrac{34}{(x+6)^2}$

31. $y' = 2\left(-\dfrac{3}{8}\right)x^{-\frac{11}{8}}+\left(-\dfrac{3}{8}\right)(2x)^{-\frac{11}{8}}(2) = -\dfrac{3}{4}x^{-\frac{11}{8}}-\dfrac{3}{4}\left(2^{-\frac{11}{8}}\right)x^{-\frac{11}{8}}$

$= -\dfrac{3}{4}x^{-\frac{11}{8}}\left(1+2^{-\frac{11}{8}}\right) = -\dfrac{3}{4}\left(1+2^{-\frac{11}{8}}\right)x^{-\frac{11}{8}}$

33. $y' = \dfrac{\left(x^2+5\right)^{\frac{1}{2}}(2x) - \left(x^2+6\right)\left(\frac{1}{2}\right)\left(x^2+5\right)^{-\frac{1}{2}}(2x)}{x^2+5}$

Multiplying the numerator and denominator by $\left(x^2+5\right)^{\frac{1}{2}}$ gives

$$y' = \frac{\left(x^2+5\right)(2x) - x\left(x^2+6\right)}{\left(x^2+5\right)^{\frac{3}{2}}} = \frac{x^3+4x}{\left(x^2+5\right)^{\frac{3}{2}}} = \frac{x\left(x^2+4\right)}{\left(x^2+5\right)^{\frac{3}{2}}}$$

35. $y' = \dfrac{3}{5}\left(x^3+6x^2+9\right)^{-\frac{2}{5}}\left(3x^2+12x\right)$

$= \dfrac{3}{5}\left(x^3+6x^2+9\right)^{-\frac{2}{5}}(3x)(x+4)$

$= \dfrac{9}{5}x(x+4)\left(x^3+6x^2+9\right)^{-\frac{2}{5}}$

37. $g(z) = -z(z-1)^2 = -z^3+2z^2-z$

$g'(z) = -3z^2+4z-1$

39. $y = x^2-6x+4$

$y' = 2x-6$

When $x = 1$, then $y = -1$ and $y' = -4$. An equation of the tangent line is $y-(-1) = -4(x-1)$, or $y = -4x+3$.

41. $y = x^{\frac{1}{3}}$

$y' = \dfrac{1}{3}x^{-\frac{2}{3}}$

When $x = 8$, then $y = 2$ and $y' = \dfrac{1}{12}$. An equation of the tangent line is $y-2 = \dfrac{1}{12}(x-8)$, or

$y = \dfrac{1}{12}x + \dfrac{4}{3}$.

43. $f(x) = 4x^2+2x+8$

$f'(x) = 8x+2$

$f(1) = 14$ and $f'(1) = 10$. The relative rate of change is $\dfrac{f'(1)}{f(1)} = \dfrac{10}{14} = \dfrac{5}{7} \approx 0.714$, so the percentage rate of change is 71.4%.

45. $r = q(20 - 0.1q) = 20q - 0.1q^2$

$$\frac{dr}{dq} = 20 - 0.2q$$

47. $\dfrac{dC}{dI} = 0.6 - 0.25\left(\dfrac{1}{2}\right)I^{-\frac{1}{2}} = 0.6 - \dfrac{1}{8\sqrt{I}}$

$$\frac{dC}{dI}\bigg|_{I=16} \approx 0.569$$

Thus the marginal propensity to consume is 0.569, so the marginal propensity to save is $1 - 0.569 = 0.431$.

49. Since $p = -0.1q + 500$, then $r = pq = -0.1q^2 + 500q$. Thus $\dfrac{dr}{dq} = 500 - 0.2q$.

51. $\dfrac{dc}{dq} = 0.125 + 0.00878q$

$$\frac{dc}{dq}\bigg|_{q=70} = 0.7396$$

53. $\dfrac{dy}{dx} = 42x^2 - 34x - 16$

$$\frac{dy}{dx}\bigg|_{x=2} = 84 \ \text{eggs/mm}$$

55. a. $\dfrac{dt}{dT}$ when $T = 38$ is

$$\frac{d}{dT}\left[\frac{4}{3}T - \frac{175}{4}\right]\bigg|_{T=38} = \frac{4}{3}\bigg|_{T=38} = \frac{4}{3}.$$

b. $\dfrac{dt}{dT}$ when $T = 35$ is

$$\frac{d}{dT}\left[\frac{1}{24}T + \frac{11}{4}\right]\bigg|_{T=35} = \frac{1}{24}\bigg|_{T=35} = \frac{1}{24}.$$

57. $V' = \dfrac{1}{2}\pi d^2$. If $d = 4$ ft, then $V' = 8\pi\dfrac{\text{ft}^3}{\text{ft}}$.

59. $c = \bar{c}q = 2q^2 + \dfrac{10,000}{q} = 2q^2 + 10,000q^{-1}$

$\dfrac{dc}{dq} = 4q - 10,000q^{-2} = 4q - \dfrac{10,000}{q^2}$

61. a. $q = 10\sqrt{m^2 + 4900} - 700$

$p = \sqrt{19,300 - 8q}; \ m = 240$

$\dfrac{dr}{dm} = \dfrac{dr}{dq} \cdot \dfrac{dq}{dm}.$

$r = pq = q\sqrt{19,300 - 8q},$ so

$\dfrac{dr}{dq} = q\left(\dfrac{1}{2}\right)(19,300 - 8q)^{-\frac{1}{2}}(-8) + \sqrt{19,300 - 8q}(1).$

If $m = 240$, then $q = 1800$, so $\dfrac{dr}{dq}\bigg|_{m=240} = -\dfrac{230}{7} \approx -32.86.$

$\dfrac{dq}{dm} = 10 \cdot \dfrac{1}{2}\left(m^2 + 4900\right)^{-\frac{1}{2}}(2m).$ $\dfrac{dq}{dm}\bigg|_{m=240} = 9.6.$ Thus $\dfrac{dr}{dm}\bigg|_{m=240} \approx (-32.86)(9.6) = -315.456$

b. $\dfrac{\frac{dr}{dm}}{r}\bigg|_{m=240} = \dfrac{-315.456}{r}\bigg|_{q=1800}$

$= \dfrac{-315.456}{1800\sqrt{4900}}$

$= -0.0025$

c. No. Since $\dfrac{dr}{dm} < 0$, there would be no additional revenue generated to offset the cost of $400.

63. 0.305

65. –0.32

Mathematical Snapshot Chapter 11

1. In Problems 63 and 64 of Sec. 11.4, the slope is ≈ 0.7. In Fig. 11.15 the slope is above 0.9. More is spent; less is saved.

3. The slope of the family consumption curve is

$\dfrac{112,040}{\sqrt{1.9667 \times 10^{10} + 224,080x}}$, which for

$x = 25,000$ equals about 0.705. You would expect the family to spend $705 and save $295.

5. Answers may vary.

Chapter 12

Principles in Practice 12.1

1. $\dfrac{dq}{dp} = \dfrac{d}{dp}\left[25 + 2\ln\left(3p^2 + 4\right)\right]$

$= 0 + 2\dfrac{d}{dp}\left[\ln\left(3p^2 + 4\right)\right]$

$= 2\left(\dfrac{1}{3p^2 + 4}\right)\dfrac{d}{dp}\left(3p^2 + 4\right) = \dfrac{2}{3p^2 + 4}(6p)$

$= \dfrac{12p}{3p^2 + 4}$

Problems 12.1

1. $\dfrac{dy}{dx} = 4 \cdot \dfrac{d}{dx}(\ln x) = 4 \cdot \dfrac{1}{x} = \dfrac{4}{x}$

3. $\dfrac{dy}{dx} = \dfrac{1}{3x - 7}(3) = \dfrac{3}{3x - 7}$

5. $y = \ln x^2 = 2\ln x$

$\dfrac{dy}{dx} = 2 \cdot \dfrac{1}{x} = \dfrac{2}{x}$

7. $\dfrac{dy}{dx} = \dfrac{1}{1 - x^2}(-2x) = -\dfrac{2x}{1 - x^2}$

9. $f'(X) = \dfrac{1}{4X^6 + 2X^3}(24X^5 + 6X^2)$

$= \dfrac{24X^5 + 6X^2}{4X^6 + 2X^3}$

$= \dfrac{6X^2(4X^3 + 1)}{2X^3(2X^3 + 1)}$

$= \dfrac{3(4X^3 + 1)}{X(2X^3 + 1)}$

11. $f'(t) = t\left(\dfrac{1}{t}\right) + (\ln t)(1) = 1 + \ln t$

13. $\dfrac{dy}{dx} = x^3\left[\dfrac{1}{2x + 5}(2)\right] + \ln(2x + 5) \cdot 3x^2$

$= \dfrac{2x^3}{2x + 5} + 3x^2\ln(2x + 5)$

15. $y = \log_3(8x - 1) = \dfrac{\ln(8x - 1)}{\ln 3}$

$\dfrac{dy}{dx} = \dfrac{1}{\ln 3} \cdot \dfrac{d}{dx}[\ln(8x - 1)]$

$= \dfrac{1}{\ln 3} \cdot \dfrac{1}{8x - 1}(8) = \dfrac{8}{(8x - 1)(\ln 3)}$

17. $y = x^2 + \log_2\left(x^2 + 4\right) = x^2 + \dfrac{\ln\left(x^2 + 4\right)}{\ln 2}$

$\dfrac{dy}{dx} = 2x + \dfrac{1}{\ln 2}\left[\dfrac{1}{x^2 + 4}(2x)\right]$

$= 2x\left[1 + \dfrac{1}{(\ln 2)\left(x^2 + 4\right)}\right]$

19. $f'(z) = \dfrac{z\left(\frac{1}{z}\right) - (\ln z)(1)}{z^2} = \dfrac{1 - \ln z}{z^2}$

21. $\dfrac{dy}{dx} = \dfrac{(\ln x)^2(2x) - (x^2 + 3)2(\ln x)\frac{1}{x}}{(\ln x)^4}$

$= \dfrac{2x^2\ln x - 2(x^2 + 3)}{x(\ln x)^3}$

23. $y = \ln\left(x^2 + 4x + 5\right)^3 = 3\ln\left(x^2 + 4x + 5\right)$

$\dfrac{dy}{dx} = 3 \cdot \dfrac{1}{x^2 + 4x + 5}(2x + 4)$

$= \dfrac{3(2x + 4)}{x^2 + 4x + 5} = \dfrac{6(x + 2)}{x^2 + 4x + 5}$

25. $y = 9 \ln \sqrt{1+x^2} = \dfrac{9}{2} \ln\left(1+x^2\right)$

$\dfrac{dy}{dx} = \dfrac{9}{2} \cdot \dfrac{1}{1+x^2}(2x) = \dfrac{9x}{1+x^2}$

27. $f(l) = \ln\left(\dfrac{1+l}{1-l}\right) = \ln(1+l) - \ln(1-l)$

$f'(l) = \dfrac{1}{1+l} - \dfrac{1}{1-l}(-1)$

$= \dfrac{(1-l)+(1+l)}{(1+l)(1-l)} = \dfrac{2}{1-l^2}$

29. $y = \ln \sqrt[4]{\dfrac{1+x^2}{1-x^2}} = \dfrac{1}{4}\left[\ln\left(1+x^2\right) - \ln\left(1-x^2\right)\right]$

$\dfrac{dy}{dx} = \dfrac{1}{4}\left[\dfrac{2x}{1+x^2} - \dfrac{-2x}{1-x^2}\right]$

$= \dfrac{1}{4}\left[\dfrac{2x\left(1-x^2\right)+2x\left(1+x^2\right)}{\left(1+x^2\right)\left(1-x^2\right)}\right] = \dfrac{x}{1-x^4}$

31. $y = \ln\left[\left(x^2+2\right)^2\left(x^3+x-1\right)\right]$

$= 2\ln\left(x^2+2\right) + \ln\left(x^3+x-1\right)$

$\dfrac{dy}{dx} = 2 \cdot \dfrac{1}{x^2+2}(2x) + \dfrac{1}{x^3+x-1}\left(3x^2+1\right)$

$= \dfrac{4x}{x^2+2} + \dfrac{3x^2+1}{x^3+x-1}$

33. $y = 13\ln\left(x^2\sqrt[3]{5x+2}\right)$

$= 13\ln x^2 + 13\ln(5x+2)^{1/3}$

$= 26\ln x + \dfrac{13}{3}\ln(5x+2)$

$\dfrac{dy}{dx} = 26\left(\dfrac{1}{x}\right) + \dfrac{13}{3} \cdot \dfrac{1}{5x+2}(5)$

$= \dfrac{26}{x} + \dfrac{65}{3(5x+2)}$

35. $\dfrac{dy}{dx} = \left(x^2+1\right)\left[\dfrac{1}{2x+1}(2)\right] + \ln(2x+1) \cdot (2x)$

$= \dfrac{2\left(x^2+1\right)}{2x+1} + 2x\ln(2x+1)$

37. $y = \ln x^3 + \ln^3 x = 3\ln x + (\ln x)^3$

$\dfrac{dy}{dx} = 3 \cdot \dfrac{1}{x} + 3(\ln x)^2 \cdot \dfrac{1}{x} = \dfrac{3}{x} + \dfrac{3(\ln x)^2}{x}$

$= \dfrac{3\left(1+\ln^2 x\right)}{x}$

39. $y = \ln^4(ax) = [\ln(ax)]^4$

$\dfrac{dy}{dx} = 4[\ln(ax)]^3\left(\dfrac{1}{ax} \cdot a\right) = \dfrac{4\ln^3(ax)}{x}$

41. $y = x\ln\sqrt{x-1} = \dfrac{1}{2}x\ln(x-1)$

$\dfrac{dy}{dx} = \dfrac{1}{2}\left[x\left(\dfrac{1}{x-1}\right) + \ln(x-1) \cdot (1)\right]$

$= \dfrac{x}{2(x-1)} + \ln\sqrt{x-1}$

43. $y = \sqrt{4+3\ln x} = (4+3\ln x)^{\frac{1}{2}}$

$\dfrac{dy}{dx} = \dfrac{1}{2}(4+3\ln x)^{-\frac{1}{2}} \cdot \dfrac{3}{x} = \dfrac{3}{2x\sqrt{4+3\ln x}}$

45. $y = \ln(x^2 - 3x - 3)$

$y' = \dfrac{2x-3}{x^2-3x-3}$

The slope of the tangent line at $x = 4$ is

$y'(4) = \dfrac{8-3}{16-12-3} = 5$. Also, if $x = 4$, then

$y = \ln(16 - 12 - 3) = \ln 1 = 0$. Thus an equation of the tangent line is

$y - 0 = 5(x - 4)$, or $y = 5x - 20$.

47. $y = \dfrac{x}{\ln x}$

$$y' = \frac{(\ln x)(1) - x\left(\frac{1}{x}\right)}{\ln^2 x} = \frac{\ln x - 1}{\ln^2 x}$$

When $x = 3$ the slope is $y'(3) = \dfrac{(\ln 3) - 1}{\ln^2 3}$.

49. $c = 25 \ln(q + 1) + 12$

$$\frac{dc}{dq} = \frac{25}{q+1}, \text{ so } \frac{dc}{dq}\bigg|_{q=6} = \frac{25}{7}.$$

51. $\dfrac{dq}{dp} = \dfrac{d}{dp}[25 + 10\ln(2p + 1)]$

$$= 0 + 10\frac{d}{dp}[\ln(2p + 1)]$$

$$= 10\left(\frac{1}{2p+1}\right)\frac{d}{dp}[2p + 1]$$

$$= \frac{10}{2p+1}(2) = \frac{20}{2p+1}$$

53. $A = 6\ln\left(\dfrac{T}{a-T} - a\right)$. Rate of change of A

with respect to T:

$$\frac{dA}{dT} = 6 \cdot \frac{1}{\frac{T}{a-T} - a}\left[\frac{(a-T)(1) - T(-1)}{(a-T)^2}\right]$$

$$= 6 \cdot \frac{1}{\frac{T - a(a-T)}{a-T}}\left[\frac{a}{(a-T)^2}\right]$$

$$= 6 \cdot \frac{a - T}{T - a^2 + aT} \cdot \frac{a}{(a-T)^2}$$

$$= \frac{6a}{\left(T - a^2 + aT\right)(a - T)}$$

55. $\dfrac{d}{dx}\left(\log_b u\right) = \dfrac{d}{dx}\left(\dfrac{\ln u}{\ln b}\right)$

$$= \frac{1}{\ln b} \cdot \frac{d}{dx}(\ln u) = \frac{1}{\ln b}\left(\frac{1}{u} \cdot \frac{du}{dx}\right)$$

$$= \left(\log_b e\right)\left(\frac{1}{u} \cdot \frac{du}{dx}\right) = \frac{1}{u}\left(\log_b e\right)\frac{du}{dx}$$

57. Note that $f(x)$ is defined for all $x \neq 0$.

$$f'(x) = \frac{x^2 \cdot \frac{1}{x^2}(2x) - \ln(x^2) \cdot 2x}{x^4}$$

$$= \frac{2 - 2\ln(x^2)}{x^3}$$

$$f'(x) = 0 \text{ for } x \approx -1.65, \ 1.65$$

Principles in Practice 12.2

1. The rate of change of temperature with respect to time is $\dfrac{dT}{dt}$. $T(t)$ has the form

Ce^u where C is a constant and $u = kt$.

$$\frac{dT}{dt} = \frac{d}{dt}\left[Ce^{kt}\right] = C\frac{d}{dt}\left[e^{kt}\right]$$

$$= C\left(e^{kt}\right)\frac{d}{dt}[kt] = Ce^{kt}(k) = Cke^{kt}$$

Problems 12.2

1. $y' = 5 \cdot \dfrac{d}{dx}(e^x) = 5e^x$

3. $y' = e^{2x^2+3}(4x) = 4xe^{2x^2+3}$

5. $y' = e^{9-5x} \cdot \dfrac{d}{dx}(9 - 5x)$

$$= e^{9-5x}(-5)$$

$$= -5e^{9-5x}$$

7. $f'(r) = e^{3r^2+4r+4}(6r + 4)$

$$= 2(3r + 2)e^{3r^2+4r+4}$$

9. $y' = x\left(e^x\right) + e^x(1) = e^x(x+1)$

11. $y' = x^2\left[e^{-x^2}(-2x)\right] + e^{-x^2}(2x)$

$= 2xe^{-x^2}\left(1 - x^2\right)$

13. $y = \dfrac{1}{3}\left(e^x + e^{-x}\right)$

$y' = \dfrac{1}{3}\left[e^x + e^{-x}(-1)\right] = \dfrac{e^x - e^{-x}}{3}$

15. $\dfrac{d}{dx}\left(5^{2x^3}\right) = \dfrac{d}{dx}\left[e^{(\ln 5)2x^3}\right]$

$= e^{(\ln 5)2x^3}[(\ln 5)6x^2]$

$= (6x^2)5^{2x^3}\ln 5$

17. $f'(w) = \dfrac{w^2\left[e^{2w}(2)\right] - e^{2w}[2w]}{w^4}$

$= \dfrac{2e^{2w}(w-1)}{w^3}$

19. $y' = e^{1+\sqrt{x}}\left(\dfrac{1}{2}x^{-\frac{1}{2}}\right) = \dfrac{e^{1+\sqrt{x}}}{2\sqrt{x}}$

21. $y = x^5 - 5^x = x^5 - e^{(\ln 5)x}$

$y' = 5x^4 - e^{(\ln 5)x}(\ln 5) = 5x^4 - 5^x\ln 5$

23. $\dfrac{dy}{dx} = \dfrac{\left(e^x + 1\right)\left[e^x\right] - \left(e^x - 1\right)\left[e^x\right]}{\left(e^x + 1\right)^2}$

$= \dfrac{2e^x}{\left(e^x + 1\right)^2}$

25. $y = \ln e^x = x$ so $y' = 1$.

27. $y' = e^{x^2\ln x^2}\left[x^2 \cdot \dfrac{1}{x^2}(2x) + (\ln x^2)(2x)\right]$

$= 2xe^{x^2\ln x^2}(1 + \ln x^2)$

29. $f(x) = ee^xe^{x^2} = e^{1+x+x^2}$

$f'(x) = e^{1+x+x^2}(1+2x) = (1+2x)e^{1+x+x^2}$

$f'(-1) = [1 + 2(-1)]e^{1+(-1)+(-1)^2} = -e$

31. $y = e^x$, $y' = e^x$. When $x = -2$, then $y = e^{-2}$ and $y' = e^{-2}$. Thus an equation of the tangent line is $y - e^{-2} = e^{-2}(x+2)$, or $y = e^{-2}x + 3e^{-2}$.

33. $\dfrac{dp}{dq} = 15e^{-0.001q}(-0.001) = -0.015e^{-0.001q}$

$\left.\dfrac{dp}{dq}\right|_{q=500} = -0.015e^{-0.5}$

35. $\bar{c} = \dfrac{7000e^{\frac{q}{700}}}{q}$, so $c = \bar{c}q = 7000e^{\frac{q}{700}}$. The marginal cost function is

$\dfrac{dc}{dq} = 7000e^{\frac{q}{700}}\left(\dfrac{1}{700}\right) = 10e^{\frac{q}{700}}$. Thus

$\left.\dfrac{dc}{dq}\right|_{q=350} = 10e^{0.5}$ and $\left.\dfrac{dc}{dq}\right|_{q=700} = 10e$.

37. $w = e^{x^3 - 4x} + x\ln(x-1)$ and $x = \dfrac{t+1}{t-1}$

By the chain rule, $\dfrac{dw}{dt} = \dfrac{dw}{dx} \cdot \dfrac{dx}{dt}$

$$= \left[e^{x^3 - 4x}\left(3x^2 - 4\right) + x\left(\frac{1}{x-1}\right) + [\ln(x-1)(1)] \right] \left[\frac{(t-1)(1) - (t+1)(1)}{(t-1)^2} \right]$$

$$= \left[\left(3x^2 - 4\right)e^{x^3 - 4x} + \frac{x}{x-1} + \ln(x-1) \right] \left[\frac{-2}{(t-1)^2} \right].$$

When $t = 3$, then $x = \dfrac{3+1}{3-1} = \dfrac{4}{2} = 2$ and $\dfrac{dw}{dt} = [8+2+0]\left[-\dfrac{1}{2}\right] = -5$.

39. $\dfrac{d}{dx}\left(c^x - x^c\right) = \dfrac{d}{dx}\left[\left(e^{\ln c}\right)^x - x^c \right]$

$$= \frac{d}{dx}\left[e^{(\ln c)x} - x^c \right]$$

$$= (\ln c)e^{(\ln c)x} - cx^{c-1} = (\ln c)c^x - cx^{c-1}$$

$$\frac{d}{dx}(c^x - x^c)\Big|_{x=1} = (\ln c)c - c$$

If this is zero, $(\ln c)c - c = 0$, or $c[\ln(c) - 1] = 0$. Since $c > 0$, we must have $\ln(c) - 1 = 0$, $\ln c = 1$, or $c = e$.

41. $q = 500\left(1 - e^{-0.2t}\right)$

$$\frac{dq}{dt} = 500\left(-e^{-0.2t}\right)(-0.2) = 100e^{-0.2t}$$

Thus $\dfrac{dq}{dt}\Big|_{t=10} = 100e^{-2}$.

43. $P = 1.92e^{0.0176t}$

$$\frac{dP}{dt} = 1.92e^{0.0176t}(0.0176) = P(0.0176)$$

$$= 0.0176P = kP \text{ for } k = 0.0176.$$

45. Since $S = Pe^{rt}$, then $\dfrac{dS}{dt} = Pe^{rt}r = rPe^{rt}$. Thus $\dfrac{\frac{dS}{dt}}{S} = \dfrac{rPe^{rt}}{Pe^{rt}} = r$.

47. $N = 10^A 10^{-bM} = 10^{A-bM} = e^{(\ln 10)(A-bM)}$

$$\frac{dN}{dM} = e^{(\ln 10)(A-bM)}(\ln 10)(-b), \text{ so } \frac{dN}{dM} = 10^{A-bM}(\ln 10)(-b) = -b\left(10^{A-bM}\right)\ln 10$$

49. $C(t) = C_0 e^{-\left(\frac{r}{V}\right)t}$

$$\frac{dC}{dt} = C_0 e^{-\left(\frac{r}{V}\right)t}\left(-\frac{r}{V}\right)$$

$$= [C(t)]\left(-\frac{r}{V}\right) = -\left(\frac{r}{V}\right)C(t)$$

51. $f(t) = 1 - e^{-0.008t}$

$$f'(t) = 0.008e^{-0.008t}$$

$$f'(100) = 0.008e^{-0.8} \approx 0.0036$$

53. $f'(x) = (6x^2 + 2x - 3)e^{2x^3 + x^2 - 3x}$

$f'(x) = 0$ for $x \approx -0.89, 0.56$

Problems 12.3

1. $\eta = \dfrac{\frac{p}{q}}{\frac{dp}{dq}} = \dfrac{\frac{p}{q}}{-2}$.

When $q = 5$ then $p = 40 - 2(5) = 30$, so

$$\eta = \frac{\frac{30}{5}}{-2} = -3$$

Because $|\eta| > 1$, demand is elastic.

3. $p = \dfrac{3500}{q} = 3500q^{-1}$

$$\frac{dp}{dq} = -3500q^{-2} = -\frac{3500}{q^2}$$

$$\eta = \frac{\frac{p}{q}}{\frac{dp}{dq}} = \frac{\frac{p}{q}}{-\frac{3500}{q^2}} = \frac{\frac{(3500/q)}{q}}{-\frac{3500}{q^2}} = -1$$

Because $|\eta| = 1$, demand has unit elasticity.

5. $\eta = \dfrac{\frac{p}{q}}{\frac{dp}{dq}} = \dfrac{\frac{p}{q}}{-\frac{500}{(q+2)^2}} = \dfrac{\frac{[500/(q+2)]}{q}}{-\frac{500}{(q+2)^2}} = -\frac{q+2}{q}$

When $q = 104$, then $\eta = -\frac{106}{104} = -\frac{53}{52}$.

Because $|\eta| > 1$, demand is elastic.

7. $\eta = \dfrac{\frac{p}{q}}{\frac{dp}{dq}} = \dfrac{\frac{p}{q}}{-\frac{e^{100}}{100}}$

When $q = 100$, then $p = 150 - e$ and

$\eta = \dfrac{\frac{150-e}{100}}{-\frac{e}{100}} = -\left(\dfrac{150}{e} - 1\right)$. Because $|\eta| > 1$,

demand is elastic.

9. $q = 1200 - 150p$

$$\eta = \frac{\frac{p}{q}}{\frac{dp}{dq}} = \frac{p}{q} \cdot \frac{dq}{dp} = \frac{p}{q}(-150)$$

If $p = 4$, then $q = 1200 - 150(4) = 600$, so

$\eta = \dfrac{4}{600}(-150) = -1$. Since $|\eta| = 1$, demand

has unit elasticity.

11. $q = \sqrt{500 - p}$

$$\eta = \frac{\frac{p}{q}}{\frac{dp}{dq}} = \frac{p}{q} \cdot \frac{dq}{dp}$$

$$\frac{dq}{dp} = \frac{1}{2}(500 - p)^{-\frac{1}{2}}(-1)$$

$$= \frac{-1}{2\sqrt{500 - p}} = -\frac{1}{2q}$$

$$\eta = \frac{p}{q}\left(-\frac{1}{2q}\right) = -\frac{p}{2q^2}$$

If $p = 400$, then $q = \sqrt{500 - 400} = 10$, so

$\eta = -\dfrac{400}{200} = -2$. $|\eta| > 1$, so demand is

elastic.

13. $q = \dfrac{(p-100)^2}{2}$

$\eta = \dfrac{\frac{p}{q}}{\frac{dp}{dq}} = \dfrac{p}{q} \cdot \dfrac{dq}{dp}$

$\dfrac{dq}{dp} = \dfrac{1}{2}(2)(p-100)(1) = p-100$, so

$\eta = \dfrac{p}{q}(p-100)$. If $p = 20$, then

$q = \dfrac{(20-100)^2}{2} = 3200$. Thus

$\eta = \dfrac{20}{3200}(20-100) = -\dfrac{1}{2}$. Demand is inelastic.

15. $p = 13 - 0.05q$

$\eta = \dfrac{\frac{p}{q}}{\frac{dp}{dq}} = -\dfrac{p}{0.05q}$

p	q	η	demand
10	60	$-\dfrac{10}{3}$	elastic
3	200	$-\dfrac{3}{10}$	inelastic
6.50	130	-1	unit elasticity

17. $q = 500 - 40p + p^2$

$\eta = \dfrac{\frac{p}{q}}{\frac{dp}{dq}} = \dfrac{p}{q} \cdot \dfrac{dq}{dp}$

$\dfrac{dq}{dp} = -40 + 2p$, so $\eta = \dfrac{p}{q}(2p-40)$.

When $p = 15$, then

$q = 500 - 40(15) + 15^2 = 125$, so

$\eta\big|_{p=15} = \dfrac{15}{125}(30-40) = -\dfrac{6}{5} = -1.2$. Now,

(% change in price) $\cdot$ (η) = % change in demand. Thus if the price of 15 increases

$\dfrac{1}{2}\%$, then the change in demand is

approximately $\left(\dfrac{1}{2}\%\right)(-1.2) = -0.6\%$. Thus demand decreases approximately 0.6%.

19. $p = 500 - 2q$

$\eta = \dfrac{\frac{p}{q}}{\frac{dp}{dq}} = \dfrac{\frac{500-2q}{q}}{-2} = \dfrac{q-250}{q}$

If demand is elastic, then $\eta = \dfrac{q-250}{q} < -1$.

For $q > 0$, we have $q - 250 < -q$, $2q < 250$, so $q < 125$. Thus, if $0 < q < 125$, demand is elastic. If demand is inelastic, then

$\eta = \dfrac{q-250}{q} > -1$. For $q > 0$, the inequality

implies $q > 125$. Thus if $125 < q < 250$, then demand is inelastic.

Since Total Revenue $= r = pq$,
$$= 500q - 2q^2,$$

then $r' = 500 - 4q = 4(125 - q)$. If $0 < q < 125$, then $r' > 0$, so r is increasing. If $125 < q < 250$, then $r' < 0$, so r is decreasing.

21. $p = \dfrac{1000}{q^2}$

$r = pq = \dfrac{1000}{q}$

$\dfrac{dr}{dq} = -1000q^{-2} = -\dfrac{1000}{q^2}$

$\eta = \dfrac{\frac{p}{q}}{\frac{dp}{dq}} = \dfrac{\frac{1000}{q^3}}{-\frac{2000}{q^3}} = -\dfrac{1}{2}$

$p\left(1 + \dfrac{1}{\eta}\right) = \dfrac{1000}{q^2}(1-2) = -\dfrac{1000}{q^2} = \dfrac{dr}{dq}$

23. a. $p = \dfrac{a}{\sqrt{b+cq^2}} = a(b+cq^2)^{-1/2}$

$\dfrac{dp}{dq} = -\dfrac{1}{2}a(b+cq^2)^{-3/2}(2cq)$

$\qquad = -acq(b+cq^2)^{-3/2}$

$\eta = \dfrac{\frac{p}{q}}{\frac{dp}{dq}} = \dfrac{a(b+cq^2)^{-1/2}}{-acq^2(b+cq^2)^{-3/2}}$

$\quad = -\dfrac{b+cq^2}{cq^2}$

Thus η does not depend on a.

b. $\eta = -\dfrac{b+cq^2}{cq^2} = -\left(\dfrac{b}{cq^2}+1\right)$

If $b,\,c>0$, then $\dfrac{b}{cq^2}+1>1$ so $|\eta|>1$

and demand is elastic.

c. If $|\eta|=1$, then $\left|\dfrac{b}{cq^2}+1\right|=1$, which can

only occur if $b=0$.

25. a. $q = \dfrac{60}{p} + \ln\left(65-p^3\right)$

$\eta = \dfrac{\frac{p}{q}}{\frac{dp}{dq}} = \dfrac{p}{q}\dfrac{dq}{dp} = \dfrac{p}{q}\left[-\dfrac{60}{p^2}-\dfrac{3p^2}{65-p^3}\right]$

If $p=4$, then $q=\dfrac{60}{4}+\ln 1 = 15$, so

$\eta = \dfrac{4}{15}\left[-\dfrac{60}{16}-\dfrac{3(16)}{65-64}\right] = -\dfrac{207}{15} \approx -13.8,$

and demand is elastic.

b. The percentage change in q is
$(-2)(-13.8) = 27.6\%$, so q increases by
approximately 27.6%.

c. Lowering the price increases revenue
because demand is elastic.

27. The percentage change in price is

$\dfrac{-5}{80}\cdot 100 = -\dfrac{25}{4}\%$ and the percentage

change in quantity is $\dfrac{50}{500}\cdot 100 = 10\%$. Thus,

since (elasticity)(% change in price)
$\approx$ % change in demand,

$(\text{elasticity})\left(-\dfrac{25}{4}\right) \approx 10.$

elasticity $\approx -\dfrac{40}{25} = -\dfrac{8}{5} = -1.6$

To estimate $\dfrac{dr}{dq}$ when $p=80$, we have

$\dfrac{dr}{dq} = p\left(1+\dfrac{1}{\eta}\right) = 80\left(1+\dfrac{1}{-\frac{8}{5}}\right) = 30.$

29. $\dfrac{dp}{dq} = 200(-1)(q+5)^{-2} = \dfrac{-200}{(q+5)^2}$

Thus $\eta = \dfrac{\frac{p}{q}}{\frac{dp}{dq}} = \dfrac{\frac{200}{q(q+5)}}{-\frac{200}{(q+5)^2}} = -\dfrac{q+5}{q}.$

For $5 \le q \le 95$, $|\eta| = \dfrac{q+5}{q} = 1+\dfrac{5}{q}$ and

$|\eta|' = -\dfrac{5}{q^2}.$

Since $|\eta|'<0$, $|\eta|$ is decreasing on [5, 95],
and thus $|\eta|$ is maximum at $q=5$ and
minimum at $q=95$.

Principles in Practice 12.4

1. Assume that P is a function of t and

differentiate both sides of $\ln\left(\dfrac{P}{1-P}\right) = 0.5t$

with respect to t.

$$\frac{d}{dt}\left[\ln\left(\frac{P}{1-P}\right)\right] = \frac{d}{dt}[0.5t]$$

$$\left(\frac{1}{\frac{P}{1-P}}\right)\frac{d}{dt}\left[\frac{P}{1-P}\right] = 0.5$$

$$\frac{1-P}{P} \cdot \frac{(1)(1-P)-P(-1)}{(1-P)^2} \cdot \frac{dP}{dt} = 0.5$$

$$\frac{1-P+P}{P(1-P)} \cdot \frac{dP}{dt} = 0.5$$

$$\frac{dP}{dt} = 0.5P(1-P)$$

3. The hypotenuse is the length of the ladder, so $x^2 + y^2 = 100$. Differentiate both sides of the equation with respect to t.

$$\frac{d}{dt}\left[x^2 + y^2\right] = \frac{d}{dt}[100]$$

$$2x\frac{dx}{dt} + 2y\frac{dy}{dt} = 0$$

When $y = 8$, we can find x by using the Pythagorean theorem.

$$x^2 + 8^2 = 100$$

$$x^2 = 100 - 64 = 36$$

$$x = 6$$

When $x = 6$, $y = 8$, and $\frac{dx}{dt} = 3$, we have

$$2(6)(3) + 2(8)\frac{dy}{dt} = 0$$

$$36 + 16\frac{dy}{dt} = 0$$

$$\frac{dy}{dt} = -\frac{36}{16} = -\frac{9}{4}$$

$$\frac{dy}{dt} = -\frac{9}{4}, \text{ thus the top of the ladder is sliding down the wall at the rate of } \frac{9}{4} \text{ feet/sec.}$$

Problems 12.4

1. $2x + 8yy' = 0$

$x + 4yy' = 0$

$4yy' = -x$

$$y' = -\frac{x}{4y}$$

3. $6y^2y' - 14x = 0$

$$y' = \frac{14x}{6y^2} = \frac{7x}{3y^2}$$

5. $x^{1/3} + y^{1/3} = 3$

$$\frac{1}{3}x^{-2/3} + \frac{1}{3}y^{-2/3}y' = 0$$

$$y^{-2/3}y' = -x^{-2/3}$$

$$y' = -\frac{x^{-2/3}}{y^{-2/3}}$$

$$= -\frac{y^{2/3}}{x^{2/3}}$$

$$= -\frac{\sqrt[3]{y^2}}{\sqrt[3]{x^2}}$$

$$= -\sqrt[3]{\frac{y^2}{x^2}}$$

7. $\left(\frac{3}{4}\right)x^{-\frac{1}{4}} + \left(\frac{3}{4}\right)y^{-\frac{1}{4}}y' = 0$

$$y' = -\frac{y^{\frac{1}{4}}}{x^{\frac{1}{4}}}$$

9. By the product rule $xy' + y(1) = 0$,

$$xy' = -y, \; y' = -\frac{y}{x}$$

11. $xy' + y(1) - y' - 11 = 0$

$y'(x-1) = 11 - y$

$y' = \dfrac{11 - y}{x - 1}$

13. $6x^2 + 3y^2 y' - 12(xy' + y) = 0$

$3y^2 y' - 12xy' = 12y - 6x^2$

$y'\left(3y^2 - 12x\right) = 12y - 6x^2$

$y'\left(y^2 - 4x\right) = 4y - 2x^2$

$y' = \dfrac{4y - 2x^2}{y^2 - 4x}$

15. $x = \sqrt{y} + \sqrt[4]{y} = y^{1/2} + y^{1/4}$

$1 = \dfrac{1}{2} y^{-1/2} y' + \dfrac{1}{4} y^{-3/4} y'$

$ = y'\left(\dfrac{1}{2y^{1/2}} + \dfrac{1}{4y^{3/4}}\right) = y'\left(\dfrac{2y^{1/4} + 1}{4y^{3/4}}\right)$

$y' = \dfrac{4y^{3/4}}{2y^{1/4} + 1}$

17. $5x^3(4y^3 y') + 15x^2 y^4 - 1 + 2yy' = 0$

$y'(20x^3 y^3 + 2y) = 1 - 15x^2 y^4$

$y' = \dfrac{1 - 15x^2 y^4}{20x^3 y^3 + 2y}$

19. $y\left(\dfrac{1}{x}\right) + (\ln x)y' = x\left(e^y y'\right) + e^y (1)$

$\left[\ln(x) - xe^y\right] y' = e^y - \dfrac{y}{x}$

$\left[\ln(x) - xe^y\right] y' = \dfrac{xe^y - y}{x}$

$y' = \dfrac{xe^y - y}{x\left[\ln(x) - xe^y\right]}$

21. $\left[x\left(e^y y'\right)+e^y(1)\right]+y'=0$

$xe^y y'+e^y+y'=0$

$\left(xe^y+1\right)y'=-e^y$

$y'=-\dfrac{e^y}{xe^y+1}$

23. $2\left(1+e^{3x}\right)\left(3e^{3x}\right)=\dfrac{1}{x+y}(1+y')$

$6e^{3x}\left(1+e^{3x}\right)(x+y)=1+y'$

$y'=6e^{3x}\left(1+e^{3x}\right)(x+y)-1$

25. $1+[xy'+y(1)]+2yy'=0$

$xy'+2yy'=-1-y$

$(x+2y)y'=-(1+y)$

$y'=-\dfrac{1+y}{x+2y}$

At the point $(1, 2)$, $y'=-\dfrac{1+2}{1+4}=-\dfrac{3}{5}$.

27. $8x+18yy'=0$

$y'=-\dfrac{8x}{18y}=-\dfrac{4x}{9y}$

Thus at $\left(0,\dfrac{1}{3}\right)$, $y'=0$; at $\left(x_0, y_0\right)$,

$y'=-\dfrac{4x_0}{9y_0}$.

29. $3x^2+xy'+y+2y'=0$

$y'=-\dfrac{3x^2+y}{x+2y}$

At $(-1, 1)$, $y'=-4$ and the tangent line is given by $y-1=-4[x-(-1)]$, or $y=-4x-3$.

31. $p=100-q^2$

$\dfrac{d}{dp}(p)=\dfrac{d}{dp}\left(100-q^2\right)$

$1=-2q\cdot\dfrac{dq}{dp}$

$\dfrac{dq}{dp}=-\dfrac{1}{2q}$

33. $p=\dfrac{20}{(q+5)^2}$

$\dfrac{d}{dp}(p)=\dfrac{d}{dp}\left[\dfrac{20}{(q+5)^2}\right]$

$\dfrac{d}{dp}(p)=\dfrac{d}{dp}\left[20(q+5)^{-2}\right]$

$1=-\dfrac{40}{(q+5)^3}\cdot\dfrac{dq}{dp}$

$\dfrac{dq}{dp}=-\dfrac{(q+5)^3}{40}$

35. $\ln\dfrac{I}{I_0}=-\lambda t$

$\ln I-\ln I_0=-\lambda t$

$\dfrac{1}{I}\dfrac{dI}{dt}=-\lambda$

$\dfrac{dI}{dt}=-\lambda I$

37. $v=f\lambda$. Differentiating implicitly with respect to λ:

$0=f(1)+\lambda\dfrac{df}{d\lambda}$, $\dfrac{df}{d\lambda}=-\dfrac{f}{\lambda}$.

Solving $v=f\lambda$ for f and differentiating:

$f=\dfrac{v}{\lambda}$, so $\dfrac{df}{d\lambda}=-\dfrac{v}{\lambda^2}=-\dfrac{f\lambda}{\lambda^2}=-\dfrac{f}{\lambda}$,

which is the same as before.

39. $S^2 + \dfrac{1}{4}I^2 = SI + I$. Differentiating implicitly with respect to I:

$$2S\frac{dS}{dI} + \frac{1}{2}I = \left[S(1) + I\frac{dS}{dI}\right] + 1, \ 2S\frac{dS}{dI} - I\frac{dS}{dI} = S + 1 - \frac{I}{2}, \ (2S - I)\frac{dS}{dI} = \frac{2S + 2 - I}{2}, \ \frac{dS}{dI} = \frac{2S + 2 - I}{2(2S - I)}.$$

Marginal propensity to consume $= \dfrac{dC}{dI} = 1 - \dfrac{dS}{dI}$. Thus $\dfrac{dC}{dI} = 1 - \dfrac{2S + 2 - I}{2(2S - I)}$. When $I = 16$ and

$S = 12$, $\dfrac{dC}{dI} = 1 - \dfrac{24 + 2 - 16}{2(24 - 16)} = 1 - \dfrac{10}{16} = \dfrac{6}{16} = \dfrac{3}{8}$.

Problems 12.5

1. $y = (x+1)^2(x-2)\left(x^2 + 3\right)$. Take natural logarithms of both sides,

$\ln y = \ln\left[(x+1)^2(x-2)\left(x^2 + 3\right)\right]$.

Using properties of logarithms on the right side gives

$\ln y = 2\ln(x+1) + \ln(x-2) + \ln\left(x^2 + 3\right)$.

Differentiating both sides with respect to x,

$\dfrac{y'}{y} = \dfrac{2}{x+1} + \dfrac{1}{x-2} + \dfrac{2x}{x^2 + 3}$.

Solving for y',

$y' = y\left[\dfrac{2}{x+1} + \dfrac{1}{x-2} + \dfrac{2x}{x^2 + 3}\right]$.

Expressing y' in terms of x,

$y' = (x+1)^2(x-2)\left(x^2 + 3\right)\left[\dfrac{2}{x+1} + \dfrac{1}{x-2} + \dfrac{2x}{x^2 + 3}\right]$

3. $\ln y = \ln\left[\left(3x^3 - 1\right)^2(2x + 5)^3\right]$

$= 2\ln\left(3x^3 - 1\right) + 3\ln(2x + 5)$

$\dfrac{y'}{y} = 2 \cdot \dfrac{9x^2}{3x^3 - 1} + 3 \cdot \dfrac{2}{2x + 5}$

$y' = y\left[\dfrac{18x^2}{3x^3 - 1} + \dfrac{6}{2x + 5}\right]$

$y' = \left(3x^3 - 1\right)^2(2x + 5)^3\left[\dfrac{18x^2}{3x^3 - 1} + \dfrac{6}{2x + 5}\right]$

5. $y = \sqrt{x+1}\sqrt{x^2 - 2}\sqrt{x+4}$

$\ln y = \ln\left(\sqrt{x+1}\sqrt{x^2 - 2}\sqrt{x+4} \right)$

$\ln y = \dfrac{1}{2}\ln(x+1) + \dfrac{1}{2}\ln\left(x^2 - 2\right) + \dfrac{1}{2}\ln(x+4)$

$\dfrac{y'}{y} = \dfrac{1}{2}\left[\dfrac{1}{x+1} + \dfrac{2x}{x^2 - 2} + \dfrac{1}{x+4} \right]$

$y' = \dfrac{y}{2}\left[\dfrac{1}{x+1} + \dfrac{2x}{x^2 - 2} + \dfrac{1}{x+4} \right]$

$= \dfrac{\sqrt{x+1}\sqrt{x^2 - 2}\sqrt{x+4}}{2}\left[\dfrac{1}{x+1} + \dfrac{2x}{x^2 - 2} + \dfrac{1}{x+4} \right]$

7. $\ln y = \ln\dfrac{\sqrt{1-x^2}}{1-2x} = \dfrac{1}{2}\ln\left(1-x^2\right) - \ln(1-2x)$

$\dfrac{y'}{y} = \dfrac{1}{2}\cdot\dfrac{-2x}{1-x^2} - \dfrac{-2}{1-2x}$

$y' = y\left[-\dfrac{x}{1-x^2} + \dfrac{2}{1-2x} \right]$

$y' = \dfrac{\sqrt{1-x^2}}{1-2x}\left[\dfrac{x}{x^2 -1} + \dfrac{2}{1-2x} \right]$

9. $y = \dfrac{\left(2x^2 + 2\right)^2}{(x+1)^2(3x+2)}$

$\ln y = \ln\left[\dfrac{\left(2x^2 + 2\right)^2}{(x+1)^2(3x+2)} \right]$

$= 2\ln\left(2x^2 + 2\right) - 2\ln(x+1) - \ln(3x+2)$

$\dfrac{y'}{y} = 2\cdot\dfrac{4x}{2x^2 + 2} - 2\cdot\dfrac{1}{x+1} - \dfrac{3}{3x+2}$

$y' = y\left[\dfrac{8x}{2x^2 + 2} - \dfrac{2}{x+1} - \dfrac{3}{3x+2} \right]$

$= \dfrac{\left(2x^2 + 2\right)^2}{(x+1)^2(3x+2)}\left[\dfrac{4x}{x^2 +1} - \dfrac{2}{x+1} - \dfrac{3}{3x+2} \right]$

11. $y = \sqrt{\dfrac{(x+3)(x-2)}{2x-1}}$

$\ln y = \ln \sqrt{\dfrac{(x+3)(x-2)}{2x-1}}$

$= \dfrac{1}{2}\ln(x+3) + \dfrac{1}{2}\ln(x-2) - \dfrac{1}{2}\ln(2x-1)$

$\dfrac{y'}{y} = \dfrac{1}{2}\cdot\dfrac{1}{x+3} + \dfrac{1}{2}\cdot\dfrac{1}{x-2} - \dfrac{1}{2}\cdot\dfrac{2}{2x-1}$

$y' = \dfrac{y}{2}\left[\dfrac{1}{x+3} + \dfrac{1}{x-2} - \dfrac{2}{2x-1}\right]$

$= \dfrac{1}{2}\sqrt{\dfrac{(x+3)(x-2)}{2x-1}}\left[\dfrac{1}{x+3} + \dfrac{1}{x-2} - \dfrac{2}{2x-1}\right]$

13. $y = x^{x^2+1}$, thus

$\ln y = \ln x^{x^2+1} = (x^2+1)\ln x.$

$\dfrac{y'}{y} = (x^2+1)\cdot\dfrac{1}{x} + (\ln x)(2x)$

$y' = y\left(\dfrac{x^2+1}{x} + 2x\ln x\right)$

$= x^{x^2+1}\left(\dfrac{x^2+1}{x} + 2x\ln x\right)$

15. $y = x^{\frac{1}{x}}$. Thus $\ln y = \dfrac{1}{x}\ln x = \dfrac{\ln x}{x}.$

$\dfrac{y'}{y} = \dfrac{x\left(\frac{1}{x}\right) - (\ln x)(1)}{x^2}$

$y' = y\left[\dfrac{1-\ln x}{x^2}\right]$

$y' = \dfrac{x^{\frac{1}{x}}(1-\ln x)}{x^2}$

17. $y = (3x+1)^{2x}$. Thus

$\ln y = \ln\left[(3x+1)^{2x}\right] = 2x\ln(3x+1)$

$\dfrac{y'}{y} = 2\left\{x\left(\dfrac{3}{3x+1}\right) + [\ln(3x+1)](1)\right\}$

$y' = 2y\left[\dfrac{3x}{3x+1} + \ln(3x+1)\right]$

$= 2(3x+1)^{2x}\left[\dfrac{3x}{3x+1} + \ln(3x+1)\right]$

19. $y = 4e^x x^{3x}$. Thus

$\ln y = \ln 4 + \ln\left(e^x x^{3x}\right) = \ln 4 + \ln e^x + \ln x^{3x}$

$= \ln 4 + x + 3x\ln x.$

$\dfrac{y'}{y} = 1 + 3\left[x\left(\dfrac{1}{x}\right) + (\ln x)(1)\right]$

$y' = y(4 + 3\ln x)$

$y' = 4e^x x^{3x}(4 + 3\ln x)$

21. $y = (4x-3)^{2x+1}$

$\ln y = \ln(4x-3)^{2x+1} = (2x+1)\ln(4x-3)$

$\dfrac{y'}{y} = (2x+1)\left[\dfrac{4}{4x-3}\right] + [\ln(4x-3)](2)$

$y' = y\left[\dfrac{4(2x+1)}{4x-3} + 2\ln(4x-3)\right]$

When $x = 1$, then $\dfrac{dy}{dx} = 1\left[\dfrac{12}{1} + 2\ln(1)\right] = 12.$

23. $y = (x+1)(x+2)^2(x+3)^2$

$\ln y = \ln(x+1) + 2\ln(x+2) + 2\ln(x+3)$

$\dfrac{y'}{y} = \dfrac{1}{x+1} + \dfrac{2}{x+2} + \dfrac{2}{x+3}$

$y' = y\left[\dfrac{1}{x+1} + \dfrac{2}{x+2} + \dfrac{2}{x+3}\right]$

When $x = 0$, then $y = 36$ and $y' = 96$. Thus an equation of the tangent line is
$y - 36 = 96(x - 0)$, or $y = 96x + 36$.

25. $y = e^x (x^2 + 1)^x$

$\ln y = \ln e^x + \ln(x^2 + 1)^x$

$\quad = x + x \ln\left(x^2 + 1\right)$

$\dfrac{y'}{y} = 1 + \left[x\left(\dfrac{2x}{x^2 + 1}\right)\right] + \left[\ln\left(x^2 + 1\right)(1)\right]$

$y' = y\left[1 + \dfrac{2x^2}{x^2 + 1} + \ln\left(x^2 + 1\right)\right]$

When $x = 1$, then $y = 2e$ and $y' = 2e[1 + 1 + \ln(2)] = 2e(2 + \ln 2)$. Thus an equation of the tangent line is $y - 2e = 2e(2 + \ln 2)(x - 1)$, or $y = (4e + 2e \ln 2)x - 2e - 2e \ln 2$.

27. $y = (3x)^{-2x}$

$\ln y = -2x \ln(3x)$

$\dfrac{y'}{y} = -2\left\{x\left[\dfrac{1}{3x}(3)\right] + [\ln(3x)](1)\right\}$

$\quad = -2[1 + \ln(3x)]$

$\dfrac{y'}{y} \cdot 100$ gives the percentage rate of change. Thus $-2[1 + \ln(3x)](100) = 60$

$1 + \ln(3x) = -0.3$

$\ln(3x) = -1.3$

$3x = e^{-1.3}$

$x = \dfrac{1}{3e^{1.3}}$

29. $\dfrac{r'}{r} \cdot 100\% = \dfrac{p'}{p} \cdot 100\% + \dfrac{q'}{q} \cdot 100\%$

$\qquad\quad = (1 + \eta)\dfrac{p'}{p}100\%$

where $\eta = \dfrac{\frac{p}{q}}{\frac{dp}{dq}} = \dfrac{p}{q} \cdot \dfrac{dq}{dp}$.

$\eta = \dfrac{p}{500 - 40p + p^2} \cdot (-40 + 2p)$

When $p = 15$, then $\eta = -1.2$ and a $\dfrac{1}{2}\%$ increase in price will result in a $(1 - 1.2)\left(\dfrac{1}{2}\%\right) = -0.1\%$ change in revenue, which is a 0.1% decrease in revenue.

Principles in Practice 12.6

1. Let $f(x) = 20x - 0.01x^2 - 850 + 3\ln x$, then $f'(x) = 20 - 0.02x + \dfrac{3}{x}$. $f(10) \approx -644$ and $f(50) \approx 137$,

so we use 50 to be the first approximation, x_1, to find the break-even quantity between 10 and 50.

$$x_{n+1} = x_n - \frac{f(x_n)}{f'(x_n)} = x_n - \frac{20x_n - 0.01x_n^2 - 850 + 3\ln x_n}{20 - 0.02x_n + 3x_n^{-1}}$$

$$= x_n - \frac{20x_n^2 - 0.01x_n^3 - 850x_n + 3x_n \ln x_n}{20x_n - 0.02x_n^2 + 3}$$

$$= \frac{20x_n^2 - 0.02x_n^3 + 3x_n - \left(20x_n^2 - 0.01x_n^3 - 850x_n + 3x_n \ln x_n\right)}{20x_n - 0.02x_n^2 + 3}$$

$$= \frac{-0.01x_n^3 + 853x_n - 3x_n \ln x_n}{20x_n - 0.02x_n^2 + 3}$$

$$x_2 = 50 - \frac{f(50)}{f'(50)} \approx 42.82602$$

$$x_3 = 42.82602 - \frac{f(42.82602)}{f'(42.82602)} \approx 42.85459$$

$$x_4 = 42.85459 - \frac{f(42.85459)}{f'(42.85459)} \approx 42.85459$$

Since the values of x_3 and x_4 differ by less than 0.0001, we take the first break-even quantity
to be $x \approx 42.85459$ or 43 televisions.
$f(1900) \approx 1073$ and $f(2000) \approx -827$, so we use 2000 to be
the first approximation, x_1, for the break-even quantity between 1900 and 2000.

$$x_2 = 2000 - \frac{f(2000)}{f'(2000)} \approx 1958.63703$$

$$x_3 = 1958.63703 - \frac{f(1958.63703)}{f'(1958.63703)} \approx 1957.74457$$

$$x_4 = 1957.74457 - \frac{f(1957.74457)}{f'(1957.74457)} \approx 1957.74415$$

$$x_5 = 1957.74415 - \frac{f(1957.74415)}{f'(1957.74415)} \approx 1957.74415$$

Since the values of x_4 and x_5 differ by less than 0.0001, we take the second break-even quantity to be
$x \approx 1957.74415$ or 1958 televisions.

Problems 12.6

1. We want a root of $f(x) = x^3 - 4x + 1 = 0$. We see that $f(0) = 1$ and $f(1) = -2$ have opposite signs, so there must be a root between 0 and 1. Moreover, $f(0)$ is closer to 0 than is $f(1)$, so we select $x_1 = 0$ as our initial estimate. Since $f'(x) = 3x^2 - 4$, the recursion formula is

$$x_{n+1} = x_n - \frac{f(x_n)}{f'(x_n)} = x_n - \frac{x_n^3 - 4x_n + 1}{3x_n^2 - 4}.$$

Simplifying gives $x_{n+1} = \dfrac{2x_n^3 - 1}{3x_n^2 - 4}$. Thus we obtain:

n	x_n	x_{n+1}
1	0.00000	0.25000
2	0.25000	0.25410
3	0.25410	0.25410

Because $|x_4 - x_3| < 0.0001$, the root is approximately $x_4 = 0.25410$.

3. Let $f(x) = x^3 - x - 1$. We have $f(1) = -1$ and $f(2) = 5$ (note the sign change). Since $f(1)$ is closer to 0 than is $f(2)$, we choose $x_1 = 1$. We have $f'(x) = 3x^2 - 1$, so the recursion formula is

$$x_{n+1} = x_n - \frac{f(x_n)}{f'(x_n)} = x_n - \frac{x_n^3 - x_n - 1}{3x_n^2 - 1}$$

$$= \frac{2x_n^3 + 1}{3x_n^2 - 1}$$

n	x_n	x_{n+1}
1	1.00000	1.50000
2	1.50000	1.34783
3	1.34783	1.32520
4	1.32520	1.32472
5	1.32472	1.32472

Since $|x_6 - x_5| < 0.0001$, the root is approximately $x_6 = 1.32472$.

5. Let $f(x) = x^3 + x + 1$. We have $f(-1) = -1$ and $f(0) = 1$ (note the sign change). Choose $x_1 = -1$. Since $f'(x) = 3x^2 + 1$, the recursion formula is

$$x_{n+1} = x_n - \frac{f(x_n)}{f'(x_n)} = x_n - \frac{x_n^3 + x_n + 1}{3x_n^2 + 1}$$

$$= \frac{2x_n^3 - 1}{3x_n^2 + 1}$$

n	x_n	x_{n+1}
1	-1	-0.75000
2	-0.75000	-0.68605
3	-0.68605	-0.68234
4	-0.68234	-0.68233

Because $|x_5 - x_4| < 0.0001$, the root is approximately $x_5 = -0.68233$.

7. $x^4 = 3x - 1$, so use $f(x) = x^4 - 3x + 1 = 0$.

Since $f(0) = 1$ and $f(1) = -1$ (note the sign change), $f(0)$ and $f(1)$ are equally close to 0. We shall choose $x_1 = 0$. Since

$f'(x) = 4x^3 - 3$, the recursion formula is

$$x_{n+1} = x_n - \frac{f(x_n)}{f'(x_n)} = x_n - \frac{x_n^4 - 3x_n + 1}{4x_n^3 - 3}$$

$$= \frac{3x_n^4 - 1}{4x_n^3 - 3}$$

n	x_n	x_{n+1}
1	0.00000	0.33333
2	0.33333	0.33766
3	0.33766	0.33767

Because $|x_4 - x_3| < 0.0001$, the root is approximately $x_4 = 0.33767$.

9. Let $f(x) = x^4 - 2x^3 + x^2 - 3$. $f(1) = -3$ and $f(2) = 1$ (note the sign change), so $f(2)$ is closer to 0 than is $f(1)$. We choose $x_1 = 2$.

Since $f'(x) = 4x^3 - 6x^2 + 2x$, the recursion formula is

$$x_{n+1} = x_n - \frac{f(x)}{f'(x_n)} = x_n - \frac{x_n^4 - 2x_n^3 + x_n^2 - 3}{4x_n^3 - 6x_n^2 + 2x_n}$$

n	x_n	x_{n+1}
1	2.00000	1.91667
2	1.91667	1.90794
3	1.90794	1.90785

Because $|x_4 - x_3| < 0.0001$, the root is approximately $x_4 = 1.90785$.

11. The desired number is x, where $x^3 = 71$, or $x^3 - 71 = 0$. Thus we want to find a root of $f(x) = x^3 - 71 = 0$. Since $4^3 = 64$, the solution should be close to 4, so we choose $x_1 = 4$ as our initial estimate. We have $f'(x) = 3x^2$, so the recursion formula is

$$x_{n+1} = x_n - \frac{f(x_n)}{f'(x_n)} = x_n - \frac{x_n^3 - 71}{3x_n^2}$$

$$= \frac{2x_n^3 + 71}{3x_n^2}$$

n	x_n	x_{n+1}
1	4	4.146
2	4.146	4.141
4	4.141	4.141

Thus to three decimal places, $\sqrt[3]{71} = 4.141$.

13. We want real solutions to $e^x = x + 5$. Thus we want to find roots of $f(x) = e^x - x - 5 = 0$. A rough sketch of the exponential function $y = e^x$ and the line $y = x + 5$ shows that there are two intersection points: one when x is near -5, and the other when x is near 3. Thus we must find two roots. Since $f'(x) = e^x - 1$, the recursion formula is

$$x_{n+1} = x_n - \frac{f(x_n)}{f'(x_n)} = x_n - \frac{e^{x_n} - x_n - 5}{e^{x_n} - 1}$$

If $x_1 = -5$, we obtain

n	x_n	x_{n+1}
1	-5	-4.99
2	-4.99	-4.99

If $x_1 = 3$, we obtain:

n	x_n	x_{n+1}
1	3	2.37
2	2.37	2.03
3	2.03	1.94
4	1.94	1.94

Thus the solutions are –4.99 and 1.94.

15. The break-even quantity is the value of q when total revenue and total cost are equal: $r = c$, or $r - c = 0$. Thus we must find a root of $3q - \left(250 + 2q - 0.1q^3\right) = 0$, or

$$f(q) = q - 250 + 0.1q^3 = 0, \text{ so}$$

$f'(q) = 1 + 0.3q^2$. The recursion formula is

$$q_{n+1} = q_n - \frac{f(q_n)}{f'(q_n)} = q_n - \frac{q_n - 250 + 0.1q_n^3}{1 + 0.3q_n^2}$$

We choose $q_1 = 13$, as suggested.

n	q_n	q_{n+1}
1	13	13.33
2	13.33	13.33

Thus $q \approx 13.33$.

17. The equilibrium quantity is the value of q for which supply and demand are equal, that is, it is a root of $2q + 5 = \dfrac{100}{q^2 + 1}$, or of

$$f(q) = 2q + 5 - \frac{100}{q^2 + 1} = 0. \text{ Since}$$

$f'(q) = 2 + \dfrac{200q}{\left(q^2 + 1\right)^2}$, the recursion formula

is

$$q_{n+1} = q_n - \frac{f(q_n)}{f'(q_n)} = q_n - \frac{2q_n + 5 - \dfrac{100}{q_n^2 + 1}}{2 + \dfrac{200q_n}{\left(q_n^2 + 1\right)^2}}$$

A rough sketch shows that the graph of the supply equation intersects the graph of the demand equation when q is near 3. Thus we select $q_1 = 3$.

n	q_n	q_{n+1}
1	3	2.875
2	2.875	2.880
3	2.880	2.880

Thus $q \approx 2.880$.

19. For a critical value of

$$f(x) = \frac{x^3}{3} - x^2 - 5x + 1, \text{ we want a root of}$$

$f'(x) = x^2 - 2x - 5 = 0$. Since

$\dfrac{d}{dx}[f'(x)] = 2x - 2$, the recursion formula is

$$x_{n+1} = x_n - \frac{x_n^2 - 2x_n - 5}{2x_n - 2}.$$

For the given interval [3, 4], note that $f'(3) = -2$ and $f'(4) = 3$ have opposite signs. Thus there is a root x between 3 and 4. Since 3 is closer to 0, we shall select $x_1 = 3$.

n	x_n	x_{n+1}
1	3.0	3.5
2	3.5	3.45
3	3.45	3.45

Thus $x \approx 3.45$.

Principles in Practice 12.7

1. $\dfrac{dh}{dt} = 0 - 16(2t) = -32t$ ft/sec

   $\dfrac{d^2h}{dt^2} = \dfrac{d}{dt}[-32t] = -32$ feet/sec^2

   The acceleration of the rock at time t is
   -32 feet/sec^2 or 32 feet/sec^2 downward.

Problems 12.7

1. $y' = 12x^2 - 24x + 6$

   $y'' = 24x - 24$

   $y''' = 24$

3. $\dfrac{dy}{dx} = -1$

   $\dfrac{d^2y}{dx^2} = 0$

5. $y' = 3x^2 + e^x$

   $y'' = 6x + e^x$

   $y''' = 6 + e^x$

   $y^{(4)} = e^x$

7. $f(x) = x^2 \ln x$

   $f'(x) = x^2\left(\dfrac{1}{x}\right) + (\ln x)(2x) = x(1 + 2\ln x)$

   $f''(x) = x\left(\dfrac{2}{x}\right) + (1 + 2\ln x)(1) = 3 + 2\ln x$

9. $f(q) = \dfrac{1}{2q^4} = \dfrac{1}{2}q^{-4}$

   $f'(q) = -2q^{-5}$

   $f''(q) = 10q^{-6}$

   $f'''(q) = -60q^{-7} = -\dfrac{60}{q^7}$

11. $f(r) = \sqrt{9-r} = (9-r)^{\frac{1}{2}}$

    $f'(r) = -\dfrac{1}{2}(9-r)^{-\frac{1}{2}}$

    $f''(r) = -\dfrac{1}{4}(9-r)^{-\frac{3}{2}} = -\dfrac{1}{4(9-r)^{\frac{3}{2}}}$

13. $y = \dfrac{1}{2x+3} = (2x+3)^{-1}$

    $\dfrac{dy}{dx} = -2(2x+3)^{-2}$

    $\dfrac{d^2y}{dx^2} = 8(2x+3)^{-3} = \dfrac{8}{(2x+3)^3}$

15. $y = \dfrac{x+1}{x-1}$

    $y' = \dfrac{(x-1)(1) - (x+1)(1)}{(x-1)^2}$

    $= -\dfrac{2}{(x-1)^2} = -2(x-1)^{-2}$

    $y'' = 4(x-1)^{-3} = \dfrac{4}{(x-1)^3}$

17. $y = \ln[x(x+6)] = \ln(x) + \ln(x+6)$

    $y' = \dfrac{1}{x} + \dfrac{1}{x+6} = x^{-1} + (x+6)^{-1}$

    $y'' = -x^{-2} + (-1)(x+6)^{-2}$

    $= -\left[\dfrac{1}{x^2} + \dfrac{1}{(x+6)^2}\right]$

19. $f(z) = z^2 e^z$

    $f'(z) = z^2\left(e^z\right) + e^z(2z) = \left(ze^z\right)(z+2)$

    $f''(z) = \left(ze^z\right)(1) + (z+2)\left[ze^z + e^z(1)\right]$

    $= e^z\left(z^2 + 4z + 2\right)$

21. $y = e^{2x} + e^{3x}$

$$\frac{dy}{dx} = 2e^{2x} + 3e^{3x}$$

$$\frac{d^2y}{dx^2} = 4e^{2x} + 9e^{3x}$$

$$\frac{d^3y}{dx^3} = 8e^{2x} + 27e^{3x}$$

$$\frac{d^4y}{dx^4} = 16e^{2x} + 81e^{3x}$$

$$\frac{d^5y}{dx^5} = 32e^{2x} + 243e^{3x}$$

$$\left.\frac{d^5y}{dx^5}\right|_{x=0} = 32e^0 + 243e^0 = 32 + 243 = 275$$

23. $x^2 + 4y^2 - 16 = 0$

$$2x + 8yy' = 0$$

$$8yy' = -2x$$

$$y' = -\frac{x}{4y}$$

$$y'' = -\frac{4y(1) - x(4y')}{16y^2}$$

$$= -\frac{4y - 4x\left(-\frac{x}{4y}\right)}{16y^2} = -\frac{4y^2 + x^2}{16y^3}$$

$$= -\frac{16}{16y^3} = -\frac{1}{y^3}$$

25. $y^2 = 4x$

$$2yy' = 4$$

$$y' = \frac{2}{y} = 2y^{-1}$$

$$y'' = -2y^{-2}y' = -2y^{-2}\left(2y^{-1}\right) = -\frac{4}{y^3}$$

27. $\sqrt{x} + 4\sqrt{y} = 4$

$$x^{\frac{1}{2}} + 4y^{\frac{1}{2}} = 4$$

$$\frac{1}{2}x^{-\frac{1}{2}} + 2y^{-\frac{1}{2}}y' = 0$$

$$2y^{-\frac{1}{2}}y' = -\frac{1}{2}x^{-\frac{1}{2}}$$

$$y' = -\frac{1}{2} \cdot \frac{x^{-\frac{1}{2}}}{2y^{-\frac{1}{2}}} = -\frac{1}{4} \cdot \frac{y^{\frac{1}{2}}}{x^{\frac{1}{2}}}$$

$$y'' = -\frac{1}{4}\left[\frac{x^{\frac{1}{2}}\left(\frac{1}{2}y^{-\frac{1}{2}}y'\right) - y^{\frac{1}{2}}\left(\frac{1}{2}x^{-\frac{1}{2}}\right)}{x}\right]$$

$$= -\frac{1}{8}\left[\frac{\frac{x^{\frac{1}{2}}}{y^{\frac{1}{2}}}\left(-\frac{y^{\frac{1}{2}}}{4x^{\frac{1}{2}}}\right) - \frac{y^{\frac{1}{2}}}{x^{\frac{1}{2}}}}{x}\right] = -\frac{1}{8}\left[\frac{-\frac{1}{4} - \frac{y^{\frac{1}{2}}}{x^{\frac{1}{2}}}}{x}\right]$$

$$= \frac{1}{8}\left[\frac{\frac{1}{4} + \frac{y^{\frac{1}{2}}}{x^{\frac{1}{2}}}}{x}\right] = \frac{1}{8}\left[\frac{x^{\frac{1}{2}} + 4y^{\frac{1}{2}}}{4x^{\frac{3}{2}}}\right]$$

$$= \frac{1}{8}\left[\frac{4}{4x^{\frac{3}{2}}}\right] = \frac{1}{8x^{\frac{3}{2}}}$$

29. $xy + y - x = 4$

$$xy' + y(1) + y' - 1 = 0$$

$$xy' + y' = 1 - y$$

$$(x+1)y' = 1 - y$$

$$y' = \frac{1-y}{1+x}$$

$$y'' = \frac{(1+x)(-y') - (1-y)(1)}{(1+x)^2}$$

$$= \frac{(1+x)\left[-\frac{(1-y)}{(1+x)}\right] - (1-y)}{(1+x)^2}$$

$$= \frac{-(1-y) - (1-y)}{(1+x)^2} = \frac{-2(1-y)}{(1+x)^2} = \frac{2(y-1)}{(1+x)^2}$$

31. $y = e^{x+y}$

$$y' = e^{x+y}(1 + y')$$

$$y' - e^{x+y}y' = e^{x+y}$$

$$y'\left(1 - e^{x+y}\right) = e^{x+y}$$

$$y' = \frac{e^{x+y}}{1 - e^{x+y}}$$

$$y' = \frac{y}{1-y}$$

$$y'' = \frac{(1-y)y' - y(-y')}{(1-y)^2} = \frac{y'}{(1-y)^2}$$

$$= \frac{\frac{y}{1-y}}{(1-y)^2} = \frac{y}{(1-y)^3}$$

33. $x^2 + 3x + y^2 = 4y$

$$2x + 3 + 2yy' = 4y'$$

$$2yy' - 4y' = -2x - 3$$

$$y' = -\frac{2x+3}{2y-4} = \frac{2x+3}{4-2y}$$

$$y'' = \frac{(4-2y)(2) - (2x+3)(-2y')}{(4-2y)^2}$$

$$= \frac{2(4-2y) + 2(2x+3)\left(\frac{2x+3}{4-2y}\right)}{(4-2y)^2}$$

$$= \frac{2(4-2y)^2 + 2(2x+3)^2}{(4-2y)^3}$$

When $x = 0$ and $y = 0$, then

$$\frac{d^2y}{dx^2} = \frac{2(4)^2 + 2(3)^2}{4^3} = \frac{25}{32}.$$

35. $f(x) = (5x-3)^4$

$$f'(x) = 20(5x-3)^3$$

$$f''(x) = 300(5x-3)^2$$

37. $\dfrac{dc}{dq} = 0.6q + 2$

$$\frac{d^2c}{dq^2} = 0.6$$

$$\left.\frac{d^2c}{dq^2}\right|_{q=100} = 0.6$$

39. $f(x) = x^4 - 6x^2 + 5x - 6$

$$f'(x) = 4x^3 - 12x + 5$$

$$f''(x) = 12x^2 - 12 = 12(x+1)(x-1)$$

Clearly $f''(x) = 0$ when $x = \pm 1$.

41. $f'(x) = 6e^x - 3x^2 - 30x$

$$f''(x) = 6\left(e^x - x - 5\right)$$

$$f''(x) = 0 \text{ when } x \approx -4.99 \text{ or } 1.94.$$

Chapter 12 Review Problems

1. $y' = 3e^x + 0 + e^{x^2}(2x) + (e^2)xe^{2}-1$

$$= 3e^x + 2xe^{x^2} + e^2 xe^{2}-1$$

3. $f'(r) = \dfrac{1}{3r^2 + 7r + 1}(6r + 7) = \dfrac{6r+7}{3r^2 + 7r + 1}$

5. $y = e^{x^2 + 4x + 5}$

$$y' = e^{x^2 + 4x + 5}(2x + 4) = 2(x+2)e^{x^2 + 4x + 5}$$

7. $y' = e^x(2x) + \left(x^2 + 2\right)e^x = e^x\left(x^2 + 2x + 2\right)$

9. $y = \sqrt{(x-6)(x+5)(9-x)}$

$\ln y = \ln \sqrt{(x-6)(x+5)(9-x)}$

$= \frac{1}{2}[\ln(x-6) + \ln(x+5) + \ln(9-x)]$

$\frac{y'}{y} = \frac{1}{2}\left[\frac{1}{x-6} + \frac{1}{x+5} + \frac{-1}{9-x}\right]$

$y' = \frac{y}{2}\left[\frac{1}{x-6} + \frac{1}{x+5} - \frac{1}{9-x}\right]$

$= \frac{\sqrt{(x-6)(x+5)(9-x)}}{2}\left[\frac{1}{x-6} + \frac{1}{x+5} + \frac{1}{x-9}\right]$

11. $y' = \dfrac{e^x\left(\frac{1}{x}\right) - (\ln x)\left(e^x\right)}{e^{2x}}$

$= \dfrac{e^x - xe^x \ln x}{xe^{2x}} = \dfrac{1 - x\ln x}{xe^x}$

13. $f(q) = \ln\left[(q+1)^2(q+2)^3\right]$

$= 2\ln(q+1) + 3\ln(q+2)$

$f'(q) = \dfrac{2}{q+1} + \dfrac{3}{q+2}$

15. $y = e^{(2x^2+2x-5)(\ln 2)}$

$y' = e^{(2x^2+2x-5)(\ln 2)}(4x+2)(\ln 2)$

$= (4x+2)(\ln 2)2^{2x^2+2x-5}$

17. $y = \dfrac{4e^{3x}}{xe^{x-1}} = \dfrac{4e^{2x+1}}{x}$

$y' = 4 \cdot \dfrac{x\left[e^{2x+1}(2)\right] - e^{2x+1}[1]}{x^2} = \dfrac{4e^{2x+1}(2x-1)}{x^2}$

19. $y = \log_2(8x+5)^2 = 2\log_2(8x+5)$

$= 2 \cdot \dfrac{\ln(8x+5)}{\ln 2}$

$y' = 2 \cdot \dfrac{1}{\ln 2} \cdot \dfrac{8}{8x+5} = \dfrac{16}{(8x+5)\ln 2}$

21. $f(l) = \ln\left(1 + l + l^2 + l^3\right)$

$$f'(l) = \frac{1}{1 + l + l^2 + l^3}\left[1 + 2l + 3l^2\right]$$

$$= \frac{1 + 2l + 3l^2}{1 + l + l^2 + l^3}$$

23. $y = (x+1)^{x+1}$

$\ln y = (x+1)\ln(x+1)$

$$\frac{y'}{y} = (x+1)\frac{1}{x+1} + \ln(x+1)[1] = 1 + \ln(x+1)$$

$$y' = y[1 + \ln(x+1)] = (x+1)^{x+1}[1 + \ln(x+1)]$$

25. $\phi(t) = \ln\left(t\sqrt{4-t^2}\right) = \ln t + \frac{1}{2}\ln(4 - t^2)$

$$\phi'(t) = \frac{1}{t} + \frac{1}{2}\cdot\frac{1}{4-t^2}\cdot(-2t) = \frac{1}{t} - \frac{t}{4-t^2}$$

27. $y = \dfrac{(x^2+1)^{1/2}(x^2+2)^{1/3}}{(2x^3+6x)^{2/5}}$

$\ln y = \dfrac{1}{2}\ln(x^2+1) + \dfrac{1}{3}\ln(x^2+2) - \dfrac{2}{5}\ln(2x^3+6x)$

$\dfrac{y'}{y} = \dfrac{1}{2}\left(\dfrac{1}{x^2+1}\right)(2x) + \dfrac{1}{3}\left(\dfrac{1}{x^2+2}\right)(2x) - \dfrac{2}{5}\left(\dfrac{1}{2x^3+6x}\right)(6x^2+6)$

$y' = y\left[\dfrac{x}{x^2+1} + \dfrac{2x}{3(x^2+2)} - \dfrac{6(x^2+1)}{5(x^3+3x)}\right]$

$= \dfrac{(x^2+1)^{1/2}(x^2+2)^{1/3}}{(2x^3+6x)^{2/5}}\left[\dfrac{x}{x^2+1} + \dfrac{2x}{3(x^2+2)} - \dfrac{6(x^2+1)}{5(x^3+3x)}\right]$

29. $y = \left(x^x\right)^x = x^{x^2}$

$\ln y = \ln x^{x^2} = x^2\ln x$

$\dfrac{y'}{y} = x^2\left(\dfrac{1}{x}\right) + (\ln x)(2x) = x + 2x\ln x$

$y' = y(x + 2x\ln x) = \left(x^x\right)^x(x + 2x\ln x)$

31. $y = (x+1)\ln x^2 = 2(x+1)\ln x$

$$y' = 2\left[(x+1)\left(\frac{1}{x}\right) + (\ln x)(1) \right]$$

$$= 2\left[\frac{x+1}{x} + \ln x \right]$$

When $x = 1$, then $y' = 2\left[\frac{2}{1} + \ln 1 \right] = 4.$

33. $y = e^{e + x \ln\left(\frac{1}{x}\right)} = e^{e - x \ln x}$

$$y' = e^{e - x \ln x}\left(-\left[x\left(\frac{1}{x}\right) + (\ln x)(1) \right] \right)$$

$$= -(1 + \ln x)e^{e - x \ln x}$$

When $x = e$, then

$$y' = -(1 + \ln e)e^{e - e \ln e} = -(2)e^0 = -2.$$

35. $y = 3e^x$

$$y' = 3e^x$$

If $x = \ln 2$, then $y = 3e^{\ln 2} = 6$ and

$$y' = 3e^{\ln 2} = 6.$$

An equation of the tangent line is
$y - 6 = 6(x - \ln 2)$, $y = 6x + 6 - 6\ln 2$,
$y = 6x + 6(1 - \ln 2)$. Alternatively, since

$6\ln 2 = \ln 2^6 = \ln 64$, the tangent line can be
written as $y = 6x + 6 - \ln 64.$

37. $y = x\left(2^{2 - x^2} \right)$. To find y' we shall use

logarithmic differentiation.

$$\ln y = \ln\left[x\left(2^{2 - x^2} \right) \right] = \ln x + \left(2 - x^2 \right)\ln 2$$

$$\frac{y'}{y} = \frac{1}{x} + (-2x)\ln 2$$

$$y' = y\left[\frac{1}{x} - 2(\ln 2)x \right]$$

When $x = 1$, then $y = 2$ and $y' = 2(1 - 2\ln 2)$.
The equation of the tangent line is
$y - 2 = 2(1 - 2\ln 2)(x - 1)$. The y-intercept
of the tangent line corresponds to the point

where $x = 0$:
$y - 2 = 2(1 - 2\ln 2)(-1) = -2 + 4\ln 2$
Thus $y = 4\ln 2$ and the y-intercept is
$(0, 4\ln 2)$.

39. $y = e^{x^2 - 2x + 1}$

$$y' = e^{x^2 - 2x + 1}[2x - 2] = (2x - 2)e^{x^2 - 2x + 1}$$

$$y'' = 2(x - 1)e^{x^2 - 2x + 1}(2x - 2) + 2e^{x^2 - 2x + 1}$$

$$= 2e^{x^2 - 2x + 1}(2(x - 1)^2 + 1)$$

At $(1, 1)$, $y'' = 2e^0(2(0) + 1) = 2.$

41. $y = \ln(2x)$

$$y' = \frac{1}{2x}(2) = x^{-1}$$

$$y'' = -1 \cdot x^{-2} = -x^{-2}$$

$$y''' = -(-2)x^{-3} = \frac{2}{x^3}$$

At $(1, \ln 2)$, $y''' = \frac{2}{1^3} = 2$

43. $2xy + y^2 = 10$

$$2(xy' + y) + 2yy' = 0$$

$$2xy' + 2yy' = -2y$$

$$(x + y)y' = -y$$

$$y' = -\frac{y}{x + y}$$

45. $\ln\left(xy^2 \right) = xy$

$$\ln x + 2\ln y = xy$$

$$\frac{1}{x} + \frac{2}{y}y' = xy' + y$$

$$y + 2xy' = x^2 yy' + xy^2$$

$$2xy' - x^2 yy' = xy^2 - y$$

$$\left(2x - x^2 y \right)y' = xy^2 - y$$

$$y' = \frac{xy^2 - y}{2x - x^2 y}$$

47. $x + xy + y = 5$

$1 + xy' + y(1) + y' = 0$

$(x+1)y' = -1 - y$

$y' = -\dfrac{1+y}{x+1}$

$y'' = -\dfrac{(x+1)y' - (1+y)}{(x+1)^2}$

At (2, 1), $y' = -\dfrac{1+1}{2+1} = -\dfrac{2}{3}$ and

$y'' = -\dfrac{3\left(-\frac{2}{3}\right) - 2}{9} = \dfrac{4}{9}$

49. $e^y = (y+1)e^x$

$e^y y' = (y+1)e^x + e^x(y')$

$e^y y' - e^x y' = (y+1)e^x$

$\left(e^y - e^x\right)y' = (y+1)e^x$

$y' = \dfrac{(y+1)e^x}{e^y - e^x} = \dfrac{(y+1)\left(\frac{e^y}{y+1}\right)}{e^y - \left(\frac{e^y}{y+1}\right)} = \dfrac{e^y}{e^y - \frac{e^y}{y+1}}$

$= \dfrac{1}{1 - \frac{1}{y+1}} = \dfrac{y+1}{y}$

$y'' = \dfrac{y(y') - (y+1)(y')}{y^2} = \dfrac{-y'}{y^2}$

$= -\dfrac{\frac{y+1}{y}}{y^2} = -\dfrac{y+1}{y^3}$

51. $f'(t)$

$= \left[-0.8e^{-0.01t}(-0.01) - 0.2e^{-0.0002t}(-0.0002)\right]$

$= 0.008e^{-0.01t} + 0.00004e^{-0.0002t}$

53. $f'(x) = (12x^3 + 6x^2 - 25)e^{3x^4 + 2x^3 - 25x}$

$f'(x) = 0$ when $x \approx 1.13$.

55. $p = \dfrac{500}{q}$

$\eta = \dfrac{\frac{p}{q}}{\frac{dp}{dq}} = \dfrac{500/q}{-\frac{500}{q^2}} = -1$

Since $|\eta| = 1,$ demand has unit elasticity when $q = 200$.

57. $p = 18 - 0.02q$

$\eta = \dfrac{\frac{p}{q}}{\frac{dp}{dq}} = \dfrac{\frac{18 - 0.02q}{q}}{-0.02} = -\dfrac{18 - 0.02q}{0.02q}$

When $q = 600$, then $\eta = -0.5$. Because $|\eta| < 1,$ demand is inelastic.

59. $\eta = \dfrac{\frac{p}{q}}{\frac{dp}{dq}} = \dfrac{p}{q} \cdot \dfrac{dq}{dp}$

$q = \sqrt{2500 - p^2}$

$\dfrac{dq}{dp} = \dfrac{-p}{\sqrt{2500 - p^2}} = \dfrac{-p}{q}$, so

$\eta = \dfrac{p}{q}\left(\dfrac{-p}{q}\right) = -\dfrac{p^2}{q^2}$. Now, if $p = 30$, then

$q = \sqrt{2500 - 30^2} = 40$, so

$\eta\big|_{p=30} = -\dfrac{(30)^2}{(40)^2} = -\dfrac{9}{16}$

If the price of 30 decreases $\dfrac{2}{3}$%, then demand would change by approximately

$\left(-\dfrac{2}{3}\right)\left(-\dfrac{9}{16}\right)$%, or $\dfrac{3}{8}$%. (That is, demand

increases by approximately $\dfrac{3}{8}$%.)

61. We want a root of $f(x) = x^3 - 2x - 2 = 0$.

We have $f(1) = -3$ and $f(2) = 2$ (note the sign change). Since $f(2)$ is closer to 0 than is $f(1)$, we choose $x_1 = 2$. We have

$f'(x) = 3x^2 - 2$, so the recursion formula is

$$x_{n+1} = x_n - \frac{f(x_n)}{f'(x_n)} = x_n - \frac{x_n^3 - 2x_n - 2}{3x_n^2 - 2}$$

$$= \frac{2x_n^3 + 2}{3x_n^2 - 2}$$

n	x_n	x_{n+1}
1	2.00000	1.80000
2	1.80000	1.76995
3	1.76995	1.76929
4	1.76929	1.76929

Because $|x_5 - x_4| < 0.0001$, the root is approximately $x_5 = 1.7693$.

Mathematical Snapshot Chapter 12

1. $F = 25, D = 3400, V = 36.5, R = 0.05$.

$$q = \sqrt{\frac{2FD}{RV}} = \sqrt{\frac{2(25)(3400)}{(0.05)(36.5)}} \approx 305.2$$

The economic order quantity is 305 units.

3. Answers may vary.

Chapter 13

Principles in Practice 13.1

1. The graph of $c(q) = 2q^3 - 21q^2 + 60q + 500$ is shown.

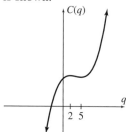

There looks to be a relative maximum at $q = 2$ and a relative minimum at $q = 5$.

$c'(q) = 6q^2 - 42q + 60 = 6(q^2 - 7q + 10)$
$= 6(q - 5)(q - 2)$

$c'(q) = 0$ when $q = 2$ or $q = 5$. If $q < 2$, then $c'(q) = 6(-)(-) = +$, so $c(q)$ is increasing. If $2 < q < 5$, then $c'(q) = 6(-)(+) = -$, so $c(q)$ is decreasing. If $5 < q$, then $c'(q) = 6(+)(+) = +$, so $c(q)$ is increasing. When $q = 2$, there is a relative maximum, since $c'(q)$ changes from $+$ to $-$. The relative maximum value is

$2(2)^3 - 21(2)^2 + 60(2) + 500 = 552$. When $q = 5$, there is a relative minimum, since $c'(q)$ changes from $-$ to $+$. The relative minimum value is

$2(5)^3 - 21(5)^2 + 60(5) + 500 = 525$.

Problems 13.1

1. Decreasing on $(-\infty, -1)$ and $(3, \infty)$; increasing on $(-1, 3)$; relative minimum $(-1, -1)$; relative maximum $(3, 4)$.

3. Decreasing on $(-\infty, -2)$ and $(0, 2)$; increasing on $(-2, 0)$ and $(2, \infty)$; relative minima $(-2, 1)$ and $(2, 1)$; no relative maximum.

In the following problems, we denote the critical value by CV.

5. $f'(x) = (x+3)(x-1)(x-2)$

$f'(x) = 0$ when $x = -3, 1, 2$

CV: $x = -3, 1, 2$

$$\begin{array}{ccccccc} & - & + & - & + \\ \hline & -3 & & 1 & & 2 \end{array}$$

Increasing on $(-3, 1)$ and $(2, \infty)$; decreasing on $(-\infty, -3)$ and $(1, 2)$; relative maximum when $x = 1$; relative minima when $x = -3, 2$.

7. $f'(x) = (x+1)(x-3)^2$

CV: $x = -1, 3$

$$\begin{array}{ccccc} & - & + & + \\ \hline & -1 & & 3 \end{array}$$

Decreasing on $(-\infty, -1)$; increasing on $(-1, 3)$ and $(3, \infty)$; relative minimum when $x = -1$.

9. $y = 2x^3 + 1$

$y' = 6x^2$

CV: $x = 0$

$$\begin{array}{ccc} & + & + \\ \hline & 0 \end{array}$$

Increasing on $(-\infty, 0)$; increasing on $(0, \infty)$; no relative maximum or minimum

11. $y = x - x^2 + 2$

$y' = 1 - 2x$

CV: $x = \dfrac{1}{2}$

$$\begin{array}{ccc} & + & - \\ \hline & \frac{1}{2} \end{array}$$

Increasing on $\left(-\infty, \dfrac{1}{2}\right)$; decreasing on $\left(\dfrac{1}{2}, \infty\right)$; relative maximum when $x = \dfrac{1}{2}$.

13. $y = -\dfrac{x^3}{3} - 2x^2 + 5x - 2$

$y' = -x^2 - 4x + 5 = -\left(x^2 + 4x - 5\right)$

$= -(x+5)(x-1)$

CV: $x = -5, 1$

$$\begin{array}{ccc} - & + & - \\ \hline & -5 & 1 \end{array}$$

Decreasing on $(-\infty, -5)$ and $(1, \infty)$; increasing on $(-5, 1)$; relative minimum when $x = -5$; relative maximum when $x = 1$.

15. $y = x^4 - 2x^2$

$y' = 4x^3 - 4x = 4x\left(x^2 - 1\right) = 4x(x+1)(x-1)$

CV: $x = 0, \pm1$

$$\begin{array}{ccccc} - & + & - & + \\ \hline & -1 & 0 & 1 \end{array}$$

Decreasing on $(-\infty, -1)$ and $(0, 1)$; increasing on $(-1, 0)$ and $(1, \infty)$; relative maximum when $x = 0$; relative minima when $x = \pm1$.

17. $y = x^3 - \dfrac{7}{2}x^2 + 2x - 5$

$y' = 3x^2 - 7x + 2 = (3x-1)(x-2)$

CV: $x = \dfrac{1}{3}, 2$

$$\begin{array}{ccc} + & - & + \\ \hline & \frac{1}{3} & 2 \end{array}$$

Increasing on $\left(-\infty, \dfrac{1}{3}\right)$ and $(2, \infty)$;

decreasing on $\left(\dfrac{1}{3}, 2\right)$; relative maximum

when $x = \dfrac{1}{3}$, relative minimum when $x = 2$.

19. $y = 2x^3 - \dfrac{11}{2}x^2 - 10x + 2$

$y' = 6x^2 - 11x - 10 = (2x-5)(3x+2)$

CV: $x = -\dfrac{2}{3}, \dfrac{5}{2}$

$$\begin{array}{ccc} + & - & + \\ \hline & -\frac{20}{3} & \frac{50}{2} \end{array}$$

Increasing on $\left(-\infty, -\dfrac{2}{3}\right)$ and $\left(\dfrac{5}{2}, \infty\right)$;

decreasing on $\left(-\dfrac{2}{3}, \dfrac{5}{2}\right)$; relative maximum

when $x = -\dfrac{2}{3}$; relative minimum when

$x = \dfrac{5}{2}$.

21. $y = \dfrac{x^3}{3} - 5x^2 + 22x + 1$

$y' = x^2 - 10x + 22$

By the quadratic formula, $y' = 0$ when

$x = \dfrac{10 \pm \sqrt{(-10)^2 - 4(1)(22)}}{2(1)}$ or $x = 5 \pm \sqrt{3}$.

CV: $x = 5 \pm \sqrt{3}$

$$\begin{array}{ccc} + & - & + \\ \hline & 5-\sqrt{3} & 5+\sqrt{3} \end{array}$$

Increasing on $\left(-\infty, 5-\sqrt{3}\right)$; decreasing on

$\left(5-\sqrt{3}, 5+\sqrt{3}\right)$; increasing on

$\left(5+\sqrt{3}, \infty\right)$; relative maximum at

$x = 5 - \sqrt{3}$; relative minimum at

$x = 5 + \sqrt{3}$.

23. $y = 3x^5 - 5x^3$

$y' = 15x^4 - 15x^2 = 15x^2(x+1)(x-1)$

CV: $x = 0, \pm1$

$$\begin{array}{ccccc} + & - & - & + \\ \hline & -1 & 0 & 1 \end{array}$$

Increasing on $(-\infty, -1)$ and $(1, \infty)$; decreasing on $(-1, 0)$ and $(0, 1)$; relative maximum when $x = -1$; relative minimum when $x = 1$.

25. $y = -x^5 - 5x^4 + 200$

$y' = -5x^4 - 20x^3 = -5x^3(x+4)$

CV: $x = 0, -4$

$$\begin{array}{c} -+- \\ \hline -40 \end{array}$$

Decreasing on $(-\infty, -4)$ and $(0, \infty)$; increasing on $(-4, 0)$; relative minimum when $x = -4$; relative maximum when $x = 0$.

27. $y = 8x^4 - x^8$

$y' = 32x^3 - 8x^7 = 8x^3\left(4 - x^4\right)$

$= 8x^3\left(2 + x^2\right)\left(2 - x^2\right)$

$= 8x^3\left(2 + x^2\right)\left(\sqrt{2} - x\right)\left(\sqrt{2} + x\right)$

CV: $x = 0, \pm\sqrt{2}$

$$\begin{array}{c} +-+- \\ \hline -\sqrt{2}0\sqrt{2} \end{array}$$

Increasing on $\left(-\infty, -\sqrt{2}\right)$ and $\left(0, \sqrt{2}\right)$;

decreasing on $\left(-\sqrt{2}, 0\right)$ and $\left(\sqrt{2}, \infty\right)$;

relative maxima when $x = \pm\sqrt{2}$, relative minimum when $x = 0$.

29. $y = (x^2 - 1)^4$

$y' = 8x(x^2 - 1)^3 = 8x(x+1)^3(x-1)^3$

CV: $0, -1, 1$

$$\begin{array}{c} -+-+ \\ \hline -101 \end{array}$$

Increasing on $(-1, 0)$ and $(1, \infty)$; decreasing on $(-\infty, -1)$ and $(0, 1)$; relative maximum when $x = 0$; relative minima when $x = \pm 1$.

31. $y = \dfrac{5}{x-1} = 5(x-1)^{-1}$

$y' = -5(x-1)^{-2} = -\dfrac{5}{(x-1)^2}$

CV: None, but $x = 1$ must be included in the sign chart because it is a point of discontinuity of y.

$$\begin{array}{c} -- \\ \hline \boxed{1} \end{array}$$

Decreasing on $(-\infty, 1)$ and $(1, \infty)$; no relative extremum.

33. $y = \dfrac{10}{\sqrt{x}} = 10x^{-\frac{1}{2}}$. [Note: $x > 0$]

$y' = -5x^{-\frac{3}{2}} = -\dfrac{5}{\sqrt{x^3}} < 0$ for $x > 0$.

Decreasing on $(0, \infty)$; no relative extremum.

35. $y = \dfrac{x^2}{2-x}$

$y' = \dfrac{(2-x)(2x) - x^2(-1)}{(2-x)^2} = \dfrac{x(4-x)}{(2-x)^2}$

CV: $x = 0, 4$, but $x = 2$ must be included in the sign chart because it is a point of discontinuity of y.

$$\begin{array}{c} -++- \\ \hline 0\boxed{2}4 \end{array}$$

Decreasing on $(-\infty, 0)$ and $(4, \infty)$; increasing on $(0, 2)$ and $(2, 4)$; relative minimum when $x = 0$; relative maximum when $x = 4$.

37. $y = \dfrac{x^2 - 3}{x+2}$

$y' = \dfrac{(x+2)(2x) - \left(x^2 - 3\right)(1)}{(x+2)^2}$

$= \dfrac{x^2 + 4x + 3}{(x+2)^2} = \dfrac{(x+1)(x+3)}{(x+2)^2}$

CV: $x = -3, -1$, but $x = -2$ must be included in the sign chart because it is a point of discontinuity of y.

$$\begin{array}{c} +--+ \\ \hline -3\boxed{-2}-1 \end{array}$$

Increasing on $(-\infty, -3)$ and $(-1, \infty)$; decreasing on $(-3, -2)$ and $(-2, -1)$; relative maximum when $x = -3$; relative minimum when $x = -1$.

39. $y = \dfrac{5x+2}{x^2+1}$

$y' = \dfrac{\left(x^2+1\right)(5)-(5x+2)(2x)}{\left(x^2+1\right)^2} = \dfrac{-5x^2-4x+5}{\left(x^2+1\right)^2}$

$y' = 0$ when $-5x^2-4x+5 = 0$; by the

quadratic formula, $x = \dfrac{-2\pm\sqrt{29}}{5}$

CV: $x = \dfrac{-2\pm\sqrt{29}}{5}$

$$\begin{array}{ccc} - & + & - \\ \hline & \frac{-2-\sqrt{29}}{5} & \frac{-2+\sqrt{29}}{5} \end{array}$$

Decreasing on $\left(-\infty, \dfrac{-2-\sqrt{29}}{5}\right)$ and

$\left(\dfrac{-2+\sqrt{29}}{5}, \infty\right)$; increasing on

$\left(\dfrac{-2-\sqrt{29}}{5}, \dfrac{-2+\sqrt{29}}{5}\right)$; relative minimum

when $x = \dfrac{-2-\sqrt{29}}{5}$; relative maximum

when $x = \dfrac{-2+\sqrt{29}}{5}$.

41. $y = (x-1)^{2/3}$

$y' = \dfrac{2}{3}(x-1)^{-1/3} = \dfrac{2}{3\sqrt[3]{x-1}}$

CV: $x = 1$

$$\begin{array}{cc} - & + \\ \hline & 1 \end{array}$$

Increasing on $(1, \infty)$; decreasing on $(-\infty, 1)$; relative minimum when $x = 1$.

43. $y = x^3(x-6)^4$

$y' = x^3\left[4(x-6)^3\right]+(x-6)^4\left(3x^2\right)$

$\quad = x^2(x-6)^3[4x+3(x-6)]$

$\quad = x^2(x-6)^3(7x-18)$

CV: $x = 0, 6, \dfrac{18}{7}$

$$\begin{array}{cccc} + & + & - & + \\ \hline 0 & \frac{18}{7} & 6 & \end{array}$$

Increasing on $(-\infty, 0)$, $\left(0, \dfrac{18}{7}\right)$, and $(6, \infty)$;

decreasing on $\left(\dfrac{18}{7}, 6\right)$; relative maximum

when $x = \dfrac{18}{7}$; relative minimum when

$x = 6$.

45. $y = e^{-\pi x}+\pi$

$y' = -\pi e^{-\pi x} < 0$ for all x. Thus decreasing

on $(-\infty, \infty)$; no relative extremum.

47. $y = x^2 - 9\ln x$. [Note: $x > 0$.]

$y' = 2x - \dfrac{9}{x} = \dfrac{2x^2-9}{x}$

CV: $x = \dfrac{3\sqrt{2}}{2}$

$$\begin{array}{cc} - & + \\ \hline 0 & \frac{3\sqrt{2}}{5} \end{array}$$

Decreasing on $\left(0, \dfrac{3\sqrt{2}}{2}\right)$; increasing on

$\left(\dfrac{3\sqrt{2}}{2}, \infty\right)$; relative minimum when

$x = \dfrac{3\sqrt{2}}{2}$.

49. $y = e^x + e^{-x}$

$y' = e^x - e^{-x}$

Setting $y' = 0$ gives $e^x - e^{-x} = 0$,

$e^x = e^{-x}$,

$x = -x, \ x = 0$

CV: $x = 0$

$$\begin{array}{c} \quad - \quad + \\ \hline 0 \end{array}$$

Decreasing on $(-\infty, 0)$; increasing on $(0, \infty)$; relative minimum when $x = 0$.

51. $y = x \ln x - x$. [Note: $x > 0$.]

$y' = \left[x \cdot \dfrac{1}{x} + (\ln x)(1) \right] - 1 = \ln x$

CV: $x = 1$

$$\begin{array}{c} \quad - \quad + \\ \hline 0 \quad 1 \end{array}$$

Decreasing on $(0, 1)$; increasing on $(1, \infty)$; relative minimum when $x = 1$; no relative maximum.

53. $y = x^2 - 3x - 10 = (x + 2)(x - 5)$

Intercepts $(-2, 0), (5, 0), (0, -10)$

$y' = 2x - 3$

CV: $x = \dfrac{3}{2}$

Decreasing on $\left(-\infty, \dfrac{3}{2} \right)$; increasing on

$\left(\dfrac{3}{2}, \infty \right)$; relative minimum when $x = \dfrac{3}{2}$.

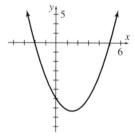

55. $y = 3x - x^3 = x\left(\sqrt{3} + x \right)\left(\sqrt{3} - x \right)$

Intercepts: $(0, 0), \left(\pm\sqrt{3}, 0 \right)$

Symmetric about origin.

$y' = 3 - 3x^2 = 3(1 + x)(1 - x)$

CV: $x = \pm 1$

Decreasing on $(-\infty, -1)$ and $(1, \infty)$; increasing on $(-1, 1)$; relative minimum when $x = -1$; relative maximum when $x = 1$.

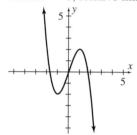

57. $y = 2x^3 - 9x^2 + 12x = x\left(2x^2 - 9x + 12 \right)$

Note that $2x^2 - 9x + 12 = 0$ has no real roots. The only intercept is $(0, 0)$.

$y' = 6x^2 - 18x + 12 = 6\left(x^2 - 3x + 2 \right)$

$= 6(x - 2)(x - 1)$

CV: $x = 1, 2$

Increasing on $(-\infty, 1)$ and $(2, \infty)$; decreasing on $(1, 2)$; relative maximum when $x = 1$; relative minimum when $x = 2$.

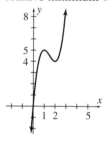

59. $y = x^4 + 4x^3 + 4x^2 = x^2(x + 2)^2$

Intercepts $(0, 0), (-2, 0)$

$y' = 4x^3 + 12x^2 + 8x = 4x(x + 1)(x + 2)$

CV: $x = 0, -1, -2$

Increasing on $(-2, -1)$ and $(0, \infty)$; decreasing on $(-\infty, -2)$ and $(-1, 0)$; relative maximum when $x = -1$; relative minima when $x = -2$ or $x = 0$.

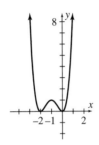

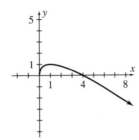

61. $y = (x-1)^2(x+2)^2$

Intercepts: $(1, 0)$, $(-2, 0)$, $(0, 4)$

$y' = (x-1)^2 \cdot 2(x+2) + (x+2)^2 \cdot 2(x-1)$
$= 2(x-1)(x+2)[(x-1) + (x+2)]$
$= 2(x-1)(x+2)(2x+1)$

CV: $x = 1, -2, -\dfrac{1}{2}$

Decreasing on $(-\infty, -2)$ and $\left(-\dfrac{1}{2}, 1\right)$;

increasing on $\left(-2, -\dfrac{1}{2}\right)$ and $(1, \infty)$; relative

minima when $x = -2$ or $x = 1$; relative

maximum when $x = -\dfrac{1}{2}$.

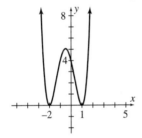

63. $y = 2\sqrt{x} - x = \sqrt{x}\left(2 - \sqrt{x}\right)$. [Note: $x \geq 0$.]

Intercepts $(0, 0)$, $(4, 0)$

$y' = \dfrac{1}{\sqrt{x}} - 1 = \dfrac{1-\sqrt{x}}{\sqrt{x}}$

CV: $x = 0, 1$

Increasing on $(0, 1)$; decreasing on $(1, \infty)$;
relative maximum when $x = 1$.

65.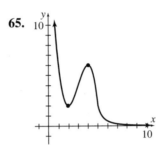

67. $c_f = 25,000$

$\overline{c}_f = \dfrac{c_f}{q} = \dfrac{25,000}{q}$

$\dfrac{d}{dq}\left(\overline{c}_f\right) = -\dfrac{25,000}{q^2} < 0$ for $q > 0$, so $\overline{c}_f$ is

a decreasing function for $q > 0$.

69. $p = 400 - 2q$
Revenue is given by
$r = pq = (400 - 2q)q$
$= 400q - 2q^2$
Marginal revenue is $r' = 400 - 4q$. Marginal
revenue is increasing when its derivative is
positive. But $(r')' = -4 < 0$. Thus marginal
revenue is never increasing.

71. $r = 240q + 57q^2 - q^3$

$r' = 240 + 114q - 3q^2 = 3(40 - q)(2 + q)$
Since $q \geq 0$, we have $q = 40$ as the only
CV. Since r is increasing on $(0, 40)$ and
decreasing on $(40, \infty)$, r is a maximum
when output is 40.

73. $E = 0.71\left(1 - \dfrac{T_c}{T_h}\right)$

$\dfrac{dE}{dT_h} = 0.71\left(\dfrac{T_c}{T_h^2}\right) > 0$, so as T_h increases, E increases.

75. $C(k) = 100\left[100 + 9k + \dfrac{144}{k}\right]$, $1 \le k \le 100$

   a. $C(1) = 25{,}300$

   b. $C'(k) = 100\left[9 - \dfrac{144}{k^2}\right]$

      $= 100\left[\dfrac{9k^2 - 144}{k^2}\right]$

      $= 100\left[\dfrac{9(k+4)(k-4)}{k^2}\right]$

      Since $k \ge 1$, the only critical value is $k = 4$. If $1 \le k < 4$, then $C'(k) < 0$ and C is decreasing. If $4 < k \le 100$, then $C'(k) > 0$ and C is increasing. Thus C has an absolute minimum for $k = 4$.

   c. $C(4) = 17{,}200$

77. Relative minimum: $(-3.83, 0.69)$

79. Relative maximum: $(2.74, 3.74)$; relative minimum: $(-2.74, -3.74)$

81. Relative minima: 0, 1.50, 2.00; relative maxima: 0.57, 1.77

83. a. $f'(x) = 4 - 6x - 3x^2$

   b.
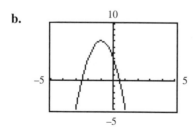

   c. $f'(x) > 0$ on $(-2.53, 0.53)$; $f'(x) < 0$ on $(-\infty, -2.53)$, $(0.53, \infty)$, f is inc. on $(-2.53, 0.53)$; f is dec. on $(-\infty, -2.53)$, $(0.53, \infty)$.

   d.
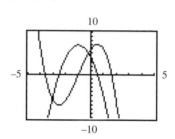

Problems 13.2

1. $f(x) = x^2 - 2x + 3$ and f is continuous over $[0, 3]$.
$f'(x) = 2x - 2 = 2(x - 1)$
The only critical value on $(0, 3)$ is $x = 1$. We evaluate f at this point and at the endpoints: $f(0) = 3$, $f(1) = 2$, and $f(3) = 6$.
Absolute maximum: $f(3) = 6$; absolute minimum: $f(1) = 2$

3. $f(x) = \dfrac{1}{3}x^3 + \dfrac{1}{2}x^2 - 2x + 1$ and f is continuous over $[-1, 0]$.
$f'(x) = x^2 + x - 2 = (x + 2)(x - 1)$
There are no critical values on $(-1, 0)$, so we only have to evaluate f at the endpoints:
$f(-1) = \dfrac{19}{6}$ and $f(0) = 1$.
Absolute maximum: $f(-1) = \dfrac{19}{6}$
Absolute minimum: $f(0) = 1$

5. $f(x) = 4x^3 + 3x^2 - 18x + 3$ and f is continuous over $\left[\dfrac{1}{2}, 3\right]$.
$f'(x) = 12x^2 + 6x - 18 = 6\left(2x^2 + x - 3\right)$
$= 6(2x + 3)(x - 1)$

The only critical value on $\left(\dfrac{1}{2}, 3\right)$ is $x = 1$.

We evaluate f at this point and the endpoints: $f\left(\dfrac{1}{2}\right) = -\dfrac{19}{4}$; $f(1) = -8$,

$f(3) = 84$.
Absolute maximum: $f(3) = 84$;
absolute minimum: $f(1) = -8$

7. $f(x) = -3x^5 + 5x^3$ and f is continuous over $[-2, 0]$.

$f'(x) = -15x^4 + 15x^2 = 15x^2\left(1 - x^2\right)$

$= 15x^2(1 + x)(1 - x)$

The only critical value on $(-2, 0)$ is $x = -1$.
We have $f(-2) = 56$, $f(-1) = -2$, and $f(0) = 0$.
Absolute maximum: $f(-2) = 56$;
absolute minimum: $f(-1) = -2$.

9. $f(x) = 3x^4 - x^6$ and f is continuous over $[-1, 2]$.

$f'(x) = 12x^3 - 6x^5 = 6x^3\left(2 - x^2\right)$

$= 6x^3\left(\sqrt{2} - x\right)\left(\sqrt{2} + x\right)$

The only critical values on $(-1, 2)$ are $x = 0$, $\sqrt{2}$. We have $f(-1) = 2$, $f(0) = 0$,

$f\left(\sqrt{2}\right) = 4$, and

$f(2) = -16$.

Absolute maximum: $f\left(\sqrt{2}\right) = 4$;

absolute minimum: $f(2) = -16$

11. $f(x) = x^4 - 9x^2 + 2$ and f is continuous over $[-1, 3]$.

$f'(x) = 4x^3 - 18x = 2x\left(2x^2 - 9\right)$

$= 2x\left(\sqrt{2}x - 3\right)\left(\sqrt{2}x + 3\right)$

The only critical values on $(-1, 3)$ are $x = 0$

and $x = \dfrac{3}{\sqrt{2}} = \dfrac{3\sqrt{2}}{2}$. We have $f(-1) = -6$,

$f(0) = 2$, $f\left(\dfrac{3\sqrt{2}}{2}\right) = -\dfrac{73}{4}$, and $f(3) = 2$.

Absolute maximum: $f(0) = f(3) = 2$;

absolute minimum: $f\left(\dfrac{3\sqrt{2}}{2}\right) = -\dfrac{73}{4}$

13. $f(x) = (x - 1)^{\frac{2}{3}}$ and f is continuous over $[-26, 28]$.

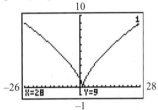

Absolute maximum: $f(-26) = f(28) = 9$;
absolute minimum: $f(1) = 0$

15.

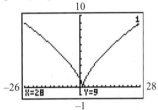

a. $-3.22, -0.78$

b. 2.75

c. 9

d. $14{,}283$

Problems 13.3

1. $f(x) = 2x^4 + 3x^3 + 2x - 3$

$f''(x) = 6x(4x + 3)$

$f''(x)$ is 0 when $x = 0$, $-\dfrac{3}{4}$. Sign chart for

f'':

+	−	+

$-\dfrac{3}{4}$ 0

Concave up on $\left(-\infty, -\dfrac{3}{4}\right)$ and $(0, \infty)$;

concave down on $\left(-\dfrac{3}{4}, 0\right)$. Inflection

points when $x = -\dfrac{3}{4}, 0.$.

3. $f(x) = \dfrac{2 + x - x^2}{x^2 - 2x + 1}$

$f''(x) = \dfrac{2(7 - x)}{(x - 1)^4}$

$f''(x)$ is 0 when $x = 7$. Although f'' is not defined when $x = 1$, f is not continuous at $x = 1$. So there is no inflection point when $x = 1$, but $x = 1$ must be considered in concavity analysis. Sign chart for f'' :

$$
\begin{array}{ccc}
+ & + & - \\
\hline
\boxed{1} & 7 &
\end{array}
$$

Concave up on $(-\infty, 1)$ and $(1, 7)$; concave down on $(7, \infty)$. Inflection point when $x = 7$.

5. $f(x) = \dfrac{x^2 + 1}{x^2 - 2}$

$f''(x) = \dfrac{6\left(3x^2 + 2\right)}{\left(x^2 - 2\right)^3} = \dfrac{6\left(3x^2 + 2\right)}{\left[\left(x - \sqrt{2}\right)\left(x + \sqrt{2}\right)\right]^3}$

$f''(x)$ is never 0. Although f'' is not defined when $x = \pm\sqrt{2}$, f is not continuous at $x = \pm\sqrt{2}$. So there is no inflection point when $x = \pm\sqrt{2}$, but $x = \pm\sqrt{2}$ must be considered in concavity analysis. Sign chart of f'' :

$$
\begin{array}{ccc}
+ & - & + \\
\hline
\boxed{-\sqrt{2}} & \boxed{\sqrt{2}} &
\end{array}
$$

Concave up on $\left(-\infty, -\sqrt{2}\right)$ and $\left(\sqrt{2}, \infty\right)$; concave down on $\left(-\sqrt{2}, \sqrt{2}\right)$. No inflection point.

7. $y = -2x^2 + 4x$

$y' = -4x + 4$

$y'' = -4 < 0$ for all x, so the graph is concave down for all x, that is, on $(-\infty, \infty)$.

9. $y = 4x^3 + 12x^2 - 12x$

$y' = 12x^2 + 24x - 12$

$y'' = 24x + 24 = 24(x + 1)$

Possible inflection point when $x = -1$.
Concave down on $(-\infty, -1)$: concave up on $(-1, \infty)$; inflection point when $x = -1$.

11. $y = 2x^3 - 5x^2 + 5x - 2$

$y' = 6x^2 - 10x + 5$

$y'' = 12x - 10 = 12\left(x - \dfrac{5}{6}\right)$

Possible inflection point when $x = \dfrac{5}{6}$.

Concave down on $\left(-\infty, \dfrac{5}{6}\right)$; concave up on

$\left(\dfrac{5}{6}, \infty\right)$; inflection point when $x = \dfrac{5}{6}$

13. $y = 2x^4 - 48x^2 + 7x + 3$

$y' = 8x^3 - 96x + 7$

$y'' = 24x^2 - 96 = 24(x^2 - 4)$
$= 24(x + 2)(x - 2)$

Possible inflection points when $x = \pm 2$.
Concave up on $(-\infty, -2)$ and $(2, \infty)$; concave down on $(-2, 2)$; inflection points when $x = \pm 2$.

15. $y = 2x^{\frac{1}{5}}$

$y' = \dfrac{2}{5}x^{-\frac{4}{5}}$

$y'' = -\dfrac{8}{25}x^{-\frac{9}{5}} = -\dfrac{8}{25x^{\frac{9}{5}}}$

y'' is not defined when $x = 0$ and y is

continuous there. Thus there is a possible inflection point when $x = 0$. Concave up on $(-\infty, 0)$; concave down on $(0, \infty)$; inflection point when $x = 0$.

17. $y = \dfrac{x^4}{2} + \dfrac{19x^3}{6} - \dfrac{7x^2}{2} + x + 5$

$y' = 2x^3 + \dfrac{19}{2}x^2 - 7x + 1$

$y'' = 6x^2 + 19x - 7 = (3x - 1)(2x + 7)$

Possible inflection points when $x = -\dfrac{7}{2}, \dfrac{1}{3}$.

Concave up on $\left(-\infty, -\dfrac{7}{2}\right)$ and $\left(\dfrac{1}{3}, \infty\right)$;

concave down on $\left(-\dfrac{7}{2}, \dfrac{1}{3}\right)$; inflection

points when $x = -\dfrac{7}{2}, \dfrac{1}{3}$.

19. $y = \dfrac{1}{20}x^5 - \dfrac{1}{4}x^4 + \dfrac{1}{6}x^3 - \dfrac{1}{2}x - \dfrac{2}{3}$

$y' = \dfrac{1}{4}x^4 - x^3 + \dfrac{1}{2}x^2 - \dfrac{1}{2}$

$y'' = x^3 - 3x^2 + x = x\left(x^2 - 3x + 1\right)$

y'' is 0 when $x = 0$ or $x^2 - 3x + 1 = 0$. Using the quadratic formula to solve

$x^2 - 3x + 1 = 0$ gives $x = \dfrac{3 \pm \sqrt{5}}{2}$. Thus possible inflection points occur when $x = 0$,

$\dfrac{3 \pm \sqrt{5}}{2}$. Concave down on $(-\infty, 0)$ and

$\left(\dfrac{3 - \sqrt{5}}{2}, \dfrac{3 + \sqrt{5}}{2}\right)$; concave up on

$\left(0, \dfrac{3 - \sqrt{5}}{2}\right)$ and $\left(\dfrac{3 + \sqrt{5}}{2}, \infty\right)$; inflection

points when $x = 0, \dfrac{3 \pm \sqrt{5}}{2}$.

21. $y = \dfrac{1}{30}x^6 - \dfrac{7}{12}x^4 + 5x^2 + 2x - 1$

$y' = \dfrac{1}{5}x^5 - \dfrac{7}{3}x^3 + 10x + 2$

$y'' = x^4 - 7x^2 + 10 = \left(x^2 - 2\right)\left(x^2 - 5\right)$

$= \left(x + \sqrt{2}\right)\left(x - \sqrt{2}\right)\left(x + \sqrt{5}\right)\left(x - \sqrt{5}\right)$

Possible inflection points when

$x = \pm\sqrt{2}, \pm\sqrt{5}$. Concave up on

$\left(-\infty, -\sqrt{5}\right), \left(-\sqrt{2}, \sqrt{2}\right)$, and $\left(\sqrt{5}, \infty\right)$;

concave down on $\left(-\sqrt{5}, -\sqrt{2}\right)$ and

$\left(\sqrt{2}, \sqrt{5}\right)$; inflection points when

$x = \pm\sqrt{5}, \pm\sqrt{2}$.

23. $y = \dfrac{x + 1}{x - 1}$

$y' = \dfrac{-2}{(x - 1)^2}$

$y'' = \dfrac{4}{(x - 1)^3}$

No possible inflection point, but we consider $x = 1$ in the concavity analysis. Concave down on $(-\infty, 1)$; concave up on $(1, \infty)$.

25. $y = \dfrac{x^2}{x^2 + 1}$

$y' = \dfrac{\left(x^2 + 1\right)(2x) - x^2(2x)}{\left(x^2 + 1\right)^2} = \dfrac{2x}{\left(x^2 + 1\right)^2}$

$$y'' = \frac{\left(x^2+1\right)^2(2) - 2x(2)\left(x^2+1\right)(2x)}{\left(x^2+1\right)^4}$$

$$= \frac{\left(x^2+1\right)(2) - 8x^2}{\left(x^2+1\right)^3}$$

$$= \frac{2\left(1-3x^2\right)}{\left(x^2+1\right)^3} = \frac{2\left(1+\sqrt{3}x\right)\left(1-\sqrt{3}x\right)}{\left(x^2+1\right)^3}$$

Possible inflection points when $x = \pm\dfrac{1}{\sqrt{3}}$. Concave down on $\left(-\infty, -\dfrac{1}{\sqrt{3}}\right)$ and $\left(\dfrac{1}{\sqrt{3}}, \infty\right)$; concave up on $\left(-\dfrac{1}{\sqrt{3}}, \dfrac{1}{\sqrt{3}}\right)$; inflection points when $x = \pm\dfrac{1}{\sqrt{3}}$.

27. $y = \dfrac{21x+40}{6(x+3)^2}$

$$y' = \frac{1}{6} \cdot \frac{(x+3)^2(21) - (21x+40)[2(x+3)]}{(x+3)^4}$$

$$= \frac{1}{6} \cdot \frac{(x+3)(21) - (21x+40)(2)}{(x+3)^3}$$

$$= \frac{1}{6} \cdot \frac{-21x-17}{(x+3)^3} = -\frac{1}{6} \cdot \frac{21x+17}{(x+3)^3}$$

$$y'' = -\frac{1}{6} \cdot \frac{(x+3)^3(21) - (21x+17)\left[3(x+3)^2\right]}{(x+3)^6}$$

$$= -\frac{1}{6} \cdot \frac{(x+3)(21) - (21x+17)(3)}{(x+3)^4}$$

$$= -\frac{1}{6} \cdot \frac{-42x+12}{(x+3)^4} = \frac{7x-2}{(x+3)^4}$$

Possible inflection point when $x = \dfrac{2}{7}$ ($x = -3$ must be considered in concavity analysis). Concave down on $(-\infty, -3)$ and $\left(-3, \dfrac{2}{7}\right)$; concave up on $\left(\dfrac{2}{7}, \infty\right)$; inflection point when $x = \dfrac{2}{7}$.

29. $y = 5e^x$

$y' = 5e^x$

$y'' = 5e^x$

Thus $y'' > 0$ for all x. Concave up on $(-\infty, \infty)$.

31. $y = 3xe^x$

$y' = 3xe^x + 3e^x = 3e^x(x+1)$

$y'' = 3e^x(1) + 3(x+1)e^x = 3e^x(x+2)$

$y'' = 0$ if $x = -2$. Concave down on $(-\infty, -2)$; concave up on $(-2, \infty)$; inflection point when $x = -2$.

33. $y = \dfrac{\ln x}{2x}$.   (Note: $x > 0$.)

$y' = \dfrac{2x \cdot \frac{1}{x} - (\ln x)(2)}{4x^2} = \dfrac{1 - \ln x}{2x^2}$

$y'' = \dfrac{2x^2\left(-\frac{1}{x}\right) - (1 - \ln x)(4x)}{4x^4}$

$\quad = \dfrac{-2x - (1 - \ln x)(4x)}{4x^4}$

$\quad = \dfrac{-1 - (1 - \ln x)(2)}{2x^3} = \dfrac{2\ln(x) - 3}{2x^3}$

y'' is 0 if $2\ln(x) - 3 = 0$, $\ln x = \dfrac{3}{2}$, $x = e^{\frac{3}{2}}$.

Concave down on $\left(0, e^{\frac{3}{2}}\right)$; concave up on $\left(e^{\frac{3}{2}}, \infty\right)$; inflection point when $x = e^{\frac{3}{2}}$.

35. $y = x^2 - x - 6 = (x-3)(x+2)$

Intercepts: $(0, -6)$, $(3, 0)$ and $(-2, 0)$

$y' = 2x - 1 = 2\left(x - \dfrac{1}{2}\right)$

CV: $x = \dfrac{1}{2}$

Decreasing on $\left(-\infty, \dfrac{1}{2}\right)$; increasing on

$\left(\dfrac{1}{2}, \infty\right)$; relative minimum at $\left(\dfrac{1}{2}, -\dfrac{25}{4}\right)$.

$y'' = 2$

No possible inflection point. Concave up on $(-\infty, \infty)$.

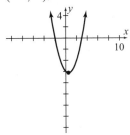

37. $y = 5x - 2x^2 = x(5 - 2x)$

Intercepts $(0, 0)$ and $\left(\dfrac{5}{2}, 0\right)$

$y' = 5 - 4x$

CV: $x = \dfrac{5}{4}$

Increasing on $\left(-\infty, \dfrac{5}{4}\right)$; decreasing

on $\left(\dfrac{5}{4}, \infty\right)$; relative maximum at $\left(\dfrac{5}{4}, \dfrac{25}{8}\right)$.

$y'' = -4$

No possible inflection point. Concave down on $(-\infty, \infty)$.

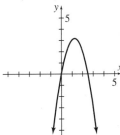

39. $y = x^3 - 9x^2 + 24x - 19$

The x-intercepts are not convenient to find; the y-intercept is $(0, -19)$.

$y' = 3x^2 - 18x + 24 = 3(x-2)(x-4)$

CV: $x = 2$, $x = 4$

Increasing on $(-\infty, 2)$ and $(4, \infty)$; decreasing

on (2, 4); relative maximum at (2, 1);
relative minimum at (4, –3).
$y'' = 6x - 18 = 6(x - 3)$
Possible inflection point when $x = 3$.
Concave down on $(-\infty, 3)$; concave up on
$(3, \infty)$; inflection point at $(3, -1)$.

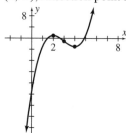

41. $y = \dfrac{x^3}{3} - 4x = \dfrac{x^3 - 12x}{3}$

$= \dfrac{1}{3}x\left(x + 2\sqrt{3}\right)\left(x - 2\sqrt{3}\right)$

Intercepts (0, 0) and $\left(\pm 2\sqrt{3},\, 0\right)$

$y' = x^2 - 4 = (x + 2)(x - 2)$
CV: $x = \pm 2$
Increasing on $(-\infty, -2)$ and $(2, \infty)$; decreasing
on $(-2, 2)$; relative maximum at $\left(-2, \dfrac{16}{3}\right)$;

relative minimum at $\left(2, -\dfrac{16}{3}\right)$.

$y'' = 2x$
Possible inflection point when $x = 0$. Concave
down on $(-\infty, 0)$; concave up on $(0, \infty)$;
inflection point at (0, 0). Symmetric about the
origin.

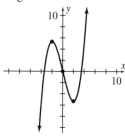

43. $y = x^3 - 3x^2 + 3x - 3$
Intercept (0, –3)
$y' = 3x^2 - 6x + 3 = 3(x - 1)^2$
CV: $x = 1$
Increasing on $(-\infty, 1)$ and $(1, \infty)$; no relative
maximum or minimum
$y'' = 6(x - 1)$
Possible inflection point when $x = 1$.
Concave down on $(-\infty, 1)$; concave up on
$(1, \infty)$; inflection point at $(1, -2)$.

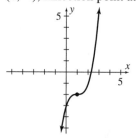

45. $y = 4x^3 - 3x^4 = x^3(4 - 3x)$

Intercepts (0, 0), $\left(\dfrac{4}{3}, 0\right)$

$y' = 12x^2 - 12x^3 = 12x^2(1 - x)$
CV: $x = 0$ and $x = 1$
Increasing on $(-\infty, 0)$ and $(0, 1)$; decreasing
on $(1, \infty)$; relative maximum at (1, 1).
$y'' = 24x - 36x^2 = 12x(2 - 3x)$
Possible inflection points at $x = 0$ and
$x = \dfrac{2}{3}$. Concave down on $(-\infty, 0)$ and
$\left(\dfrac{2}{3}, \infty\right)$; concave up on $\left(0, \dfrac{2}{3}\right)$; inflection
points at (0, 0) and $\left(\dfrac{2}{3}, \dfrac{16}{27}\right)$

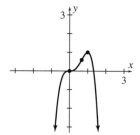

47. $y = -2 + 12x - x^3$

Intercept $(0, -2)$

$y' = 12 - 3x^2 = 3(2 + x)(2 - x)$

CV: $x = \pm 2$

Decreasing on $(-\infty, -2)$ and $(2, \infty)$; increasing on $(-2, 2)$; relative minimum at $(-2, -18)$; relative maximum at $(2, 14)$.

$y'' = -6x$

Possible inflection point when $x = 0$. Concave up on $(-\infty, 0)$; concave down on $(0, \infty)$; inflection point at $(0, -2)$.

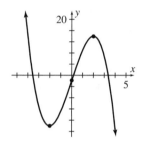

49. $y = 2x^3 - 6x^2 + 6x - 2 = 2(x-1)^3$

Intercepts $(0, -2)$, $(1, 0)$

$y' = 6(x-1)^2$

CV: $x = 1$

Increasing on $(-\infty, 1)$ and $(1, \infty)$; no relative maximum or minimum.

$y'' = 12(x-1)$

Possible inflection point when $x = 1$. Concave down on $(-\infty, 1)$; concave up on $(1, \infty)$; inflection point at $(1, 0)$.

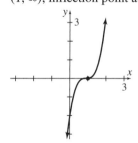

51. $y = 5x - x^5 = x\left(5 - x^4\right)$

$= x\left(\sqrt{5} + x^2\right)\left(\sqrt{5} - x^2\right)$

$= x\left(\sqrt{5} + x^2\right)\left(\sqrt[4]{5} + x\right)\left(\sqrt[4]{5} - x\right)$

Intercepts $(0, 0)$ and $\left(\pm\sqrt[4]{5}, 0\right)$

Symmetric about the origin.

$y' = 5 - 5x^4 = 5\left(1 - x^4\right) = 5\left(1 - x^2\right)\left(1 + x^2\right)$

$= 5(1-x)(1+x)\left(1 + x^2\right)$

CV: $x = \pm 1$

Decreasing on $(-\infty, -1)$ and $(1, \infty)$; increasing on $(-1, 1)$; relative minimum at $(-1, -4)$; relative maximum at $(1, 4)$.

$y'' = -20x^3$

Possible inflection point when $x = 0$. Concave up on $(-\infty, 0)$; concave down on $(0, \infty)$; inflection point at $(0, 0)$.

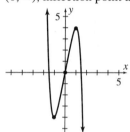

53. $y = 3x^4 - 4x^3 + 1$

Intercepts $(0, 1)$ and $(1, 0)$ [the latter is found by inspection of the equation]. No symmetry.

$y' = 12x^3 - 12x^2 = 12x^2(x-1)$

CV: $x = 0$ and $x = 1$

Decreasing on $(-\infty, 0)$ and $(0, 1)$; increasing on $(1, \infty)$; relative minimum at $(1, 0)$.

$y'' = 36x^2 - 24x = 12x(3x - 2)$

Possible inflection points at $x = 0$ and $x = \dfrac{2}{3}$. Concave up on $(-\infty, 0)$ and $\left(\dfrac{2}{3}, \infty\right)$;

concave down on $\left(0, \dfrac{2}{3}\right)$; inflection points at $(0, 1)$ and $\left(\dfrac{2}{3}, \dfrac{11}{27}\right)$.

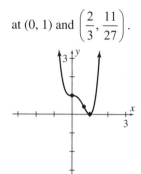

55. $y = 4x^2 - x^4 = x^2(2 + x)(2 - x)$

Intercepts $(0, 0)$ and $(\pm 2, 0)$

Symmetric about the y-axis.

$y' = 8x - 4x^3 = 4x\left(2 - x^2\right)$

$\quad = 4x\left(\sqrt{2} + x\right)\left(\sqrt{2} - x\right)$

CV: $x = 0,\ \pm\sqrt{2}$

Increasing on $\left(-\infty, -\sqrt{2}\right)$ and $\left(0, \sqrt{2}\right)$;

decreasing on $\left(-\sqrt{2}, 0\right)$ and $\left(\sqrt{2}, \infty\right)$;

relative maxima at $\left(\pm\sqrt{2}, 4\right)$; relative

minimum at $(0, 0)$.

$y'' = 8 - 12x^2 = 12\left[\dfrac{2}{3} - x^2\right]$

$\quad = 12\left(\sqrt{\dfrac{2}{3}} - x\right)\left(\sqrt{\dfrac{2}{3}} + x\right)$

Possible inflection points when $x = \pm\sqrt{\dfrac{2}{3}}$.

Concave down on $\left(-\infty, -\sqrt{\dfrac{2}{3}}\right)$ and

$\left(\sqrt{\dfrac{2}{3}}, \infty\right)$; concave up on $\left(-\sqrt{\dfrac{2}{3}}, \sqrt{\dfrac{2}{3}}\right)$;

inflection points at $\left(\pm\sqrt{\dfrac{2}{3}}, \dfrac{20}{9}\right)$.

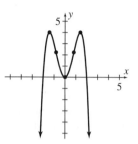

57. $y = x^{1/3}(x - 8) = x^{4/3} - 8x^{1/3}$

Intercepts $(0, 0)$ and $(8, 0)$

$y' = \dfrac{4}{3}x^{1/3} - \dfrac{8}{3}x^{-2/3}$

$\quad = \dfrac{4}{3}\left[x^{1/3} - \dfrac{2}{x^{2/3}}\right] = \dfrac{4(x - 2)}{3x^{2/3}}$

CV: $x = 0, 2$

Decreasing on $(-\infty, 0)$ and $(0, 2)$; increasing

on $(2, \infty)$; relative minimum at

$\left(2, -6\sqrt[3]{2}\right) \approx (2, -7.56)$.

$y'' = \dfrac{4}{9}x^{-2/3} + \dfrac{16}{9}x^{-5/3}$

$\quad = \dfrac{4}{9}\left[\dfrac{1}{x^{2/3}} + \dfrac{4}{x^{5/3}}\right] = \dfrac{4(x + 4)}{9x^{5/3}}$

Possible inflection points when $x = -4, 0$.

Concave up on $(-\infty, -4)$ and $(0, \infty)$; concave

down on $(-4, 0)$; inflection points at

$\left(-4, 12\sqrt[3]{4}\right)$ and $(0, 0)$. Observe that at the

origin the tangent line exists but it is

vertical.

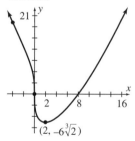

59. $y = 4x^{1/3} + x^{4/3} = x^{1/3}(4 + x)$

Intercepts $(0, 0)$ and $(-4, 0)$

$y' = \dfrac{4}{3}x^{-2/3} + \dfrac{4}{3}x^{1/3} = \dfrac{4}{3}\left[\dfrac{1}{x^{2/3}} + x^{1/3}\right]$

$= \dfrac{4(1 + x)}{3x^{2/3}}$

CV: $x = 0, -1$

Decreasing on $(-\infty, -1)$; increasing on $(-1, 0)$ and $(0, \infty)$; rel. min at $(-1, -3)$

$y'' = -\dfrac{8}{9}x^{-5/3} + \dfrac{4}{9}x^{-2/3} = \dfrac{4}{9}\left[\dfrac{1}{x^{2/3}} - \dfrac{2}{x^{5/3}}\right]$

$= \dfrac{4(x - 2)}{9x^{5/3}}$

Possible inflection points when $x = 0, 2$.
Concave up on $(-\infty, 0)$ and $(2, \infty)$; concave down on $(0, 2)$; inflection point at $(0, 0)$ and $\left(2, 6\sqrt[3]{2}\right)$. Observe that at the origin the tangent line exists but it is vertical.

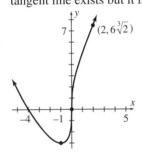

61. $y = 6x^{2/3} - \dfrac{x}{2} = 6x^{2/3}\left(1 - \dfrac{x^{1/3}}{12}\right)$

Intercepts $(0, 0)$ and $(1728, 0)$

$y' = 4x^{-1/3} - \dfrac{1}{2} = \dfrac{1}{2}\left(\dfrac{8}{\sqrt[3]{x}} - 1\right) = \dfrac{1}{2}\left(\dfrac{8 - \sqrt[3]{x}}{\sqrt[3]{x}}\right)$

CV: $x = 0, 512$

Increasing on $(0, 512)$; decreasing on $(-\infty, 0)$ and $(512, \infty)$; relative maximum at $(512, 128)$; relative minimum at $(0, 0)$.

$y'' = -\dfrac{4}{3}x^{-4/3} = -\dfrac{4}{3x^{4/3}}$

Possible inflection point at $x = 0$. Concave down on $(-\infty, 0)$ and $(0, \infty)$. Observe that at the origin the tangent line exists but it is

vertical.

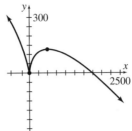

63.

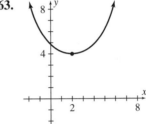

65.

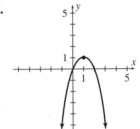

67. $p = \dfrac{100}{q + 2}$

$\dfrac{dp}{dq} = -\dfrac{100}{(q + 2)^2} < 0$ for $q > 0$, so p is

decreasing. Since $\dfrac{d^2 p}{dq^2} = \dfrac{200}{(q + 2)^3} > 0$ for

$q > 0$, the demand curve is concave up.

69. $S = f(A) = 12\sqrt[4]{A}$, $0 \le A \le 625$. For the

given values of A we have $S' = 3A^{-\frac{3}{4}} > 0$

and $S'' = -\left(\dfrac{9}{4}\right)A^{-\frac{7}{4}} < 0$. Thus y is

increasing and concave down.

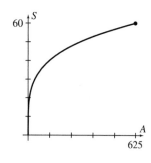

71. $y = 12.5 + 5.8(0.42)^x$

$y' = 5.8(0.42)^x \ln(0.42)$

Since $\ln(0.42) < 0$, we have $y' < 0$, so the function is decreasing.

$y'' = 5.8(0.42)^x \ln^2(0.42) > 0$, so the function is concave up.

73. $n = f(r) = 0.1\ln(r) + \dfrac{7}{r} - 0.8$, $1 \le r \le 10$

a. $\dfrac{dn}{dr} = \dfrac{0.1}{r} - \dfrac{7}{r^2} = \dfrac{0.1r - 7}{r^2}$

$= \dfrac{0.1(r - 70)}{r^2} < 0$

for $1 \le r \le 10$. Thus the graph of f is always falling. Also,

$\dfrac{d^2 n}{dr^2} = -\dfrac{0.1}{r^2} + \dfrac{14}{r^3} = \dfrac{14 - 0.1r}{r^3}$

$= \dfrac{0.1(140 - r)}{r^3} > 0$

for $1 \le r \le 10$. Thus the graph is concave up.

b.

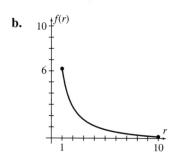

c. $\left.\dfrac{dn}{dr}\right|_{r=5} = -0.26$, so the rate of decrease is 0.26.

75.

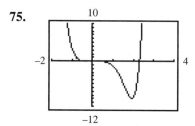

Two inflection points

$y = x^5(x - a) = x^6 - ax^5$

$y' = 6x^5 - 5ax^4$

$y'' = 30x^4 - 20ax^3 = 10x^3(3x - 2a)$

Possible inflection points when $x = 0$ and $x = \dfrac{2a}{3}$. If $a > 0$, y is concave up on $(-\infty, 0)$ and $\left(\dfrac{2a}{3}, \infty\right)$; concave down on $\left(0, \dfrac{2a}{3}\right)$.

If $a < 0$, y is concave up on $\left(-\infty, \dfrac{2a}{3}\right)$ and $(0, \infty)$; concave down on $\left(\dfrac{2a}{3}, 0\right)$. In either case, y has two points of inflection, when $x = 0$ and $x = \dfrac{2a}{3}$.

77. $y = x^3 - 2x^2 + x + 3$

$y' = 3x^2 - 4x + 1$

When $x = 2$, then $y = 5$ and $y' = 5$. Thus an equation of the tangent line at $x = 2$ is $y - 5 = 5(x - 2)$, or $y = 5x - 5$. Graphing the curve and the tangent line indicates that the curve lies above the tangent line around $x = 2$. Thus the curve is concave up at $x = 2$.

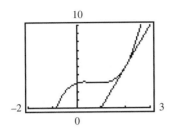

79. $f(x) = x^6 + 3x^5 - 4x^4 + 2x^2 + 1$

$f'(x) = 6x^5 + 15x^4 - 16x^3 + 4x$

$f''(x) = 30x^4 + 60x^3 - 48x^2 + 4$

Inflection points of f when $x \approx -2.61, -0.26$.

Problems 13.4

1. $y = x^2 - 5x + 6$

$y' = 2x - 5$

CV: $x = \dfrac{5}{2}$

$y'' = 2$

$y''\left(\dfrac{5}{2}\right) = 2 > 0$

Thus there is a relative minimum when

$x = \dfrac{5}{2}$. Because there is only one relative

extremum and f is continuous, the relative minimum is an absolute minimum.

3. $y = -4x^2 + 2x - 8$

$y' = -8x + 2$

CV: $x = \dfrac{1}{4}$

$y'' = -8$

$y''\left(\dfrac{1}{4}\right) = -8 < 0$

Thus there is a relative maximum when

$x = \dfrac{1}{4}$. Because there is only one relative

extremum and f is continuous, the relative maximum is an absolute maximum.

5. $y = \dfrac{1}{3}x^3 + 2x^2 - 5x + 1$

$y' = x^2 + 4x - 5 = (x + 5)(x - 1)$

CV: $x = -5,\ 1$

$y'' = 2x + 4$

$y''(-5) = -6 < 0 \Rightarrow$ relative maximum when

$x = -5$

$y''(1) = 6 > 0 \Rightarrow$ relative minimum when

$x = 1$

7. $y = -x^3 + 3x^2 + 1$

$y' = -3x^2 + 6x = -3x(x - 2)$

CV: $x = 0,\ 2$

$y'' = -6x + 6$

$y''(0) = 6 > 0 \Rightarrow$ relative minimum when

$x = 0$

$y''(2) = -6 < 0 \Rightarrow$ relative maximum when

$x = 2$

9. $y = 7 - 2x^4$

$y' = -8x^3$

CV: $x = 0$

$y'' = -24x^2$

Since $y''(0) = 0$, the second-derivative test fails. Using the first-derivative test, we see that f increases for $x < 0$ and f decreases for $x > 0$, so there is a relative maximum when $x = 0$.

11. $y = 81x^5 - 5x$

$y' = 81 \cdot 5x^4 - 5 = 5\left(81x^4 - 1\right)$

$= 5\left(9x^2 - 1\right)\left(9x^2 + 1\right)$

$= 5(3x + 1)(3x - 1)\left(9x^2 + 1\right)$

CV: $x = \pm\dfrac{1}{3}$

$y'' = 81 \cdot 5 \cdot 4x^3$

$y''\left(-\dfrac{1}{3}\right) = -60 < 0 \Rightarrow$ relative maximum when $x = -\dfrac{1}{3}$

$y''\left(\dfrac{1}{3}\right) = 60 > 0 \Rightarrow$ relative minimum when

$x = \dfrac{1}{3}$

13. $y = \left(x^2 + 7x + 10\right)^2$

$y' = 2\left(x^2 + 7x + 10\right)(2x + 7)$

$= 2(x + 2)(x + 5)(2x + 7)$

CV: $x = -2, -5, -\dfrac{7}{2}$

$y'' = 2\left[\left(x^2 + 7x + 10\right)(2) + (2x + 7)(2x + 7)\right]$

$y''(-5) = 18 > 0 \Rightarrow$ relative minimum when

$x = -5$

$y''\left(-\dfrac{7}{2}\right) = -9 < 0 \Rightarrow$ relative maximum

when $x = -\dfrac{7}{2}$

$y''(-2) = 18 > 0 \Rightarrow$ relative minimum when

$x = -2$

Problems 13.5

1. $y = f(x) = \dfrac{x}{x - 1}$

When $x = 1$ the denominator is zero but the numerator is not zero. Thus $x = 1$ is a vertical asymptote.

$\lim\limits_{x \to \infty} \dfrac{x}{x - 1} = \lim\limits_{x \to \infty} \dfrac{x}{x} = \lim\limits_{x \to \infty} 1 = 1$.

Similarly $\lim\limits_{x \to -\infty} f(x) = 1$. Thus the line

$y = 1$ is a horizontal asymptote.

3. $f(x) = \dfrac{x + 2}{3x - 5}$

When $x = \dfrac{5}{3}$ the denominator is zero but the

numerator is not. Thus $x = \dfrac{5}{3}$ is a vertical asymptote.

$\lim\limits_{x \to \infty} f(x) = \lim\limits_{x \to \infty} \dfrac{x}{3x} = \lim\limits_{x \to \infty} \dfrac{1}{3} = \dfrac{1}{3}$.

Similarly $\lim\limits_{x \to -\infty} f(x) = \dfrac{1}{3}$. Thus $y = \dfrac{1}{3}$ is a horizontal asymptote.

5. $y = f(x) = \dfrac{4}{x}$

When $x = 0$ the denominator is zero but the numerator is not zero, so $x = 0$ is a vertical asymptote.

$\lim\limits_{x \to \infty} \left(\dfrac{4}{x}\right) = 0$. Similarly, $\lim\limits_{x \to -\infty} \left(\dfrac{4}{x}\right) = 0$, so

$y = 0$ is a horizontal asymptote.

7. $y = f(x) = \dfrac{1}{x^2 - 1} = \dfrac{1}{(x - 1)(x + 1)}$

Vertical asymptotes are $x = 1$ and $x = -1$.

$\lim\limits_{x \to \infty} \dfrac{1}{x^2 - 1} = \lim\limits_{x \to \infty} \dfrac{1}{x^2} = 0$. Similarly,

$\lim\limits_{x \to -\infty} f(x) = 0$. Thus $y = 0$ is a horizontal asymptote.

9. $y = f(x) = x^2 - 5x + 5$ is a polynomial function, so there are no horizontal or vertical asymptotes.

11. $f(x) = \dfrac{2x^2}{x^2 + x - 6} = \dfrac{2x^2}{(x + 3)(x - 2)}$

Vertical asymptotes are $x = -3$ and $x = 2$.

$\lim\limits_{x \to \infty} f(x) = \lim\limits_{x \to \infty} \dfrac{2x^2}{x^2} = \lim\limits_{x \to \infty} 2 = 2$, and

$\lim\limits_{x \to -\infty} f(x) = 2$. Thus $y = 2$ is a horizontal asymptote.

13. $y = \dfrac{2x^2 + 3x + 1}{x^2 - 5} = \dfrac{2x^2 + 3x + 1}{\left(x - \sqrt{5}\right)\left(x + \sqrt{5}\right)}$

Vertical asymptotes are $x = -\sqrt{5}$ and $x = \sqrt{5}$.

$\displaystyle\lim_{x \to \infty} f(x) = \lim_{x \to \infty} \frac{2x^2}{x^2} = \lim_{x \to \infty} 2 = 2$

Similarly, $\displaystyle\lim_{x \to -\infty} = 2$. Thus $y = 2$ is a horizontal asymptote.

15. $y = f(x) = \dfrac{2}{x - 3} + 5 = \dfrac{5x - 13}{x - 3}$

From the denominator, $x = 3$ is a vertical asymptote.

$\displaystyle\lim_{x \to \infty} f(x) = \lim_{x \to \infty} \frac{5x}{x} = \lim_{x \to \infty} 5 = 5,$ and

$\displaystyle\lim_{x \to -\infty} f(x) = 5.$ Thus, $y = 5$ is a horizontal asymptote.

17. $f(x) = \dfrac{3 - x^4}{x^3 + x^2} = \dfrac{3 - x^4}{x^2(x + 1)}$

Vertical asymptotes are $x = 0$ and $x = -1$.

$\dfrac{3 - x^4}{x^3 + x^2} = -x + 1 + \dfrac{3 - x^2}{x^3 + x^2}$ so the line

$y = -x + 1$ is an oblique asymptote.

19. $y = f(x) = \dfrac{x^2 - 3x - 4}{1 + 4x + 4x^2} = \dfrac{x^2 - 3x - 4}{(1 + 2x)^2}$

From the denominator, $x = -\dfrac{1}{2}$ is a vertical asymptote.

$\displaystyle\lim_{x \to \infty} f(x) = \lim_{x \to \infty} \frac{x^2}{4x^2} = \lim_{x \to \infty} \frac{1}{4} = \frac{1}{4},$ and

$\displaystyle\lim_{x \to -\infty} f(x) = \frac{1}{4},$ so $y = \dfrac{1}{4}$ is a horizontal asymptote.

21. $y = f(x) = \dfrac{9x^2 - 16}{2(3x + 4)^2} = \dfrac{(3x + 4)(3x - 4)}{2(3x + 4)^2}$

When $x = -\dfrac{4}{3}$, both the numerator and denominator are zero. Since

$\displaystyle\lim_{x \to -4/3^+} f(x) = \lim_{x \to -4/3^+} \frac{3x - 4}{2(3x + 4)} = -\infty,$

the line $x = -\dfrac{4}{3}$ is a vertical asymptote.

$\displaystyle\lim_{x \to \infty} \frac{9x^2 - 16}{2(3x + 4)^2} = \lim_{x \to \infty} \frac{9x^2}{18x^2} = \lim_{x \to \infty} \frac{1}{2} = \frac{1}{2}.$

Similarly, $\displaystyle\lim_{x \to -\infty} f(x) = \frac{1}{2}.$ Thus $y = \dfrac{1}{2}$ is a horizontal asymptote.

23. $y = f(x) = 2e^{x+2} + 4$

We have $\displaystyle\lim_{x \to \infty} f(x) = +\infty$ and

$\displaystyle\lim_{x \to -\infty} f(x) = 2 \cdot \lim_{x \to -\infty} e^x + \lim_{x \to -\infty} 4$
$= 2(0) + 4 = 4$

Thus $y = 4$ is a horizontal asymptote. There is no vertical asymptote because $f(x)$ neither increases nor decreases without bound around any fixed value of x.

25. $y = \dfrac{3}{x}$

Symmetric about the origin. Vertical asymptote is $x = 0$. $\displaystyle\lim_{x \to \infty} \frac{3}{x} = 0 = \lim_{x \to -\infty} \frac{3}{x},$ so $y = 0$ is a horizontal asymptote.

$y' = -\dfrac{3}{x^2}$

CV: None, however $x = 0$ must be included in the inc.-dec. analysis. Decreasing on $(-\infty, 0)$ and $(0, \infty)$.

$y'' = \dfrac{6}{x^3}$

No possible inflection point, but we include $x = 0$ in the concavity analysis. Concave down on $(-\infty, 0)$; concave up on $(0, \infty)$.

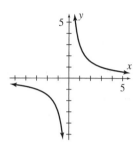

27. $y = \dfrac{x}{x-1}$

Intercept $(0, 0)$

Vertical asymptote is $x = 1$

$\displaystyle\lim_{x \to \infty} y = 1 = \lim_{x \to -\infty} y$, so $y = 1$ is a horizontal asymptote.

$y' = \dfrac{(x-1)(1) - x(1)}{(x-1)^2} = -\dfrac{1}{(x-1)^2}$

CV: None, but $x = 1$ must be included in the inc.-dec. analysis. Decreasing on $(-\infty, 1)$ and $(1, \infty)$.

$y'' = \dfrac{2}{(x-1)^3}$

No possible inflection point, but $x = 1$ must be included in concavity analysis. Concave up on $(1, \infty)$, concave down on $(-\infty, 1)$.

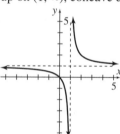

29. $y = x^2 + \dfrac{1}{x^2} = \dfrac{x^4 + 1}{x^2}$

$x \neq 0$, so there is no y-intercept. Setting $y = 0 \Rightarrow$ no x-intercept. Replacing x by $-x$ yields symmetry about the y-axis. Setting $x^2 = 0$ gives $x = 0$ as the only vertical asymptote. Because the degree of the numerator is greater than the degree of the denominator, no horizontal asymptote

exists.

$y = x^2 + x^{-2}$

$y' = 2x - 2x^{-3} = 2x - \dfrac{2}{x^3} = \dfrac{2x^4 - 2}{x^3} = \dfrac{2\left(x^4 - 1\right)}{x^3}$

$= \dfrac{2\left(x^2 + 1\right)(x+1)(x-1)}{x^3}.$

CV: $x = \pm 1$, but $x = 0$ must be included in the inc.-dec. analysis. Decreasing on $(-\infty, -1)$ and $(0, 1)$; increasing on $(-1, 0)$ and $(1, \infty)$; relative minima at $(-1, 2)$ and $(1, 2)$,

$y'' = 2 + \dfrac{6}{x^4} > 0$

for all $x \neq 0$. Concave up on $(-\infty, 0)$ and $(0, \infty)$.

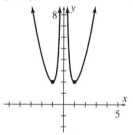

31. $y = \dfrac{1}{x^2 - 1} = \dfrac{1}{(x+1)(x-1)}$

Intercept $(0, -1)$

Symmetric about the y-axis.

Vertical asymptotes are $x = -1$ and $x = 1$.

$\displaystyle\lim_{x \to \infty} \dfrac{1}{x^2 - 1} = 0 = \lim_{x \to -\infty} \dfrac{1}{x^2 - 1}$, so $y = 0$ is a horizontal asymptote.

$y' = -\dfrac{2x}{\left(x^2 - 1\right)^2}$

CV: $x = 0$, but $x = \pm 1$ must be included in the inc.-dec. analysis. Increasing on $(-\infty, -1)$ and $(-1, 0)$; decreasing on $(0, 1)$ and $(1, \infty)$; relative maximum at $(0, -1)$.

$$y'' = -2 \cdot \frac{\left(x^2-1\right)^2 (1) - x\left[4x\left(x^2-1\right)\right]}{\left(x^2-1\right)^4}$$

$$= -2 \cdot \frac{\left(x^2-1\right)\left[\left(x^2-1\right)-4x^2\right]}{\left(x^2-1\right)^4}$$

$$= \frac{2\left(3x^2+1\right)}{\left(x^2-1\right)^3} = \frac{2\left(3x^2+1\right)}{[(x+1)(x-1)]^3}$$

No possible inflection point, but $x = \pm 1$ must be considered in the concavity analysis. Concave up on $(-\infty, -1)$ and $(1, \infty)$; concave down on $(-1, 1)$.

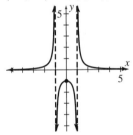

33. $y = \dfrac{1+x}{1-x}$

Intercepts: $(0, 1)$ and $(-1, 0)$.

$x = 1$ is the only vertical asymptote. Since
$$\lim_{x \to \infty} \frac{1+x}{1-x} = \lim_{x \to \infty} \frac{x}{-x} = \lim_{x \to \infty} -1 = -1$$
$$= \lim_{x \to -\infty} \frac{1+x}{1-x}$$

the only horizontal asymptote is $y = -1$.

$$y' = \frac{(1-x)(1) - (1+x)(-1)}{(1-x)^2} = \frac{2}{(1-x)^2}$$

No critical values, but $x = 1$ must be considered in the ind.-dec. analysis. Increasing on $(-\infty, 1)$ and $(1, \infty)$.

$$y'' = \frac{4}{(1-x)^3}$$

No possible inflection point, but $x = 1$ must be included in the concavity analysis. Concave up on $(-\infty, 1)$; concave down on $(1, \infty)$.

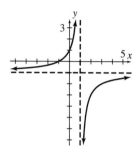

35. $y = \dfrac{x^2}{7x+4}$

Intercept: (0, 0)

Vertical asymptote is $x = -\dfrac{4}{7}$. $\dfrac{x^2}{7x+4} = \dfrac{1}{7}x - \dfrac{4}{49} + \dfrac{16}{49(7x+4)}$ so $y = \dfrac{1}{7}x - \dfrac{4}{49}$ is an oblique asymptote.

$$y' = \dfrac{(7x+4)(2x) - x^2(7)}{(7x+4)^2}$$

$$= \dfrac{7x^2 + 8x}{(7x+4)^2} = \dfrac{x(7x+8)}{(7x+4)^2}$$

CV: $x = 0, -\dfrac{8}{7}$, but $x = -\dfrac{4}{7}$ must be included in the inc.-dec. analysis. Increasing on $\left(-\infty, -\dfrac{8}{7}\right)$ and

$(0, \infty)$; decreasing on $\left(-\dfrac{8}{7}, -\dfrac{4}{7}\right)$ and $\left(-\dfrac{4}{7}, 0\right)$; relative maximum at $\left(-\dfrac{8}{7}, -\dfrac{16}{49}\right)$; relative minimum

at (0, 0).

$$y'' = \dfrac{\left(7x^2+4\right)^2 (14x+8) - \left(7x^2+8x\right)[14(7x+4)]}{(7x+4)^4}$$

$$= \dfrac{(7x+4)\left[(7x+4)(14x+8) - 14\left(7x^2+8x\right)\right]}{(7x+4)^4}$$

$$= \dfrac{32}{(7x+4)^3}$$

No possible inflection point but $x = -\dfrac{4}{7}$ must be included in concavity analysis. Concave down on

$\left(-\infty, -\dfrac{4}{7}\right)$; concave up on $\left(-\dfrac{4}{7}, \infty\right)$.

37. $y = \dfrac{9}{9x^2 - 6x - 8} = \dfrac{9}{(3x+2)(3x-4)}$

Intercept: $\left(0, -\dfrac{9}{8}\right)$

Vertical asymptotes: $x = -\dfrac{2}{3}$, $x = \dfrac{4}{3}$

$\displaystyle \lim_{x \to \infty} y = \lim_{x \to \infty} \dfrac{9}{9x^2} = \lim_{x \to \infty} \dfrac{1}{x^2} = 0 = \lim_{x \to -\infty} y$

Thus $y = 0$ is a horizontal asymptote. Since $y = 9\left(9x^2 - 6x - 8\right)^{-1}$,

$y' = 9(-1)\left(9x^2 - 6x - 8\right)^{-2}(18x - 6)$

$= -\dfrac{54(3x - 1)}{[(3x+2)(3x-4)]^2}$

CV: $x = \dfrac{1}{3}$, but $x = -\dfrac{2}{3}$ and $x = \dfrac{4}{3}$ must be included in inc.-dec. analysis.

Increasing on $\left(-\infty, -\dfrac{2}{3}\right)$ and $\left(-\dfrac{2}{3}, \dfrac{1}{3}\right)$; decreasing on $\left(\dfrac{1}{3}, \dfrac{4}{3}\right)$ and $\left(\dfrac{4}{3}, \infty\right)$;

relative maximum at $\left(\dfrac{1}{3}, -1\right)$. Finding y'' gives:

$y'' = -54 \cdot \dfrac{\left(9x^2 - 6x - 8\right)^2(3) - (3x-1)\left[2\left(9x^2 - 6x - 8\right)(18x - 6)\right]}{\left(9x^2 - 6x - 8\right)^4}$

$= -54 \cdot \dfrac{3\left(9x^2 - 6x - 8\right)\left[\left(9x^2 - 6x - 8\right) - 4(3x-1)(3x-1)\right]}{\left(9x^2 - 6x - 8\right)^4}$

$= \dfrac{-162\left(-27x^2 + 18x - 12\right)}{\left(9x^2 - 6x - 8\right)^3} = \dfrac{486\left(9x^2 - 6x + 4\right)}{[(3x+2)(3x-4)]^3}$

Since $9x^2 - 6x + 4 = 0$ has no real roots, y'' is never zero. No possible inflection points,

but $x = -\dfrac{2}{3}$ and $x = \dfrac{4}{3}$ must be included in concavity analysis. Concave up on $\left(-\infty, -\dfrac{2}{3}\right)$ and $\left(\dfrac{4}{3}, \infty\right)$;

concave down on $\left(-\dfrac{2}{3}, \dfrac{4}{3}\right)$.

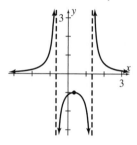

39. $y = \dfrac{3x+1}{(3x-2)^2}$

Intercepts: $\left(-\dfrac{1}{3}, 0\right), \left(0, \dfrac{1}{4}\right)$

Vertical asymptote is $x = \dfrac{2}{3}$.

$$\lim_{x \to \infty} y = \lim_{x \to \infty} \frac{3x}{9x^2} = \lim_{x \to \infty} \frac{1}{3x} = 0 = \lim_{x \to -\infty} y$$

Thus $y = 0$ is a horizontal asymptote.

$$\begin{aligned}
y' &= \frac{(3x-2)^2(3) - (3x+1)(2)(3x-2)(3)}{(3x-2)^4} \\
&= \frac{3(3x-2)[(3x-2) - 2(3x+1)]}{(3x-2)^4} \\
&= -\frac{3(3x+4)}{(3x-2)^3}
\end{aligned}$$

CV: $x = -\dfrac{4}{3}$, but $x = \dfrac{2}{3}$ must be included in inc.-dec. analysis.

Decreasing on $\left(-\infty, -\dfrac{4}{3}\right)$ and $\left(\dfrac{2}{3}, \infty\right)$; increasing on $\left(-\dfrac{4}{3}, \dfrac{2}{3}\right)$; relative minimum at $\left(-\dfrac{4}{3}, -\dfrac{1}{12}\right)$.

$$y'' = -3 \cdot \frac{(3x-2)^3(3)-(3x+4)(3)(3x-2)^2(3)}{(3x-2)^6}$$

$$= -3 \cdot \frac{3(3x-2)^2[(3x-2)-3(3x+4)]}{(3x-2)^6}$$

$$= -3 \cdot \frac{3(-6x-14)}{(3x-2)^4} = \frac{18(3x+7)}{(3x-2)^4}$$

Possible inflection point when $x = -\dfrac{7}{3}$, but $x = \dfrac{2}{3}$ must be included in concavity analysis.

Concave down on $\left(-\infty, -\dfrac{7}{3}\right)$; concave up on $\left(-\dfrac{7}{3}, \dfrac{2}{3}\right)$ and $\left(\dfrac{2}{3}, \infty\right)$; inflection point at $\left(-\dfrac{7}{3}, -\dfrac{2}{27}\right)$.

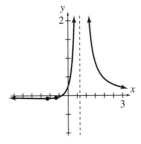

41. $y = \dfrac{x^2-1}{x^3} = \dfrac{(x+1)(x-1)}{x^3}$

Intercepts are $(-1, 0)$ and $(1, 0)$.
Symmetric about the origin.
Vertical asymptote $x = 0$.

$$\lim_{x\to\infty} \frac{x^2-1}{x^3} = \lim_{x\to\infty}\frac{x^2}{x^3} = \lim_{x\to\infty}\frac{1}{x}$$

$$= 0 = \lim_{x\to-\infty} \frac{1-x}{x^2}, \text{ so } y = 0 \text{ is the only horizontal asymptote. Since } y = x^{-1} - x^{-3}, \text{ then}$$

$$y' = -x^{-2} + 3x^{-4} = x^{-4}\left(-x^2 + 3\right) = \frac{3-x^2}{x^4}$$

CV: $x = \pm\sqrt{3}$, but $x = 0$ must be included in the inc.-dec. analysis. Increasing on $\left(-\sqrt{3}, 0\right)$ and

$\left(0, \sqrt{3}\right)$; decreasing on $\left(-\infty, -\sqrt{3}\right)$ and $\left(\sqrt{3}, \infty\right)$; relative maximum at $\left(\sqrt{3}, \dfrac{2\sqrt{3}}{9}\right)$; relative minimum

at $\left(-\sqrt{3}, -\dfrac{2\sqrt{3}}{9}\right)$.

$$y'' = 2x^{-3} - 12x^{-5} = 2x^{-5}\left(x^2 - 6\right) = \frac{2\left(x^2-6\right)}{x^5}$$

Possible inflection points when $x = \pm\sqrt{6}$, but $x = 0$ must be included in the concavity analysis. Concave down on $\left(-\infty, -\sqrt{6}\right)$ and $\left(0, \sqrt{6}\right)$; concave up on $\left(-\sqrt{6}, 0\right)$ and $\left(\sqrt{6}, \infty\right)$; inflection points at $\left(\sqrt{6}, \dfrac{5\sqrt{6}}{36}\right)$ and $\left(-\sqrt{6}, \dfrac{-5\sqrt{6}}{36}\right)$.

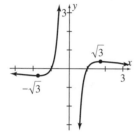

43. $y = x + \dfrac{1}{x+1} = \dfrac{x^2 + x + 1}{x+1}$

Intercept: $(0, 1)$. $x = -1$ is the only vertical asymptote. $y = x$ is an oblique asymptote.

$$y' = \dfrac{(x+1)(2x+1) - \left(x^2 + x + 1\right)}{(x+1)^2}$$

$$= \dfrac{x^2 + 2x}{(x+1)^2} = \dfrac{x(x+2)}{(x+1)^2}$$

CV: 0 and -2, but $x = -1$ must be included in the inc.-dec. analysis. Increasing on $(-\infty, -2)$ and $(0, \infty)$; decreasing on $(-2, -1)$ and $(-1, 0)$; relative maximum at $(-2, -3)$; relative minimum at $(0, 1)$.

$$y'' = \dfrac{(x+1)^2(2x+2) - \left(x^2 + 2x\right)[2(x+1)]}{(x+1)^4}$$

$$= \dfrac{(x+1)(2x+2) - \left(x^2 + 2x\right)[2]}{(x+1)^3} = \dfrac{2}{(x+1)^3}$$

No possible inflection point, but $x = -1$ must be included in the concavity analysis. Concave down on $(-\infty, -1)$; concave up on $(-1, \infty)$.

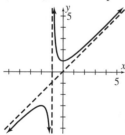

45. $y = \dfrac{-3x^2 + 2x - 5}{3x^2 - 2x - 1} = \dfrac{-3x^2 + 2x - 5}{(3x+1)(x-1)}$

Note that $-3x^2 + 2x - 5$ is never zero.
Intercept: $(0, 5)$

Vertical asymptotes are $x = -\dfrac{1}{3}$ and $x = 1$.

$\lim\limits_{x \to \infty} y = \lim\limits_{x \to \infty} \dfrac{-3x^2}{3x^2} = \lim\limits_{x \to \infty} -1 = -1 = \lim\limits_{x \to -\infty} y$

Thus $y = -1$ is horizontal asymptote.

$y' = \dfrac{\left(3x^2 - 2x - 1\right)(-6x+2) - \left(-3x^2 + 2x - 5\right)(6x-2)}{\left(3x^2 - 2x - 1\right)^2}$

$= \dfrac{2(3x-1)\left[\left(3x^2 - 2x - 1\right)(-1) - \left(-3x^2 + 2x - 5\right)\right]}{\left(3x^2 - 2x - 1\right)^2}$

$= \dfrac{12(3x-1)}{\left(3x^2 - 2x - 1\right)^2} = \dfrac{12(3x-1)}{[(3x+1)(x-1)]^2}$

CV: $x = \dfrac{1}{3}$, but $x = -\dfrac{1}{3}$ and $x = 1$ must be included in inc.-dec. analysis.

Decreasing on $\left(-\infty, -\dfrac{1}{3}\right)$ and $\left(-\dfrac{1}{3}, \dfrac{1}{3}\right)$; increasing on $\left(\dfrac{1}{3}, 1\right)$ and $(1, \infty)$; relative minimum at

$\left(\dfrac{1}{3}, \dfrac{7}{2}\right)$.

$y'' = 12 \cdot \dfrac{\left(3x^2 - 2x - 1\right)^2 (3) - (3x-1)\left[2\left(3x^2 - 2x - 1\right)(6x-2)\right]}{\left(3x^2 - 2x - 1\right)^4}$

$= 12 \cdot \dfrac{\left(3x^2 - 2x - 1\right)\left[3\left(3x^2 - 2x - 1\right) - 2(3x-1)(6x-2)\right]}{\left(3x^2 - 2x - 1\right)^4}$

$= 12 \cdot \dfrac{-27x^2 + 18x - 7}{\left(3x^2 - 2x - 1\right)^3} = \dfrac{-12\left(27x^2 - 18x + 7\right)}{[(3x+1)(x-1)]^3}$

Since $27x^2 - 18x + 7$ is never zero, there is no possible inflection point, but $x = -\dfrac{1}{3}$ and $x = 1$ must be included in concavity analysis. Concave down on $\left(-\infty, -\dfrac{1}{3}\right)$ and $(1, \infty)$; concave up on $\left(-\dfrac{1}{3}, 1\right)$.

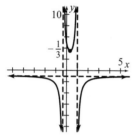

47.

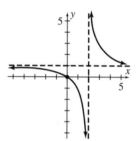

49.

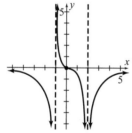

51. When $x = -\dfrac{a}{b}$, then $a + bx = 0$ so $x = -\dfrac{a}{b}$ is a vertical asymptote.

$$\lim_{x \to \infty} \frac{x}{a + bx} = \lim_{x \to \infty} \frac{x}{bx} = \lim_{x \to \infty} \frac{1}{b} = \frac{1}{b}$$

Thus $y = \dfrac{1}{b}$ is a horizontal asymptote.

53. $\lim_{t \to \infty} \left(150 - 76e^{-t}\right) = \lim_{t \to \infty} \left(150 - \dfrac{76}{e^t}\right)$
$= 150 - 0 = 150$
Thus $y = 150$ is a horizontal asymptote.

55.

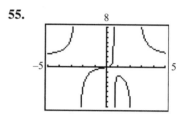

$x \approx \pm 2.45,\ x \approx 0.67,\ y = 2$

57.

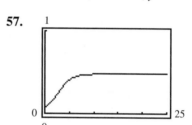

From the graph, it appears that $\lim_{x \to \infty} y \approx 0.48$. Thus a horizontal asymptote is $y \approx 0.48$. Algebraically, we have

$$\lim_{x \to \infty} \frac{0.34e^{0.7x}}{4.2 + 0.71e^{0.7x}} = \lim_{x \to \infty} \frac{\dfrac{0.34e^{0.7x}}{e^{0.7x}}}{\dfrac{4.2 + 0.71e^{0.7x}}{e^{0.7x}}}$$

$$= \lim_{x \to \infty} \frac{0.34}{\dfrac{4.2}{e^{0.7x}} + 0.71} = \frac{0.34}{0 + 0.71} \approx 0.48$$

Problems 13.6

1. Let the numbers be x and $82 - x$. Then if $P = x(82 - x) = 82x - x^2$, we have $P' = 82 - 2x$. Setting $P' = 0 \Rightarrow x = 41$. Since $P'' = -2 < 0$, there is a maximum when $x = 41$. Because $82 - x = 41$, the required numbers are 41 and 41.

3. We are given that $15x + 9(2y) = 9000$, or
$y = \dfrac{9000 - 15x}{18}$. We want to maximize area
A, where $A = xy$.

$$A = xy = x\left(\frac{9000 - 15x}{18}\right)$$

$$= \frac{1}{18}\left(9000x - 15x^2\right)$$

$$A' = \frac{1}{18}(9000 - 30x)$$

Setting $A' = 0 \Rightarrow x = 300$. Since

$A''(300) = \dfrac{1}{18}(-30) < 0$, we have a

maximum at $x = 300$. Thus
$y = \dfrac{9000 - 15(300)}{18} = 250$. The dimensions
are 300 ft by 250 ft.

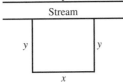

5. $c = 0.05q^2 + 5q + 500$

Avg. cost per unit

$$= \overline{c} = \frac{c}{q} = 0.05q + 5 + \frac{500}{q}$$

$\overline{c}' = 0.05 - \dfrac{500}{q^2}$. Setting $\overline{c}' = 0$ yields

$0.05 = \dfrac{500}{q^2}$, $q^2 = 10,000$, $q = \pm 100$. We

exclude $q = -100$ because q represents the

number of units. Since $\overline{c}'' = \dfrac{1000}{q^3} > 0$ for

$q > 0$, $\overline{c}$ is an absolute minimum when
$q = 100$ units.

7. $p = -5q + 30$
Since total revenue = (price)(quantity),
$r = pq = (-5q + 30)q = -5q^2 + 30q$
Setting $r' = -10q + 30 = 0 \Rightarrow q = 3$. Since
$r'' = -10 < 0$, r is maximum at $q = 3$ units,
for which the corresponding price is
$p = -5(3) + 30 = \$15$.

9. $f(p) = 160 - p - \dfrac{900}{p + 10}$, where
$0 \le p \le 100$.

   a. Setting $f'(p) = 0$ gives

$$-1 + \frac{900}{(p+10)^2} = 0, \quad \frac{900}{(p+10)^2} = 1,$$

$(p+10)^2 = 900, \quad p + 10 = \pm 30$, from
which $p = 20$.

Since $f''(p) = \dfrac{-1800}{(p+10)^3} < 0$ for $p = 20$,

we have an absolute maximum of
$f(20) = 110$ grams.

   b. $f(0) = 70$ and $f(100) = 51\dfrac{9}{11}$, so we
have an absolute minimum of
$f(100) = 51\dfrac{9}{11}$ grams.

11. $p = 85 - 0.05q$
$c = 600 + 35q$
Profit = Total Revenue – Total Cost
$P = pq - c = (85 - 0.05q)q - (600 + 35q)$
$\quad = -(0.05q^2 - 50q + 600)$
Setting $P' = -(0.1q - 50) = 0$ yields
$q = 500$.
Since $P''(500) = -0.1 < 0$, P is a maximum
when $q = 500$ units. This corresponds to a
price of $p = 85 - 0.05(500) = \$60$ and a
profit of $P = \$11,900$.

13. $p = 42 - 4q$

$$\overline{c} = 2 + \frac{80}{q}$$

Total Cost $= c = \overline{c}q = 2q + 80$

Profit = Total Revenue − Total Cost

$P = pq - c = (42 - 4q)q - (2q + 80)$

$= -\left(4q^2 - 40q + 80\right)$

$P' = -(8q - 40)$

Setting $P' = -(8q - 40) = 0$ gives $q = 5$. We find that $P'' = -8 < 0$, so P has a maximum value when $q = 5$. The corresponding price p is $42 - 4(5) = \$22$.

15. $p = q^2 - 100q + 3200$ on $[0, 120]$

$$\overline{c} = \frac{2}{3}q^2 - 40q + \frac{10,000}{q}$$

Profit = Total Revenue − Total Cost

Since total revenue $r = pq$ and total cost $= c = \overline{c}q$,

$P = pq - \overline{c}q$

$= q^3 - 100q^2 + 3200q - \left(\frac{2}{3}q^3 - 40q^2 + 10,000\right)$

$= \frac{1}{3}q^3 - 60q^2 + 3200q - 10,000$

$P' = q^2 - 120q + 3200 = (q - 40)(q - 80)$

Setting $P' = 0$ gives $q = 40$ or 80. Evaluating profit at $q = 0, 40, 80,$ and 120 gives

$P(0) = -10,000$

$$P(40) = \frac{130,000}{3} = 43,333\frac{1}{3}$$

$$P(80) = \frac{98,000}{3} = 32,666\frac{2}{3}$$

$P(120) = 86,000$

Thus the profit maximizing output is $q = 120$ units, and the corresponding maximum profit is \$86,000.

17. Total fixed costs = \$1200,

material-labor costs/unit = \$2, and the demand equation is $p = \dfrac{100}{\sqrt{q}}$.

Profit = Total Revenue − Total Cost

$P = pq - c$

$$P = \frac{100}{\sqrt{q}} \cdot q - (2q + 1200)$$
$$= 100\sqrt{q} - 2q - 1200$$
$$= 2\left(50\sqrt{q} - q - 600\right)$$

Setting $P' = 2\left(\dfrac{25}{\sqrt{q}} - 1\right) = 0$ yields $q = 625$. We see that $P'' = -25q^{-\frac{3}{2}} < 0$ for $q > 0$, so P is maximum when $q = 625$. When

$q = 625$, $\text{MR} = \dfrac{50}{\sqrt{625}} = 2 = \text{MC}$. When

$q = 625$, then $p = \$4$.

19. If $x = $ number of $\$0.50$ decreases, where $0 \le x \le 36$, then the monthly fee for each subscriber is $18 - 0.50x$, and the total number of subscribers is $4800 + 150x$. Let r be the total (monthly) revenue.
revenue = (monthly rate)(number of subscribers)
$r = (18 - 0.50x)(4800 + 150x)$
$r' = (18 - 0.50x)(150) + (4800 + 150x)(-0.50)$
$= 300 - 150x = 150(2 - x)$
Setting $r' = 0$ yields $x = 2$.
Evaluating r when $x = 0$, 2, and 36, we find that r is a maximum when $x = 2$. This corresponds to a monthly fee of $18 - 0.50(2) = \$17$ and a monthly revenue r of $\$86,700$.

21. See the figure in the text. Given that $x^2 y = 32$, we want to minimize $S = 4(xy) + x^2$. Since $y = \dfrac{32}{x^2}$,

where $x > 0$, we have $S = 4x\left(\dfrac{32}{x^2}\right) + x^2 = \dfrac{128}{x} + x^2$, from which $S' = -\dfrac{128}{x^2} + 2x$. Setting $S' = 0$ gives

$2x^3 = 128$, $x^3 = 64$, $x = 4$. Since $S'' = \dfrac{256}{x^3} + 2$, we get $S''(4) > 0$, so $x = 4$ gives a minimum. If $x = 4$,

then $y = \dfrac{32}{16} = 2$. The dimensions are 4 ft $\times$ 4 ft $\times$ 2 ft.

23. $V = x(L - 2x)^2 = L^2 x - 4Lx^2 + 4x^3$

where $0 < x < \dfrac{L}{2}$.

$V' = L^2 - 8Lx + 12x^2$
$= 12x^2 - 8Lx + L^2$
$= (2x - L)(6x - L)$

For $0 < x < \dfrac{L}{2}$, setting $V' = 0$ gives $x = \dfrac{L}{6}$. Since $V' > 0$ on $\left(0, \dfrac{L}{6}\right)$ and $V' < 0$ on $\left(\dfrac{L}{6}, \dfrac{L}{2}\right)$, V is maximum when $x = \dfrac{L}{6}$. Thus the length of

the side of the square must be $\dfrac{L}{6}$ in., which

results in a volume of

$$\frac{L}{6}\left(L - \frac{L}{3}\right)^2 = \frac{2L^3}{27} \text{ in}^3.$$

25. See the figure in the text.

$$V = K = \pi r^2 h \qquad (1)$$

$$S = 2\pi rh + \pi r^2 \qquad (2)$$

From Equation (1) $h = \dfrac{K}{\pi r^2}$. Thus Equation

(2) becomes

$$S = \frac{2K}{r} + \pi r^2$$

$$\frac{dS}{dr} = -\frac{2K}{r^2} + 2\pi r = \frac{2\left(\pi r^3 - K\right)}{r^2}.$$

If $S' = 0$, then $\pi r^3 - K = 0$, $\pi r^3 = K$,

$r = \sqrt[3]{\dfrac{K}{\pi}}$. Thus

$$h = \frac{K}{\pi\left(\frac{K}{\pi}\right)^{\frac{2}{3}}} = \left(\frac{K}{\pi}\right)^{\frac{1}{3}} = \sqrt[3]{\frac{K}{\pi}}.$$

Note that since $S'' = 2\pi + \dfrac{4K}{r^3} > 0$ for $r > 0$,

we have a minimum.

27. $p = 600 - 2q$

$c = 0.2q^2 + 28q + 200$

Profit = Total Revenue – Total Cost

$P = pq - c$

$P = (600 - 2q)q - \left(0.2q^2 + 28q + 200\right)$

$= -\left(2.2q^2 - 572q + 200\right)$

$P' = -(4.4q - 572)$

Setting $P' = 0$ yields $q = 130$. Since

$P'' = -4.4 < 0$, P is maximum when

$q = 130$ units. The corresponding price is

$p = 600 - 2(130) = \$340$, and the profit is
$P = \$36,980$. If a tax of $\$22$/unit is imposed
on the manufacturer, then the cost equation
is

$$c_1 = 0.2q^2 + 28q + 200 + 22q$$

$$= 0.2q^2 + 50q + 200.$$

The demand equation remains the same.
Thus

$P_1 = pq - c_1$

$= (600 - 2q)q - \left(0.2q^2 + 50q + 200\right)$

$= -\left(2.2q^2 - 550q + 200\right)$

$P_1' = -(4.4q - 550)$

Setting $P_1' = 0$ yields $q = 125$. Since

$P_1'' = -4.4 < 0$, P_1 is maximum when

$q = 125$ units. The corresponding price is

$p = \$350$ and the profit is $P_1 = \$34,175$.

29. Let q = number of units in a production run.
Since inventory is depleted at a uniform
rate, assume that the average inventory is

$\dfrac{q}{2}$. The value of average inventory is

$10\left(\dfrac{q}{2}\right)$, and carrying costs are

$0.128\left[10\left(\dfrac{q}{2}\right)\right]$. The number of production

runs per year is $\dfrac{1000}{q}$, and total set-up costs

are $40\left(\dfrac{1000}{q}\right)$. We want to minimize the

sum C of carrying costs and set-up costs.

$$C = 0.128\left[10\left(\frac{q}{2}\right)\right] + 40\left(\frac{1000}{q}\right)$$

$$= 0.64q + \frac{40,000}{q}$$

$$C' = 0.64 - \frac{40,000}{q^2}$$

Setting $C' = 0$ yields

$$q^2 = \frac{40,000}{0.64} = 62,500,$$

$q = 250$ (since $q > 0$). Since

$$C'' = \frac{80,000}{q^3} > 0,$$

C is minimum when $q = 250$. Thus the economic lot size is 250/lot (4 lots).

31. Let x = number of people over the 30.
Note: $0 \le x \le 10$.
Revenue = r
= (number attending)(charge/person)
= $(30 + x)(50 - 1.25x)$

$$= 1500 + 12.5x - 1.25x^2$$

$r' = 12.5 - 2.5x$
Setting $r' = 0$ yields $x = 5$. Since
$r'' = -2.5 < 0$, r is maximum when $x = 5$,
that is, when 35 attend.

33. The cost per mile of operating the truck is

$0.165 + \dfrac{s}{200}$. Driver's salary is \$18/hr. The

number of hours for 700 mi trip is $\dfrac{700}{s}$.

Driver's salary for trip is $18\left(\dfrac{700}{s}\right)$, or

$\dfrac{12,600}{s}$. The cost of operating the truck for

the trip is $700\left[0.165 + \dfrac{s}{200}\right]$.

Total cost of trip is

$$C = \frac{12,600}{s} + 700\left(0.165 + \frac{s}{200}\right)$$

Setting $C' = -\dfrac{12,600}{s^2} + \dfrac{7}{2} = 0$ yields

$s^2 = 3600$, or $s = 60$ (since $s > 0$). Since

$C'' = \dfrac{25,200}{s^3} > 0$ for $s > 0$, C is a minimum

when $s = 60$ mi/h.

35. Profit P is given by
P = Total revenue – Total cost
= Total revenue – (salaries + fixed cost)
= $50q - (1000m + 3000)$

$$= 50\left(m^3 - 15m^2 + 92m\right) - 1000m - 3000$$

$$= 50\left(m^3 - 15m^2 + 72m - 60\right), \text{ where}$$

$$0 \le m \le 8$$

$$P' = 50\left(3m^2 - 30m + 72\right)$$

$$= 150\left(m^2 - 10m + 24\right) = 150(m - 4)(m - 6)$$

Setting $P' = 0$ gives the critical values 4 and 6. We now evaluate P at these critical values and also at the endpoints 0 and 8.
$P(0) = -3000$
$P(4) = 2600$
$P(6) = 2400$
$P(8) = 3400$
Thus Ms. Jones should hire 8 salespeople to obtain a maximum weekly profit of \$3400.

37. x = tons of chemical A ($x \le 4$),

$$y = \frac{24 - 6x}{5 - x} = \text{ tons of chemical B, profit on}$$

A = \$2000/ton, and profit on B = \$1000/ton.

Total Profit $= P_T = 2000x + 1000\left(\dfrac{24 - 6x}{5 - x}\right)$

$$= 2000\left[x + \frac{12 - 3x}{5 - x}\right]$$

$$P'_T = 2000\left[1 + \frac{(5 - x)(-3) - (12 - 3x)(-1)}{(5 - x)^2}\right]$$

$$= 2000\left[1 - \frac{3}{(5 - x)^2}\right]$$

$$= 2000\left[\frac{x^2 - 10x + 22}{(5 - x)^2}\right]$$

Setting $P'_T = 0$ yields (by the quadratic formula)

$$x = \frac{10 \pm 2\sqrt{3}}{2} = 5 \pm \sqrt{3}$$

Because $x \le 4$, choose $x = 5 - \sqrt{3}$. Since

P_T is increasing on $\left[0, 5-\sqrt{3}\right)$ and decreasing on $\left(5-\sqrt{3}, 4\right]$, P_T is a maximum for $x = 5-\sqrt{3}$ tons. If profit on A is P/ton and profit on B is $\dfrac{P}{2}$/ton, then

$$P_T = Px + \frac{P}{2}\left(\frac{24-6x}{5-x}\right) = P\left[x + \frac{12-3x}{5-x}\right]$$

$$P'_T = P\left[\frac{x^2 - 10x + 22}{(5-x)^2}\right]$$

Setting $P'_T = 0$ and using an argument similar to that above, we find that P_T is a maximum when $x = 5-\sqrt{3}$ tons.

39. $P(j) = Aj\dfrac{L^4}{V} + B\dfrac{V^3 L^2}{1+j}$

$$\frac{dP}{dj} = \frac{AL^4}{V} - \frac{BV^3 L^2}{(1+j)^2} = 0$$

Solving for $(1+j)^2$ gives $(1+j)^2 = \dfrac{BV^4}{AL^2}$

41. $\bar{c} = \dfrac{c}{q} = 3q + 50 - 18\ln(q) + \dfrac{120}{q}$, $q > 0$

$$\frac{d\bar{c}}{dq} = 3 - \frac{18}{q} - \frac{120}{q^2} = \frac{3q^2 - 18q - 120}{q^2}$$

$$= \frac{3\left(q^2 - 6q - 40\right)}{q^2}$$

$$= \frac{3(q-10)(q+4)}{q^2}$$

Critical value is $q = 10$ since $q \geq 0$.

Since $\dfrac{d\bar{c}}{dq} < 0$ for $0 < q < 10$, and $\dfrac{d\bar{c}}{dq} > 0$ for $q > 10$, we have a minimum when $q = 10$ cases. This minimum average cost is $3(10) + 50 - 18\ln 10 + 12 \approx \$50.55$.

Chapter 13 Review Problems

1. $y = \dfrac{3x^2}{x^2 - 16} = \dfrac{3x^2}{(x+4)(x-4)}$

When $x = \pm 4$ the denominator is zero and the numerator is not zero. Thus $x = 4$ and $x = -4$ are vertical asymptotes.

$$\lim_{x \to \infty} \frac{3x^2}{x^2 - 16} = \lim_{x \to \infty} \frac{3x^2}{x^2} = \lim_{x \to \infty} 3 = 3$$

Similarly, $\lim_{x \to -\infty} y = 3$. Thus $y = 3$ is the only horizontal asymptote.

3. $y = \dfrac{5x^2 - 3}{(3x+2)^2} = \dfrac{5x^2 - 3}{9x^2 + 12x + 4}$

When $x = -\dfrac{2}{3}$, the denominator is zero and the numerator is not zero. Thus $x = -\dfrac{2}{3}$ is a vertical asymptote.

$$\lim_{x \to \infty} y = \lim_{x \to \infty} \frac{5x^2}{9x^2} = \lim_{x \to \infty} \frac{5}{9} = \frac{5}{9}$$

Similarly, $\lim_{x \to -\infty} y = \dfrac{5}{9}$. Thus $y = \dfrac{5}{9}$ is the only horizontal asymptote.

5. $f(x) = \dfrac{5x^2}{3 - x^2}$

$$f'(x) = \frac{(3-x^2)(10x) - 5x^2(-2x)}{(3-x^2)^2}$$

$$= \frac{10x(3 - x^2 + x^2)}{(3-x^2)^2}$$

$$= \frac{30x}{(3-x^2)^2}$$

Thus $x = 0$ is the only critical value.
Note: Although $f'\left(\pm\sqrt{3}\right)$ is not defined, $\pm\sqrt{3}$ are not critical values because $\pm\sqrt{3}$ are not in the domain of f.

7. $f(x) = \dfrac{\sqrt[3]{x+1}}{3-4x}$

$$f'(x) = \frac{(3-4x)\left[\frac{1}{3}(x+1)^{-\frac{2}{3}}\right]-(x+1)^{\frac{1}{3}}(-4)}{(3-4x)^2} = \frac{\frac{1}{3}(x+1)^{-\frac{2}{3}}[(3-4x)+12(x+1)]}{(3-4x)^2} = \frac{8x+15}{3(x+1)^{\frac{2}{3}}(3-4x)^2}$$

$f'(x)$ is zero when $x = -\dfrac{15}{8}$; $f'(x)$ is not defined when $x = -1$ or $x = \dfrac{3}{4}$. However $\dfrac{3}{4}$ is not in the

domain of f. Thus $x = -\dfrac{15}{8}$ and $x = -1$ are critical values.

9. $f(x) = -\dfrac{5}{3}x^3 + 15x^2 + 35x + 10$

$f'(x) = -5x^2 + 30x + 35$

$\qquad = -5(x^2 - 6x - 7) = -5(x-7)(x+1)$

CV: $x = -1$ and $x = 7$. Decreasing on $(-\infty, -1)$ and $(7, \infty)$; increasing on $(-1, 7)$

11. $f(x) = \dfrac{6x^4}{x^2-3}$

$$f'(x) = 6 \cdot \frac{\left(x^2-3\right)\left(4x^3\right)-x^4(2x)}{\left(x^2-3\right)^2}$$

$$= \frac{12x^3\left[2\left(x^2-3\right)-x^2\right]}{\left(x^2-3\right)^2} = \frac{12x^3\left(x^2-6\right)}{\left(x^2-3\right)^2}$$

$$= \frac{12x^3\left(x+\sqrt{6}\right)\left(x-\sqrt{6}\right)}{\left[\left(x+\sqrt{3}\right)\left(x-\sqrt{3}\right)\right]^2}$$

CV: $x = 0$, $\pm\sqrt{6}$, but $x = \pm\sqrt{3}$ must also be considered in the inc.-dec. analysis. Decreasing on

$\left(-\infty, -\sqrt{6}\right), \left(0, \sqrt{3}\right)$, and $\left(\sqrt{3}, \sqrt{6}\right)$; increasing on $\left(-\sqrt{6}, -\sqrt{3}\right), \left(-\sqrt{3}, 0\right)$ and $\left(\sqrt{6}, \infty\right)$.

13. $f(x) = x^4 - x^3 - 14$

$f'(x) = 4x^3 - 3x^2$

$f''(x) = 12x^2 - 6x = 6x(2x-1)$

$f''(x) = 0$ when $x = 0$ or $x = \dfrac{1}{2}$. Concave up on $(-\infty, 0)$ and $\left(\dfrac{1}{2}, \infty\right)$; concave down on $\left(0, \dfrac{1}{2}\right)$.

15. $f(x) = \dfrac{1}{2x-1} = (2x-1)^{-1}$

$f'(x) = -2(2x-1)^{-2}$

$f'' = 8(2x-1)^3 = \dfrac{8}{(2x-1)^3}$

$f''(x)$ is not defined when $x = \dfrac{1}{2}$. Concave down on $\left(-\infty, \dfrac{1}{2}\right)$; concave up on $\left(\dfrac{1}{2}, \infty\right)$.

17. $f(x) = (2x+1)^3(3x+2)$

$\begin{aligned}
f'(x) &= (2x+1)^3(3) + (3x+2)[3(2x+1)^2(2)] \\
&= 3(2x+1)^2(2x+1+6x+4) \\
&= 3(2x+1)^2(8x+5)
\end{aligned}$

$\begin{aligned}
f''(x) &= 3\{(2x+1)^2(8) + (8x+5)[2(2x+1)(2)]\} \\
&= 12(2x+1)[2(2x+1) + 8x+5] \\
&= 12(2x+1)(12x+7)
\end{aligned}$

$f''(x) = 0$ when $x = -\dfrac{1}{2}$ or $x = -\dfrac{7}{12}$. Concave up on $\left(-\infty, -\dfrac{7}{12}\right)$ and $\left(-\dfrac{1}{2}, \infty\right)$; concave down on

$\left(-\dfrac{7}{12}, -\dfrac{1}{2}\right)$.

19. $f(x) = 2x^3 - 9x^2 + 12x + 7$

$f'(x) = 6x^2 - 18x + 12 = 6\left(x^2 - 3x + 2\right)$

$= 6(x-1)(x-2)$

CV: $x = 1$ and $x = 2$

Increasing on $(-\infty, 1)$ and $(2, \infty)$; decreasing on $(1, 2)$. Relative maximum when $x = 1$; relative minimum when $x = 2$.

21. $f(x) = \dfrac{x^{10}}{10} + \dfrac{x^5}{5}$

$f'(x) = x^9 + x^4 = x^4(x^5 + 1)$

CV: $x = 0$ and $x = -1$

Decreasing on $(-\infty, -1)$; increasing on $(-1, 0)$ and $(0, \infty)$; relative minimum when $x = -1$

23. $f(x) = x^{\frac{2}{3}}(x+1) = x^{\frac{5}{3}} + x^{\frac{2}{3}}$

$f'(x) = \frac{5}{3}x^{\frac{2}{3}} + \frac{2}{3}x^{-\frac{1}{3}}$

$= \frac{1}{3}x^{-\frac{1}{3}}(5x+2)$

$= \frac{5x+2}{3x^{\frac{1}{3}}}$

CV: $x = 0$ and $x = -\frac{2}{5}$

Increasing on $\left(-\infty, -\frac{2}{5}\right)$ and $(0, \infty)$;

decreasing on $\left(-\frac{2}{5}, 0\right)$. Relative maximum

when $x = -\frac{2}{5}$; relative minimum when

$x = 0$.

25. $y = x^5 - 5x^4 + 3x$

$y' = 5x^4 - 20x^3 + 3$

$y'' = 20x^3 - 60x^2 = 20x^2(x-3)$

Possible inflection points occur when $x = 0$
or $x = 3$. Concave down on $(-\infty, 0)$ and
$(0, 3)$; concave up on $(3, \infty)$. Concavity
changes at $x = 3$, so there is an inflection
point when $x = 3$.

27. $y = 4(3x-5)\left(x^4 + 2\right)$

$= 12x^5 - 20x^4 + 24x - 40$

$y' = 60x^4 - 80x^3 + 24$

$y'' = 240x^3 - 240x^2 = 240x^2(x-1)$

Possible inflection points occur when $x = 0$
or $x = 1$. Concave down on $(-\infty, 0)$ and
$(0, 1)$; concave up on $(1, \infty)$. Inflection point
when $x = 1$.

29. $y = \frac{x^3}{e^x} = x^3 e^{-x}$

$y' = x^3(-e^{-x}) + e^{-x}(3x^2) = -e^{-x}(x^3 - 3x^2)$

$y'' = -e^{-x}(3x^2 - 6x) - (x^3 - 3x^2)(-e^{-x})$

$= e^{-x}(x^3 - 6x^2 + 6x)$

$= xe^{-x}(x^2 - 6x + 6)$

y'' is defined for all x and y'' is zero only

when $x = 0$ or $x^2 - 6x + 6 = 0$. Using the
quadratic formula on the second equation, the
possible points of inflection occur when $x = 0$,
$3 \pm \sqrt{3}$. Concave up on $\left(0, 3 - \sqrt{3}\right)$ and

$\left(3 + \sqrt{3}, \infty\right)$; concave down on $(-\infty, 0)$ and

$\left(3 - \sqrt{3}, 3 + \sqrt{3}\right)$. Inflection points when

$x = 0, \ 3 \pm \sqrt{3}$.

31. $f(x) = 3x^4 - 4x^3$ and f is continuous on
[0, 2].
$f'(x) = 12x^3 - 12x^2 = 12x^2(x-1)$
The only critical value on $(0, 2)$ is $x = 1$.
Evaluating f at this value and at the endpoints
gives $f(0) = 0$, $f(1) = -1$, and $f(2) = 16$.
Absolute maximum: $f(2) = 16$; absolute
minimum: $f(1) = -1$.

33. $f(x) = \frac{x}{(5x-6)^2}$ and f is continuous on

$[-2, 0]$.

$f'(x) = \frac{(5x-6)^2(1) - x[10(5x-6)]}{(5x-6)^4}$

$= \frac{(5x-6)[(5x-6) - 10x]}{(5x-6)^4} = \frac{-5x-6}{(5x-6)^3}$

$= -\frac{5x+6}{(5x-6)^3}$

The only critical value on $(-2, 0)$ is $x = -\frac{6}{5}$.

Evaluating f at this value and at the
endpoints gives

$f(-2) = -\dfrac{1}{128}$, $f\left(-\dfrac{6}{5}\right) = -\dfrac{1}{120}$ and $f(0) = 0$. Absolute maximum: $f(0) = 0$; absolute minimum:

$f\left(-\dfrac{6}{5}\right) = -\dfrac{1}{120}$.

35. $f(x) = \left(x^2 + 1\right)e^{-x}$

    a. $f'(x) = \left(x^2 + 1\right)\left(-e^{-x}\right) + e^{-x}(2x)$

        $= -e^{-x}\left[\left(x^2 + 1\right) - 2x\right] = -e^{-x}\left(x^2 - 2x + 1\right)$

        $= -e^{-x}(x - 1)^2$

    CV: $x = 1$

    Decreasing on $(-\infty, 1)$ and $(1, \infty)$. No relative extrema.

    b. $f''(x) = -\left\{e^{-x}[2(x - 1)] + (x - 1)^2\left(-e^{-x}\right)\right\}$

        $= e^{-x}(x - 1)[-2 + (x - 1)]$

        $= e^{-x}(x - 1)(x - 3)$

    Possible inflection points when $x = 1, 3$. Concave up on $(-\infty, 1)$ and $(3, \infty)$; concave down on $(1, 3)$.

    Inflection points at $(1, f(1)) = \left(1, 2e^{-1}\right)$ and $(3, f(3)) = \left(3, 10e^{-3}\right)$.

37. $y = x^2 - 2x - 24 = (x + 4)(x - 6)$

Intercepts: $(-4, 0)$, $(6, 0)$, $(0, -24)$

No symmetry. No asymptotes.

$y' = 2x - 2 = 2(x - 1)$

CV: $x = 1$

Increasing on $(1, \infty)$; decreasing on $(-\infty, 1)$; relative minimum at $(1, -25)$.

$y'' = 2$

No possible inflection point. Concave up on $(-\infty, \infty)$.

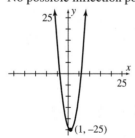

39. $y = x^3 - 12x + 20$

Intercept: (0, 20)

No symmetry; no asymptotes

$y' = 3x^2 - 12$

$= 3\left(x^2 - 4\right) = 3(x+2)(x-2)$

CV: $x = \pm 2$

Increasing on $(-\infty, -2)$ and $(2, \infty)$;
decreasing on $(-2, 2)$; relative maximum at
$(-2, 36)$; relative minimum at $(2, 4)$.

$y'' = 6x$

Possible inflection point when $x = 0$.
Concave up on $(0, \infty)$; concave down on
$(-\infty, 0)$; inflection point at $(0, 20)$.

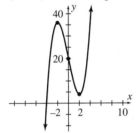

41. $y = x^3 - x = x\left(x^2 - 1\right) = x(x+1)(x-1)$

Intercepts $(0, 0)$, $(-1, 0)$, and $(1, 0)$

Symmetric about the origin. No asymptotes.

$y' = 3x^2 - 1 = \left(\sqrt{3}x + 1\right)\left(\sqrt{3}x - 1\right)$

CV: $\pm \dfrac{\sqrt{3}}{3}$

Increasing on $\left(-\infty, -\dfrac{\sqrt{3}}{3}\right)$ and $\left(\dfrac{\sqrt{3}}{3}, \infty\right)$;

decreasing on $\left(-\dfrac{\sqrt{3}}{3}, \dfrac{\sqrt{3}}{3}\right)$.

$y'' = 6x$

Possible inflection point when $x = 0$.
Concave down on $(-\infty, 0)$; concave up on
$(0, \infty)$; inflection point at $(0, 0)$.

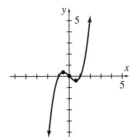

43. $y = f(x) = \dfrac{100(x+5)}{x^2}$

Intercept: $(-5, 0)$

No symmetry.

$x = 0$ is the only vertical asymptote.

$\lim_{x \to \infty} y = 100 \lim_{x \to \infty} \dfrac{x}{x^2} = 100 \lim_{x \to \infty} \dfrac{1}{x} = 0$, and

$\lim_{x \to -\infty} y = 0$, so $y = 0$ is the only horizontal

asymptote.

$y = 100\left[x^{-1} + 5x^{-2}\right]$

$y' = 100\left[-x^{-2} - 10x^{-3}\right] = -100\left[\dfrac{1}{x^2} + \dfrac{10}{x^3}\right]$

$= \dfrac{-100(x+10)}{x^3}$

CV: $x = -10$ but $x = 0$ must be included in
inc.-dec. analysis. Increasing on $(-10, 0)$;
decreasing on $(-\infty, -10)$ and $(0, \infty)$; relative
minimum at $(-10, -5)$.

$y'' = 100\left[2x^{-3} + 30x^{-4}\right] = 200\left[\dfrac{1}{x^3} + \dfrac{15}{x^4}\right]$

$= \dfrac{200(x+15)}{x^4}$

Possible inflection point when $x = -15$, but
$x = 0$ must also be considered in concavity
analysis. Concave up on $(-15, 0)$ and $(0, \infty)$;
concave down on $(-\infty, -15)$; inflection point

at $\left(-15, -\dfrac{40}{9}\right)$

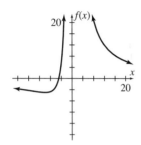

45. $y = \dfrac{2x}{(3x-1)^3}$

Intercept: $(0, 0)$
No symmetry

Vertical asymptote is $x = \dfrac{1}{3}$.

$$\lim_{x \to \infty} y = \lim_{x \to \infty} \frac{2x}{27x^3} = \frac{2}{27} \lim_{x \to \infty} \frac{1}{x^2} = 0$$
$$= \lim_{x \to -\infty} y,$$

so $y = 0$ is a horizontal asymptote.

$$y' = \frac{(3x-1)^3(2) - 2x[3(3x-1)^2(3)]}{(3x-1)^6}$$

$$= \frac{2(3x-1)^2[(3x-1) - 9x]}{(3x-1)^6}$$

$$= \frac{2(-6x-1)}{(3x-1)^4} = \frac{-2(6x+1)}{(3x-1)^4}$$

CV: $x = -\dfrac{1}{6}$, but $x = \dfrac{1}{3}$ must be considered in inc.-dec. analysis. Increasing on $\left(-\infty, -\dfrac{1}{6}\right)$; decreasing

on $\left(-\dfrac{1}{6}, \dfrac{1}{3}\right)$ and $\left(\dfrac{1}{3}, \infty\right)$; relative maximum at $\left(-\dfrac{1}{6}, \dfrac{8}{81}\right)$.

$$y'' = -2 \cdot \frac{(3x-1)^4(6) - (6x+1)[4(3x-1)^3(3)]}{(3x-1)^8}$$

$$= -2 \cdot \frac{6(3x-1)^3[(3x-1) - 2(6x+1)]}{(3x-1)^8}$$

$$= \frac{-12(-9x-3)}{(3x-1)^5} = \frac{36(3x+1)}{(3x-1)^5}$$

Possible inflection point when $x = -\dfrac{1}{3}$, but $x = \dfrac{1}{3}$ must be considered in concavity analysis. Concave up

on $\left(-\infty, -\dfrac{1}{3}\right)$ and $\left(\dfrac{1}{3}, \infty\right)$; concave down on $\left(-\dfrac{1}{3}, \dfrac{1}{3}\right)$; inflection point at $\left(-\dfrac{1}{3}, \dfrac{1}{12}\right)$.

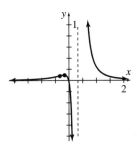

47. $f(x) = \dfrac{e^x + e^{-x}}{2}$

Intercept: (0, 1)

Symmetric about the *y*-axis. No asymptotes.

$f'(x) = \dfrac{e^x - e^{-x}}{2}$

Setting

$f'(x) = 0 \Rightarrow e^x = e^{-x} \Rightarrow x = -x \Rightarrow x = 0$

CV: $x = 0$

Increasing on $(0, \infty)$; decreasing on $(-\infty, 0)$; relative minimum at (0, 1). Finding $f''(x)$

gives: $f''(x) = \dfrac{e^x + e^{-x}}{2}$. $f''(x) > 0$ for all

x. No possible inflection point. Concave up on $(-\infty, \infty)$.

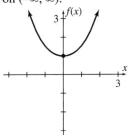

49. a. False. $f'(x_0) = 0$ only indicates the possibility of a relative extremum at x_0, For example, if $f(x) = x^3$, then $f'(x) = 3x^2$ and $f'(0) = 0$. However there is no relative extremum at $x = 0$.

b. False. For example, let $x_1 = -1$ and $x_2 = 1$. Then $x_1 < x_2$ and $f(x_1) = -1 < f(x_2) = 1$.

c. True. The absolute minimum is $f(0) = 0$ and the absolute maximum is $f(1) = 1$.

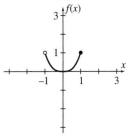

d. False. If concavity does not change around x_0, then $(x_0, f(x_0))$ is not an inflection point. For example, consider $f(x) = x^4$. If $x_0 = 0$, then $f''(x_0) = 0$, but $(x_0, f(x_0))$ is not an inflection point. See graph in part (c).

e. False. Consider the function *f* whose graph is shown. On $(-2, 2)$ it has exactly one relative maximum [at the point (0, 1)] but no absolute maximum.

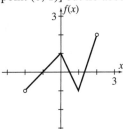

51. $c = q^3 - 6q^2 + 12q + 18$

Marginal cost $= \dfrac{dc}{dq} = 3q^2 - 12q + 12$.

Marginal cost is increasing when its

derivative, which is $\dfrac{d^2c}{dq^2}$, is positive.

$\dfrac{d^2c}{dq^2} = 6q - 12 = 6(q - 2)$

$\dfrac{d^2c}{dq^2} > 0$ for $q > 2$. Thus marginal cost is

increasing for $q > 2$.

53. $p = 200 - \dfrac{\sqrt{q}}{5}$, $q > 0$. The revenue function

r is given by

$$r = pq = \left(200 - \frac{\sqrt{q}}{5}\right)q = 200q - \frac{q^{\frac{3}{2}}}{5}.$$

$$r' = 200 - \frac{3}{10}q^{\frac{1}{2}}$$

$$r'' = -\frac{3}{20}q^{-\frac{1}{2}} = -\frac{3}{20\sqrt{q}}$$

Since $r'' < 0$ for $q > 0$, the graph of the revenue function is concave down for $q > 0$.

55. $f(t) = At^3 + Bt^2 + Ct + D$

$f'(t) = 3At^2 + 2Bt + C$

$f''(t) = 6At + 2B$, which gives an inflection point when $6At + 2B = 0$, that is for

$a = -\dfrac{B}{3A}$. This value of a must be such that

$f'(a) = 0$.

$$3A\left(-\frac{B}{3A}\right)^2 + 2B\left(-\frac{B}{3A}\right) + C = 0$$

$$\frac{1}{3}\left(\frac{B^2}{A}\right) - \frac{2}{3}\left(\frac{B^2}{A}\right) + C = 0$$

$$C = \frac{1}{3}\left(\frac{B^2}{A}\right)$$

$3AC = B^2$,

which was to be shown.

57.

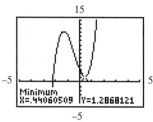

Relative maximum $(-1.32, 12.28)$; relative minimum $(0.44, 1.29)$

59.

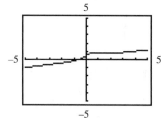

The x-value of the inflection point of f corresponds to the x-intercept of f''. Thus the x-value of the inflection point is $x \approx -0.60$.

61. $q = 80m^2 - 0.1m^4$

$$\frac{dq}{dm} = 160m - 0.4m^3 = 0.4m\left(400 - m^2\right)$$

$= 0.4m(20 + m)(20 - m)$. Setting $\dfrac{dq}{dm} = 0$

yields $m = 0$ or $m = 20$ (for $m \geq 0$). We find that q is increasing on $(0, 20)$ and decreasing on $(20, \infty)$, so q is maximum at $m = 20$.

63. $p = \sqrt{500 - q}$, where $100 \leq q \leq 200$.

Total revenue $= r = pq = q\sqrt{500 - q}$

$$r' = q\left(\frac{1}{2}\right)(500 - q)^{-1/2}(-1) + \sqrt{500 - q}\,(1)$$

$$= \frac{1}{2}(500 - q)^{-1/2}[-q + 2(500 - q)]$$

$$= \frac{1000 - 3q}{2\sqrt{500 - q}}$$

$$= \frac{3\left(\frac{1000}{3} - q\right)}{2\sqrt{500 - q}}$$

No critical values on $(100, 200)$.
$r(100) = 2000$; $r(200) \approx 3464$, so 200 units should be produced for maximum revenue.

65. $p = 500 - 3q$

$$\bar{c} = q + 200 + \frac{1000}{q}$$

Total Cost $= c = \bar{c}q = q^2 + 200q + 1000$

Profit = Total Revenue – Total Cost

$P = pq - c$

$ = (500 - 3q)q - (q^2 + 200q + 1000)$

$ = -4(q^2 - 75q + 250)$

$P' = -4(2q - 75)$

Setting $P' = 0$ yields $q = 37.5$. Since $P'' = -8 < 0$, P is maximum when $q = 37.5$. In reality, whole units are likely. Since $P(37) = P(38) = 4624$, the maximum profit is \$4624.

67. $2x + 4y = 800$; thus $x = 400 - 2y$

Area $= A = xy = (400 - 2y)y$

$$= 400y - 2y^2$$

$$\frac{dA}{dy} = 400 - 4y = 4(100 - y)$$

Setting $\frac{dA}{dy} = 0$ gives $y = 100$. Since

$$\frac{d^2A}{dy^2} = -4 < 0,\ A \text{ is maximum when}$$

$y = 100$. When $y = 100$, then $x = 200$. The dimensions are 200 ft by 100 ft.

69. a. $c = 2q^3 - 9q^2 + 12q + 20$, where

$$\frac{3}{4} \le q \le 6.$$

$$\frac{dc}{dq} = 6q^2 - 18q + 12 = 6\left(q^2 - 3q + 2\right)$$

$$= 6(q - 1)(q - 2)$$

Setting $\frac{dc}{dq} = 0$ gives $q = 1$ or 2.

Evaluating c at these critical values and

the endpoints: $c\left(\dfrac{3}{4}\right) = \dfrac{793}{32} \approx 24.78$,

$c(1) = 25$, $c(2) = 24$, $c(6) = 200$. Thus a minimum occurs at $q = 2$, which corresponds to 200 stands and a total cost of \$24,000. This gives an average

cost per stand of $\dfrac{24,000}{200} = \$120$.

b. There are no critical values of c in $3 \le q \le 6$, so we only evaluate c at the endpoints: $c(3) = 29$, $c(6) = 200$. Thus a minimum occurs at $q = 3$, which gives 300 stands.

Mathematical Snapshot Chapter 13

1. Figure 13.74 does not readily show how long it takes for the population to reach its final size. Figure 13.75 shows that this takes about 45 days.

3. Even if the graph starts out exactly coinciding with the ideal curve, a line segment tangent to the curve at one end must (in general) lie slightly off the curve at the other end. This introduces errors that accumulate over successive iterations. The amount of cumulative error could be reduced by taking smaller time steps, such as 1 month instead of 1 year, and correspondingly drawing shorter line segments.

Chapter 14

Problems 14.1

1. $y = 5x - 7$

$$dy = \frac{d}{dx}(5x - 7)dx = 5 \ dx$$

3. $d[f(x)] = f'(x)dx = \frac{1}{2}(x^4 - 9)^{-\frac{1}{2}}(4x^3)dx$

$$= \frac{2x^3}{\sqrt{x^4 - 9}}dx$$

5. $u = x^{-2}$

$$du = \frac{d}{dx}\left(x^{-2}\right)dx = -2x^{-3}dx = -\frac{2}{x^3}dx$$

7. $dp = \frac{d}{dx}\left[\ln\left(x^2 + 7\right)\right]dx = \frac{1}{x^2 + 7}(2x)dx$

$$= \frac{2x}{x^2 + 7}dx$$

9. $dy = y'dx$

$$= \left[(9x + 3)e^{2x^2+3}(4x) + e^{2x^2+3}(9)\right]dx$$

$$= 3e^{2x^2+3}[(3x + 1)(4x) + 3]dx$$

$$= 3e^{2x^2+3}\left(12x^2 + 4x + 3\right)dx$$

11. $\Delta y = [4 - 7(3.02)] - [4 - 7(3)] = -0.14$
$dy = -7 \ dx = -7(0.02) = -0.14$

13. $\Delta y = [2(-1.9)^2 + 5(-1.9) - 7] - [2(-2)^2 + 5(-2) - 7]$
$\quad = -0.28$
$dy = (4x + 5)dx = [4(-2) + 5](0.1) = -0.3$

15. $\Delta y = \sqrt{32 - (3.95)^2} - \sqrt{32 - (4^2)} \approx 0.049$

$$dy = \frac{-x}{\sqrt{32 - x^2}}dx = \frac{-4}{\sqrt{16}}(-0.05) = 0.050$$

17. a. $f(x) = \dfrac{x+5}{x+1}$

$f'(x) = \dfrac{(x+1)(1) - (x+5)(1)}{(x+1)^2} = \dfrac{-4}{(x+1)^2}$

$f'(1) = \dfrac{-4}{4} = -1$

b. We use $f(x + dx) \approx f(x) + dy$ with $x = 1$, $dx = 0.1$.

$f(1.1) = f(1 + 0.1) \approx f(1) + f'(1)dx$

$= \dfrac{6}{2} + (-1)(0.1) = 2.9$

19. Let $y = f(x) = \sqrt{x}$

$f(x + dx) \approx f(x) + dy = \sqrt{x} + \dfrac{1}{2\sqrt{x}} dx$

If $x = 289$ and $dx = -1$, then

$\sqrt{288} = f(289 - 1)$

$\approx \sqrt{289} + \dfrac{1}{2\sqrt{289}}(-1)$

$= \dfrac{577}{34}$

≈ 16.97

21. Let $y = f(x) = \sqrt[3]{x}$

$f(x + dx) \approx f(x) + dy = \sqrt[3]{x} + \dfrac{1}{3x^{\frac{2}{3}}} dx$

If $x = 64$ and $dx = 1.5$, then

$\sqrt[3]{65.5} = f(64 + 1.5) \approx \sqrt[3]{64} + \dfrac{1}{3\left(\sqrt[3]{64}\right)^2}(1.5)$

$= 4 + \dfrac{1.5}{3 \cdot 4^2} = 4\dfrac{1}{32}$

23. Let $y = f(x) = \ln x$

$f(x + dx) \approx f(x) + dy = \ln(x) + \dfrac{1}{x} dx$

If $x = 1$ and $dx = -0.03$, then

$\ln(0.97) = f(1 + (-0.03))$

$\approx \ln(1) + \dfrac{1}{1}(-0.03) = -0.03$

25. Let $y = f(x) = e^x$

$f(x + dx) \approx f(x) + dy = e^x + e^x dx$

If $x = 0$ and $dx = 0.001$, then

$e^{0.001} = f(0 + 0.001)$

$\approx e^0 + e^0(0.001) = 1.001$

27. $\dfrac{dy}{dx} = 2$, so $\dfrac{dx}{dy} = \dfrac{1}{\frac{dy}{dx}} = \dfrac{1}{2}$

29. $\dfrac{dq}{dp} = 6p\left(p^2 + 5\right)^2$, so $\dfrac{dp}{dq} = \dfrac{1}{6p\left(p^2 + 5\right)^2}$

31. $q = p^{-1}$, $\dfrac{dq}{dp} = -1p^{-2} = \dfrac{-1}{p^2}$, so $\dfrac{dp}{dq} = -p^2$

33. $\dfrac{dx}{dy} = \dfrac{1}{\frac{dy}{dx}} = \dfrac{1}{14x - 6}$

If $x = 3$, $\dfrac{dx}{dy} = \dfrac{1}{36}$

35. $p = \dfrac{500}{q + 2}$

$\dfrac{dp}{dq} = \dfrac{-500}{(q + 2)^2}$

$\dfrac{dq}{dp} = -\dfrac{(q + 2)^2}{500}$

$\left.\dfrac{dq}{dp}\right|_{q=18} = -\dfrac{(q + 2)^2}{500}\Bigg|_{q=18} = -\dfrac{4}{5}$

37. $P = 397q - 2.3q^2 - 400$, q changes from 90 to 91.

$\Delta P \approx dP = P'dq = (397 - 4.6q)dq$

Choosing $q = 90$ and $dq = 1$,

$\Delta P \approx [397 - 4.6(90)](1) = -17$.

True change is

$P(91) - P(90) = 16,680.7 - 16,700 = -19.3$.

39. $p = \dfrac{10}{\sqrt{q}}$. We approximate p when $q = 24$.

$$p(q + dq) \approx p + dp = \frac{10}{\sqrt{q}} - \frac{5}{\sqrt{q^3}} dq$$

If $q = 25$ and $dq = -1$, then

$$p(24) = p(25 + (-1)) \approx \frac{10}{\sqrt{25}} - \frac{5}{\sqrt{(25)^3}}(-1)$$

$$= 2 + \frac{1}{25} = \frac{51}{25} = 2.04$$

41. $c = \dfrac{q^4}{2} + 3q + 400$

If $q = 10$ and $dq = 2$,

$$\frac{dc}{c} = \frac{\left(2q^3 + 3\right) dq}{\frac{q^4}{2} + 3q + 400} = \frac{(2003)(2)}{5430} \approx 0.7$$

43. $V = \dfrac{4}{3}\pi r^3$

$$\Delta V \approx dV = V' dr = 4\pi r^2 dr$$

$$dr = \left(6.6 \times 10^{-4}\right) - \left(6.5 \times 10^{-4}\right)$$

$$= 0.1 \times 10^{-4} = 10^{-5}$$

$$\Delta V \approx 4\pi \left(6.5 \times 10^{-4}\right)^2 \left(10^{-5}\right)$$

$$= \left(1.69 \times 10^{-11}\right)\pi \text{ cm}^3.$$

45. a. We substitute $q = 40$ and $p = 20$

$$2 + \frac{40^2}{200} = \frac{4000}{20^2}$$

$$2 + 8 = 10$$

$$10 = 10$$

b. We differentiate implicitly with respect to p.

$$0 + \frac{1}{200}\left(2q\frac{dq}{dp}\right) = -\frac{8000}{p^3}$$

From part (a) $q = 40$ when $p = 20$. Substituting gives

$$\frac{1}{200}\left(2 \cdot 40 \frac{dq}{dp}\right) = -\frac{8000}{20^3}$$

$$\frac{dq}{dp} = -2.5$$

c. $q(p + dp) \approx q(p) + dq = q(p) + q'(p)dp$

$$q(19.20) = q(20 + (-0.8))$$

$$\approx q(20) + q'(20)dp$$

$$= 40 + (-2.5)(-0.8)$$

$$= 42 \text{ units}$$

Principles in Practice 14.2

1. $\displaystyle\int 28.3 \, dq = 28.3q + C$

The form of the cost function is $28.3q + C$.

3. Let $S(t)$ = the number of subscribers t months after the competition entered the market, then $S'(t) = -\dfrac{480}{t^3}$.

$$S(t) = \int -\frac{480}{t^3} dt = -480 \int t^{-3} dt$$

$$= -480\left(\frac{t^{-2}}{-2}\right) + C = 240t^{-2} + C = \frac{240}{t^2} + C$$

The number of subscribers is

$$S(t) = \frac{240}{t^2} + C.$$

5. The amount of money saved is $\displaystyle\int \frac{dS}{dt} dt$.

$$\int \left(2.1t^2 - 65.4t + 491.6\right) dt$$

$$= 2.1\left(\frac{t^3}{3}\right) - 65.4\left(\frac{t^2}{2}\right) + 491.6t + C$$

$$= 0.7t^3 - 32.7t^2 + 491.6t + C$$

The amount of money saved is

$$S(t) = 0.7t^3 - 32.7t^2 + 491.6t + C$$

Problems 14.2

1. $\int 7\,dx = 7x + C$

3. $\int x^8\,dx = \dfrac{x^{8+1}}{8+1} + C = \dfrac{x^9}{9} + C$

5. $\int 5x^{-7}\,dx = 5\int x^{-7}\,dx = 5\cdot\dfrac{x^{-7+1}}{-7+1} + C$

$\qquad = 5\cdot\dfrac{x^{-6}}{-6} + C = -\dfrac{5}{6x^6} + C$

7. $\int\dfrac{2}{x^{10}}\,dx = 2\int x^{-10}\,dx = 2\cdot\dfrac{x^{-10+1}}{-10+1} + C$

$\qquad = \dfrac{2x^{-9}}{-9} + C = -\dfrac{2}{9x^9} + C$

9. $\int\dfrac{1}{t^{7/4}}\,dt = \int t^{-7/4}\,dt = \dfrac{t^{-7/4+1}}{-\frac{7}{4}+1} + C$

$\qquad = \dfrac{t^{-3/4}}{-\frac{3}{4}} + C$

$\qquad = -\dfrac{4}{3t^{3/4}} + C$

11. $\int(4+t)\,dt = \int 4\,dt + \int t\,dt = 4t + \dfrac{t^{t+1}}{1+1} + C$

$\qquad = 4t + \dfrac{t^2}{2} + C$

13. $\int\left(y^5 - 5y\right)dy = \int y^5\,dy - \int 5y\,dy$

$\qquad = \dfrac{y^{5+1}}{5+1} - 5\cdot\dfrac{y^{1+1}}{1+1} + C$

$\qquad = \dfrac{y^6}{6} - 5\cdot\dfrac{y^2}{2} + C = \dfrac{y^6}{6} - \dfrac{5y^2}{2} + C$

15. $\int\left(3t^2 - 4t + 5\right)dt = 3\int t^2\,dt - 4\int t\,dt + \int 5\,dt$

$\qquad = 3\cdot\dfrac{t^3}{3} - 4\cdot\dfrac{t^2}{2} + 5t + C = t^3 - 2t^2 + 5t + C$

17. Since $7 + e$ is a constant,

$\qquad \int(7+e)\,dx = (7+e)x + C$.

19. $\int\left(\dfrac{x}{7} - \dfrac{3}{4}x^4\right)dx = \dfrac{1}{7}\int x\,dx - \dfrac{3}{4}\int x^4\,dx$

$\qquad = \dfrac{1}{7}\cdot\dfrac{x^2}{2} - \dfrac{3}{4}\cdot\dfrac{x^5}{5} + C$

$\qquad = \dfrac{x^2}{14} - \dfrac{3x^5}{20} + C$

21. $\int\pi e^x\,dx = \pi\int e^x\,dx = \pi e^x + C$

23. $\int\left(x^{8.3} - 9x^6 + 3x^{-4} + x^{-3}\right)dx$

$\qquad = \dfrac{x^{9.3}}{9.3} - 9\cdot\dfrac{x^7}{7} + 3\cdot\dfrac{x^{-3}}{-3} + \dfrac{x^{-2}}{-2} + C$

$\qquad = \dfrac{x^{9.3}}{9.3} - \dfrac{9x^7}{7} - \dfrac{1}{x^3} - \dfrac{1}{2x^2} + C$

25. $\int\dfrac{-2\sqrt{x}}{3}\,dx = -\dfrac{2}{3}\int x^{\frac{1}{2}}\,dx = -\dfrac{2}{3}\cdot\dfrac{x^{\frac{1}{2}+1}}{\frac{1}{2}+1} + C$

$\qquad = -\dfrac{2}{3}\cdot\dfrac{x^{\frac{3}{2}}}{\frac{3}{2}} + C = -\dfrac{4x^{\frac{3}{2}}}{9} + C$

27. $\int\dfrac{1}{4\sqrt[8]{x^2}}\,dx = \dfrac{1}{4}\int x^{-\frac{1}{4}}\,dx = \dfrac{1}{4}\cdot\dfrac{x^{-\frac{1}{4}+1}}{-\frac{1}{4}+1} + C$

$\qquad = \dfrac{1}{4}\cdot\dfrac{x^{\frac{3}{4}}}{\frac{3}{4}} + C = \dfrac{x^{\frac{3}{4}}}{3} + C$

29. $\int\left(\dfrac{x^3}{3} - \dfrac{3}{x^3}\right)dx = \dfrac{1}{3}\int x^3\,dx - 3\int x^{-3}\,dx$

$\qquad = \dfrac{1}{3}\cdot\dfrac{x^{3+1}}{3+1} - 3\cdot\dfrac{x^{-3+1}}{-3+1} + C$

$\qquad = \dfrac{1}{3}\cdot\dfrac{x^4}{4} - 3\cdot\dfrac{x^{-2}}{-2} + C = \dfrac{x^4}{12} + \dfrac{3}{2x^2} + C$

31. $\displaystyle\int\left(\frac{3w^2}{2}-\frac{2}{3w^2}\right)dw=\frac{3}{2}\int w^2\,dw-\frac{2}{3}\int w^{-2}\,dw$

$\displaystyle=\frac{3}{2}\cdot\frac{w^3}{3}-\frac{2}{3}\cdot\frac{w^{-1}}{-1}+C=\frac{w^3}{2}+\frac{2}{3w}+C$

33. $\displaystyle\int\frac{3u-4}{5}\,du=\frac{1}{5}\int(3u-4)\,du$

$\displaystyle=\frac{1}{5}\left(3\int u\,du-4\int du\right)$

$\displaystyle=\frac{1}{5}\left(3\frac{u^2}{2}-4u\right)+C=\frac{3}{10}u^2-\frac{4}{5}u+C$

$\displaystyle=\frac{1}{7}\left(2\int z\,dz-\int 5\,dz\right)$

$\displaystyle=\frac{1}{7}\left(2\cdot\frac{z^2}{2}-5z\right)+C=\frac{1}{7}\left(z^2-5z\right)+C$

35. $\displaystyle\int(u^e+e^u)\,du=\int u^e\,du+\int e^u\,du$

$\displaystyle=\frac{u^{e+1}}{e+1}+e^u+C$

37. $\displaystyle\int\left(2\sqrt{x}-3\sqrt[4]{x}\right)dx=\int\left(2x^{\frac{1}{2}}-3x^{\frac{1}{4}}\right)dx$

$\displaystyle=2\int x^{\frac{1}{2}}\,dx-3\int x^{\frac{1}{4}}\,dx$

$\displaystyle=2\cdot\frac{x^{\frac{3}{2}}}{\frac{3}{2}}-3\cdot\frac{x^{\frac{5}{4}}}{\frac{5}{4}}+C=\frac{4x^{\frac{3}{2}}}{3}-\frac{12x^{\frac{5}{4}}}{5}+C$

39. $\displaystyle\int\left(-\frac{\sqrt[3]{x^2}}{5}-\frac{7}{2\sqrt{x}}+6x\right)dx$

$\displaystyle=\int\left(-\frac{x^{\frac{2}{3}}}{5}-\frac{7x^{-\frac{1}{2}}}{2}+6x\right)dx$

$\displaystyle=-\frac{1}{5}\int x^{\frac{2}{3}}\,dx-\frac{7}{2}\int x^{-\frac{1}{2}}\,dx+6\int x\,dx$

$\displaystyle=-\frac{1}{5}\cdot\frac{x^{\frac{5}{3}}}{\frac{5}{3}}-\frac{7}{2}\cdot\frac{x^{\frac{1}{2}}}{\frac{1}{2}}+6\cdot\frac{x^2}{2}+C$

$\displaystyle=-\frac{3x^{\frac{5}{3}}}{25}-7x^{\frac{1}{2}}+3x^2+C$

41. $\displaystyle\int\left(x^2+5\right)(x-3)\,dx$

$\displaystyle=\int\left(x^3-3x^2+5x-15\right)dx$

$\displaystyle=\frac{x^4}{4}-3\cdot\frac{x^3}{3}+5\cdot\frac{x^2}{2}-15x+C$

$\displaystyle=\frac{x^4}{4}-x^3+\frac{5x^2}{2}-15x+C$

43. $\displaystyle\int\sqrt{x}(x+3)\,dx=\int\left(x^{\frac{3}{2}}+3x^{\frac{1}{2}}\right)dx$

$\displaystyle=\frac{x^{\frac{5}{2}}}{\frac{5}{2}}+3\cdot\frac{x^{\frac{3}{2}}}{\frac{3}{2}}+C$

$\displaystyle=\frac{2x^{\frac{5}{2}}}{5}+2x^{\frac{3}{2}}+C$

45. $\displaystyle\int(3u+2)^3\,du=\int(27u^3+54u^2+36u+8)\,du$

$\displaystyle=27\cdot\frac{u^4}{4}+54\cdot\frac{u^3}{3}+36\cdot\frac{u^2}{2}+8u+C$

$\displaystyle=\frac{27}{4}u^4+18u^3+18u^2+8u+C$

47. $\int v^{-2}\left(2v^4 + 3v^2 - 2v^{-3}\right)dv$

$= \int\left(2v^2 + 3 - 2v^{-5}\right)dv$

$= 2 \cdot \dfrac{v^3}{3} + 3v - 2 \cdot \dfrac{v^{-4}}{-4} + C$

$= \dfrac{2v^3}{3} + 3v + \dfrac{1}{2v^4} + C$

49. $\int\dfrac{z^4 + 10z^3}{2z^2}\,dz = \dfrac{1}{2}\int\left(\dfrac{z^4}{z^2} + \dfrac{10z^3}{z^2}\right)dz$

$= \dfrac{1}{2}\int\left(z^2 + 10z\right)dz$

$= \dfrac{1}{2}\left(\dfrac{z^3}{3} + 10 \cdot \dfrac{z^2}{2}\right) + C$

$= \dfrac{z^3}{6} + \dfrac{5z^2}{2} + C$

51. $\int\dfrac{e^x + e^{2x}}{e^x}\,dx = \int\left(\dfrac{e^x}{e^x} + \dfrac{e^{2x}}{e^x}\right)dx$

$= \int\left(1 + e^x\right)dx$

$= x + e^x + C$

53. No, $F(x) - G(x)$ might be a nonzero constant.

55. Because an antiderivative of the derivative of a function is the function itself, we have

$\int\dfrac{d}{dx}\left[\dfrac{1}{\sqrt{x^2+1}}\right]dx = \dfrac{1}{\sqrt{x^2+1}} + C.$

Principles in Practice 14.3

1. $N(t) = \int\dfrac{dN}{dt}\,dt = \int\left(800 + 200e^t\right)dt$

$= 800t + 200e^t + C$

Since $N(5) = 40,000$, we have

$40,000 = 800(5) + 200e^5 + C$, so

$C = 40,000 - \left(4000 + 200e^5\right)$

$= 36,000 - 200e^5 \approx 6317.37$

$N(t) = 800t + 200e^t + 6317.37$

Problems 14.3

1. $\dfrac{dy}{dx} = 3x - 4$

$y = \int(3x-4)dx = \dfrac{3x^2}{2} - 4x + C$

Using $y(-1) = \dfrac{13}{2}$ gives

$\dfrac{13}{2} = \dfrac{3(-1)^2}{2} - 4(-1) + C$

$\dfrac{13}{2} = \dfrac{11}{2} + C$

Thus $C = 1$, so $y = \dfrac{3x^2}{2} - 4x + 1$.

3. $y' = \dfrac{5}{\sqrt{x}}$

$y = \int\dfrac{5}{\sqrt{x}}\,dx = \int 5x^{-\frac{1}{2}}dx$

$= 5 \cdot \dfrac{x^{\frac{1}{2}}}{\frac{1}{2}} + C = 10\sqrt{x} + C$

$y(9) = 50$ implies $50 = 10\sqrt{9} + C$,

$50 = 30 + C$, $C = 20$.

Thus $y = 10\sqrt{x} + 20$.

$y(16) = 10 \cdot 4 + 20 = 60$

5. $y'' = -3x^2 + 4x$

$y' = \int (-3x^2 + 4x)dx = -x^3 + 2x^2 + C_1$

$y'(1) = 2$ implies $2 = -1 + 2 + C_1$, so

$C_1 = 1$.

$y = \int (-x^3 + 2x^2 + 1)dx$

$= -\dfrac{x^4}{4} + \dfrac{2x^3}{3} + x + C_2$

$y(1) = 3$ implies $3 = -\dfrac{1}{4} + \dfrac{2}{3} + 1 + C_2$, so

$C_2 = \dfrac{19}{12}$. Thus $y = -\dfrac{x^4}{4} + \dfrac{2x^3}{3} + x + \dfrac{19}{12}$.

7. $y''' = 2x$

$y'' = \int 2x\, dx = x^2 + C_1$

$y''(-1) = 3$ implies that $3 = 1 + C_1$, so

$C_1 = 2$.

$y' = \int \left(x^2 + 2\right)dx = \dfrac{x^3}{3} + 2x + C_2$

$y'(3) = 10$ implies $10 = 9 + 6 + C_2$, so

$C_2 = -5$.

$y = \int \left(\dfrac{x^3}{3} + 2x - 5\right)dx = \dfrac{x^4}{12} + x^2 - 5x + C_3$.

$y(0) = 13$ implies that $13 = 0 + 0 - 0 + C_3$, so

$C_3 = 13$. Therefore $y = \dfrac{x^4}{12} + x^2 - 5x + 13$.

9. $\dfrac{dr}{dq} = 0.7$

$r = \int 0.7 dq = 0.7q + C$

If $q = 0$, r must be 0, so $0 = 0 + C$, $C = 0$.

Thus $r = 0.7q$. Since $r = pq$, we have

$p = \dfrac{r}{q} = \dfrac{0.7q}{q} = 0.7$. The demand function

is $p = 0.7$.

11. $\dfrac{dr}{dq} = 275 - q - 0.3q^2$

Thus $r = \int \left(275 - q - 0.3q^2\right)dq$

$= 275q - 0.5q^2 - 0.1q^3 + C$. When $q = 0$, r

must be 0, so $C = 0$ and

$r = 275q - 0.5q^2 - 0.1q^3$. Since $r = pq$, then

$p = \dfrac{r}{q} = 275 - 0.5q - 0.1q^2$. Thus the

demand function is $p = 275 - 0.5q - 0.1q^2$.

13. $\dfrac{dc}{dq} = 1.35$

$c = \int 1.35 dq = 1.35q + C$

When $q = 0$, then $c = 200$, so $200 = 0 + C$,

or $C = 200$. Thus $c = 1.35q + 200$.

15. $\dfrac{dc}{dq} = 0.08q^2 - 1.6q + 6.5$

$c = \int \left(0.08q^2 - 1.6q + 6.5\right)dq$

$\dfrac{0.08}{3}q^3 - 0.8q^2 + 6.5q + C$. If $q = 0$, then

$c = 8000$, from which $C = 8000$. Hence

$c = \dfrac{0.08}{3}q^3 - 0.8q^2 + 6.5q + 8000$. If $q = 25$,

substituting gives $c(25) = 8079\dfrac{1}{6}$ or

$\$8079.17$.

17. $G = \int \left[-\dfrac{P}{25} + 2\right]dP = -\dfrac{P^2}{50} + 2P + C$

When $P = 10$, then $G = 38$, so

$38 = -2 + 20 + C$, from which $C = 20$. Thus

$G = -\dfrac{1}{50}P^2 + 2P + 20$.

19. $v = \int -\dfrac{(P_1 - P_2)r}{2l\eta}\,dr = -\dfrac{(P_1 - P_2)r^2}{4l\eta} + C$

Since $v = 0$ when $r = R$, then

$0 = -\dfrac{(P_1 - P_2)R^2}{4l\eta} + C$, so

$C = \dfrac{(P_1 - P_2)R^2}{4l\eta}$. Thus

$v = -\dfrac{(P_1 - P_2)r^2}{4l\eta} + \dfrac{(P_1 - P_2)R^2}{4l\eta}$

$= \dfrac{(P_1 - P_2)\left(R^2 - r^2\right)}{4l\eta}$.

21. $\dfrac{dc}{dq} = 0.003q^2 - 0.4q + 40$

$c = \int\left(0.003q^2 - 0.4q + 40\right)dq$

$= 0.001q^3 - 0.2q^2 + 40q + C$

When $q = 0$, then $c = 5000$, so
$5000 = 0 - 0 + 0 + C$, or $C = 5000$. Thus
$c = 0.001q^3 - 0.2q^2 + 40q + 5000$. When
$q = 100$, then $c = 8000$. Since

Avg. Cost $= \bar{c} = \dfrac{\text{Total Cost}}{\text{Quantity}} = \dfrac{c}{q}$, when

$q = 100$, we have $\bar{c} = \dfrac{8000}{100} = \$80$. (Observe

that knowing $\dfrac{dc}{dq} = 27.50$ when $q = 50$ is

not relevant to the problem.)

Principles in Practice 14.4

1. Using the values given,

$\dfrac{dT}{dt} = -0.5(70 - 60)e^{-0.5t} = -5e^{-0.5t}$

$T(t) = \int \dfrac{dT}{dt}\,dt = \int -5e^{-0.5t}\,dt = 10e^{-0.5t} + C$

Problems 14.4

1. Let $u = x + 5 \Rightarrow du = 1dx = dx$

$\int (x + 5)^7\,[dx] = \int u^7\,du = \dfrac{u^8}{8} + C$

$= \dfrac{(x + 5)^8}{8} + C$

3. Let $u = x^2 + 3 \Rightarrow du = 2x\,dx$

$\int 2x\left(x^2 + 3\right)^5 dx$

$= \int \left(x^2 + 3\right)^5 [2x\,dx] = \int u^5\,du = \dfrac{u^6}{6} + C$

$= \dfrac{\left(x^2 + 3\right)^6}{6} + C$

5. Let $u = y^3 + 3y^2 + 1 \Rightarrow du = \left(3y^2 + 6y\right)dy$

$\int \left(3y^2 + 6y\right)\left(y^3 + 3y^2 + 1\right)^{\frac{2}{3}}\,dy$

$= \int \left(y^3 + 3y^2 + 1\right)^{\frac{2}{3}}\left[\left(3y^2 + 6y\right)dy\right]$

$= \int u^{\frac{2}{3}}\,du = \dfrac{u^{\frac{5}{3}}}{\frac{5}{3}} + C$

$= \dfrac{3}{5}\left(y^3 + 3y^2 + 1\right)^{\frac{5}{3}} + C$

7. Let $u = 3x - 1 \Rightarrow du = 3\,dx$

$\int \dfrac{5}{(3x - 1)^3}\,dx = \dfrac{5}{3}\int \dfrac{1}{(3x - 1)^3}[3\,dx]$

$= \dfrac{5}{3}\int \dfrac{1}{u^3}\,du = \dfrac{5}{3}\int u^{-3}\,du$

$= \dfrac{5}{3}\cdot\dfrac{u^{-2}}{-2} + C = -\dfrac{5(3x - 1)^{-2}}{6} + C$

9. Let $u = 2x - 1 \Rightarrow du = 2\ dx$.

$$\int \sqrt{2x-1}\,dx = \int (2x-1)^{\frac{1}{2}}\,dx$$

$$= \frac{1}{2}\int (2x-1)^{\frac{1}{2}}[2\ dx]$$

$$= \frac{1}{2}\int u^{\frac{1}{2}}\,du = \frac{1}{2}\cdot\frac{u^{\frac{3}{2}}}{\frac{3}{2}} + C = \frac{1}{3}(2x-1)^{\frac{3}{2}} + C$$

11. Let $u = 7x - 6 \Rightarrow du = 7\ dx$

$$\int (7x-6)^4\,dx = \frac{1}{7}\int (7x-6)^4[7\ dx]$$

$$= \frac{1}{7}\int u^4\,du = \frac{1}{7}\cdot\frac{u^5}{5} + C$$

$$= \frac{(7x-6)^5}{35} + C$$

13. Let $v = 5u^2 - 9 \Rightarrow dv = 10u\ du$

$$\int u(5u^2-9)^{14}\,du = \frac{1}{10}\int (5u^2-9)^{14}[10u\ du]$$

$$\frac{1}{10}\int v^{14}\,dv = \frac{1}{10}\cdot\frac{v^{15}}{15} + C = \frac{(5u^2-9)^{15}}{150} + C$$

15. Let $u = 27 + x^5 \Rightarrow du = 5x^4\,dx$

$$\int 4x^4\left(27+x^5\right)^{\frac{1}{3}}\,dx$$

$$= \frac{4}{5}\int\left(27+x^5\right)^{\frac{1}{3}}\left[5x^4\,dx\right]$$

$$= \frac{4}{5}\int u^{\frac{1}{3}}\,du = \frac{4}{5}\cdot\frac{u^{\frac{4}{3}}}{\frac{4}{3}} + C$$

$$= \frac{3}{5}\left(27+x^5\right)^{\frac{4}{3}} + C$$

17. Let $u = 3x \Rightarrow du = 3\ dx$

$$\int 3e^{3x}\,dx = \int e^{3x}[3\ dx]$$

$$= \int e^u\,du = e^u + C = e^{3x} + C$$

19. Let $u = t^2 + t \Rightarrow du = (2t+1)dt$

$$\int (2t+1)e^{t^2+t}\,dt = \int e^{t^2+t}[(2t+1)\ dt]$$

$$= \int e^u\,du = e^u + C = e^{t^2+t} + C$$

21. Let $u = 7x^2 \Rightarrow du = 14x\,dx$

$$\int xe^{7x^2}\,dx = \frac{1}{14}\int e^{7x^2}[14x\ dx] = \frac{1}{14}\int e^u\,du$$

$$= \frac{1}{14}e^u + C = \frac{1}{14}e^{7x^2} + C$$

23. Let $u = -3x \Rightarrow du = -3dx$.

$$\int 4e^{-3x}\,dx = -\frac{4}{3}\int e^{-3x}[-3\ dx]$$

$$= -\frac{4}{3}\int e^u\,du = -\frac{4}{3}e^u + C = -\frac{4}{3}e^{-3x} + C$$

25. Let $u = x + 5 \Rightarrow du = dx$

$$\int\frac{1}{x+5}[dx] = \int\frac{1}{u}\,du = \ln|u| + C$$

$$= \ln|x+5| + C$$

27. Let $u = x^3 + x^4 \Rightarrow du = (3x^2+4x^3)dx$

$$\int\frac{3x^2+4x^3}{x^3+x^4}\,dx = \int\frac{1}{x^3+x^4}\left[\left(3x^2+4x^3\right)dx\right]$$

$$= \int\frac{1}{u}\,du = \ln|u| + C$$

$$= \ln|x^3+x^4| + C$$

29. Let $u = z^2 - 6 \Rightarrow du = 2z\ dz$

$$\int\frac{6z}{(z^2-6)^5} = 3\int(z^2-6)^{-5}[2z\ dz]$$

$$= 3\int u^{-5}\,du = 3\frac{u^{-4}}{-4} + C = -\frac{3}{4}(z^2-6)^{-4} + C$$

31. $\int\frac{4}{x}\,dx = 4\int\frac{1}{x}\,dx = 4\ln|x| + C$

33. Let $u = s^3 + 5 \Rightarrow du = 3s^2\,ds$

$$\int \frac{s^2}{s^3+5}\,ds = \frac{1}{3}\int \frac{1}{s^3+5}\left[3s^2\,ds\right]$$

$$= \frac{1}{3}\int \frac{1}{u}\,du = \frac{1}{3}\ln|u| + C = \frac{1}{3}\ln\left|s^3+5\right| + C$$

35. Let $u = 4 - 2x \Rightarrow du = -2\,dx$

$$\int \frac{5}{4-2x}\,dx = -\frac{5}{2}\int \frac{1}{4-2x}[-2\,dx]$$

$$= -\frac{5}{2}\int \frac{1}{u}\,du = -\frac{5}{2}\ln|u| + C$$

$$= -\frac{5}{2}\ln|4-2x| + C$$

37. $\int \sqrt{5x}\,dx = \sqrt{5}\int x^{1/2}\,dx = \sqrt{5}\,\dfrac{x^{\frac{3}{2}}}{\frac{3}{2}} + C$

$$= \frac{2\sqrt{5}}{3}x^{\frac{3}{2}} + C$$

39. Let $u = x^2 - 4 \Rightarrow du = 2x\,dx$

$$\int \frac{x}{\sqrt{x^2-4}}\,dx = \frac{1}{2}\int \left(x^2-4\right)^{-\frac{1}{2}}[2x\,dx]$$

$$= \frac{1}{2}\int u^{-\frac{1}{2}}\,du = \frac{1}{2}\cdot\frac{u^{\frac{1}{2}}}{\frac{1}{2}} + C$$

$$= \sqrt{x^2-4} + C$$

41. Let $u = y^4 + 1 \Rightarrow du = 4y^3\,dy$

$$\int 2y^3 e^{y^4+1}\,dy = 2\int y^3 e^{y^4+1}\,dy$$

$$= 2\cdot\frac{1}{4}\int e^{y^4+1}\left[4y^3\,dy\right]$$

$$= \frac{1}{2}\int e^u\,du = \frac{1}{2}e^u + C$$

$$= \frac{1}{2}e^{y^4+1} + C$$

43. Let $u = -2v^3 + 1 \Rightarrow du = -6v^2\,dv$

$$\int v^2 e^{-2v^3+1}\,dv = -\frac{1}{6}\int e^{-2v^3+1}\left[-6v^2\,dv\right]$$

$$= -\frac{1}{6}\int e^u\,du = -\frac{1}{6}e^u + C$$

$$= -\frac{1}{6}e^{-2v^3+1} + C$$

45. $\int \left(e^{-5x} + 2e^x\right)dx = \int e^{-5x}\,dx + 2\int e^x\,dx$

$$= -\frac{1}{5}\int e^{-5x}[-5\,dx] + 2\int e^x\,dx$$

$$= -\frac{1}{5}e^{-5x} + 2e^x + C$$

47. $\int (8x+10)(7-2x^2-5x)^3\,dx$

$$= -2\int (7-2x^2-5x)^3[(-4x-5)dx]$$

$$= -2\cdot\frac{(7-2x^2-5x)^4}{4} + C$$

$$= -\frac{1}{2}(7-2x^2-5x)^4 + C$$

49. $\int \dfrac{x^2+2}{x^3+6x}\,dx = \dfrac{1}{3}\int \dfrac{1}{x^3+6x}\left[\left(3x^2+6\right)dx\right]$

$$= \frac{1}{3}\ln\left|x^3+6x\right| + C$$

51. $\int \dfrac{16s-4}{3-2s+4s^2}\,dx$

$$= 2\int \frac{1}{3-2s+4s^2}[(8s-2)ds]$$

$$= 2\ln\left|3-2s+4s^2\right| + C$$

53. $\int x\left(2x^2+1\right)^{-1}dx = \int \dfrac{x}{2x^2+1}\,dx$

$$= \frac{1}{4}\int \frac{1}{2x^2+1}[4x\,dx]$$

$$= \frac{1}{4}\ln\left(2x^2+1\right) + C$$

55. $\int -\left(x^2 - 2x^5\right)\left(x^3 - x^6\right)^{-10} dx$

$$= -\frac{1}{3} \int \left(x^3 - x^6\right)^{-10} \left[\left(3x^2 - 6x^5\right) dx\right]$$

$$= -\frac{1}{3} \cdot \frac{\left(x^3 - x^6\right)^{-9}}{-9} + C = \frac{1}{27}\left(x^3 - x^6\right)^{-9} + C$$

57. $\int \left(2x^3 + x\right)\left(x^4 + x^2\right) dx$

$$= \frac{1}{2} \int \left(x^4 + x^2\right)^1 \left[\left(4x^3 + 2x\right) dx\right]$$

$$= \frac{1}{2} \cdot \frac{\left(x^4 + x^2\right)^2}{2} + C = \frac{1}{4}\left(x^4 + x^2\right)^2 + C$$

59. $\int \dfrac{7 + 14x}{(4 - x - x^2)^5} dx$

$$= -7 \int (4 - x - x^2)^{-5} [(-1 - 2x) dx]$$

$$= -7 \frac{(4 - x - x^2)^{-4}}{-4} + C$$

$$= \frac{7}{4}(4 - x - x^2)^{-4} + C$$

61. $u = 4x^3 + 3x^2 - 4$

$du = \left(12x^2 + 6x\right) dx = 6x(2x + 1) dx$

$\int x(2x + 1)e^{4x^3 + 3x^2 - 4} dx$

$$= \frac{1}{6} \int e^{4x^3 + 3x^2 - 4} [6x(2x + 1) dx]$$

$$= \frac{1}{6} \int e^u du = \frac{1}{6} e^u + C = \frac{1}{6} e^{4x^3 + 3x^2 - 4} + C$$

63. $\int x\sqrt{\left(8 - 5x^2\right)^3} dx$

$$= -\frac{1}{10} \int \left(8 - 5x^2\right)^{\frac{3}{2}} [-10x \, dx]$$

$$= -\frac{1}{10} \cdot \frac{\left(8 - 5x^2\right)^{\frac{5}{2}}}{\frac{5}{2}} + C = -\frac{1}{25}\left(8 - 5x^2\right)^{\frac{5}{2}} + C$$

65. $\int \left(\sqrt{2x} - \dfrac{1}{\sqrt{2x}}\right) dx = \int \sqrt{2x} \, dx - \int \dfrac{1}{\sqrt{2x}} dx$

$$= \frac{1}{2} \int (2x)^{\frac{1}{2}} [2 \, dx] - \frac{1}{2} \int (2x)^{-\frac{1}{2}} [2 \, dx]$$

$$= \frac{1}{2} \cdot \frac{(2x)^{\frac{3}{2}}}{\frac{3}{2}} - \frac{1}{2} \cdot \frac{(2x)^{\frac{1}{2}}}{\frac{1}{2}} + C$$

$$= \frac{(2x)^{\frac{3}{2}}}{3} - \sqrt{2x} + C$$

$$= \frac{2\sqrt{2}}{3} x^{\frac{3}{2}} - \sqrt{2} x^{\frac{1}{2}} + C$$

67. $\int \left(x^2 + 1\right)^2 dx = \int \left(x^4 + 2x^2 + 1\right) dx$

$$= \frac{x^5}{5} + \frac{2x^3}{3} + x + C$$

69. $\int \left[\dfrac{x}{x^2 + 1} + \dfrac{x^5}{(x^6 + 1)^2}\right] dx$

$$= \int \frac{x}{x^2 + 1} dx + \int \frac{x^5}{\left(x^6 + 1\right)^2} dx$$

$$= \frac{1}{2} \int \frac{1}{x^2 + 1} [2x \, dx] + \frac{1}{6} \int \left(x^6 + 1\right)^{-2} \left[6x^5 dx\right]$$

$$= \frac{1}{2} \ln\left(x^2 + 1\right) + \frac{1}{6} \cdot \frac{\left(x^6 + 1\right)^{-1}}{-1} + C$$

$$= \frac{1}{2} \ln\left(x^2 + 1\right) - \frac{1}{6\left(x^6 + 1\right)} + C$$

71. $\int \left[\dfrac{2}{4x+1} - (4x^2 - 8x^5)(x^3 - x^6)^{-8} \right] dx$

$= \dfrac{1}{2} \int \dfrac{1}{4x+1} [4\,dx] - \dfrac{4}{3} \int (x^3 - x^6)^{-8} [(3x^2 - 6x^5)dx]$

$= \dfrac{1}{2} \ln|4x+1| - \dfrac{4}{3} \cdot \dfrac{(x^3 - x^6)^{-7}}{-7} + C$

$= \dfrac{1}{2} \ln|4x+1| + \dfrac{4}{21}(x^3 - x^6)^{-7} + C$

73. $\int \left[\sqrt{3x+1} - \dfrac{x}{x^2+3} \right] dx = \int (3x+1)^{\frac{1}{2}}\,dx - \int \dfrac{x}{x^2+3}\,dx = \dfrac{1}{3} \int (3x+1)^{\frac{1}{2}} [3\,dx] - \dfrac{1}{2} \int \dfrac{1}{x^2+3} [2x\,dx]$

$= \dfrac{1}{3} \cdot \dfrac{(3x+1)^{\frac{3}{2}}}{\frac{3}{2}} - \dfrac{1}{2} \ln\left(x^2 + 3\right) + C = \dfrac{2}{9}(3x+1)^{\frac{3}{2}} - \ln\sqrt{x^2+3} + C$

75. Let $u = \sqrt{x} \Rightarrow du = \dfrac{1}{2} x^{-\frac{1}{2}}\,dx = \dfrac{1}{2\sqrt{x}}\,dx$.

$\int \dfrac{e^{\sqrt{x}}}{\sqrt{x}}\,dx = 2 \int e^{\sqrt{x}} \left[\dfrac{1}{2\sqrt{x}}\,dx \right]$

$= 2 \int e^u\,du = 2e^u + C = 2e^{\sqrt{x}} + C$

77. $\int \dfrac{1 + e^{2x}}{4e^x}\,dx = \dfrac{1}{4} \int \left(\dfrac{1}{e^x} + \dfrac{e^{2x}}{e^x} \right) dx$

$= \dfrac{1}{4} \int \left(e^{-x} + e^x \right) dx$

$= -\dfrac{1}{4} \int e^{-x} [-1\,dx] + \dfrac{1}{4} \int e^x\,dx$

$= -\dfrac{1}{4} e^{-x} + \dfrac{1}{4} e^x + C$

79. Let $u = \ln\left(x^2 + 2x\right) \Rightarrow du = \dfrac{1}{x^2+2x}(2x+2)dx$

$\int \dfrac{x+1}{x^2+2x} \ln\left(x^2 + 2x\right) dx$

$= \dfrac{1}{2} \int \ln\left(x^2 + 2x\right) \left[\dfrac{2x+2}{x^2+2x}\,dx \right]$

$= \dfrac{1}{2} \int u\,du = \dfrac{1}{2} \cdot \dfrac{u^2}{2} + C = \dfrac{1}{4} \ln^2\left(x^2 + 2x\right) + C$

81. $y = \int (3-2x)^2 \, dx = -\frac{1}{2} \int (3-2x)^2 [-2 \, dx]$

$= -\frac{1}{2} \cdot \frac{(3-2x)^3}{3} + C = -\frac{1}{6}(3-2x)^3 + C$

$y(0) = 1$ implies $1 = -\frac{1}{6}(27) + C$, so

$C = \frac{11}{2}$. Thus $y = -\frac{1}{6}(3-2x)^3 + \frac{11}{2}$.

83. $y'' = \frac{1}{x^2}$

$y' = \int x^{-2} \, dx = -x^{-1} + C_1$

$y'(-2) = 3$ implies $3 = \frac{1}{2} + C_1$, so $C_1 = \frac{5}{2}$.

Thus $y' = -x^{-1} + \frac{5}{2}$.

$y = \int \left(-x^{-1} + \frac{5}{2} \right) dx = -\int \frac{1}{x} \, dx + \int \frac{5}{2} \, dx$

$= -\ln|x| + \frac{5}{2}x + C_2$

$y(1) = 2$ implies that $2 = 0 + \frac{5}{2} + C_2$, so

$C_2 = -\frac{1}{2}$. Thus

$y = -\ln|x| + \frac{5}{2}x - \frac{1}{2} = \ln\left|\frac{1}{x}\right| + \frac{5}{2}x - \frac{1}{2}$.

85. $V(t) = \int \frac{dV}{dt} \, dt = \int 8e^{0.05t} \, dt$

$= \frac{8}{0.05} \int e^{0.05t} [0.05 \, dt]$

$= 160e^{0.05t} + C$

The house cost \$350,000 to build, so
$V(0) = 350$.

$350 = 160e^0 + C = 160 + C$

$190 = C$

$V(t) = 160e^{0.05t} + 190$

87. Note that $r > 0$.

$C = \int \left[\frac{Rr}{2K} + \frac{B_1}{r} \right] dr = \int \frac{Rr}{2K} \, dr + \int \frac{B_1}{r} \, dr$

$= \frac{R}{2K} \int r \, dr + B_1 \int \frac{1}{r} \, dr$

$= \frac{R}{2K} \cdot \frac{r^2}{2} + B_1 \ln|r| + B_2$

Thus we obtain $C = \frac{Rr^2}{4K} + B_1 \ln|r| + B_2$.

Problems 14.5

1. $\int \frac{2x^6 + 8x^4 - 4x}{2x^2} \, dx$

$= \int \left(\frac{2x^6}{2x^2} + \frac{8x^4}{2x^2} - \frac{4x}{2x^2} \right) dx$

$= \int x^4 \, dx + 4 \int x^2 \, dx - 2 \int \frac{1}{x} \, dx$

$= \frac{x^5}{5} + \frac{4}{3}x^3 - 2\ln|x| + C$

3. $\int (3x^2 + 2)\sqrt{2x^3 + 4x + 1} \, dx$

$= \frac{1}{2} \int (2x^3 + 4x + 1)^{\frac{1}{2}} \left[(6x^2 + 4) dx \right]$

$= \frac{1}{2} \cdot \frac{(2x^3 + 4x + 1)^{\frac{3}{2}}}{\frac{3}{2}} + C$

$= \frac{1}{3}(2x^3 + 4x + 1)^{\frac{3}{2}} + C$

5. $\int \frac{9}{\sqrt{2-3x}} \, dx = 9 \int (2-3x)^{-1/2} \, dx$

$= 9\left(-\frac{1}{3} \right) \int (2-3x)^{-1/2} [-3 \, dx]$

$= -3 \frac{(2-3x)^{1/2}}{\frac{1}{2}} + C = -6\sqrt{2-3x} + C$

7. $\int 4^{7x}\,dx = \int\left(e^{\ln 4}\right)^{7x}\,dx = \int e^{(\ln 4)(7x)}\,dx$

$= \dfrac{1}{7\ln 4}\int e^{(\ln 4)(7x)}[7\ln 4\,dx]$

$= \dfrac{1}{7\ln 4}\cdot e^{(\ln 4)(7x)} + C$

$= \dfrac{1}{7\ln 4}\left(e^{\ln 4}\right)^{7x} + C = \dfrac{4^{7x}}{7\ln 4} + C$

9. $\int 2x\left(7 - e^{\frac{x^2}{4}}\right)dx = \int\left(14x - 2xe^{\frac{x^2}{4}}\right)dx$

$= 14\int x\,dx - 2\int xe^{\frac{x^2}{4}}\,dx$

$= 14\int x\,dx - 2\cdot 2\int e^{\frac{x^2}{4}}\left[\dfrac{1}{2}x\,dx\right]$

$= 14\cdot\dfrac{x^2}{2} - 4\cdot e^{\frac{x^2}{4}} + C = 7x^2 - 4e^{\frac{x^2}{4}} + C$

11. By long division,

$\dfrac{6x^2 - 11x + 5}{3x - 1} = 2x - 3 + \dfrac{2}{3x - 1}$. Thus

$\int\dfrac{6x^2 - 11x + 5}{3x - 1}\,dx = \int\left(2x - 3 + \dfrac{2}{3x - 1}\right)dx$

$= 2\int x\,dx - \int 3\,dx + 2\cdot\dfrac{1}{3}\int\dfrac{1}{3x - 1}[3\,dx]$

$= x^2 - 3x + \dfrac{2}{3}\ln|3x - 1| + C$

13. $\int\dfrac{5e^{2x}}{7e^{2x} + 4}\,dx = \dfrac{5}{14}\int\dfrac{1}{7e^{2x} + 4}[7e^{2x}(2)dx]$

$= \dfrac{5}{14}\ln(7e^{2x} + 4) + C$

15. $\int\dfrac{e^{\frac{7}{x}}}{x^2}\,dx = \int e^{\frac{7}{x}}\cdot\dfrac{1}{x^2}\,dx = -\dfrac{1}{7}\int e^{\frac{7}{x}}\left[-\dfrac{7}{x^2}\,dx\right]$

$= -\dfrac{1}{7}e^{\frac{7}{x}} + C$

17. By using long division on the integrand,

$\int\dfrac{5x^3}{x^2 + 9}\,dx = \int\left(5x - \dfrac{45x}{x^2 + 9}\right)dx$

$= \int 5x\,dx - \dfrac{45}{2}\int\dfrac{1}{x^2 + 9}[2x\,dx]$

$= \dfrac{5}{2}x^2 - \dfrac{45}{2}\ln(x^2 + 9) + C$

Note that since $x^2 + 9 > 0$ for all values of x, the absolute value bars are not needed.

19. $\int\dfrac{\left(\sqrt{x} + 2\right)^2}{3\sqrt{x}}\,dx = \dfrac{2}{3}\int\left(\sqrt{x} + 2\right)^2\left[\dfrac{1}{2\sqrt{x}}\,dx\right]$

$= \dfrac{2}{3}\cdot\dfrac{\left(\sqrt{x} + 2\right)^3}{3} + C = \dfrac{2}{9}\left(\sqrt{x} + 2\right)^3 + C$

21. $\int\dfrac{5\left(x^{\frac{1}{3}} + 2\right)^4}{\sqrt[3]{x^2}}\,dx = 3\int 5\left(x^{\frac{1}{3}} + 2\right)^4\left[\dfrac{1}{3}x^{-\frac{2}{3}}\,dx\right]$

$= 3\left(x^{\frac{1}{3}} + 2\right)^5 + C$

23. $\int\dfrac{\ln x}{x}\,dx = \int(\ln x)\left[\dfrac{1}{x}\,dx\right] = \dfrac{(\ln x)^2}{2} + C$

$= \dfrac{1}{2}\left(\ln^2 x\right) + C$

25. $\int\dfrac{\ln^2(r + 1)}{r + 1}\,dr = \int[\ln(r + 1)]^2\left(\dfrac{1}{r + 1}\,dr\right)$

$= \dfrac{1}{3}\ln^3(r + 1) + C$

27. $\displaystyle\int \frac{3^{\ln x}}{x}\,dx = \int \frac{\left(e^{\ln 3}\right)^{\ln x}}{x}\,dx$

$\displaystyle = \frac{1}{\ln 3}\int e^{(\ln 3)\ln x}\left[\frac{\ln 3}{x}\,dx\right]$

$\displaystyle = \frac{1}{\ln 3}\cdot e^{(\ln 3)\ln x} + C$

$\displaystyle = \frac{1}{\ln 3}\left(e^{\ln 3}\right)^{\ln x} + C = \frac{3^{\ln x}}{\ln 3} + C$

29. $\displaystyle\int x^2\sqrt{e^{x^3+1}}\,dx = \int x^2 (e^{x^3+1})^{1/2}\,dx$

$\displaystyle = \frac{2}{3}\int e^{\frac{x^3+1}{2}}\left[\frac{3}{2}x^2\,dx\right] = \frac{2}{3}e^{\frac{x^3+1}{2}}\,dx$

31. $\displaystyle\int \frac{8}{(x+3)\ln(x+3)}\,dx = 8\int\frac{1}{\ln(x+3)}\left[\frac{1}{x+3}\,dx\right]$

$\displaystyle = 8\ln\big|\ln(x+3)\big| + C$

33. By using long division on the integrand,

$\displaystyle\int\frac{x^3+x^2-x-3}{x^2-3}\,dx = \int\left(x+1+\frac{2x}{x^2-3}\right)dx$

$\displaystyle = \int(x+1)dx + \int\frac{1}{x^2-3}[2x\,dx]$

$\displaystyle = \frac{x^2}{2}+x+\ln\big|x^2-3\big| + C$

35. $\displaystyle\int\frac{6x^2\sqrt{\ln(x^3+1)^2}}{x^3+1}\,dx$

$\displaystyle = \int[2\ln(x^3+1)]^{1/2}\left[\frac{6x^2}{x^3+1}\,dx\right]$

$\displaystyle = \frac{[2\ln(x^3+1)]^{3/2}}{\frac{3}{2}} + C$

$\displaystyle = \frac{2}{3}\ln^{3/2}(x^3+1)^2 + C$

37. $\displaystyle\int\left(\frac{x^3-1}{\sqrt{x^4-4x}} - \ln 7\right)dx$

$\displaystyle = \frac{1}{4}\int(x^4-4x)^{\frac{1}{2}}[(4x^3-4)dx] - \ln 7\int dx$

$\displaystyle = \frac{1}{6}(x^4-4x)^{\frac{3}{2}} - (\ln 7)x + C$

39. $\displaystyle\int\frac{2x^4-8x^3-6x^2+4}{x^3}\,dx$

$\displaystyle = \int\left(2x-8-\frac{6}{x}+\frac{4}{x^3}\right)dx$

$\displaystyle = 2\int x\,dx - \int 8\,dx - 6\int\frac{1}{x}\,dx + 4\int x^{-3}\,dx$

$\displaystyle = 2\cdot\frac{x^2}{2} - 8x - 6\ln|x| + 4\cdot\frac{x^{-2}}{-2} + C$

$\displaystyle = x^2 - 8x - 6\ln|x| - \frac{2}{x^2} + C$

41. By using long division on the integrand,

$\displaystyle\int\frac{x}{x+1}\,dx = \int\left(1-\frac{1}{x+1}\right)dx = x - \ln|x+1| + C$

43. $\displaystyle\int\frac{xe^{x^2}}{\sqrt{e^{x^2}+2}}\,dx = \frac{1}{2}\int\left(e^{x^2}+2\right)^{-\frac{1}{2}}\left[2xe^{x^2}\,dx\right]$

$\displaystyle = \frac{1}{2}\cdot\frac{\left(e^{x^2}+2\right)^{\frac{1}{2}}}{\frac{1}{2}} + C = \sqrt{e^{x^2}+2} + C$

45. $\displaystyle\int\frac{\left(e^{-x}+6\right)^2}{e^x}\,dx = -\int\left(e^{-x}+6\right)^2\left[-e^{-x}\,dx\right]$

$\displaystyle = -\frac{\left(e^{-x}+6\right)^3}{3} + C$

47. $\int \left(x^3 + ex\right)\sqrt{x^2 + e}\, dx$

$$= \int x\left(x^2 + e\right)\left(x^2 + e\right)^{\frac{1}{2}} dx$$

$$= \frac{1}{2}\int \left(x^2 + e\right)^{\frac{3}{2}} [2x\, dx] = \frac{1}{2}\cdot\frac{\left(x^2 + e\right)^{\frac{5}{2}}}{\frac{5}{2}} + C$$

$$= \frac{1}{5}\left(x^2 + e\right)^{\frac{5}{2}} + C$$

49. $\int \sqrt{x}\sqrt{(8x)^{\frac{3}{2}} + 3}\, dx = \int \left(8^{\frac{3}{2}} x^{\frac{3}{2}} + 3\right)^{\frac{1}{2}} \cdot x^{\frac{1}{2}} dx$

$$= \frac{2}{3\cdot 8^{\frac{3}{2}}}\int \left(8^{\frac{3}{2}} x^{\frac{3}{2}} + 3\right)^{\frac{1}{2}}\left[8^{\frac{3}{2}}\cdot\frac{3}{2}\cdot x^{\frac{1}{2}} dx\right]$$

$$= \frac{2}{3\cdot 16\sqrt{2}}\cdot\frac{\left(8^{\frac{3}{2}} x^{\frac{3}{2}} + 3\right)^{\frac{3}{2}}}{\frac{3}{2}} + C$$

$$= \frac{1}{36\sqrt{2}}\left[(8x)^{\frac{3}{2}} + 3\right]^{\frac{3}{2}} + C$$

51. $\int \dfrac{\sqrt{s}}{e^{\sqrt{s^3}}}\, ds = -\dfrac{2}{3}\int e^{-s^{\frac{3}{2}}}\left[-\dfrac{3}{2}s^{\frac{1}{2}} ds\right]$

$$= -\frac{2}{3}e^{-\sqrt{s^3}} + C$$

53. $e^{\ln(x^2+1)}$ is simply $x^2 + 1$. Thus

$$\int e^{\ln(x^2+1)}\, dx = \int (x^2 + 1)\, dx = \frac{1}{3}x^3 + x + C$$

55. $\int \dfrac{\ln\left(xe^x\right)}{x}\, dx = \int \dfrac{\ln x + \ln e^x}{x}\, dx$

$$= \int \frac{\ln x + x}{x}\, dx = \int \left(\frac{\ln x}{x} + 1\right) dx$$

$$= \int (\ln x)\left[\frac{1}{x}dx\right] + \int 1\, dx = \frac{\ln^2 x}{2} + x + C$$

57. $\dfrac{dr}{dq} = \dfrac{200}{(q+2)^2}$

$$r = \int 200(q+2)^{-2}\, dq = 200\cdot\frac{(q+2)^{-1}}{-1} + C$$

$$= -\frac{200}{q+2} + C$$

When $q = 0$, then $r = 0$, so $0 = -100 + C$, or

$C = 100$. Hence $r = -\dfrac{200}{q+2} + 100 = \dfrac{100q}{q+2}$.

Since $r = pq$, then $p = \dfrac{r}{q} = \dfrac{100}{q+2}$.

The demand function is $p = \dfrac{100}{q+2}$.

59. $\dfrac{dc}{dq} = \dfrac{20}{q+5}$

$$c = \int \frac{20}{q+5}\, dq = 20\int \frac{1}{q+5}\, dq$$

$$= 20\ln|q+5| + C$$

When $q = 0$, then $c = 2000$, so
$2000 = 20\ln(5) + C$, or $C = 2000 - 20\ln 5$.
Hence $c = 20\ln|q+5| + 2000 - 20\ln 5$

$$= 20\left(\ln|q+5| - \ln 5\right) + 2000$$

$$= 20\ln\left|\frac{q+5}{5}\right| + 2000$$

The cost function is $c = 20\ln\left|\dfrac{q+5}{5}\right| + 2000$.

61. $\dfrac{dC}{dI} = \dfrac{1}{\sqrt{I}}$

$$C = \int I^{-\frac{1}{2}}\, dI = \frac{I^{\frac{1}{2}}}{\frac{1}{2}} + C_1 = 2\sqrt{I} + C_1$$

$C(9) = 8$ implies that $8 = 2\cdot 3 + C_1$, or

$C_1 = 2$. Thus $C = 2\sqrt{I} + 2 = 2\left(\sqrt{I} + 1\right)$.

The consumption function is $C = 2\left(\sqrt{I} + 1\right)$.

63. $\dfrac{dC}{dI} = \dfrac{3}{4} - \dfrac{1}{6\sqrt{I}}$

$C = \int\left[\dfrac{3}{4} - \dfrac{I^{-\frac{1}{2}}}{6}\right]dI = \int\dfrac{3}{4}dI - \dfrac{1}{6}\int I^{-\frac{1}{2}}dI$

$= \dfrac{3}{4}I - \dfrac{1}{6}\cdot\dfrac{I^{\frac{1}{2}}}{\frac{1}{2}} + C_1 = \dfrac{3}{4}I - \dfrac{\sqrt{I}}{3} + C_1$

Thus $C = \dfrac{3}{4}I - \dfrac{\sqrt{I}}{3} + C_1$.

$C(25) = 23$ implies that $23 = \dfrac{3}{4}\cdot 25 - \dfrac{5}{3} + C_1$,

so $C_1 = \dfrac{71}{12}$.

The consumption function is

$C = \dfrac{3}{4}I - \dfrac{1}{3}\sqrt{I} + \dfrac{71}{12}$.

65. $\dfrac{dc}{dq} = \dfrac{100q^2 - 3998q + 60}{q^2 - 40q + 1}$

a. $\left.\dfrac{dc}{dq}\right|_{q=40} = \dfrac{100(40)^2 - 3998(40) + 60}{(40)^2 - 40(40) + 1}$

$= \$140$ per unit

b. To find c, we integrate $\dfrac{dc}{dq}$ by using long division:

$c = \int\dfrac{100q^2 - 3998q + 60}{q^2 - 40q + 1}dq$

$= \int\left(100 + \dfrac{2q - 40}{q^2 - 40q + 1}\right)dq$

$= \int 100\,dq + \int\dfrac{1}{q^2 - 40q + 1}[(2q - 40)dq]$

Thus $c = 100q + \ln\left|q^2 - 40q + 1\right| + C$.

When $q = 0$, then $c = 10,000$, so
$10,000 = 0 + \ln(1) + C$, so $C = 10,000$.

Hence

$c = 100q + \ln\left|q^2 - 40q + 1\right| + 10,000$.

When $q = 40$, then
$c = 4000 + \ln(1) + 10,000 = \$14,000$.

c. If $c = f(q)$, then

$f(q + dq) \approx f(q) + dc = f(q) + \dfrac{dc}{dq}dq$

Letting $q = 40$ and $dq = 2$, we have

$f(42) = f(40 + 2) \approx f(40) + \left.\dfrac{dc}{dq}\right|_{q=40}\cdot(2)$

$= 14,000 + 140(2) = \$14,280$

67. $\dfrac{dV}{dt} = \dfrac{8t^3}{\sqrt{0.2t^4 + 8000}}$

$V = \int\dfrac{8t^3}{\sqrt{0.2t^4 + 8000}}dt$

$= 10\int\left(0.2t^4 + 8000\right)^{-\frac{1}{2}}\left[0.8t^3\right]dt$

$= 10\dfrac{\left(0.2t^4 + 8000\right)^{\frac{1}{2}}}{\frac{1}{2}} + C$

Thus $V = 20\sqrt{0.2t^4 + 8000} + C$. If $t = 0$,
then $V = 500$, so $500 = 20\sqrt{8000} + C$,
$500 = 20\sqrt{1600\cdot 5} + C$, $500 = 800\sqrt{5} + C$,
or $C = 500 - 800\sqrt{5}$. Hence
$V = 20\sqrt{0.2t^4 + 8000} + 500 - 800\sqrt{5}$.
When $t = 10$, then
$V = 20\sqrt{10,000} + 500 - 800\sqrt{5}$
$= 20(100) + 500 - 800\sqrt{5} \approx \$711$ per acre.

69. $S = \int\dfrac{dS}{dI}dI = \int\dfrac{5}{(I + 2)^2}dI$

$= 5\int(I + 2)^{-2}dI = 5\cdot\dfrac{(I + 2)^{-1}}{-1} + C_1$

Thus $S = -\dfrac{5}{I + 2} + C_1$. If C is the total
national consumption (in billions of dollars),

then $C + S = I$, or $C = I - S$. Hence $C = I + \dfrac{5}{I+2} - C_1$. When $I = 8$, then $C = 7.5$, so $7.5 = 8 + \dfrac{1}{2} - C_1$, or

$C_1 = 1$. Thus $S = 1 - \dfrac{5}{I+2}$. If $S = 0$, then $0 = 1 - \dfrac{5}{I+2} \Rightarrow \dfrac{5}{I+2} = 1 \Rightarrow 5 = I + 2 \Rightarrow I = 3$

Principles in Practice 14.6

1. Divide the interval $[0, 10]$ into n subintervals of equal length Δx, so $\Delta x = \dfrac{10}{n}$. The endpoints of the

subintervals are $0, \dfrac{10}{n}, 2\left(\dfrac{10}{n}\right), 3\left(\dfrac{10}{n}\right), \dots, (n-1)\left(\dfrac{10}{n}\right)$, and $n\left(\dfrac{10}{n}\right) = 10$. Letting S_n denote the sum of

the areas of the rectangles corresponding to right-hand endpoints, we have

$$S_n = \frac{10}{n} R\left(\frac{10}{n}\right) + \frac{10}{n} R\left[2\left(\frac{10}{n}\right)\right] + \dots + \frac{10}{n} R\left[n\left(\frac{10}{n}\right)\right]$$

$$= \frac{10}{n}\left[\left\{600 - 0.5\left(\frac{10}{n}\right)\right\} + \left\{600 - 0.5(2)\left(\frac{10}{n}\right)\right\} + \dots + \left\{600 - 0.5(n)\left(\frac{10}{n}\right)\right\}\right]$$

$$= \frac{10}{n}\left[600n - 0.5\left(\frac{10}{n}\right)\{1 + 2 + \dots + n\}\right]$$

$$= \frac{10}{n}\left[600n - 0.5\left(\frac{10}{n}\right)\frac{n(n+1)}{2}\right]$$

$$= \frac{10}{n}[600n - 2.5(n+1)]$$

$$= 6000 - 25\left(\frac{n+1}{n}\right)$$

Now take the limit of S_n as $n \to \infty$

$$\lim_{n\to\infty} S_n = \lim_{n\to\infty}\left[6000 - 25\left(\frac{n+1}{n}\right)\right] = \lim_{n\to\infty}\left[6000 - 25\left(1 + \frac{1}{n}\right)\right] = 6000 - 25 = 5975$$

The total revenue for selling 10 units is $5975.

Problems 14.6

1. $f(x) = x,\ y = 0,\ x = 1$

$S_3,\ \Delta x = \dfrac{1}{3}$

$$S_3 = \frac{1}{3} f\left(\frac{1}{3}\right) + \frac{1}{3} f\left(\frac{2}{3}\right) + \frac{1}{3} f\left(\frac{3}{3}\right) = \frac{1}{3}\left[\frac{1}{3} + \frac{2}{3} + \frac{3}{3}\right] = \frac{1}{3} \cdot \frac{6}{3} = \frac{2}{3}$$

The area is approximately $\dfrac{2}{3}$ sq unit.

3. $f(x) = x^2$, $y = 0$, $x = 1$

S_4, $\Delta x = \dfrac{1}{4}$

$S_4 = \dfrac{1}{4} f\left(\dfrac{1}{4}\right) + \dfrac{1}{4} f\left(\dfrac{2}{4}\right) + \dfrac{1}{4} f\left(\dfrac{3}{4}\right) + \dfrac{1}{4} f\left(\dfrac{4}{4}\right)$

$ = \dfrac{1}{4}\left[\dfrac{1}{16} + \dfrac{4}{16} + \dfrac{9}{16} + \dfrac{16}{16}\right]$

$ = \dfrac{1}{4} \cdot \dfrac{30}{16} = \dfrac{15}{32}$

The area is approximately $\dfrac{15}{32}$ sq units.

5. $f(x) = 4x$; $[0, 1]$

$\Delta x = \dfrac{1}{n}$

$S_n = \dfrac{1}{n} f\left(\dfrac{1}{n}\right) + \ldots + \dfrac{1}{n} f\left(n \cdot \dfrac{1}{n}\right)$

$ = \dfrac{1}{n}\left[f\left(\dfrac{1}{n}\right) + \ldots + f\left(n \cdot \dfrac{1}{n}\right)\right]$

$ = \dfrac{1}{n}\left[4 \cdot \dfrac{1}{n} + \ldots + 4 \cdot \dfrac{n}{n}\right]$

$ = \dfrac{4}{n^2}[1 + \ldots + n] = \dfrac{4}{n^2} \cdot \dfrac{n(n+1)}{2}$

$ = \dfrac{2(n+1)}{n}$

7. a. $S_n = \dfrac{1}{n}\left[\left(\dfrac{1}{n} + 1\right) + \left(\dfrac{2}{n} + 1\right) + \ldots + \left(\dfrac{n}{n} + 1\right)\right]$

$ = \dfrac{1}{n}\left[\dfrac{1}{n}(1 + 2 + \ldots + n) + n\right]$

$ = \dfrac{1}{n}\left[\dfrac{1}{n} \cdot \dfrac{n(n+1)}{2} + n\right]$

$ = \dfrac{n+1}{2n} + 1$

b. $\displaystyle\lim_{n\to\infty} S_n = \lim_{n\to\infty}\left[\frac{n+1}{2n}+1\right]$

$\displaystyle = \lim_{n\to\infty}\left[\frac{1}{2}+\frac{1}{2n}+1\right]$

$\displaystyle = \frac{1}{2}+0+1 = \frac{3}{2}$

9. $f(x) = x, \ y = 0, \ x = 1$

$\displaystyle \Delta x = \frac{1}{n}$

$\displaystyle S_n = \frac{1}{n}f\left(\frac{1}{n}\right)+\ldots+\frac{1}{n}f\left(n\cdot\frac{1}{n}\right) = \frac{1}{n}\left[\frac{1}{n}+\ldots+\frac{n}{n}\right] = \frac{1}{n^2}[1+\ldots+n] = \frac{1}{n^2}\cdot\frac{n(n+1)}{2}$

$\displaystyle = \frac{1}{2}\cdot\frac{n+1}{n} = \frac{1}{2}\left[1+\frac{1}{n}\right]$

$\displaystyle \lim_{n\to\infty} S_n = \frac{1}{2}$

The area is $\dfrac{1}{2}$ sq unit.

11. $f(x) = x^2, \ y = 0, \ x = 1$

$\displaystyle \Delta x = \frac{1}{n}$

$\displaystyle S_n = \frac{1}{n}f\left(\frac{1}{n}\right)+\ldots+\frac{1}{n}f\left(n\cdot\frac{1}{n}\right) = \frac{1}{n}\left[\left(\frac{1}{n}\right)^2+\ldots+\left(n\cdot\frac{1}{n}\right)^2\right]$

$\displaystyle = \frac{1}{n^3}\left[1^2+\ldots+n^2\right] = \frac{1}{n^3}\cdot\frac{n(n+1)(2n+1)}{6}$

$\displaystyle = \frac{1}{6}\cdot\frac{2n^2+3n+1}{n^2} = \frac{1}{6}\left[2+\frac{3}{n}+\frac{1}{n^2}\right]$

$\displaystyle \lim_{n\to\infty} S_n = \frac{1}{3}$

The area is $\dfrac{1}{3}$ sq unit.

13. $f(x) = 3x^2$, $y = 0$, $x = 1$

$$\Delta x = \frac{1}{n}$$

$$S_n = \frac{1}{n} f\left(\frac{1}{n}\right) + \ldots + \frac{1}{n} f\left(n \cdot \frac{1}{n}\right)$$

$$= \frac{1}{n}\left[3\left(\frac{1}{n}\right)^2 + \ldots + 3\left(n \cdot \frac{1}{n}\right)^2 \right]$$

$$= \frac{3}{n^3}\left[1^2 + \ldots + n^2 \right]$$

$$= \frac{3}{n^3} \cdot \frac{n(n+1)(2n+1)}{6} = \frac{1}{2} \cdot \frac{2n^2 + 3n + 1}{n^2}$$

$$= \frac{1}{2}\left[2 + \frac{3}{n} + \frac{1}{n^2} \right]$$

$$\lim_{n \to \infty} S_n = 1$$

The area is 1 sq unit.

15. $\int_1^3 5x\, dx$

Let $f(x) = 5x$.

$$\Delta x = \frac{2}{n}$$

$$S_n = \frac{2}{n} f\left(1 + \frac{2}{n}\right) + \cdots + \frac{2}{n} f\left(1 + n \cdot \frac{2}{n}\right)$$

$$= \frac{2}{n}\left[5\left(1 + \frac{2}{n}\right) + \cdots + 5\left(1 + n \cdot \frac{2}{n}\right) \right]$$

$$= \frac{10}{n}\left[(1 + \cdots + 1) + \frac{2}{n}(1 + \cdots + n) \right]$$

$$= \frac{10}{n}\left[n + \frac{2}{n} \cdot \frac{n(n+1)}{2} \right]$$

$$= \frac{10}{n}\left[n + n + 1 \right]$$

$$= \frac{10}{n}(2n + 1)$$

$$= 20 + \frac{10}{n}$$

$$\int_1^3 5x\, dx = \lim_{n \to \infty} S_n = 20$$

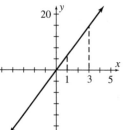

17. $\int_0^3 -4x\, dx$

Let $f(x) = -4x$.

$$\Delta x = \frac{3}{n}$$

$$S_n = \frac{3}{n} f\left(\frac{3}{n}\right) + \ldots + \frac{3}{n} f\left(n \cdot \frac{3}{n}\right)$$

$$= \frac{3}{n}\left[-4\left(\frac{3}{n}\right) - \ldots - 4\left(n \cdot \frac{3}{n}\right) \right]$$

$$= -\frac{36}{n^2}[1 + \ldots + n]$$

$$= -\frac{36}{n^2} \cdot \frac{n(n+1)}{2} = -18 \cdot \frac{n+1}{n} = -18\left[1 + \frac{1}{n}\right]$$

$$\int_0^3 -4x\, dx = \lim_{n \to \infty} S_n = -18$$

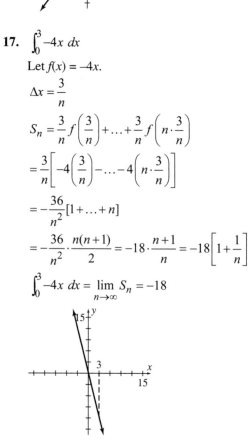

19. $\int_0^1 \left(x^2 + x \right) dx$

Let $f(x) = x^2 + x$.

$\Delta x = \dfrac{1}{n}$

$S_n = \dfrac{1}{n} f\!\left(\dfrac{1}{n} \right) + \ldots + \dfrac{1}{n} f\!\left(n \cdot \dfrac{1}{n} \right)$

$= \dfrac{1}{n} \left\{ \left[\left(\dfrac{1}{n} \right)^2 + \dfrac{1}{n} \right] + \ldots + \left[\left(n \cdot \dfrac{1}{n} \right)^2 + n \cdot \dfrac{1}{n} \right] \right\}$

$= \dfrac{1}{n} \left\{ \left(\dfrac{1}{n} \right)^2 \left[1^2 + \ldots + n^2 \right] + \dfrac{1}{n} [1 + \ldots + n] \right\}$

$= \dfrac{1}{n^3} \cdot \dfrac{n(n+1)(2n+1)}{6} + \dfrac{1}{n^2} \cdot \dfrac{n(n+1)}{2}$

$= \dfrac{1}{6} \cdot \dfrac{2n^2 + 3n + 1}{n^2} + \dfrac{1}{2} \cdot \dfrac{n+1}{n}$

$= \dfrac{1}{6} \left[2 + \dfrac{3}{n} + \dfrac{1}{n^2} \right] + \dfrac{1}{2} \left[1 + \dfrac{1}{n} \right]$

$\int_0^1 \left(x^2 + x \right) dx = \lim_{n \to \infty} S_n = \dfrac{1}{3} + \dfrac{1}{2} = \dfrac{5}{6}$

21. $\int_2^3 \sqrt{x^2 + 1}\, dx$ is simply a real number. Thus

$D_x \left[\int_2^3 \sqrt{x^2 + 1}\, dx \right] = D_x \,(\text{real number}) = 0.$

23. $f(x) = \begin{cases} 1 & \text{if } x \le 1 \\ 2 - x & \text{if } 1 \le x \le 2 \\ -1 + \frac{x}{2} & \text{if } x > 2 \end{cases}$

f is continuous and $f(x) \ge 0$ on $[-1, 3]$. Thus

$\int_{-1}^3 f(x)\, dx$ gives the area A bounded by

$y = f(x)$, $y = 0$, $x = -1$, and $x = 3$. From the diagram, this area is composed of three subareas, A_1, A_2, and A_3 and

$A = A_1 + A_2 + A_3$.

$A_1 = $ area of rectangle $= (2)(1) = 2$ sq units

$A_2 = $ area of triangle $= \dfrac{1}{2}(1)(1) = \dfrac{1}{2}$ sq unit

$A_3 = $ area of triangle $= \dfrac{1}{2}(1)\!\left(\dfrac{1}{2} \right) = \dfrac{1}{4}$ sq unit

Since

$A = A_1 + A_2 + A_3 = 2 + \dfrac{1}{2} + \dfrac{1}{4} = \dfrac{11}{4}$ sq units,

we have $\int_{-1}^3 f(x)\, dx = \dfrac{11}{4}$.

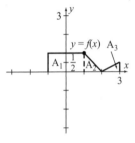

25. 14.77 sq units

27. 2.4

29. −25.5

Principles in Practice 14.7

1. $\int_3^6 10{,}000 e^{0.02t}\, dt = \left(10{,}000 \dfrac{e^{0.02t}}{0.02} \right) \Big|_3^6$

$= \left(500{,}000 e^{0.02t} \right) \Big|_3^6$

$= 500{,}000 \left(e^{0.02(6)} - e^{0.02(3)} \right)$

$= 500{,}000 \left(e^{0.12} - e^{0.06} \right) \approx 32{,}830$

The total income for the chain between the third and sixth years was about \$32,830.

Problems 14.7

1. $\int_0^3 5\,dx = 5x\Big|_0^3 = 5(3) - 5(0) = 15 - 0 = 15$

3. $\int_1^2 5x\,dx = 5 \cdot \dfrac{x^2}{2}\Big|_1^2 = 10 - \dfrac{5}{2} = \dfrac{15}{2}$

5. $\int_{-3}^1 (2x-3)dx = \left(x^2 - 3x\right)\Big|_{-3}^1 = -2 - 18 = -20$

7. $\int_2^3 \left(y^2 - 2y + 1\right)dy = \int_2^3 (y-1)^2\,dy$

$= \dfrac{1}{3}(y-1)^3\Big|_2^3$

$= \dfrac{8}{3} - \dfrac{1}{3} = \dfrac{7}{3}$

9. $\int_{-2}^{-1}\left(3w^2 - w - 1\right)dw = \left(w^3 - \dfrac{w^2}{2} - w\right)\Big|_{-2}^{-1}$

$= -\dfrac{1}{2} - (-8) = \dfrac{15}{2}$

11. $\int_1^3 3t^{-3}\,dt = -\dfrac{3}{2}\cdot t^{-2}\Big|_1^3 = -\dfrac{1}{6} - \left(-\dfrac{3}{2}\right) = \dfrac{4}{3}$

13. $\int_{-8}^8 \sqrt[3]{x^4}\,dx = \int_{-8}^8 x^{4/3}\,dx$

$= \dfrac{3x^{7/3}}{7}\Big|_{-8}^8$

$= \dfrac{3\cdot 128}{7} - \dfrac{3(-128)}{7}$

$= \dfrac{768}{7}$

15. $\int_{1/2}^3 \dfrac{1}{x^2}\,dx = -\dfrac{1}{x}\Big|_{1/2}^3 = -\dfrac{1}{3} - (-2) = \dfrac{5}{3}$

17. $\int_{-1}^1 (z+1)^5\,dz = \dfrac{(z+1)^6}{6}\Big|_{-1}^1 = \dfrac{32}{3} - 0 = \dfrac{32}{3}$

19. $\int_0^1 2x^2\left(x^3 - 1\right)^3 dx = \dfrac{2}{3}\int_0^1 \left(x^3 - 1\right)^3\left[3x^2 dx\right]$

$= \dfrac{1}{6}\left(x^3 - 1\right)^4\Big|_0^1 = 0 - \dfrac{1}{6} = -\dfrac{1}{6}$

21. $\int_1^8 \dfrac{4}{y}\,dy = 4\ln|y|\Big|_1^8 = 4(\ln 8 - \ln 1)$

$= 4(\ln 8 - 0) = 4\ln 8$

23. $\int_0^1 e^5\,dx = e^5 x\Big|_0^1 = e^5 - 0 = e^5$

25. $\int_0^1 5x^2 e^{x^3}\,dx = \dfrac{5}{3}\int_0^1 e^{x^3}\left[3x^2 dx\right] = \dfrac{5}{3}e^{x^3}\Big|_0^1$

$= \dfrac{5}{3}\left(e^1 - e^0\right) = \dfrac{5}{3}(e-1)$

27. $\int_4^5 \dfrac{2}{(x-3)^3}\,dx = 2\int_4^5 (x-3)^{-3}\,dx$

$= 2\cdot \dfrac{(x-3)^{-2}}{-2}\Big|_4^5$

$= -\dfrac{1}{(x-3)^2}\Big|_4^5 = -\dfrac{1}{4} - (-1) = \dfrac{3}{4}$

29. $\int_{1/3}^2 \sqrt{10 - 3p}\,dp = -\dfrac{1}{3}\int_{1/3}^2 (10-3p)^{\frac{1}{2}}[-3\,dp]$

$= -\dfrac{2}{9}(10 - 3p)^{\frac{3}{2}}\Big|_{1/3}^2 = -\dfrac{2}{9}(8 - 27) = \dfrac{38}{9}$

31. $\int_0^1 x^2 \sqrt[3]{7x^3+1}\,dx = \frac{1}{21}\int_0^1 \left(7x^3+1\right)^{\frac{1}{3}}\left[21x^2\,dx\right]$

$= \frac{1}{21}\cdot\left.\frac{\left(7x^3+1\right)^{\frac{4}{3}}}{\frac{4}{3}}\right|_0^1 = \left.\frac{\left(7x^3+1\right)^{\frac{4}{3}}}{28}\right|_0^1$

$= \frac{16}{28} - \frac{1}{28} = \frac{15}{28}$

33. $\int_0^1 \frac{2x^3+x}{x^2+x^4+1}\,dx$

$= \frac{1}{2}\int_0^1 \frac{1}{x^4+x^2+1}\left[\left(4x^3+2x\right)dx\right]$

$= \left.\frac{1}{2}\ln\left(x^4+x^2+1\right)\right|_0^1 = \frac{1}{2}[\ln 3 - \ln 1] = \frac{1}{2}\ln 3$

35. $\int_0^1 \frac{e^x - e^{-x}}{2}\,dx = \left.\frac{1}{2}(e^x+e^{-x})\right|_0^1$

$= \frac{1}{2}[(e+e^{-1})+(1+1)]$

$= \frac{1}{2}\left(e+\frac{1}{e}+2\right)$

37. $\int_\pi^e 3(x^{-2}+x^{-3}-x^{-4})\,dx$

$= 3\left.\left(\frac{x^{-1}}{-1}+\frac{x^{-2}}{-2}-\frac{x^{-3}}{-3}\right)\right|_\pi^e$

$= 3\left(-\frac{1}{e}-\frac{1}{2e^2}+\frac{1}{3e^3}\right)-3\left(-\frac{1}{\pi}-\frac{1}{2\pi^2}+\frac{1}{3\pi^3}\right)$

$= \frac{1}{e^3}-\frac{3}{e}-\frac{3}{2e^2}+\frac{3}{\pi}+\frac{3}{2\pi^2}-\frac{1}{\pi^3}$

39. $\int_1^3 (x+1)e^{x^2+2x}\,dx = \frac{1}{2}\int_1^3 e^{x^2+2x}[(2x+2)dx]$

$= \left.\frac{1}{2}e^{x^2+2x}\right|_1^3 = \frac{1}{2}\left(e^{15}-e^3\right) = \frac{e^3}{2}\left(e^{12}-1\right)$

41. Using long division on the integrand

$\int_0^2 \frac{x^6+6x^4+x^3+8x^2+x+5}{x^3+5x+1}\,dx$

$= \int_0^2\left[x^3+x+\frac{3x^2+5}{x^3+5x+1}\right]dx$

$= \left.\left[\frac{x^4}{4}+\frac{x^2}{2}+\ln\left|x^3+5x+1\right|\right]\right|_0^2$

$= (6+\ln 19) - 0 = 6 + \ln 19$

43. $\int_0^2 f(x)\,dx = \int_0^{1/2} 4x^2\,dx + \int_{1/2}^2 2x\,dx$

$= \left.\frac{4x^3}{3}\right|_0^{1/2} + \left.x^2\right|_{1/2}^2 = \left(\frac{1}{6}-0\right)+\left(4-\frac{1}{4}\right) = \frac{47}{12}$

45. $f(x) = \int_1^x 3\frac{1}{t^2}\,dt = \left.-\frac{3}{t}\right|_1^x = -\frac{3}{x}+3 = 3-\frac{3}{x}$

$\int_e^1 f(x)\,dx = \int_e^1\left(3-\frac{3}{x}\right)dx = \left.(3x-3\ln|x|)\right|_e^1$

$= (3-0)-(3e-3) = 6-3e$

47. $\int_1^3 f(x)\,dx = \int_1^2 f(x)\,dx + \int_2^3 f(x)\,dx$, so

$\int_1^2 f(x)\,dx = \int_1^3 f(x)\,dx - \int_2^3 f(x)\,dx$

$= \int_1^3 f(x)\,dx + \int_3^2 f(x)\,dx = 4+3 = 7$

49. $\int_2^3 e^{x^3}\,dx$ is a constant, so $\frac{d}{dx}\int_2^3 e^{x^3}\,dx = 0.$

Thus

$\int_2^3\left(\frac{d}{dx}\int_2^3 e^{x^3}\,dx\right)dx = \int_2^3 0\,dx$

$= \left.C\right|_2^3 = C-C = 0$

51. $\int_0^T \alpha^{\frac{5}{2}}\,dt = \left.\alpha^{\frac{5}{2}}t\right|_0^T = \alpha^{\frac{5}{2}}T - 0 = \alpha^{\frac{5}{2}}T$

53. The total number receiving between a and b dollars equals the number $N(a)$ receiving a or more dollars minus the number $N(b)$ receiving b or more dollars. Thus

$$N(a) - N(b) = \int_b^a -Ax^{-B} dx.$$

55. $\displaystyle\int_0^5 2000e^{-0.06t} dt$

$$= 2000 \cdot \frac{1}{-0.06} \int_0^5 e^{-0.06t} [-0.06 \, dt]$$

$$= -\frac{2000}{0.06} e^{-0.06t} \Big|_0^5 = -\frac{2000}{0.06}\left(e^{-0.03} - 1\right)$$

$$\approx \$8639$$

57. $\displaystyle\int_{36}^{64} 10,000\sqrt{100 - t}\,dt$

$$= (-1)(10,000)\int_{36}^{64}(100 - t)^{\frac{1}{2}}[(-1)\,dt]$$

$$= -\frac{2}{3}(10,000)(100 - t)^{\frac{3}{2}}\Big|_{36}^{64}$$

$$= -\frac{2}{3}(10,000)[216 - 512]$$

$$\approx 1,973,333$$

59. $\displaystyle\int_{65}^{75}(0.2q + 8)dq = \left(0.1q^2 + 8q\right)\Big|_{65}^{75}$

$$= 1162.5 - 942.5 = \$220$$

61. $\displaystyle\int_{500}^{800} \frac{2000}{\sqrt{300q}}\,dq = \int_{500}^{800} \frac{2000}{10\sqrt{3q}}\,dq$

$$= \frac{200}{\sqrt{3}}\int_{500}^{800} q^{-1/2}\,dq = \frac{200}{\sqrt{3}} \cdot \frac{q^{1/2}}{\frac{1}{2}}\Big|_{500}^{800}$$

$$= \frac{400}{\sqrt{3}}\sqrt{q}\Big|_{500}^{800} = \frac{400}{\sqrt{3}}\left(\sqrt{800} - \sqrt{500}\right) \approx \$1367.99$$

63. $\displaystyle\int_0^{12}(8t + 10)dt = \left(4t^2 + 10t\right)\Big|_0^{12}$

$$= 696 - 0 = 696$$

$$\int_6^{12}(8t + 10)dt = \left(4t^2 + 10t\right)\Big|_6^{12}$$

$$= 696 - 204 = 492$$

65. $\displaystyle G = \int_{-R}^{R} i \, dx = ix\Big|_{-R}^{R} = iR - (-iR) = 2Ri$

67. $\displaystyle A = \frac{\int_0^R (m + x)[1 - (m + x)]dx}{\int_0^R [1 - (m + x)]dx}$

$$= \frac{\int_0^R \left(m + x - m^2 - 2mx - x^2\right)dx}{\int_0^R (1 - m - x)dx}$$

$$= \frac{\left[mx + \frac{x^2}{2} - m^2x - mx^2 - \frac{x^3}{3}\right]\Big|_0^R}{\left[x - mx - \frac{x^2}{2}\right]\Big|_0^R}$$

$$= \frac{\left[mR + \frac{R^2}{2} - m^2R - mR^2 - \frac{R^3}{3}\right] - 0}{\left[R - mR - \frac{R^2}{2}\right] - 0}$$

$$= \frac{R\left[m + \frac{R}{2} - m^2 - mR - \frac{R^2}{3}\right]}{R\left[1 - m - \frac{R}{2}\right]}$$

$$= \frac{m + \frac{R}{2} - m^2 - mR - \frac{R^2}{3}}{1 - m - \frac{R}{2}}$$

69. $\displaystyle\int_0^4 \frac{1}{(4x + 4)^2} dx = \frac{1}{4}\int_0^4 (4x + 4)^{-2}[4 \, dx]$

$$= \frac{1}{4} \cdot \frac{(4x + 4)^{-1}}{-1}\Big|_0^4$$

$$= -\frac{1}{4} \cdot \frac{1}{4x + 4}\Big|_0^4 = -\frac{1}{16} \cdot \frac{1}{x + 1}\Big|_0^4 = -\frac{1}{16}\left(\frac{1}{5} - 1\right)$$

$$= \frac{1}{20} = 0.05$$

71. 3.52

73. 14.34

Principles in Practice 14.8

1. In this case, $f(t) = \dfrac{60}{\sqrt{t^2 + 9}}$, $n = 5$, $a = 0$,

and $b = 5$. Thus $h = \dfrac{b - a}{n} = \dfrac{5 - 0}{5} = 1$. The

terms to be added are

$f(0) = \dfrac{60}{\sqrt{0^2 + 9}} = \dfrac{60}{3} = 20$

$2f(1) = \dfrac{2(60)}{\sqrt{1^2 + 9}} = \dfrac{120}{\sqrt{10}} \approx 37.9473$

$2f(2) = \dfrac{2(60)}{\sqrt{2^2 + 9}} = \dfrac{120}{\sqrt{13}} \approx 33.2820$

$2f(3) = \dfrac{2(60)}{\sqrt{3^2 + 9}} = \dfrac{120}{\sqrt{18}} \approx 28.2843$

$2f(4) = \dfrac{2(60)}{\sqrt{4^2 + 9}} = \dfrac{120}{5} = 24$

$f(5) = \dfrac{60}{\sqrt{5^2 + 9}} = \dfrac{60}{\sqrt{34}} \approx 10.2899$

The sum of the above terms is 153.8035.
The estimate of the radius after 5 seconds is

$\displaystyle\int_0^5 \dfrac{60}{\sqrt{t^2 + 9}}\, dt \approx \dfrac{1}{2}(153.8035) \approx 76.90$ feet.

Problems 14.8

1. $f(x) = \dfrac{170}{1 + x^2}$, $n = 6$, $a = -2$, $b = 4$.

Trapezoidal

$h = \dfrac{b - a}{n} = \dfrac{4 - (-2)}{6} = \dfrac{6}{6} = 1$

$\begin{aligned}
f(-2) &= 34 = 34 \\
2f(-1) &= 2(85) = 170 \\
2f(0) &= 2(170) = 340 \\
2f(1) &= 2(85) = 170 \\
2f(2) &= 2(34) = 68 \\
2f(3) &= 2(17) = 34 \\
f(4) &= 10 = \underline{10} \\
& 826
\end{aligned}$

$\displaystyle\int_{-2}^4 \dfrac{170}{1 + x^2}\, dx \approx \dfrac{1}{2}(826) = 413$

3. $f(x) = x^2$, $n = 5$, $a = 0$, $b = 1$

Trapezoidal

$h = \dfrac{b - a}{n} = \dfrac{1 - 0}{5} = \dfrac{1}{5} = 0.2$

$\begin{aligned}
f(0) &= 0.0000 \\
2f(0.2) &= 0.0800 \\
2f(0.4) &= 0.3200 \\
2f(0.6) &= 0.7200 \\
2f(0.8) &= 1.2800 \\
f(1) &= \underline{1.0000} \\
& 3.4000
\end{aligned}$

$\displaystyle\int_0^1 x^2\, dx \approx \dfrac{0.2}{2}(3.4000) = 0.340$

Actual value: $\displaystyle\int_0^1 x^2\, dx = \dfrac{x^3}{3}\bigg|_0^1 = \dfrac{1}{3} \approx 0.333$

5. $f(x) = \dfrac{1}{x^2}$, $n = 4$, $a = 1$, $b = 4$

Simpson's

$h = \dfrac{b - a}{n} = \dfrac{4 - 1}{4} = 0.75$

$\begin{aligned}
f(1) &= 1.0000 \\
4f(1.75) &= 1.3061 \\
2f(2.50) &= 0.3200 \\
4f(3.25) &= 0.3787 \\
f(4) &= \underline{0.0625} \\
& 3.0673
\end{aligned}$

$\displaystyle\int_1^4 \dfrac{1}{x^2}\, dx \approx \dfrac{0.75}{3}(3.0673) \approx 0.767$

Actual value:

$$\int_1^4 \frac{1}{x^2}\,dx = -\frac{1}{x}\Big|_1^4 = -\frac{1}{4} - (-1) = 0.750$$

7. $f(x) = \dfrac{x}{x+1}$, $n = 4$, $a = 0$, $b = 2$

Trapezoidal

$$h = \frac{b-a}{n} = \frac{2-0}{4} = 0.5$$

$$
\begin{aligned}
f(0) &= 0.0000 \\
2f(0.5) &= 0.6667 \\
2f(1) &= 1.0000 \\
2f(1.5) &= 1.2000 \\
f(2) &= 0.6667 \\
\hline
&\ \ 3.5334
\end{aligned}
$$

Thus

$$\int_0^2 \frac{x}{x+1}\,dx \approx \frac{0.5}{2}(3.5334) \approx 0.883$$

9. $\displaystyle\int_{45}^{70} l(t)\,dt$, males, $n = 5$, $a = 45$, $b = 70$

$$h = \frac{70-45}{5} = 5$$

$$
\begin{aligned}
l(45) &= \ \ 93,717 \\
2l(50) &= 183,232 \\
2l(55) &= 177,292 \\
2l(60) &= 168,376 \\
2l(65) &= 155,094 \\
l(70) &= \ \ 68,375 \\
\hline
&\ 846,086
\end{aligned}
$$

$$\int_{45}^{70} l(t)\,dt \approx \frac{5}{2}(846,086) = 2,115,215$$

11. $a = 1$, $b = 5$, $h = 1$

$$
\begin{aligned}
f(1) = 0.4 &\quad = 0.4 \\
4f(2) = 4(0.6) &= 2.4 \\
2f(3) = 2(1.2) &= 2.4 \\
4f(4) = 4(0.8) &= 3.2 \\
f(5) = 0.5 &\quad = 0.5 \\
\hline
&\quad\ \ 8.9
\end{aligned}
$$

$$\int_1^5 f(x)\,dx \approx \frac{1}{3}(8.9) \approx 3.0$$

The area is about 3.0 square units.

13. $\displaystyle\int_1^3 f(x)\,dx$, $n = 4$, $a = 1$, $b = 3$

$$h = \frac{3-1}{4} = 0.5$$

$$
\begin{aligned}
f(1) = 1 &\quad = \ 1 \\
4f(1.5) = 4(2) &= \ 8 \\
2f(2) = 2(2) &= \ 4 \\
4f(2.5) = 4(0.5) &= \ 2 \\
f(3) = 1 &\quad = \ \frac{1}{16}
\end{aligned}
$$

$$\int_1^3 f(x)\,dx \approx \frac{0.5}{3}(16) = \frac{8}{3}$$

15. $f(x) = \sqrt{1-x^2}$, $a = 0$, $b = 1$, $n = 4$

$$h = \frac{1-0}{4} = 0.25$$

Simpson's

$$
\begin{aligned}
f(0) &= 1.0000 \\
4f(0.25) &= 3.8730 \\
2f(0.50) &= 1.7321 \\
4f(0.75) &= 2.6458 \\
f(1) &= 0.0000 \\
\hline
&\ \ 9.2509
\end{aligned}
$$

$$\int_0^1 \sqrt{1-x^2}\,dx \approx \frac{0.25}{3}(9.2509) \approx 0.771$$

17. Let $f(x)$ = distance from near to far shore at point x on highway. Then area $\approx \displaystyle\int_0^4 f(x)\,dx$.

Using Simpson's rule with $h = 0.5$:

$$
\begin{aligned}
f(0) &= 0.5 - 0.5 &&= 0 &&= 0 \\
4f(0.5) &= 4(2.3 - 0.3) = 4(2) &&= 8 \\
2f(1) &= 2(2.2 - 0.7) = 2(1.5) &&= 3 \\
4f(1.5) &= 4(3 - 1) &&= 4(2) &&= 8 \\
2f(2) &= 2(2.5 - 0.5) = 2(2) &&= 4 \\
4f(2.5) &= 4(2.2 - 0.2) = 4(2) &&= 8 \\
2f(3) &= 2(1.5 - 0.5) = 2(1) &&= 2 \\
4f(3.5) &= 4(1.3) - 0.8) = 4(0.5) &&= 2 \\
f(4) &= 1 - 1 &&= 0 &&= 0 \\
&&&&&\overline{35}
\end{aligned}
$$

$$
\text{Area} \approx \int_0^4 f(x)dx \approx \frac{0.5}{3}(35) = \frac{35}{6} \text{ km}^2
$$

Problems 14.9

In Problems 1–33, answers are assumed to be expressed in square units.

1. $y = 4x$, $x = 2$

$$
\text{Area} = \int_0^2 4x\ dx = 2x^2 \Big|_0^2 = 8 - 0 = 8
$$

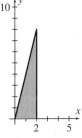

3. $y = 5x + 2$, $x = 1$, $x = 4$

$$
\text{Area} = \int_1^4 (5x + 2)dx
$$

$$
= \left(\frac{5x^2}{2} + 2x \right) \Big|_1^4 = 48 - \frac{9}{2} = \frac{87}{2}
$$

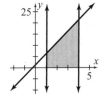

5. $y = x - 1$, $x = 5$

$$
\text{Area} = \int_1^5 (x - 1)dx = \left(\frac{x^2}{2} - x \right) \Big|_1^5
$$

$$
= \frac{15}{2} - \left(-\frac{1}{2} \right) = \frac{16}{2} = 8
$$

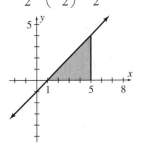

7. $y = x^2$, $x = 2$, $x = 3$

$$
\text{Area} = \int_2^3 x^2 dx = \frac{x^3}{3} \Big|_2^3 = 9 - \frac{8}{3} = \frac{19}{3}
$$

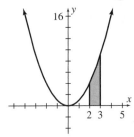

9. $y = x^2 + 2$, $x = -1$, $x = 2$

$$
\text{Area} = \int_{-1}^2 \left(x^2 + 2 \right) dx = \left(\frac{x^3}{3} + 2x \right) \Big|_{-1}^2
$$

$$
= \frac{20}{3} - \left(-\frac{7}{3} \right) = \frac{27}{3} = 9
$$

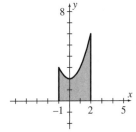

11. $y = x^2 - 2x$, $x = -3$, $x = -1$

$$\text{Area} = \int_{-3}^{-1} \left(x^2 - 2x \right) dx = \left(\frac{x^3}{3} - x^2 \right) \Bigg|_{-3}^{-1}$$

$$= -\frac{4}{3} - (-18) = \frac{50}{3}$$

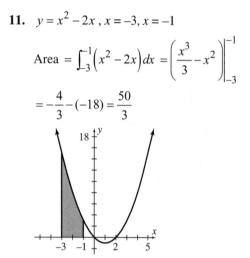

13. $y = 2 - x - x^2$

$$\text{Area} = \int_{-2}^{1} (2 - x - x^2) dx$$

$$= \left(2x - \frac{x^2}{2} - \frac{x^3}{3} \right) \Bigg|_{-2}^{1}$$

$$= \frac{7}{6} - \left(-\frac{10}{3} \right)$$

$$= \frac{9}{2}$$

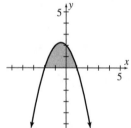

15. $y = 2 - x - x^3$, $x = -3$, $x = 0$

Area

$$= \int_{-3}^{0} \left(2 - x - x^3 \right) dx = \left(2x - \frac{x^2}{2} - \frac{x^4}{4} \right) \Bigg|_{-3}^{0}$$

$$= 0 - \left(-\frac{123}{4} \right) = \frac{123}{4}$$

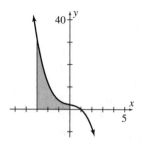

17. $A = 3 + 2x - x^2$

$$\text{Area} = \int_{-1}^{3} \left(3 + 2x - x^2 \right) dx$$

$$= \left(3x + x^2 - \frac{x^3}{3} \right) \Bigg|_{-1}^{3}$$

$$= 9 - \left(-\frac{5}{3} \right) = \frac{32}{3}$$

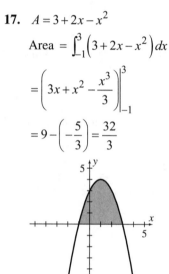

19. $y = \frac{1}{x}$, $x = 1$, $x = e$

$$\text{Area} = \int_{1}^{e} \frac{1}{x} dx = \ln|x| \Big\|_{1}^{e} = \ln e - \ln 1$$

$$= 1 - 0 = 1$$

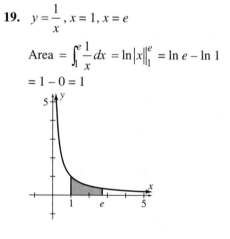

21. $y = \sqrt{x+9}$, $x = -9$, $x = 0$

Area $= \int_{-9}^{0} \sqrt{x+9}\,dx = \int_{-9}^{0} (x+9)^{\frac{1}{2}}\,dx$

$= \dfrac{(x+9)^{\frac{3}{2}}}{\frac{3}{2}}\Bigg|_{-9}^{0} = \dfrac{2(x+9)^{\frac{3}{2}}}{3}\Bigg|_{-9}^{0}$

$= 18 - 0 = 18$

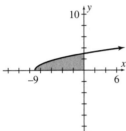

23. $y = \sqrt{2x-1}$, $x = 1$, $x = 5$

Area $= \int_{1}^{5} \sqrt{2x-1}\,dx$

$= \dfrac{1}{2} \int_{1}^{5} (2x-1)^{\frac{1}{2}} [2\ dx]$

$= \dfrac{(2x-1)^{\frac{3}{2}}}{3}\Bigg|_{1}^{5} = 9 - \dfrac{1}{3} = \dfrac{26}{3}$

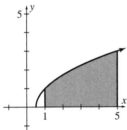

25. $y = \sqrt[3]{x}$, $x = 2$

Area

$= \int_{0}^{2} \sqrt[3]{x}\,dx = \int_{0}^{2} x^{\frac{1}{3}}\,dx = \dfrac{3x^{\frac{4}{3}}}{4}\Bigg|_{0}^{2} = \dfrac{3(2)^{\frac{4}{3}}}{4} - 0$

$= \dfrac{3\left(2\sqrt[3]{2}\right)}{4} = \dfrac{3}{2}\sqrt[3]{2}$

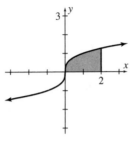

27. $y = e^{x} + 1$, $x = 0$, $x = 1$

Area $= \int_{0}^{1} (e^{x}+1)\,dx = (e^{x}+x)\Big|_{0}^{1}$

$= (e^{1}+1) - 1 = e$

29. $y = x + \dfrac{2}{x}$, $x = 1$, $x = 2$

Area $= \int_{1}^{2}\left(x + \dfrac{2}{x}\right)dx = \left(\dfrac{x^{2}}{2} + 2\ln|x|\right)\Bigg|_{1}^{2}$

$= (2 + 2\ln 2) - \dfrac{1}{2} = \dfrac{3}{2} + 2\ln 2 = \dfrac{3}{2} + \ln 4$

31. $y = x^3$, $x = -2$, $x = 4$

$$\text{Area} = \int_{-2}^{0} -x^3 \, dx + \int_{0}^{4} x^3 \, dx = -\frac{x^4}{4}\Big|_{-2}^{0} + \frac{x^4}{4}\Big|_{0}^{4}$$

$$= [0 - (-4)] + [64 - 0] = 68$$

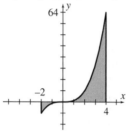

33. $y = 2x - x^2$, $x = 1$, $x = 3$

$$\text{Area} = \int_{1}^{2} \left(2x - x^2\right) dx + \int_{2}^{3} -\left(2x - x^2\right) dx$$

$$= \left(x^2 - \frac{x^3}{3}\right)\Big|_{1}^{2} - \left(x^2 - \frac{x^3}{3}\right)\Big|_{2}^{3}$$

$$= \left[\frac{4}{3} - \frac{2}{3}\right] - \left[0 - \frac{4}{3}\right] = \frac{6}{3} = 2$$

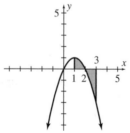

35. $f(x) = \begin{cases} 3x^2 & \text{if } 0 \le x < 2 \\ 16 - 2x & \text{if } x \ge 2 \end{cases}$

$$\text{Area} = \int_{0}^{3} f(x) \, dx$$

$$= \int_{0}^{2} 3x^2 \, dx + \int_{2}^{3} (16 - 2x) \, dx$$

$$= x^3 \Big|_{0}^{2} + \left(16x - x^2\right)\Big|_{2}^{3}$$

$$= [8 - 0] + [39 - 28] = 19 \text{ sq units}$$

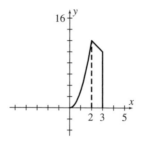

37. a. $P(0 \le x \le 1) = \int_{0}^{1} \frac{1}{8} x \, dx = \frac{x^2}{16}\Big|_{0}^{1} = \frac{1}{16} - 0$

$$= \frac{1}{16}$$

b. $P(2 \le x \le 4) = \int_{2}^{4} \frac{1}{8} x \, dx = \frac{x^2}{16}\Big|_{2}^{4}$

$$= 1 - \frac{1}{4} = \frac{3}{4}$$

c. $P(x \ge 3) = \int_{3}^{4} \frac{1}{8} x \, dx = \frac{x^2}{16}\Big|_{3}^{4} = 1 - \frac{9}{16} = \frac{7}{16}$

39. a. $P(3 \le x \le 7) = \int_{3}^{7} \frac{1}{x} \, dx = \ln|x| \Big|_{3}^{7}$

$$= \ln 7 - \ln 3 = \ln \frac{7}{3}$$

b. $P(x \le 5) = \int_{e}^{5} \frac{1}{x} \, dx = \ln|x| \Big|_{e}^{5}$

$$= \ln(5) - \ln e = \ln(5) - 1$$

c. $P(x \ge 4) = \int_{4}^{e^2} \frac{1}{x} \, dx = \ln|x| \Big|_{4}^{e^2}$

$$= \ln e^2 - \ln 4 = 2 - \ln 4$$

d. $P\left(e \le x \le e^2\right) = \int_{e}^{e^2} \frac{1}{x} \, dx$

$$= \ln|x| \Big|_{e}^{e^2} = \ln e^2 - \ln e$$

$$= 2 - 1 = 1$$

41. 1.89 sq units

43. The *x*-intercept on [1, 3] is $A \approx 2.190327947$

$$\text{Area} = \int_1^A -\left(x^4 - 2x^3 - 2\right)dx + \int_A^3 \left(x^4 - 2x^3 - 2\right)dx$$

$$\approx 11.41 \text{ sq units}$$

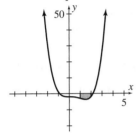

Problems 14.10

1. $\text{Area} = \int_a^b \left(y_{\text{UPPER}} - y_{\text{LOWER}}\right)dx = \int_{-2}^3 \left[(x+6) - x^2\right]dx$

3. Intersection points:

$x^2 - x = 2x,\ x^2 - 3x = 0,\ x(x-3) = 0 \Rightarrow x = 0 \text{ or } x = 3$

$$\text{Area} = \int_0^3 \left(y_{\text{UPPER}} - y_{\text{LOWER}}\right)dx + \int_3^4 \left(y_{\text{UPPER}} - y_{\text{LOWER}}\right)dx$$

$$= \int_0^3 \left[2x - \left(x^2 - x\right)\right]dx + \int_3^4 \left[\left(x^2 - x\right) - 2x\right]dx$$

5. The graphs of $y = 1 - x^2$ and $y = x - 1$ intersect when $1 - x^2 = x - 1$, $0 = x^2 + x - 2$,

$0 = (x-1)(x+2) \Rightarrow x = 1$ or

$x = -2$. When $x = 1$, then $y = 0$. We use horizontal elements, where *y* ranges from 0 to 1. Solving $y = x - 1$ for *x* gives $x = y + 1$, and solving $y = 1 - x^2$ for *x* gives $x^2 = 1 - y$, $x = \pm\sqrt{1-y}$. We must choose

$x = \sqrt{1-y}$ because *x* is not negative over the given region.

$$\text{Area} = \int_0^1 \left(x_{\text{RIGHT}} - x_{\text{LEFT}}\right)dy = \int_0^1 \left[(y+1) - \sqrt{1-y}\right]dy$$

7. The graphs of $y = x^2 - 5$ and $y = 7 - 2x^2$ intersect when $x^2 - 5 = 7 - 2x^2$, $3x^2 = 12$, $x^2 = 4$, so

$x = \pm\sqrt{4} = \pm2$. We use vertical elements.

$$\text{Area} = \int_{-2}^2 \left(y_{\text{UPPER}} - y_{\text{LOWER}}\right)dx = \int_{-2}^2 [(7 - 2x^2) - (x^2 - 5)]dx$$

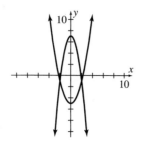

In Problems 10–34, the answers are assumed to be expressed in square units.

9. $y = x^2$, $y = 2x$

Region appears below.

Intersection: $x^2 = 2x$, $x^2 - 2x = 0$,

$x(x - 2) = 0$, so $x = 0$ or 2.

$$\text{Area} = \int_0^2 \left(2x - x^2\right)dx = \left(x^2 - \frac{x^3}{3}\right)\Bigg|_0^2$$

$$= \left(4 - \frac{8}{3}\right) - 0 = \frac{4}{3}$$

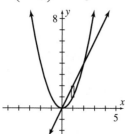

11. $y = x^2 + 1$, $x \geq 0$, $x = 0$, $y = 3$. Region appears below.

Intersection: $x^2 + 1 = 3$, so $x = \pm\sqrt{2}$

$$\text{Area} = \int_0^{\sqrt{2}} [3 - (x^2 + 1)]dx = \int_0^{\sqrt{2}} (2 - x^2)dx$$

$$= \left(2x - \frac{x^3}{3}\right)\Bigg|_0^{\sqrt{2}} = \frac{4\sqrt{2}}{3} - 0 = \frac{4\sqrt{2}}{3}$$

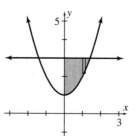

13. $y = 10 - x^2$, $y = 4$. Region appears below.

Intersection: $10 - x^2 = 4$, $x^2 = 6$, so

$x = \pm\sqrt{6}$

$$\text{Area} = \int_{-\sqrt{6}}^{\sqrt{6}} [(10 - x^2) - 4]\,dx$$

$$= \int_{-\sqrt{6}}^{\sqrt{6}} (6 - x^2)\,dx$$

$$= \left(6x - \frac{x^3}{3}\right)\Bigg|_{-\sqrt{6}}^{\sqrt{6}}$$

$$= \left(6\sqrt{6} - \frac{6\sqrt{6}}{3}\right) - \left(-6\sqrt{6} + \frac{6\sqrt{6}}{3}\right)$$

$$= 8\sqrt{6}$$

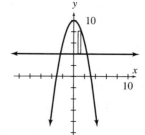

15. $x = 8 + 2y$, $x = 0$, $y = -1$, $y = 3$. Region appears below.

$$\text{Area} = \int_{-1}^{3} (8 + 2y)dy = \left(8y + y^2\right)\Bigg|_{-1}^{3}$$

$$= (24 + 9) - (-8 + 1) = 40$$

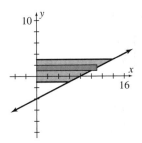

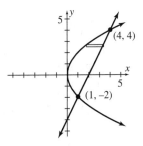

17. $y = 4 - x^2$, $y = -3x$. Region appears below.

Intersection: $-3x = 4 - x^2$, $x^2 - 3x - 4 = 0$,

$(x + 1)(x - 4) = 0$, so $x = -1$ or 4.

Area $\int_{-1}^{4} \left[\left(4 - x^2 \right) - (-3x) \right] dx$

$= \left(4x - \dfrac{x^3}{3} + \dfrac{3x^2}{2} \right) \Bigg|_{-1}^{4}$

$= \left(16 - \dfrac{64}{3} + 24 \right) - \left(-4 + \dfrac{1}{3} + \dfrac{3}{2} \right) = \dfrac{125}{6}$

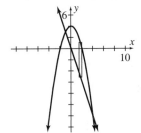

19. $y^2 = 4x$, $y = 2x - 4$. Region appears below.

Intersection: $y^2 = 4 \left(\dfrac{y}{2} + 2 \right)$,

$y^2 - 2y - 8 = 0$,

$(y + 2)(y - 4) = 0$, so $y = -2$ or 4.

Area $= \int_{-2}^{4} \left[\left(\dfrac{y}{2} + 2 \right) - \dfrac{y^2}{4} \right] dy$

$= \left(\dfrac{y^2}{4} + 2y - \dfrac{y^3}{12} \right) \Bigg|_{-2}^{4}$

$= \left(4 + 8 - \dfrac{16}{3} \right) - \left(1 - 4 + \dfrac{2}{3} \right)$

$= 9$

21. $2y = 4x - x^2$, $2y = x - 4$. Region appears below.

Intersection: $x - 4 = 4x - x^2$,

$x^2 - 3x - 4 = 0$,

$(x + 1)(x - 4) = 0$, so $x = -1$ or 4. Note that the y-values of the curves are given by

$y = \dfrac{4x - x^2}{2}$ and $y = \dfrac{x - 4}{2}$.

Area $= \int_{-1}^{4} \left[\left(\dfrac{4x - x^2}{2} \right) - \left(\dfrac{x - 4}{2} \right) \right] dx$

$= \int_{-1}^{4} \left(\dfrac{3}{2} x - \dfrac{x^2}{2} + 2 \right) dx$

$= \left(\dfrac{3x^2}{4} - \dfrac{x^3}{6} + 2x \right) \Bigg|_{-1}^{4}$

$= \left(12 - \dfrac{64}{6} + 8 \right) - \left(\dfrac{3}{4} + \dfrac{1}{6} - 2 \right)$

$= \dfrac{125}{12}$

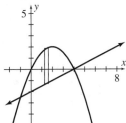

23. $y^2 = 3x$, $3x - 2y = 15$ $\left(\text{or } x = \dfrac{2y+15}{3}\right)$.

Region appears below.

Intersection: $y^2 = 3\left(\dfrac{2y+15}{3}\right)$,

$y^2 - 2y - 15 = 0$, $(y+3)(y-5) = 0$, so
$y = -3$ or 5.

$\text{Area} = \displaystyle\int_{-3}^{5}\left[\left(\frac{2}{3}y+5\right)-\frac{y^2}{3}\right]dy$

$\qquad = \left(\dfrac{1}{3}y^2 + 5y - \dfrac{y^3}{9}\right)\Bigg|_{-3}^{5}$

$\qquad = \left(\dfrac{25}{3} + 25 - \dfrac{125}{9}\right) - (3 - 15 + 3)$

$\qquad = \dfrac{256}{9}$

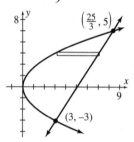

25. $y = 8 - x^2$, $y = x^2$, $x = -1$, $x = 1$. Region appears below.

Intersection: $x^2 = 8 - x^2$, $2x^2 = 8$, $x^2 = 4$,
so
$x = \pm 2$.
Area

$= \displaystyle\int_{-1}^{1}\left[\left(8 - x^2\right) - x^2\right]dx = \int_{-1}^{1}\left(8 - 2x^2\right)dx$

$= \left(8x - \dfrac{2x^3}{3}\right)\Bigg|_{-1}^{1} = \left(8 - \dfrac{2}{3}\right) - \left(-8 + \dfrac{2}{3}\right) = \dfrac{44}{3}$

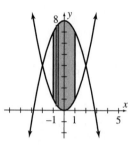

27. $y = x^2$, $y = 2$, $y = 5$. Region appears below.

Area

$= \displaystyle\int_{2}^{5}\left[\sqrt{y} - \left(-\sqrt{y}\right)\right]dy = \int_{2}^{5} 2\sqrt{y}\,dy$

$= 2 \cdot \dfrac{y^{\frac{3}{2}}}{\frac{3}{2}}\Bigg|_{2}^{5}$

$= \dfrac{4y^{\frac{3}{2}}}{3}\Bigg|_{2}^{5} = \dfrac{4 \cdot 5\sqrt{5}}{3} - \dfrac{4 \cdot 2\sqrt{2}}{3} = \dfrac{4}{3}\left(5\sqrt{5} - 2\sqrt{2}\right)$

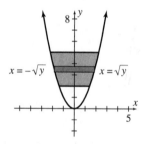

29. $y = x^3 - 1$, $y = x - 1$. Region appears below. Intersection: $x^3 - 1 = x - 1$, $x^3 - x = 0$, $x(x^2 - 1) = 0$,

$x(x+1)(x-1) = 0$, so $x = 0$ or $x = \pm 1$.

Area $= \int_{-1}^{0}[x^3 - 1 - (x-1)]dx + \int_{0}^{1}[x-1-(x^3-1)]dx$

$= \int_{-1}^{0}\left(x^3 - x\right)dx + \int_{0}^{1}\left(x - x^3\right)dx$

$= \left(\dfrac{x^4}{4} - \dfrac{x^2}{2}\right)\Big|_{-1}^{0} + \left(\dfrac{x^2}{2} - \dfrac{x^4}{4}\right)\Big|_{0}^{1}$

$= \left[0 - \left(\dfrac{1}{4} - \dfrac{1}{2}\right)\right] + \left[\left(\dfrac{1}{2} - \dfrac{1}{4}\right) - 0\right] = \dfrac{1}{2}$

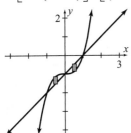

31. $4x + 4y + 17 = 0$, $y = \dfrac{1}{x}$. Region appears below. Intersection: $\dfrac{-17 - 4x}{4} = \dfrac{1}{x}$, $-17x - 4x^2 = 4$,

$4x^2 + 17x + 4 = 0$, $(4x + 1)(x + 4) = 0$, so $x = -\dfrac{1}{4}$ or -4.

Area $= \int_{-4}^{-1/4}\left[\dfrac{1}{x} - \left(\dfrac{-17-4x}{4}\right)\right]dx = \left(\ln|x| + \dfrac{17}{4}x + \dfrac{x^2}{2}\right)\Big|_{-4}^{-1/4}$

$= \left(\ln\dfrac{1}{4} - \dfrac{17}{16} + \dfrac{1}{32}\right) - (\ln 4 - 17 + 8) = \dfrac{255}{32} - 4\ln 2$

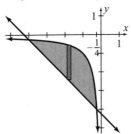

33. $y = x - 1$, $y = 5 - 2x$. Region appears below.

Intersection: $x - 1 = 5 - 2x$, $3x = 6$, so $x = 2$.

$$\text{Area} = \int_0^2 [(5-2x)-(x-1)]dx + \int_2^4 [(x-1)-(5-2x)]dx = \int_0^2 (6-3x)dx + \int_2^4 (3x-6)dx$$

$$= -\frac{1}{3}\int_0^2 (6-3x)[-3\ dx] + \frac{1}{3}\int_2^4 (3x-6)[3\ dx] = -\frac{(6-3x)^2}{6}\Big|_0^2 + \frac{(3x-6)^2}{6}\Big|_2^4$$

$$= -[0-6] + [6-0] = 6 + 6 = 12$$

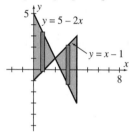

35. $\dfrac{\text{Area between curve and diag.}}{\text{Area under diagonal}} = \dfrac{\int_0^1 \left[x - \left(\frac{14}{15}x^2 + \frac{1}{15}x\right)\right]dx}{\int_0^1 x\ dx}$

$$\text{Numerator} = \int_0^1 \left[\frac{14}{15}x - \frac{14}{15}x^2\right]dx = \frac{14}{15}\int_0^1 \left(x - x^2\right)dx = \frac{14}{15}\left(\frac{x^2}{2} - \frac{x^3}{3}\right)\Big|_0^1 = \frac{14}{15}\left[\left(\frac{1}{2} - \frac{1}{3}\right) - 0\right] = \frac{14}{15}\cdot\frac{1}{6} = \frac{7}{45}$$

$$\text{Denominator} = \int_0^1 x\ dx = \frac{x^2}{2}\Big|_0^1 = \frac{1}{2}$$

$$\text{Coefficient of inequality} = \frac{\frac{7}{45}}{\frac{1}{2}} = \frac{14}{45}$$

37. $y^2 = 3x$, $y = mx$

Intersection: $(mx)^2 = 3x$, $m^2 x^2 = 3x$

$m^2 x^2 - 3x = 0$, $x(m^2 x - 3) = 0$, $x = 0$ or $x = \dfrac{3}{m^2}$.

If $x = 0$, then $y = 0$; if $x = \dfrac{3}{m^2}$, then $y = \dfrac{3}{m}$.

With horizontal elements,

$$\text{Area} = \int_0^{3/m} \left(\frac{y}{m} - \frac{y^2}{3}\right)dy = \left(\frac{y^2}{2m} - \frac{y^3}{9}\right)\Big|_0^{\frac{3}{m}}$$

$$= \frac{9}{2m^3} - \frac{3}{m^3} = \frac{3}{2m^3}\ \text{square units}$$

Note: With vertical elements,

Area $= \int_0^{3/m^2} \left(\sqrt{3}\sqrt{x} - mx \right) dx$.

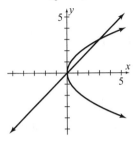

39. $y = x^2$ and $y = k$ intersect when

$x^2 = k$, $x = \pm\sqrt{k}$. Equating areas gives

$$\int_{-\sqrt{k}}^{\sqrt{k}} \left(k - x^2 \right) dx = \frac{1}{2} \int_{-2}^{2} \left(4 - x^2 \right) dx$$

$$\left(kx - \frac{x^3}{3} \right) \Bigg|_{-\sqrt{k}}^{\sqrt{k}} = \frac{1}{2} \left(4x - \frac{x^3}{3} \right) \Bigg|_{-2}^{2}$$

$$\frac{4}{3} k^{\frac{3}{2}} = \frac{16}{3}$$

$$k^{\frac{3}{2}} = 4 \Rightarrow k = 4^{\frac{2}{3}} = \left(2^2 \right)^{\frac{2}{3}} = 2^{\frac{4}{3}} \approx 2.52$$

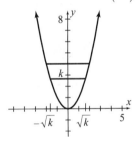

41. 4.76 sq units

43. Two integrals are involved.
Answer: 7.26 sq units

Problems 14.11

1. $D: p = 22 - 0.8q$
$S: p = 6 + 1.2q$

Equilibrium pt. $= (q_0, p_0) = (8, 15.6)$

$$CS = \int_0^{q_0} \left[f(q) - p_0 \right] dq$$

$$= \int_0^8 [(22 - 0.8q) - 15.6] = \int_0^8 (6.4 - 0.8q) dq$$

$$= \left(6.4q - 0.4q^2 \right) \Bigg|_0^8 = (51.2 - 25.6) - 0 = 25.6$$

$$PS = \int_0^{q_0} \left[p_0 - g(q) \right] dq$$

$$= \int_0^8 [15.6 - (6 + 1.2q)] dq$$

$$= \int_0^8 (9.6 - 1.2q) dq$$

$$= \left(9.6q - 0.6q^2 \right) \Bigg|_0^8 = (76.8 - 38.4) - 0 = 38.4$$

3. $D: p = \frac{50}{q+5}$
$S: p = \frac{q}{10} + 4.5$

Equilibrium pt. $= (q_0, p_0) = (5, 5)$

$$CS = \int_0^{q_0} \left[f(q) - p_0 \right] dq$$

$$= \int_0^5 \left[\frac{50}{q+5} - 5 \right] dq = \left(50 \ln|q+5| - 5q \right) \Big|_0^5$$

$$= [50 \ln(10) - 25] - [50 \ln(5)]$$
$$= 50[\ln(10) - \ln(5)] - 25 = 50 \ln(2) - 25$$

$$PS = \int_0^{q_0} \left[p_0 - g(q) \right] dq$$

$$= \int_0^5 \left[5 - \left(\frac{q}{10} + 4.5 \right) \right] dq = \int_0^5 \left(0.5 - \frac{q}{10} \right) dq$$

$$= \left(0.5q - \frac{q^2}{20} \right) \Bigg|_0^5 = (2.5 - 1.25) - 0 = 1.25$$

5. $D : q = 100(10 - 2p)$
$S : q = 50(2p - 1)$

Equilibrium pt. $= (q_0, \ p_0) = (300, \ 3.5)$

We use horizontal strips and integrate with respect to p.

$CS = \int_{3.5}^{5} 100(10 - 2p)dp$

$= 100(10p - p^2)\Big|_{3.5}^{5}$

$= 100[(50 - 25) - (35 - 12.25)]$

$= 225$

$PS = \int_{0.5}^{3.5} 50(2p - 1)dp = 50(p^2 - p)\Big|_{0.5}^{3.5}$

$= 50[(12.25 - 3.5) - (0.25 - 0.5)]$

$= 450$

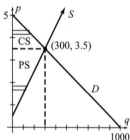

7. We integrate with respect to p. From the demand equation, when $q = 0$, then $p = 100$.

$CS = \int_{84}^{100} 10\sqrt{100 - p}\,dp$

$= \int_{84}^{100} -10(100 - p)^{\frac{1}{2}}[-dp]$

$= -\frac{20}{3}(100 - p)^{\frac{3}{2}}\Big|_{84}^{100}$

$= -\frac{20}{3}\left[0 - (16)^{\frac{3}{2}}\right] = -\frac{20}{3}(-64)$

$= 426\frac{2}{3} \approx \$426.67$

9. At equilibrium,

$2^{11-q} = 2^{q+1} \Rightarrow 11 - q = q + 1 \Rightarrow q = 5$, so

$p = 2^{11-5} = 64$

$CS = \int_{0}^{5}\left(2^{11-q} - 64\right)dq = \left(-\frac{2^{11-q}}{\ln 2} - 64q\right)\Big|_{0}^{5}$

$= -\frac{64}{\ln 2} - 320 - \left(-\frac{2^{11}}{\ln 2} - 0\right)$

≈ 2542.307 hundred $\approx \$254,000$

11. CS ≈ 1197; PS ≈ 477

Chapter 14 Review Problems

1. $\int\left(x^3 + 2x - 7\right)dx = \frac{x^4}{4} + 2 \cdot \frac{x^2}{2} - 7x + C$

$= \frac{x^4}{4} + x^2 - 7x + C$

3. $\int_{0}^{8}\left(\sqrt{2x} + 2x\right)dx = \int_{0}^{8}\left(\sqrt{2}x^{\frac{1}{2}} + 2x\right)dx$

$= \left(\sqrt{2} \cdot \frac{2x^{\frac{3}{2}}}{3} + x^2\right)\Big|_{0}^{8}$

$= \left(\sqrt{2} \cdot \frac{2}{3}\left(2\sqrt{2}\right)^3 + 64\right) - 0$

$= \frac{64}{3} + 64 = \frac{256}{3}$

5. $\int\frac{6}{(x + 5)^3}dx = 6\int(x + 5)^{-3}dx$

$= \frac{6(x + 5)^{-2}}{-2} + C$

$= -3(x + 5)^{-2} + C$

7. $\displaystyle\int \frac{6x^2-12}{x^3-6x+1}\,dx$

$\displaystyle = 2\int \frac{1}{x^3-6x+1}\Big[\big(3x^2-6\big)dx\Big]$

$\displaystyle = 2\ln\big|x^3-6x+1\big|+C$

9. $\displaystyle\int_0^1 \sqrt[3]{3t+8}\,dt = \frac{1}{3}\int_0^1 (3t+8)^{\frac{1}{3}}[3\ dt]$

$\displaystyle = \frac{1}{3}\cdot\frac{(3t+8)^{\frac{4}{3}}}{\frac{4}{3}}\Bigg|_0^1$

$\displaystyle = \frac{(3t+8)^{\frac{4}{3}}}{4}\Bigg|_0^1 = \frac{11\sqrt[3]{11}}{4}-4$

11. $\displaystyle\int y(y+1)^2\,dy = \int\big(y^3+2y^2+y\big)dy$

$\displaystyle = \frac{y^4}{4}+\frac{2y^3}{3}+\frac{y^2}{2}+C$

13. $\displaystyle\int \frac{\sqrt[5]{t}-\sqrt[3]{t}}{\sqrt{t}}\,dt = \int\left(\frac{t^{1/5}}{t^{1/2}}-\frac{t^{1/3}}{t^{1/2}}\right)dt$

$\displaystyle = \int(t^{-3/10}-t^{-1/6})dt = \frac{t^{7/10}}{\frac{7}{10}}-\frac{t^{5/6}}{\frac{5}{6}}+C$

$\displaystyle = \frac{10}{7}t^{7/10}-\frac{6}{5}t^{5/6}+C$

15. $\displaystyle\int_1^3 \frac{2t^2}{3+2t^3}\,dt = \frac{1}{3}\int_1^3 \frac{1}{3+2t^3}[6t^2\ dt]$

$\displaystyle = \frac{1}{3}\ln(3+2t^3)\Bigg|_1^3$

$\displaystyle = \frac{1}{3}[\ln(57)-\ln(5)] = \frac{1}{3}\ln\left(\frac{57}{5}\right)$

17. $\displaystyle\int x^2\sqrt{3x^3+2}\,dx = \frac{1}{9}\int\big(3x^3+2\big)^{\frac{1}{2}}\Big[9x^2dx\Big]$

$\displaystyle = \frac{1}{9}\cdot\frac{\big(3x^3+2\big)^{\frac{3}{2}}}{\frac{3}{2}}+C = \frac{2}{27}\big(3x^3+2\big)^{\frac{3}{2}}+C$

19. $\displaystyle\int\big(e^{2y}-e^{-2y}\big)dy$

$\displaystyle = \frac{1}{2}\int e^{2y}[2\ dy]-\left(-\frac{1}{2}\right)\int e^{-2y}[-2\ dy]$

$\displaystyle = \frac{1}{2}e^{2y}+\frac{1}{2}e^{-2y}+C = \frac{1}{2}\big(e^{2y}+e^{-2y}\big)+C$

21. $\displaystyle\int\left(\frac{1}{x}+\frac{2}{x^2}\right)dx = \int\frac{1}{x}\,dx+2\int x^{-2}dx$

$\displaystyle = \ln|x|+2\cdot\frac{x^{-1}}{-1}+C$

$\displaystyle = \ln|x|-\frac{2}{x}+C$

23. $\displaystyle\int_{-2}^1 10\big(y^4-y+1\big)dy = 10\left(\frac{y^5}{5}-\frac{y^2}{2}+y\right)\Bigg|_{-2}^1$

$\displaystyle = 10\left(\frac{1}{5}-\frac{1}{2}+1\right)-10\left(-\frac{32}{5}-2-2\right) = 111$

25. $\displaystyle\int_1^2 5x\sqrt{5-x^2}\,dx = -\frac{5}{2}\int_1^2(5-x^2)^{1/2}[-2x\,dx]$

$\displaystyle = -\frac{5}{2}\cdot\frac{(5-x^2)^{3/2}}{\frac{3}{2}}\Bigg|_1^2 = -\frac{5}{3}(5-x^2)^{3/2}\Bigg|_1^2$

$\displaystyle = -\frac{5}{3}(1^{3/2}-4^{3/2}) = -\frac{5}{3}(1-8) = \frac{35}{3}$

27.　$\displaystyle\int_0^1\left[2x-\frac{1}{(x+1)^{\frac{2}{3}}}\right]dx$

$\displaystyle=2\int_0^1 x\,dx-\int_0^1(x+1)^{-\frac{2}{3}}[dx]$

$\displaystyle=\left[2\cdot\frac{x^2}{2}-\frac{(x+1)^{\frac{1}{3}}}{\frac{1}{3}}\right]\Bigg|_0^1=\left[x^2-3(x+1)^{\frac{1}{3}}\right]\Bigg|_0^1$

$\displaystyle=\left[1-3\sqrt[3]{2}\right]-\left[0-3\right]=4-3\sqrt[3]{2}$

29.

$\displaystyle\int\frac{\sqrt{t}-3}{t^2}\,dt=\int\left[\frac{t^{\frac{1}{2}}}{t^2}-\frac{3}{t^2}\right]dt$

$\displaystyle=\int\left(t^{-\frac{3}{2}}-3t^{-2}\right)dt$

$\displaystyle=\frac{t^{-\frac{1}{2}}}{-\frac{1}{2}}-3\cdot\frac{t^{-1}}{-1}+C=-2t^{-\frac{1}{2}}+3t^{-1}+C$

$\displaystyle=\frac{3}{t}-\frac{2}{\sqrt{t}}+C$

31.

$\displaystyle\int_{-1}^0\frac{x^2+4x-1}{x+2}\,dx$

$\displaystyle=\int_{-1}^0\left(x+2-\frac{5}{x+2}\right)dx$

$\displaystyle=\left(\frac{x^2}{2}+2x-5\ln|x+2|\right)\Bigg|_{-1}^0$

$\displaystyle=(-5\ln2)-\left(\frac{1}{2}-2-0\right)=\frac{3}{2}-5\ln2$

33.　$\displaystyle\int 9\sqrt{x}\sqrt{x^{\frac{3}{2}}+1}\,dx=9\cdot\frac{2}{3}\int\left(x^{\frac{3}{2}}+1\right)^{\frac{1}{2}}\left[\frac{3}{2}x^{\frac{1}{2}}dx\right]$

$\displaystyle=6\cdot\frac{\left(x^{\frac{3}{2}}+1\right)^{\frac{3}{2}}}{\frac{3}{2}}+C=4\left(x^{\frac{3}{2}}+1\right)^{\frac{3}{2}}+C$

35.　$\displaystyle\int_1^e\frac{e^{\ln x}}{x^2}\,dx=\int_1^e\frac{x}{x^2}\,dx=\int_1^e\frac{1}{x}\,dx=\ln|x|\Big|_1^e$

$\displaystyle=\ln e-\ln 1$

$\displaystyle=1-0=1$

37.　$\displaystyle\int\frac{(1+e^{2x})^3}{e^{-2x}}\,dx=\frac{1}{2}\int(1+e^{2x})^3[2e^{2x}dx]$

$\displaystyle=\frac{(1+e^{2x})^3}{8}+C$

39.　$\displaystyle\int 3\sqrt{10^{3x}}\,dx=3\int e^{\frac{3x}{2}\ln 10}\,dx$

$\displaystyle=3\cdot\frac{2}{3\ln 10}\int e^{\frac{3x}{2}\ln 10}\left[\frac{3\ln 10}{2}\,dx\right]$

$\displaystyle=\frac{2}{\ln 10}e^{\frac{3x}{2}\ln 10}+C=\frac{2}{\ln 10}10^{\frac{3x}{2}}+C$

$\displaystyle=\frac{2\sqrt{10^{3x}}}{\ln 10}+C$

41.　$\displaystyle y=\int\left(e^{2x}+3\right)dx=\int e^{2x}dx+\int 3\,dx$

$\displaystyle=\frac{1}{2}\int e^{2x}[2\,dx]+\int 3\,dx$

$\displaystyle=\frac{1}{2}e^{2x}+3x+C$

$\displaystyle y(0)=-\frac{1}{2}$ implies that $\displaystyle -\frac{1}{2}=\frac{1}{2}+0+C$, so

$C=-1$. Thus $\displaystyle y=\frac{1}{2}e^{2x}+3x-1$

In Problems 43–57, answers are assumed to be expressed in square units.

43. $y = x^2 - 1$, $x = 2$, $y \geq 0$. Region appears below.

$$\text{Area} = \int_1^2 \left(x^2 - 1 \right) dx$$

$$= \left(\frac{x^3}{3} - x \right) \Bigg|_1^2$$

$$= \left(\frac{8}{3} - 2 \right) - \left(\frac{1}{3} - 1 \right) = \frac{4}{3}$$

45. $y = \sqrt{x+4}$, $x = 0$. Region appears below.

$$\text{Area} = \int_{-4}^0 \sqrt{x+4}\, dx$$

$$= \int_{-4}^0 (x+4)^{\frac{1}{2}} [dx]$$

$$= \frac{(x+4)^{\frac{3}{2}}}{\frac{3}{2}} \Bigg|_{-4}^0 = \frac{2(x+4)^{\frac{3}{2}}}{3} \Bigg|_{-4}^0 = \frac{16}{3} - 0 = \frac{16}{3}$$

47. $y = 5x - x^2$. Region appears below.

$$\text{Area} = \int_0^5 \left(5x - x^2 \right) dx = \left(\frac{5x^2}{2} - \frac{x^3}{3} \right) \Bigg|_0^5$$

$$= \left(\frac{125}{2} - \frac{125}{3} \right) - 0 = \frac{125}{6}$$

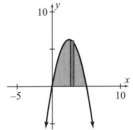

49. $y = \frac{1}{x} + 2$, $x = 1$, $x = 4$. Region appears below.

$$\text{Area} = \int_1^4 \left(\frac{1}{x} + 2 \right) dx = \left(\ln|x| + 2x \right) \Big|_1^4$$

$$= [\ln(4) + 8] - [0 + 2] = 6 + \ln 4$$

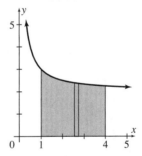

51. $y^2 = 4x$, $x = 0$, $y = 2$. Region appears below.

$$\text{Area} = \int_0^2 \frac{y^2}{4} dy = \frac{y^3}{12}\bigg|_0^2 = \frac{8}{12} - 0 = \frac{2}{3}$$

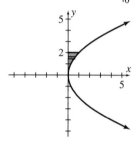

53. $y = x^2 + 4x - 5$, $y = 0$. Region appears below.

$x^2 + 4x - 5 = 0$, $(x + 5)(x - 1) = 0$, so $x = -5, 1$.

$$\text{Area} = \int_{-5}^1 -\left(x^2 + 4x - 5\right) dx$$

$$= -\left(\frac{x^3}{3} + 2x^2 - 5x\right)\bigg|_{-5}^1$$

$$= -\left(\frac{1}{3} + 2 - 5\right) + \left(-\frac{125}{3} + 50 + 25\right) = 36$$

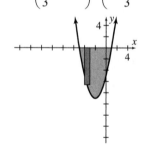

55. $y = x^2 - x$, $y = 10 - x^2$. Region appears below.

$x^2 - x = 10 - x^2$, $2x^2 - x - 10 = 0$,

$(x + 2)(2x - 5) = 0$, so $x = -2$ or $\dfrac{5}{2}$.

$$\text{Area} = \int_{-2}^{5/2} [(10 - x^2) - (x^2 - x)] dx$$

$$= \int_{-2}^{5/2} (10 + x - 2x^2) dx$$

$$= \left(10x + \frac{x^2}{2} - \frac{2x^3}{3}\right)\bigg|_{-2}^{5/2}$$

$$= \left(25 + \frac{25}{8} - \frac{125}{12}\right) - \left(-20 + 2 + \frac{16}{3}\right)$$

$$= \frac{243}{8}$$

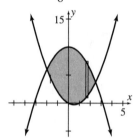

57. $y = \ln x$, $x = 0$, $y = 0$, $y = 1$. Region appears below.

$y = \ln x \Rightarrow x = e^y$

$$\text{Area} = \int_0^1 e^y dy = e^y\bigg|_0^1 = e - 1$$

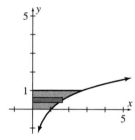

59. $r = \int \left(100 - \frac{3}{2}\sqrt{2q}\right) dq$

$= \int 100\, dq - \frac{3}{2}\sqrt{2} \int q^{\frac{1}{2}}\, dq$

$= 100q - \frac{3}{2}\sqrt{2} \cdot \frac{q^{\frac{3}{2}}}{\frac{3}{2}} + C = 100q - \sqrt{2}q^{\frac{3}{2}} + C$

When $q = 0$, then $r = 0$. Thus $0 = 0 - 0 + C$, so $C = 0$. Hence $r = 100q - \sqrt{2}q^{\frac{3}{2}}$. Since

$r = pq$, then $p = \frac{r}{q} = 100 - \sqrt{2}q^{\frac{1}{2}} = 100 - \sqrt{2q}$. Thus $p = 100 - \sqrt{2q}$.

61. $\int_{15}^{25} (250 - q - 0.2q^2)\, dq$

$= \left(250q - \frac{q^2}{2} - \frac{0.2q^3}{3}\right)\Bigg|_{15}^{25}$

$= \left(6250 - 312.5 - \frac{3125}{3}\right) - (3750 - 112.5 - 225)$

$\approx \$1483.33$

63. $\int_{0}^{100} 0.008e^{-0.008t}\, dt = -\int_{0}^{100} e^{0.008t}[-0.008\ dt]$

$= -e^{-0.008t}\Big|_{0}^{100} = -e^{-0.8} + 1 \approx 0.5507$

65. $y = 9 - 2x$, $y = x$; from $x = 0$ to $x = 4$. Region appears below. Intersection: $x = 9 - 2x$, $3x = 9$, so $x = 3$.

Area $= \int_{0}^{3} [(9-2x) - x]dx + \int_{3}^{4} [x - (9-2x)]dx$

$= \int_{0}^{3} (9 - 3x)dx + \int_{3}^{4} (3x - 9)dx$

$= \left(9x - \frac{3x^2}{2}\right)\Bigg|_{0}^{3} + \left(\frac{3x^2}{2} - 9x\right)\Bigg|_{3}^{4}$

$= \left[\left(27 - \frac{27}{2}\right) - 0\right] + \left[(24 - 36) - \left(\frac{27}{2} - 27\right)\right]$

$= 15$ square units

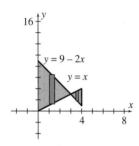

67. $D: p = 0.01q^2 - 1.1q + 30$
$\left.\begin{array}{l}\end{array}\right\}$
$S: p = 0.01q^2 + 8$

Equilibrium pt. $= (q_0, \, p_0) = (20, \, 12)$

$CS = \displaystyle\int_0^{q_0} [f(q) - p_0] \, dq$

$= \displaystyle\int_0^{20} \left[\left(0.01q^2 - 1.1q + 30\right) - 12\right] dq$

$= \displaystyle\int_0^{20} \left(0.01q^2 - 1.1q + 18\right) dq$

$= \left(\dfrac{0.01q^3}{3} - \dfrac{1.1q^2}{2} + 18q\right)\Bigg|_0^{20}$

$= \left(\dfrac{80}{3} - 220 + 360\right) - 0 = 166\dfrac{2}{3}$

$PS = \displaystyle\int_0^{q_0} [p_0 - g(q)] \, dq$

$= \displaystyle\int_0^{20} \left[12 - \left(0.01q^2 + 8\right)\right] dq$

$= \displaystyle\int_0^{20} \left(4 - 0.01q^2\right) dq = \left(4q - \dfrac{0.01q^3}{3}\right)\Bigg|_0^{20}$

$= \left(80 - \dfrac{80}{3}\right) - 0 = 53\dfrac{1}{3}$

69. $\displaystyle\int_{q_0}^{q_n} \dfrac{dq}{q - \hat{q}} = -(u+v) \int_0^n dt$

$\ln |q - \hat{q}| \Big\|_{q_0}^{q_n} = -(u+v)t \Big|_0^n$

$\ln |q_n - \hat{q}| - \ln |q_0 - \hat{q}| = -(u+v)n$

$\ln |q_0 - \hat{q}| - \ln |q_n - \hat{q}| = (u+v)n$

$\ln \left|\dfrac{q_0 - \hat{q}}{q_n - \hat{q}}\right| = (u+v)n$

$n = \dfrac{1}{u+v} \ln \left|\dfrac{q_0 - \hat{q}}{q_n - \hat{q}}\right|$

as was to be shown.

71. Case 1. $r \neq -1$

$g(x) = \dfrac{1}{k} \displaystyle\int_1^{1/x} k u^r \, du = \int_1^{1/x} u^r \, du = \dfrac{u^{r+1}}{r+1}\Bigg|_1^{1/x}$

$= \dfrac{1}{r+1}\left(x^{-r-1} - 1\right)$

$g'(x) = \dfrac{1}{r+1}\left[-(r+1)x^{-r-2}\right] = -\dfrac{1}{x^{r+2}}$

Case 2. $r = -1$

$g(x) = \dfrac{1}{k} \displaystyle\int_1^{1/x} k u^{-1} \, du = \int_1^{1/x} \dfrac{1}{u} \, du$

$= \ln |u| \Big\|_1^{1/x} = \ln\left(\dfrac{1}{x}\right) - 0 = -\ln x$

$g'(x) = -\dfrac{1}{x} = -\dfrac{1}{x^{r+2}}$

73. Two integrals are involved.
Answer: 15.08 sq units

75. $CS \approx 1148; \, PS \approx 251$

Mathematical Snapshot Chapter 14

1. a. $\displaystyle\int_0^5 f(t)dt = \int_0^5 (100-2t)dt = \left(100t-t^2\right)\Big|_0^5$

   $= (500-25)-0 = 475$

 b. $\displaystyle\int_{20}^{25} f(t)dt = \int_{20}^{25} (100-2t)dt = (100t-t^2)\Big|_{20}^{25}$

   $= (2500-625)-(2000-400) = 275$

3. a. Total revenue

   $\displaystyle = \int_0^R (m+st)f(t)dt = \int_0^{30} (100+t)(900-t^2)dt = \int_0^{30}\left(90,000+900t-100t^2-t^3\right)dt$

   $\displaystyle = 90,000t+450t^2-\frac{100}{3}t^3-\frac{1}{4}t^4\Big|_0^{30}$

   $= 2,700,000+405,000-900,000-202,500$

   $= \$2,002,500$

 b. Total number of units sold

   $\displaystyle = \int_0^R f(t)dt = \int_0^{30}\left(900-t^2\right)dt$

   $\displaystyle = \left(900t-\frac{1}{3}t^3\right)\Big|_0^{30} = 27,000-9000 = 18,000$

 c. Average delivered price $=\dfrac{\text{total revenue}}{\text{total number of units sold}}$

   $\displaystyle = \frac{2,002,500}{18,000} = \$111.25$

Chapter 15

1. $S(t) = \int -4te^{0.1t} dt$

Let $u = -4t$ and $dv = e^{0.1t} dt$, so $du = -4\, dt$,

and $v = \int e^{0.1t} dt = \frac{1}{0.1} e^{0.1t} = 10e^{0.1t}$.

$$\int -4te^{0.1t} dt$$
$$= (-4t)\left(10e^{0.1t}\right) - \int \left(10e^{0.1t}\right)(-4)dt$$
$$= -40te^{0.1t} + \int 40e^{0.1t} dt$$
$$= -40te^{0.1t} + 40\frac{e^{0.1t}}{0.1} + C$$
$$= -40te^{0.1t} + 400e^{0.1t} + C$$
$$S(t) = -40te^{0.1t} + 400e^{0.1t} + C \text{ and }$$
$$S(0) = 5000$$
$$5000 = 0 + 400e^0 + C$$
$$C = 4600$$
$$S(t) = -40te^{0.1t} + 400e^{0.1t} + 4600$$

Problems 15.1

1. $\int f(x) dx = uv - \int v\, du$

$$= x \cdot \frac{2}{3}(x+5)^{\frac{3}{2}} - \int \frac{2}{3}(x+5)^{\frac{3}{2}} dx$$
$$= \frac{2}{3} x(x+5)^{\frac{3}{2}} - \frac{2}{3} \cdot \frac{2}{5}(x+5)^{\frac{5}{2}} + C$$
$$= \frac{2}{3} x(x+5)^{\frac{3}{2}} - \frac{4}{15}(x+5)^{\frac{5}{2}} + C$$

3. $\int xe^{-x} dx$

Letting $u = x$, $dv = e^{-x} dx$, then $du = dx$,

$v = -e^{-x}$.

$$\int xe^{-x} dx = -xe^{-x} - \int -e^{-x} dx$$
$$= -xe^{-x} - \int e^{-x}[-dx] = -xe^{-x} - e^{-x} + C$$
$$= -e^{-x}(x+1) + C$$

5. $\int y^3 \ln y\, dy$

Letting $u = \ln y$, $dv = y^3 dy$, then

$$du = \left(\frac{1}{y}\right) dy, \quad v = \frac{y^4}{4}$$

$$\int y^3 \ln y\, dy = \frac{y^4 \ln y}{4} - \int \frac{y^4}{4}\left(\frac{1}{y}\right) dy$$
$$= \frac{y^4 \ln y}{4} - \int \frac{y^3}{4} dy = \frac{y^4 \ln y}{4} - \frac{y^4}{16} + C$$
$$= \frac{y^4}{4}\left[\ln(y) - \frac{1}{4}\right] + C$$

7. $\int \ln(4x)\, dx$

Letting $u = \ln(4x)$, $dv = dx$, then

$$du = \left(\frac{1}{x}\right) dx,$$

$v = x$.

$$\int \ln(4x) dx = x\ln(4x) - \int x\left(\frac{1}{x} dx\right)$$
$$= x\ln(4x) - \int dx = x\ln(4x) - x + C$$
$$= x[\ln(4x) - 1] + C$$

9. $\int 3x\sqrt{2x+3}\, dx$

Letting $u = 3x$, $dv = \sqrt{2x+3}\, dx$, then $du = 3dx$, $v = \frac{1}{3}(2x+3)^{3/2}$.

$$\int 3x\sqrt{2x+3}\, dx$$
$$= 3x \cdot \frac{1}{3}(2x+3)^{3/2} - \int \frac{1}{3}(2x+3)^{3/2} \cdot 3\, dx$$
$$= x(2x+3)^{3/2} - \frac{1}{5}(2x+3)^{5/2} + C$$
$$= \frac{1}{5}(2x+3)^{3/2}[5x - (2x+3)] + C$$
$$= \frac{1}{5}(2x+3)^{3/2}(3x-3) + C$$
$$= \frac{3}{5}(2x+3)^{3/2}(x-1) + C$$

11. $\displaystyle\int \frac{x}{(5x+2)^3}\,dx$

Letting $u = x$, $dv = (5x+3)^{-3}\,dx$, then

$du = dx$ and $v = -\dfrac{1}{10}(5x+3)^{-2}$.

$\displaystyle\int \frac{x}{(5x+2)^3}\,dx$

$\displaystyle = -\frac{x}{10(5x+3)^2} - \int -\frac{1}{10}(5x+3)^{-2}\,dx$

$\displaystyle = -\frac{x}{10(5x+3)^2} + \frac{1}{10}\cdot\frac{(5x+3)^{-1}}{5(-1)} + C$

$\displaystyle = -\frac{x}{10(5x+3)^2} - \frac{1}{50(5x+3)} + C$

13. $\displaystyle\int \frac{\ln x}{x^2}\,dx$

Letting $u = \ln x$, $dv = x^{-2}dx$, then

$du = \dfrac{1}{x}dx$, $v = -x^{-1}$.

$\displaystyle\int \frac{\ln x}{x^2}\,dx = -\frac{\ln x}{x} - \int -x^{-1}\left(\frac{1}{x}dx\right)$

$\displaystyle = -\frac{\ln x}{x} + \int x^{-2}\,dx = -\frac{\ln x}{x} - \frac{1}{x} + C$

$\displaystyle = -\frac{1}{x}(1 + \ln x) + C$

15. $\displaystyle\int_1^2 4xe^{2x}\,dx$

Letting $u = 4x$, $dv = e^{2x}dx$, then $du = 4dx$,

$v = \dfrac{1}{2}e^{2x}$

$\displaystyle\int_1^2 4xe^{2x}\,dx = \left[2xe^{2x} - \int 2e^{2x}\,dx\right]\Big|_1^2$

$\displaystyle = \left[2xe^{2x} - e^{2x}\right]\Big|_1^2 = e^{2x}(2x-1)\Big|_1^2$

$\displaystyle = e^4(3) - e^2(1) = e^2\left(3e^2 - 1\right)$

17. $\displaystyle\int_0^1 xe^{-x^2}\,dx = -\frac{1}{2}\int_0^1 e^{-x^2}(-2x\,dx)$

(Form: $\int e^u\,du$)

$\displaystyle = -\frac{1}{2}e^{-x^2}\Big|_0^1 = -\frac{1}{2}\left(e^{-1} - 1\right) = \frac{1}{2}\left(1 - e^{-1}\right)$

19. $\displaystyle\int_1^2 \frac{3x}{\sqrt{4-x}}\,dx$

Letting $u = 3x$, $dv = (4-x)^{-\frac{1}{2}}dx$, then

$du = 3dx$, $v = -2(4-x)^{\frac{1}{2}}$.

$\displaystyle\int_1^2 \frac{3x}{\sqrt{4-x}}\,dx$

$\displaystyle = \left[-6x(4-x)^{\frac{1}{2}} - \int -2(4-x)^{\frac{1}{2}}(3\,dx)\right]\Big|_1^2$

$\displaystyle = \left[-6x(4-x)^{\frac{1}{2}} - 4(4-x)^{\frac{3}{2}}\right]\Big|_1^2$

$\displaystyle = \left\{-2\sqrt{4-x}\,[3x + 2(4-x)]\right\}\Big|_1^2$

$\displaystyle = \left\{-2\sqrt{4-x}\,(x+8)\right\}\Big|_1^2 = -2\left(10\sqrt{2} - 9\sqrt{3}\right)$

$\displaystyle = 2\left(9\sqrt{3} - 10\sqrt{2}\right)$

21. $\displaystyle\int 3(2x-2)\ln(x-2)\,dx$

Letting $u = 3\ln(x-2)$, $dv = (2x-2)dx$, then

$du = \dfrac{3}{x-2}\,dx$ and $v = x^2 - 2x = x(x-2)$.

$\displaystyle\int 3(2x-2)\ln(x-2)\,dx$

$\displaystyle = 3x(x-2)\ln(x-2) - \int x(x-2)\cdot\frac{3}{x-2}\,dx$

$\displaystyle = 3x(x-2)\ln(x-2) - \int 3x\,dx$

$\displaystyle = 3x(x-2)\ln(x-2) - \frac{3}{2}x^2 + C$

23. $\int x^2 e^x dx$

Letting $u = x^2$, $dv = e^x dx$, then $du = 2x\, dx$ and $v = e^x$.

$\int x^2 e^x dx = x^2 e^x - \int e^x (2x\, dx) = x^2 e^x - 2\int xe^x dx$

For $\int xe^x dx$, let $u = x$, $dv = e^x dx$. Then

$du = dx$, $v = e^x$ and

$\int xe^x dx = xe^x - \int e^x dx = xe^x - e^x + C_1$

$= e^x (x-1) + C_1$.

Thus $\int x^2 e^x dx = x^2 e^x - 2\left[e^x (x-1) \right] + C$

$= e^x \left(x^2 - 2x + 2 \right) + C$.

25. $\int \left(x - e^{-x} \right)^2 dx = \int \left(x^2 - 2xe^{-x} + e^{-2x} \right) dx = \dfrac{x^3}{3} - \dfrac{e^{-2x}}{2} - 2\int xe^{-x} dx$

Using Problem 3 for $\int xe^{-x} dx$,

$\int \left(x - e^{-x} \right)^2 dx$

$= \dfrac{x^3}{3} - \dfrac{e^{-2x}}{2} + 2e^{-x}(x+1) + C$

27. $\int x^3 e^{x^2} dx$

Letting $u = x^2$, $dv = xe^{x^2} dx$, then $du = 2x\, dx$, $v = \left(\dfrac{1}{2} \right) e^{x^2}$.

$\int x^3 e^{x^2} dx = \dfrac{x^2 e^{x^2}}{2} - \int \dfrac{e^{x^2}}{2} (2x\, dx)$

$= \dfrac{x^2 e^{x^2}}{2} - \dfrac{e^{x^2}}{2} + C = \dfrac{e^{x^2}}{2} \left(x^2 - 1 \right) + C$

29. $\int \left(2^x + x \right)^2 dx = \int \left(2^{2x} + 2x2^x + x^2 \right) dx$

$= \int 2^{2x} dx + \int x2^{x+1} dx + \int x^2 dx$

For $\int x2^{x+1} dx$, let $u = x$, $dv = 2^{x+1} dx$. Then

$du = dx$, $v = \dfrac{1}{\ln 2} \cdot 2^{x+1}$ and

$\int x2^{x+1} dx = \dfrac{x}{\ln 2} \cdot 2^{x+1} - \dfrac{1}{\ln 2} \int 2^{x+1} dx = \dfrac{x}{\ln 2} \cdot 2^{x+1} - \dfrac{1}{\ln^2 2} \cdot 2^{x+1} + C_1$. Thus

$$\int\left(2^x+x\right)^2 dx = \int 2^{2x} dx + \int x 2^{x+1} dx + \int x^2 dx$$

$$= \frac{1}{2}\int 2^{2x}[2\ dx] + \int x 2^{x+1} dx + \int x^2 dx$$

$$= \frac{1}{2\ln 2}\cdot 2^{2x} + \frac{x}{\ln 2}\cdot 2^{x+1} - \frac{1}{\ln^2 2}\cdot 2^{x+1} + \frac{x^3}{3} + C = \frac{1}{\ln 2}\cdot 2^{2x-1} + \frac{x}{\ln 2}\cdot 2^{x+1} - \frac{1}{\ln^2 2}\cdot 2^{x+1} + \frac{x^3}{3} + C$$

31. Area $= \int_1^{e^3} (\ln x) dx$. Letting $u = \ln x$, $dv = dx$, then $du = \left(\frac{1}{x}\right) dx$, $v = x$.

$$\int_1^{e^3} (\ln x) dx = \left[(x\ln x) - \int x\cdot\frac{1}{x} dx\right]\Bigg|_1^{e^3}$$

$$= \left[(x\ln x) - \int dx\right]\Bigg|_1^{e^3} = [x\ln(x) - x]\Bigg|_1^{e^3}$$

$$= \left[e^3\cdot 3 - e^3\right] - [1\cdot 0 - 1] = 2e^3 + 1$$

The area is $(2e^3 + 1)$ sq units.

33. Area $= \int_1^2 x^2 \ln x\, dx$.

Letting $u = \ln x$, $dv = x^2 dx$, then $du = \frac{1}{x} dx$, $v = \frac{x^3}{3}$.

$$\int_1^2 x^2 \ln x\, dx = \left(\frac{x^3}{3}\ln x - \int\frac{x^3}{3}\cdot\frac{1}{x} dx\right)\Bigg|_1^2$$

$$= \left(\frac{x^3}{3}\ln x - \frac{1}{3}\int x^2 dx\right)\Bigg|_1^2$$

$$= \left(\frac{x^3}{3}\ln x - \frac{1}{9}x^3\right)\Bigg|_1^2$$

$$= \left(\frac{8}{3}\ln 2 - \frac{8}{9}\right) - \left(0 - \frac{1}{9}\right)$$

$$= \frac{8}{3}\ln(2) - \frac{7}{9}$$

The area is $\left(\frac{8}{3}\ln(2) - \frac{7}{9}\right)$ sq units.

35. a. Consider $\int p \, dq$. Letting $u = p$,

$dv = dq$, then $du = \dfrac{dp}{dq} dq$, $v = q$. Thus

$$\int p \, dq = pq - \int q \frac{dp}{dq} dq = r - \int q \frac{dp}{dq} dq$$

(since $r = pq$).

b. From (a), $r = \int p \, dq + \int q \dfrac{dp}{dq} dq$.

Combining the integrals gives

$$r = \int \left(p + q \frac{dp}{dq} \right) dq \, .$$

c. From (b), $\dfrac{dr}{dq} = p + q \dfrac{dp}{dq}$. Thus

$$\int_0^{q_0} \left(p + q \frac{dp}{dq} \right) dq$$

$$= \int_0^{q_0} \frac{dr}{dq} dq = r(q_0) - r(0) = r(q_0)$$

[since $r(0) = 0$].

37. f and its inverse f^{-1} satisfy the equation

$f(f^{-1}(x)) = x$. Differentiating this equation
using the Chain Rule we get:
$f'(f^{-1}(x)) \cdot (f^{-1})'(x) = 1$. Thus

$(f^{-1})'(x) = \dfrac{1}{f'(f^{-1}(x))}$. Now to evaluate

$\int f^{-1}(x) \, dx$ we will use integration by

parts, letting $u = f^{-1}(x)$ and $dv = dx$. Then

$du = \dfrac{1}{f'(f^{-1}(x))} dx$ and $v = x$.

So $\int f^{-1}(x) \, dx = xf^{-1}(x) - \int \dfrac{x}{f'(f^{-1}(x))} dx$.

To evaluate $\int \dfrac{x}{f'(f^{-1}(x))} dx$ we will use the

fact that $x = f(f^{-1}(x))$ and

$(f^{-1})'(x) = \dfrac{1}{f'(f^{-1}(x))}$.

Hence

$$\int \frac{x}{f'(f^{-1}(x))} dx = \int f(f^{-1}(x)) \cdot (f^{-1})'(x) \, dx$$

$$= F(f^{-1}(x))$$

since $F' = f$. Finally,

$$\int f^{-1}(x) \, dx = xf^{-1}(x) - F(f^{-1}(x)) + C.$$

Principles in Practice 15.2

1. $r(q) = \int r'(q) dq = \int \dfrac{5(q+4)}{q^2 + 4q + 3} dq$

Express $\dfrac{5(q+4)}{q^2 + 4q + 3}$ as a sum of partial

fractions.

$$\frac{5(q+4)}{q^2 + 4q + 3} = \frac{5(q+4)}{(q+1)(q+3)} = \frac{A}{q+1} + \frac{B}{q+3}$$

$5(q+4) = A(q+3) + B(q+1)$
When $q = -3$, we get $5(1) = -2B$, so

$B = -\dfrac{5}{2}$. When $q = -1$, we get $5(3) = A(2)$,

so $A = \dfrac{15}{2}$.

$$r(q) = \int \frac{5(q+4)}{q^2 + 4q + 3} dx$$

$$= \int \frac{\frac{15}{2}}{q+1} dq - \int \frac{\frac{5}{2}}{q+3} dq$$

$$= \frac{15}{2} \ln|q+1| - \frac{5}{2} \ln|q+3| + C$$

$$= \frac{5}{2} \ln \left| \frac{(q+1)^3}{q+3} \right| + C$$

Since $r(0) = 0$, $0 = \dfrac{5}{2} \ln \left| \dfrac{1}{3} \right| + C$ so

$$C = \frac{5}{2} \ln 3 \text{ and } r(q) = \frac{5}{2} \ln \left| \frac{3(q+1)^3}{q+3} \right|.$$

Problems 15.2

1. $\dfrac{10x}{x^2+7x+6} = \dfrac{10x}{(x+6)(x+1)} = \dfrac{A}{x+6} + \dfrac{B}{x+1}$

$10x = A(x+1) + B(x+6)$

If $x = -1$, then $-10 = 5B$, or $B = -2$. If $x = -6$, then $-60 = -5A$, or $A = 12$.

Answer $\dfrac{12}{x+6} - \dfrac{2}{x+1}$

3. $\dfrac{2x^2}{x^2+5x+6} = 2 + \dfrac{-10x-12}{x^2+5x+6}$

$\dfrac{-10x-12}{x^2+5x+6} = \dfrac{-10x-12}{(x+2)(x+3)} = \dfrac{A}{x+2} + \dfrac{B}{x+3}$

$-10x - 12 = A(x+3) + B(x+2)$

If $x = -3$, then $18 = -B$, or $B = -18$.

If $x = -2$, then $8 = A$.

Answer: $2 + \dfrac{8}{x+2} - \dfrac{18}{x+3}$

5. $f(x) = \dfrac{x+4}{x^2+4x+4} = \dfrac{x+4}{(x+2)^2}$

$= \dfrac{A}{x+2} + \dfrac{B}{(x+2)^2}$

$x + 4 = A(x+2) + B$

If $x = -2$, then $2 = B$. If $x = 0$, then

$4 = 2A + B$, $2A = 4 - B = 4 - 2 = 2$, or $A = 1$.

Answer: $\dfrac{1}{x+2} + \dfrac{2}{(x+2)^2}$

7. $\dfrac{x^2+3}{x^3+x} = \dfrac{x^2+3}{x\left(x^2+1\right)} = \dfrac{A}{x} + \dfrac{Bx+C}{x^2+1}$

$x^2 + 3 = A\left(x^2+1\right) + (Bx+C)x$

$x^2 + 3 = (A+B)x^2 + Cx + A$

Thus $A + B = 1$, $C = 0$, $A = 3$. This gives
$A = 3$, $B = -2$, $C = 0$.

Answer: $\dfrac{3}{x} - \dfrac{2x}{x^2+1}$

9. $\dfrac{5x-2}{x^2-x} = \dfrac{5x-2}{x(x-1)} = \dfrac{A}{x} + \dfrac{B}{x-1}$

$5x - 2 = A(x-1) + Bx$

If $x = 1$, then $3 = B$. If $x = 0$, then $-2 = -A$,
or $A = 2$.

$\displaystyle\int \dfrac{5x-2}{x^2-x}\,dx = \int\left(\dfrac{2}{x} + \dfrac{3}{x-1}\right)dx$

$= 2\ln|x| + 3\ln|x-1| + C = \ln\left|x^2(x-1)^3\right| + C$

11. $\dfrac{x+10}{x^2-x-2} = \dfrac{x+10}{(x+1)(x-2)} = \dfrac{A}{x+1} + \dfrac{B}{x-2}$

$x + 10 = A(x-2) + B(x+1)$

If $x = 2$, then $12 = 3B$, or $B = 4$. If $x = -1$,
then $9 = -3A$, or $A = -3$.

$\displaystyle\int \dfrac{x+10}{x^2-x-2}\,dx = \int\left(\dfrac{-3}{x+1} + \dfrac{4}{x-2}\right)dx$

$= -3\ln|x+1| + 4\ln|x-2| + C$

$= \ln\left|\dfrac{(x-2)^4}{(x+1)^3}\right| + C$

13. $\dfrac{3x^3-3x+4}{4x^2-4} = \dfrac{1}{4}\cdot\dfrac{3x^3-3x+4}{x^2-1}$

$= \dfrac{1}{4}\left(3x + \dfrac{4}{x^2-1}\right)$

$\dfrac{4}{x^2-1} = \dfrac{4}{(x-1)(x+1)} = \dfrac{A}{x-1} + \dfrac{B}{x+1}$

$4 = A(x+1) + B(x-1)$

If $x = -1$, then $4 = -2B$, or $B = -2$. If $x = 1$,
then $4 = 2A$, or $A = 2$.

$\displaystyle\int \dfrac{3x^3-3x+4}{4x^2-4}\,dx = \dfrac{1}{4}\int\left(3x + \dfrac{2}{x-1} + \dfrac{-2}{x+1}\right)dx$

$= \left(\dfrac{1}{4}\right)\left[\dfrac{3x^2}{2} + 2\ln|x-1| - 2\ln|x+1|\right] + C$

$= \left(\dfrac{1}{4}\right)\left[\dfrac{3x^2}{2} + \ln\left|\dfrac{x-1}{x+1}\right|^2\right] + C$

15. $\dfrac{3x-4}{x^3-x^2-2x} = \dfrac{3x-4}{x(x+1)(x-2)} = \dfrac{A}{x} + \dfrac{B}{x+1} + \dfrac{C}{x-2}$

$3x - 4 = A(x + 1)(x - 2) + Bx(x - 2) + Cx(x + 1)$

If $x = 0$, then $-4 = -2A$, or $A = 2$.

If $x = -1$, then $-7 = 3B$, or $B = -\dfrac{7}{3}$.

If $x = 2$, then $2 = 6C$, or $C = \dfrac{1}{3}$.

$\displaystyle\int \dfrac{3x-4}{x^3-x^2-2x}\,dx = \int \left(\dfrac{2}{x} + \dfrac{-\frac{7}{3}}{x+1} + \dfrac{\frac{1}{3}}{x-2} \right) dx$

$\qquad\qquad = 2\ln|x| - \dfrac{7}{3}\ln|x+1| + \dfrac{1}{3}\ln|x-2| + C$

$\qquad\qquad = \ln\left| \dfrac{x^2 \sqrt[3]{x-2}}{\sqrt[3]{(x+1)^7}} \right| + C$

17. $\displaystyle\int \dfrac{2(3x^5 + 4x^3 - x)}{x^6 + 2x^4 - x^2 - 2}\,dx = \int \dfrac{1}{x^6 + 2x^4 - x^2 - 2}\left[\left(6x^5 + 8x^3 - 2x\right)dx \right]$

$\left(\text{Form: } \displaystyle\int \left(\dfrac{1}{u} \right) du \right)$ (Partial fractions not required.)

Answer: $\ln\left| x^6 + 2x^4 - x^2 - 2 \right| + C$

19. $\dfrac{2x^2 - 5x - 2}{(x-2)^2(x-1)} = \dfrac{A}{x-1} + \dfrac{B}{x-2} + \dfrac{C}{(x-2)^2}$

$2x^2 - 5x - 2 = A(x-2)^2 + B(x-1)(x-2) + C(x-1)$

If $x = 1$, then $-5 = A$. If $x = 2$, then $-4 = C$.

If $x = 0$, then $-2 = 4A + 2B - C$, $-2 = -20 + 2B + 4$, or $B = 7$.

$\displaystyle\int \dfrac{2x^2 - 5x - 2}{(x-2)^2(x-1)}\,dx = \int \left[\dfrac{-5}{x-1} + \dfrac{7}{x-2} + \dfrac{-4}{(x-2)^2} \right] dx$

$= -5\ln|x-1| + 7\ln|x-2| + \dfrac{4}{x-2} + C = \dfrac{4}{x-2} + \ln\left| \dfrac{(x-2)^7}{(x-1)^5} \right| + C$

21. $\dfrac{2(x^2 + 8)}{x^3 + 4x} = \dfrac{2x^2 + 16}{x\left(x^2 + 4\right)} = \dfrac{A}{x} + \dfrac{Bx+C}{x^2+4}$

$2x^2 + 16 = A\left(x^2 + 4\right) + (Bx + C)x$

$2x^2 + 16 = (A + B)x^2 + Cx + 4A$

402

Thus $A + B = 2$, $C = 0$, $4A = 16$. This gives $A = 4$, $B = -2$, $C = 0$.

$$\int \frac{2(x^2 + 8)}{x^3 + 4x} dx = \int \left(\frac{4}{x} + \frac{-2x}{x^2 + 4} \right) dx = 4 \int \frac{1}{x} dx - \int \frac{1}{x^2 + 4} [2x\ dx]$$

$$= 4\ln|x| - \ln\left(x^2 + 4\right) + C = \ln\left[\frac{x^4}{x^2 + 4} \right] + C$$

23. $\dfrac{-x^3 + 8x^2 - 9x + 2}{\left(x^2 + 1\right)(x - 3)^2} = \dfrac{Ax + B}{x^2 + 1} + \dfrac{C}{x - 3} + \dfrac{D}{(x - 3)^2}$

$-x^3 + 8x^2 - 9x + 2 = (Ax + B)(x - 3)^2 + C(x - 3)\left(x^2 + 1\right) + D\left(x^2 + 1\right)$

$$= (Ax + B)\left(x^2 - 6x + 9\right) + C\left(x^3 - 3x^2 + x - 3\right) + D\left(x^2 + 1\right)$$

$$= (A + C)x^3 + (B - 6A - 3C + D)x^2 + (9A - 6B + C)x + (9B - 3C + D)$$

Thus $A + C = -1$, $B - 6A - 3C + D = 8$, $9A - 6B + C = -9$, $9B - 3C + D = 2$. This gives $A = -1$, $B = 0$, $C = 0$,
$D = 2$.

$$\int \frac{-x^3 + 8x^2 - 9x + 2}{\left(x^2 + 1\right)(x - 3)^2} dx = \int \left(\frac{-x}{x^2 + 1} + \frac{0}{x - 3} + \frac{2}{(x - 3)^2} \right) dx = -\frac{1}{2}\ln\left(x^2 + 1\right) - \frac{2}{x - 3} + C$$

25. $\dfrac{14x^3 + 24x}{\left(x^2 + 1\right)\left(x^2 + 2\right)} = \dfrac{Ax + B}{x^2 + 1} + \dfrac{Cx + D}{x^2 + 2}$

$14x^3 + 24x = \left(x^2 + 2\right)(Ax + B) + \left(x^2 + 1\right)(Cx + D) = (A + C)x^3 + (B + D)x^2 + (2A + C)x + (2B + D)$

Thus $A + C = 14$, $B + D = 0$, $2A + C = 24$, $2B + D = 0$.
This gives $A = 10$, $B = 0$, $C = 4$, $D = 0$.

$$\int \frac{14x^3 + 24x}{\left(x^2 + 1\right)\left(x^2 + 2\right)} dx = \int \left(\frac{10x}{x^2 + 1} + \frac{4x}{x^2 + 2} \right) dx$$

$$= 5\int \frac{1}{x^2 + 1}[2\ dx] + 2\int \frac{1}{x^2 + 2}[2\ dx]$$

$$= 5\ln\left(x^2 + 1\right) + 2\ln\left(x^2 + 2\right) + C$$

$$= \ln\left[\left(x^2 + 1\right)^5 \left(x^2 + 2\right)^2 \right] + C$$

27. $\dfrac{3x^3 + 8x}{(x^2 + 2)^2} = \dfrac{Ax + B}{x^2 + 2} + \dfrac{Cx + D}{(x^2 + 2)^2}$

$3x^3 + 8x = (Ax + B)(x^2 + 2) + Cx + D$

$$= Ax^3 + Bx^2 + (2A + C)x + (2B + D)$$

Thus, $A = 3$, $B = 0$, $2A + C = 8$, $2B + D = 0$.
This gives $A = 3$, $B = 0$, $C = 2$, $D = 0$.

$$\int \frac{3x^3 + 8x}{(x^2 + 2)^2} \, dx = \int \left(\frac{3x}{x^2 + 2} + \frac{2x}{(x^2 + 2)^2} \right) dx$$

$$= \frac{3}{2} \ln(x^2 + 2) - \frac{1}{x^2 + 2} + C$$

29. $\dfrac{2 - 2x}{x^2 + 7x + 12} = \dfrac{2 - 2x}{(x + 3)(x + 4)} = \dfrac{A}{x + 3} + \dfrac{B}{x + 4}$

$2 - 2x = A(x + 4) + B(x + 3)$
If $x = -4$, then $10 = -B$, or $B = -10$. If
$x = -3$, then $8 = A$.

$$\int_0^1 \frac{2 - 2x}{x^2 + 7x + 12} \, dx = \int_0^1 \left(\frac{8}{x + 3} + \frac{-10}{x + 4} \right) dx$$

$$= \left[8 \ln|x + 3| - 10 \ln|x + 4| \right]\Big|_0^1$$

$= 8 \ln 4 - 10 \ln 5 - (8 \ln 3 - 10 \ln 4)$
$= 18 \ln(4) - 10 \ln(5) - 8 \ln(3)$

31. Note that $\dfrac{6(x^2 + 1)}{(x + 2)^2} \geq 0$ on $[0, 1]$.

Area $= \displaystyle\int_0^1 \frac{6(x^2 + 1)}{(x + 2)^2} \, dx$

$\dfrac{6(x^2 + 1)}{(x + 2)^2} = 6 + \dfrac{-24x - 18}{(x + 2)^2}$ (by long division)

$\dfrac{-24x - 18}{(x + 2)^2} = \dfrac{A}{x + 2} + \dfrac{B}{(x + 2)^2}$

$-24x - 18 = A(x + 2) + B$
If $x = -2$, then $30 = B$. If $x = 0$, then
$-18 = 2A + B$, $-18 = 2A + 30$, or $A = -24$.

$$\int_0^1 \frac{6(x^2 + 1)}{(x + 2)^2} \, dx$$

$$= \int_0^1 \left[6 + \frac{-24}{x + 2} + \frac{30}{(x + 2)^2} \right] dx$$

$$= \left[6x - 24 \ln|x + 2| - \frac{30}{x + 2} \right]\Big|_0^1$$

$= 6 - 24 \ln 3 - 10 - (-24 \ln 2 - 15)$

$= 11 + 24 \ln \dfrac{2}{3}$

The area is $11 + 24 \ln \dfrac{2}{3}$ sq units.

Problems 15.3

1. Let $u = x$, $a^2 = 9$. Then $du = dx$.

$$\int \frac{dx}{(9 - x^2)^{3/2}} = \frac{x}{9\sqrt{9 - x^2}} + C$$

3. Let $u = 4x$, $a^2 = 3$. Then $du = 4 \, dx$.

$$\int \frac{dx}{x^2 \sqrt{16x^2 + 3}} = 4 \int \frac{(4 \, dx)}{(4x)^2 \sqrt{(4x)^2 + 3}}$$

$$= 4 \left[-\frac{\sqrt{(4x)^2 + 3}}{3(4x)} \right] + C$$

$$= -\frac{\sqrt{16x^2 + 3}}{3x} + C$$

5. Formula 5 with $u = x$, $a = 6$, $b = 7$. Then
$du = dx$.

$$\int \frac{dx}{x(6 + 7x)} = \frac{1}{6} \ln\left| \frac{x}{6 + 7x} \right| + C$$

7. Formula 28 with $u = x$, $a = 3$. Then $du = dx$.

$$\int \frac{dx}{x\sqrt{x^2 + 9}} = \frac{1}{3} \ln\left| \frac{\sqrt{x^2 + 9} - 3}{x} \right| + C$$

9. Formula 12 with $u = x$, $a = 2$, $b = 3$, $c = 4$, $k = 5$. Then $du = dx$.

$$\int \frac{x\,dx}{(2+3x)(4+5x)}$$

$$= \frac{1}{2}\left[\frac{4}{5}\ln|4+5x| - \frac{2}{3}\ln|2+3x|\right] + C$$

11. Formula 45 with $u = x$, $a = 5$, $b = 2$, $c = 3$. Then $du = dx$.

$$\int \frac{dx}{5+2e^{3x}} = \frac{1}{15}\left(3x - \ln\left|5x + 2e^{3x}\right|\right) + C$$

13. Formula 9 with $u = x$, $a = 5$, $b = 2$. Then $du = dx$.

$$\int \frac{7\,dx}{x(5+2x)^2} = 7\left[\int \frac{dx}{x(5+2x)^2}\right]$$

$$= 7\left[\frac{1}{5(5+2x)} + \frac{1}{25}\ln\left|\frac{x}{5+2x}\right|\right] + C$$

15. Formula 3 with $u = x$, $a = 2$, $b = 1$. Then $du = dx$.

$$\int_0^1 \frac{x\,dx}{2+x} = \left(x - 2\ln|2+x|\right)\Big|_0^1 = 1 - 2\ln 3 + 2\ln 2$$

$$= 1 - \ln 9 + \ln 4 = 1 + \ln\left(\frac{4}{9}\right)$$

17. Formula 23 with $u = x$, $a^2 = 3$. Then $du = dx$.

$$\int \sqrt{x^2 - 3}\,dx$$

$$= \frac{1}{2}\left(x\sqrt{x^2-3} - 3\ln\left|x + \sqrt{x^2-3}\right|\right) + C$$

19. Formula 38 with $u = x$, $a = 12$. Then $du = dx$.

$$\int_0^{1/12} xe^{12x}\,dx = \frac{e^{12x}}{144}(12x-1)\Big|_0^{1/12}$$

$$= \frac{1}{144}[e(0) - 1(-1)]$$

$$= \frac{1}{144}$$

21. Formula 39 with $u = x$, $n = 2$, $a = 1$. Then $du = dx$.

$$\int x^2 e^x\,dx = x^2 e^x - 2\int xe^x\,dx$$

Applying Formula 38 on $\int xe^x\,dx$ with $u = x$, $a = 1$ (so $du = dx$) gives

$$\int xe^x\,dx = e^x(x-1) + C_1. \text{ Thus}$$

$$\int x^2 e^x\,dx = x^2 e^x - 2\left[e^x(x-1)\right] + C$$

$$= e^x\left[x^2 - 2(x-1)\right] + C$$

$$= e^x\left(x^2 - 2x + 2\right) + C$$

23. Formula 26 with $u = \sqrt{5}x$, $a^2 = 1$. Then $du = \sqrt{5}\,dx$.

$$\int \frac{\sqrt{5x^2+1}}{2x^2}\,dx$$

$$= \frac{5}{2\sqrt{5}}\int \frac{\sqrt{5x^2+1}}{5x^2}\left(\sqrt{5}\,dx\right)$$

$$= \frac{\sqrt{5}}{2}\left(-\frac{\sqrt{5x^2+1}}{\sqrt{5}x} + \ln\left|\sqrt{5}x + \sqrt{5x^2+1}\right|\right) + C$$

25. Formula 7 with $u = x$, $a = 1$, $b = 3$. Then $du = dx$.

$$\int \frac{x\,dx}{(1+3x)^2} = \frac{1}{9}\left(\ln|1+3x| + \frac{1}{1+3x}\right) + C$$

27. Formula 34 with $u = \sqrt{5}x$, $a = \sqrt{7}$. Then $du = \sqrt{5}\,dx$

$$\int \frac{dx}{7-5x^2} = \frac{1}{\sqrt{5}}\int \frac{1}{\left(\sqrt{7}\right)^2 - \left(\sqrt{5}x\right)^2}\left(\sqrt{5}\,dx\right)$$

$$= \frac{1}{\sqrt{5}}\left(\frac{1}{2\sqrt{7}}\ln\left|\frac{\sqrt{7}+\sqrt{5}x}{\sqrt{7}-\sqrt{5}x}\right|\right) + C$$

29. Formula 42 with $u = 3x$, $n = 5$. Then $du = 3\ dx$.

$$\int 36x^5 \ln(3x)dx = 36 \int x^5 \ln(3x)dx$$

$$= \frac{36}{3^6} \int (3x)^5 \ln(3x)(3\ dx)$$

$$= \frac{4}{81} \left[\frac{(3x)^6 \ln(3x)}{6} - \frac{(3x)^6}{36} \right] + C$$

$$= x^6 [6\ln(3x) - 1] + C$$

31. Formula 13 with $u = x$, $a = 1$, $b = 3$. Then $du = dx$.

$$\int 270x\sqrt{1+3x}dx = 270 \int x\sqrt{1+3x}\ dx$$

$$= 270 \left[\frac{2(9x-2)(1+3x)^{\frac{3}{2}}}{15 \cdot 9} \right] + C$$

$$= 4(9x-2)(1+3x)^{\frac{3}{2}} + C$$

33. Formula 27 with $u = 2x$, $a^2 = 13$. Then $du = 2\ dx$.

$$\int \frac{dx}{\sqrt{4x^2-13}} = \frac{1}{2} \int \frac{1}{\sqrt{(2x)^2-13}}(2\ dx)$$

$$= \frac{1}{2} \ln \left| 2x + \sqrt{4x^2-13} \right| + C$$

35. Formula 21 with $u = 3x$, $a^2 = 16$. Then $du = 3\ dx$.

$$\int \frac{2\ dx}{x^2\sqrt{16-9x^2}} = 2(3) \int \frac{(3\ dx)}{(3x)^2\sqrt{16-(3x)^2}}$$

$$= 6 \left(-\frac{\sqrt{16-9x^2}}{16(3x)} \right) + C$$

$$= -\frac{\sqrt{16-9x^2}}{8x} + C$$

37. Formula 45 with
$u = \sqrt{x}$, $a = \pi$, $b = 7$, $c = 4$. Then

$$du = \frac{1}{2\sqrt{x}}dx$$

$$\int \frac{dx}{\sqrt{x}\left(\pi + 7e^{4\sqrt{x}}\right)} = 2 \int \frac{1}{\pi + 7e^{4\sqrt{x}}} \left(\frac{1}{2\sqrt{x}}dx \right)$$

$$= 2 \left[\frac{1}{4\pi} \left(4\sqrt{x} - \ln \left| \pi + 7e^{4\sqrt{x}} \right| \right) \right] + C$$

$$= \frac{1}{2\pi} \left(4\sqrt{x} - \ln \left| \pi + 7e^{4\sqrt{x}} \right| \right) + C$$

39. Can be put in the form $\int \frac{1}{u}du$.

$$\int \frac{x\ dx}{x^2+1} = \frac{1}{2} \int \frac{1}{x^2+1}(2x\ dx)$$

$$= \frac{1}{2} \ln \left(x^2+1 \right) + C$$

41. Can be put in the form $\int u^n du$.

$$\int 6x\sqrt{2x^2+1}dx = \frac{3}{2} \int \left(2x^2+1 \right)^{\frac{1}{2}} (4x\ dx)$$

$$= \frac{3}{2} \cdot \frac{\left(2x^2+1 \right)^{\frac{3}{2}}}{\frac{3}{2}} + C$$

$$= \left(2x^2+1 \right)^{\frac{3}{2}} + C$$

43. $\int \dfrac{1}{x^2-5x+6}dx = \int \dfrac{1}{(x-3)(x-2)}dx$

Formula 11 with $u = x$, $a = -3$, $b = 1$, $c = -2$, and $k = 1$. Then $du = dx$.

$$\int \frac{1}{x^2-5x+6}dx = \int \frac{1}{(x-3)(x-2)}dx$$

$$= \ln \left| \frac{x-3}{x-2} \right| + C$$

45. Formula 42 with $u = x$ and $n = 3$. Then $du = dx$.

$$\int x^3 \ln x \, dx = \frac{x^4}{4} \left[\ln(x) - \frac{1}{4} \right] + C$$

47. Formula 38 with $u = x^2$ and $a = 3$. Then $du = 2x \, dx$.

$$\int 4x^3 e^{3x^2} dx = 2 \int x^2 e^{3x^2} [2x \, dx]$$

$$= 2 \left[\frac{e^{3x^2}}{9} (3x^2 - 1) \right] + C$$

$$= \frac{2}{9} e^{3x^2} (3x^2 - 1) + C$$

49. Formula 43 and then Formula 41. For Formula 43, let $u = x$, $n = 0$, and $m = 2$. Then $du = dx$.

$$\int \ln^2 x \, dx = x \ln^2 x - 2 \int \ln x \, dx$$

Now we apply Formula 41 to the last integral with $u = x$ (so $du = dx$).

$$\int \ln^2 x \, dx = x(\ln x)^2 - 2x(\ln x) + 2x + C$$

51. Formula 15 with $u = x$, $a = 4$ and $b = -1$. Then $du = dx$.

$$\int_1^2 \frac{x \, dx}{\sqrt{4-x}} = \frac{2(-x-8)\sqrt{4-x}}{3} \Bigg|_1^2$$

$$= \frac{2}{3} \left(9\sqrt{3} - 10\sqrt{2} \right)$$

53. Can be put in the form $\int u^n \, du$.

$$\int_0^1 \frac{2x \, dx}{\sqrt{8-x^2}} = -\int_0^1 \left(8 - x^2 \right)^{-\frac{1}{2}} (-2x \, dx)$$

$$= -\frac{\left(8 - x^2 \right)^{\frac{1}{2}}}{\frac{1}{2}} \Bigg|_0^1$$

$$= -2 \left(8 - x^2 \right)^{\frac{1}{2}} \Bigg|_0^1 = -2 \left(\sqrt{7} - \sqrt{8} \right)$$

$$= -2 \left(\sqrt{7} - 2\sqrt{2} \right)$$

$$= 2 \left(2\sqrt{2} - \sqrt{7} \right)$$

55. Integration by parts or Formula 42. For Formula 42, let $u = 2x$, $n = 1$. Then $du = 2 \, dx$.

$$\int_1^2 x \ln(2x) dx = \frac{1}{4} \int_1^2 (2x) \ln(2x) [2 \, dx]$$

$$= \frac{1}{4} \left[\frac{(2x)^2 \ln(2x)}{2} - \frac{(2x)^2}{4} \right] \Bigg|_1^2$$

$$= 2 \ln(4) - 1 - \frac{1}{2} \ln(2) + \frac{1}{4}$$

$$= 2 \ln \left(2^2 \right) - \frac{1}{2} \ln(2) - \frac{3}{4}$$

$$= 4 \ln(2) - \frac{1}{2} \ln(2) - \frac{3}{4}$$

$$= \frac{7}{2} (\ln 2) - \frac{3}{4}$$

57. Formula 5 with $u = q$, $a = 1$, and $b = -1$. Then $du = dq$.

$$\int_{q_0}^{q_n} \frac{dq}{q(1-q)} = \ln \left| \frac{q}{1-q} \right| \Bigg\|_{q_0}^{q_n}$$

$$= \ln \left| \frac{q_n}{1-q_n} \right| - \ln \left| \frac{q_0}{1-q_0} \right|$$

$$= \ln \left| \frac{q_n (1 - q_0)}{q_0 (1 - q_n)} \right|$$

59. a. For $\int_0^9 1000e^{-0.04t}\,dt$, the form $\int e^u\,du$

can be applied.

$$\int_0^9 1000e^{-0.04t}\,dt$$

$$= \frac{1000}{-0.04}\int_0^9 e^{-0.04t}(-0.04\,dt)$$

$$= -\frac{1000}{0.04}e^{-0.04t}\Big|_0^9$$

$$= -\frac{1000}{0.04}(e^{-0.36}-1)$$

$$\approx \$7558.09$$

b. For $\int_0^{10} 500te^{-0.06t}\,dt$ use Formula 38

with $t = u$ and $a = -0.06$, so $du = dt$.

$$\int_0^{10} 500te^{-0.06t}\,dt$$

$$= 500\int_0^{10} te^{-0.06t}\,dt$$

$$= 500\left[\frac{e^{-0.06t}}{0.0036}(-0.06t-1)\right]_0^{10}$$

$$= \frac{500}{0.0036}[e^{-0.6}(-1.6)-(-1)]$$

$$\approx \$16{,}930.75$$

61. a. $\int_0^{10} 400e^{0.06(10-t)}\,dt$

$$= 400\int_0^{10} e^{0.6-0.06t}\,dt$$

$$= 400\int_0^{10} e^{0.6}e^{-0.06t}\,dt$$

$$= 400e^{0.6}\int_0^{10} e^{-0.06t}\,dt$$

$$= 400e^{0.6}\left(\frac{1}{-0.06}\right)\int_0^{10} e^{-0.06t}(-0.06\,dt)$$

$$= \frac{400e^{0.6}}{-0.06}e^{-0.06t}\Big|_0^{10}$$

$$= \frac{400e^{0.6}}{-0.06}\left[e^{-0.6}-1\right]$$

$$\approx \$5481$$

b. Use Formula 38 with $u = t$ and $a = -0.04$, so $du = dt$.

$$\int_0^5 40te^{0.04(5-t)}\,dt = 40\int_0^5 te^{0.2}e^{-0.04t}\,dt$$

$$= 40e^{0.2}\int_0^5 te^{-0.04t}\,dt$$

$$= 40e^{0.2}\left[\frac{e^{-0.04t}}{0.0016}(-0.04t-1)\right]_0^5$$

$$= \frac{40e^{0.2}}{0.0016}\left[e^{-0.2}(-0.2-1)-1(-1)\right]$$

$$\approx \$535$$

Problems 15.4

1. $\bar{f} = \dfrac{1}{3-(-1)}\int_{-1}^3 x^2\,dx = \dfrac{1}{4}\cdot\dfrac{x^3}{3}\Big|_{-1}^3$

$$= \frac{1}{4}\left(9-\frac{-1}{3}\right) = \frac{7}{3}$$

3. $\bar{f} = \dfrac{1}{2-(-1)}\int_{-1}^2\left(2-3x^2\right)dx$

$$= \frac{1}{3}\left(2x-x^3\right)\Big|_{-1}^2 = -1$$

5. $\bar{f} = \dfrac{1}{3-(-3)}\int_{-3}^3 2t^5\,dt$

$$= \frac{1}{6}\cdot\frac{t^6}{3}\Big|_{-3}^3$$

$$= \frac{1}{18}[3^6-(-3)^6]$$

$$= 0$$

7. $\bar{f} = \dfrac{1}{9-1}\int_1^9 6\sqrt{x}\,dx = \dfrac{1}{8}\left(4x^{\frac{3}{2}}\right)\Big|_1^9 = 13$

9. $\bar{P} = \dfrac{1}{100-0} \displaystyle\int_0^{100} \left(369q - 2.1q^2 - 400\right) dq$

$= \dfrac{1}{100}\left(184.5q^2 - 0.7q^3 - 400q\right)\Big|_0^{100}$

$= \dfrac{1}{100}(1,845,000 - 700,000 - 40,000) - 0$

$= 11,050$

Answer: $11,050

11. $\dfrac{1}{2-0}\displaystyle\int_0^2 3000e^{0.05t}\,dt$

$= \dfrac{3000}{2}\cdot\dfrac{1}{0.05}\displaystyle\int_0^2 e^{0.05t}[0.05\ dt]$

$= 30,000e^{0.05t}\Big|_0^2$

$= 30,000\left(e^{0.1} - 1\right) \approx 3155.13$

Answer: $3155.13

13. Average value $= \dfrac{1}{q_0 - 0}\displaystyle\int \dfrac{dr}{dq}\,dq$.

$= \dfrac{1}{q_0}\left[r(q_0) - r(0)\right]$

But $r(0) = 0$, so avg. value $= \dfrac{r(q_0)}{q_0}$. Since

$r(q_0) = $ [price per unit when q_0 units are sold] $\cdot q_0$,
we have

avg. value $= \dfrac{\left[\begin{array}{c}\text{price per unit}\\ \text{when } q_0 \text{ units}\\ \text{are sold}\end{array}\right]\cdot q_0}{q_0}$

$= $ price per unit when $\cdot q_0$ units are sold.

Principles in Practice 15.5

1. Separating variables, we have

$$\frac{dI}{dx} = -0.0085I$$

$$\frac{dI}{I} = -0.0085dx$$

$$\int \frac{1}{I} dI = -\int 0.0085 \, dx$$

$$\ln|I| = -0.0085x + C_1$$

To solve for I, we convert to exponential

Formula

$I = e^{-0.0085x + C_1} = Ce^{-0.0085x}$. Since $I = I_0$

when $x = 0$, $I_0 = Ce^0 = C$, so

$I(x) = I_0 e^{-0.0085x}$.

Problems 15.5

1. $y' = 2xy^2$

$$\frac{dy}{dx} = 2xy^2$$

$$\frac{dy}{y^2} = 2x \, dx$$

$$\int y^{-2} dy = \int 2x \, dx$$

$$-\frac{1}{y} = x^2 + C$$

$$y = -\frac{1}{x^2 + C}$$

3. $\dfrac{dy}{dx} - 3x\sqrt{x^2 + 1} = 0$

$$dy = 3x\left(x^2 + 1\right)^{\frac{1}{2}} dx$$

$$\int dy = 3\int x\left(x^2 + 1\right)^{\frac{1}{2}} dx$$

$$\int dy = \frac{3}{2} \int \left(x^2 + 1\right)^{\frac{1}{2}} [2x \, dx]$$

$$y = \frac{3}{2} \cdot \frac{\left(x^2 + 1\right)^{\frac{3}{2}}}{\frac{3}{2}} + C$$

$$y = \left(x^2 + 1\right)^{\frac{3}{2}} + C$$

5. $\dfrac{dy}{dx} = y$, where $y > 0$.

$$\frac{dy}{y} = dx$$

$$\int \frac{dy}{y} = \int dx$$

$$\ln y = x + C_1$$

$$y = e^{x + C_1} = e^{C_1}e^x = Ce^x, \text{ where } C = e^{C_1}.$$

Thus $y = Ce^x$, where $C > 0$.

7. $y' = \dfrac{y}{x}$, where $x, y > 0$.

$$\frac{dy}{dx} = \frac{y}{x}$$

$$\frac{dy}{y} = \frac{dx}{x}$$

$$\int \frac{dy}{y} = \int \frac{dx}{x}$$

$\ln y = \ln x + C_1$

$\ln y = \ln x + \ln C$, where $C > 0$.

$\ln y = \ln(Cx) \Rightarrow y = Cx$, where $C > 0$.

9. $y' = \dfrac{1}{y^2}$ where $y(1) = 1$.

$$\frac{dy}{dx} = \frac{1}{y^2}$$

$$y^2 dy = dx$$

$$\int y^2 \, dy = \int dx$$

$$\frac{y^3}{3} = x + C$$

Given $y(1) = 1$, we obtain $\dfrac{1^3}{3} = 1 + C$, so

$C = -\dfrac{2}{3}$. Thus $y^3 = 3\left(x - \dfrac{2}{3}\right) = 3x - 2$,

$y = \sqrt[3]{3x - 2}$.

11. $e^y y' - x^2 = 0$, where $y = 0$ when $x = 0$.

$$e^y \frac{dy}{dx} = x^2$$

$$e^y dy = x^2 dx$$

$$\int e^y dy = \int x^2 dx$$

$$e^y = \frac{x^3}{3} + C$$

Given that $y(0) = 0$, we have $e^0 = 0 + C$, so

$1 = C \Rightarrow e^y = \dfrac{x^3}{3} + 1$, $e^y = \dfrac{x^3 + 3}{3}$, so

$y = \ln \dfrac{x^3 + 3}{3}$.

13. $(3x^2 + 2)^3 \, y' - xy^2 = 0$, where $y(0) = \dfrac{3}{2}$.

$$(3x^2 + 2)^3 \frac{dy}{dx} = xy^2$$

$$\frac{dy}{y^2} = \frac{x}{(3x^2 + 2)^3}$$

$$\int \frac{dy}{y^2} = \int \frac{x}{(3x^2 + 2)^3} \, dx$$

$$\int y^{-2} dy = \frac{1}{6} \int (3x^2 + 2)^{-3} [6x \, dx]$$

$$-\frac{1}{y} = -\frac{1}{12(3x^2 + 2)^2} + C$$

Given that $y(0) = \dfrac{3}{2}$ we have

$$-\frac{1}{\frac{3}{2}} = -\frac{1}{2(2)^2} + C, \quad -\frac{2}{3} = -\frac{1}{48} + C, \text{ so}$$

$C = -\dfrac{31}{48}$. Thus,

$$-\frac{1}{y} = -\frac{1}{12(3x^2 + 2)^2} - \frac{31}{48}$$

$$= -\frac{4 + 31(3x^2 + 2)^2}{48(3x^2 + 2)^2}.$$

Hence, $y = \dfrac{48(3x^2 + 2)^2}{4 + 31(3x^2 + 2)^2}$.

15. $\dfrac{dy}{dx} = \dfrac{3x\sqrt{1 + y^2}}{y}$, where $y > 0$ and

$y(1) = \sqrt{8}$.

$$\frac{y \, dy}{\sqrt{1 + y^2}} = 3x \, dx$$

$$\frac{1}{2} \int (1 + y^2)^{-\frac{1}{2}} [2y \, dy] = 3 \int x \, dx$$

$$(1 + y^2)^{\frac{1}{2}} = \frac{3x^2}{2} + C$$

$$y(1) = \sqrt{8} \Rightarrow (1 + 8)^{\frac{1}{2}} = \frac{3}{2} + C$$

$$C = \frac{3}{2}$$

Thus

$$\left(1+y^2\right)^{\frac{1}{2}} = \frac{3x^2}{2}+\frac{3}{2}$$

$$1+y^2 = \left[\frac{3x^2}{2}+\frac{3}{2}\right]^2$$

$$y^2 = \left[\frac{3x^2}{2}+\frac{3}{2}\right]^2 - 1$$

Since $y > 0$, $y = \sqrt{\left[\frac{3x^2}{2}+\frac{3}{2}\right]^2 - 1}$.

17. $2\dfrac{dy}{dx} = \dfrac{xe^{-y}}{\sqrt{x^2+3}}$, where $y(1) = 0$.

$$e^y\,dy = \frac{1}{2}x\left(x^2+3\right)^{-\frac{1}{2}}\,dx$$

$$\int e^y dy = \frac{1}{2}\cdot\frac{1}{2}\int\left(x^2+3\right)^{-\frac{1}{2}}[2x\,dx]$$

$$e^y = \frac{1}{2}\left(x^2+3\right)^{\frac{1}{2}} + C$$

Now, $y(1) = 0 \Rightarrow e^0 = \dfrac{1}{2}(2) + C$, so $C = 0$.

Thus

$$e^y = \frac{1}{2}\left(x^2+3\right)^{\frac{1}{2}} \Rightarrow y = \ln\left(\frac{1}{2}\sqrt{x^2+3}\right).$$

19. $(q+1)^2\dfrac{dc}{dq} = cq$

$$\int\frac{1}{c}dc = \int\frac{q}{(q+1)^2}\,dq$$

Using partial fractions or Formula 7 for

$$\int\frac{q}{(q+1)^2}\,dq\text{, we obtain}$$

$\ln c = \ln(q+1) + \dfrac{1}{q+1} + C$. Now, fixed cost is given to be e, which means that $c = e$ when $q = 0$. This implies $1 = 0 + 1 + C$, so $C = 0$. Thus

$$\ln c = \ln(q+1) + \frac{1}{q+1} \Rightarrow c = e^{\ln(q+1)+\frac{1}{q+1}},$$

$$c = e^{\ln(q+1)}e^{\frac{1}{q+1}},\text{ or }c = (q+1)e^{\frac{1}{q+1}}.$$

21. $\dfrac{dy}{dt} = -0.025y$

$$\int\frac{1}{y}dy = -0.025\int dt$$

$$\ln|y| = -0.025t + C$$

Given that $y = 1000$ when $t = 0$, we have $\ln 1000 = -0 + C = C$. Thus $\ln|y| = -0.025t + \ln 1000$. To find t when money is 95% new, we note that y would be $5\%(1000) = 50$. Solving $\ln 50 = -0.025t + \ln 1000$ gives

$$t = \frac{\ln 1000 - \ln 50}{0.025} \approx 120\text{ weeks.}$$

23. Let N be the population at time t, where $t = 0$ corresponds to 1985. Since N follows exponential growth, $N = N_0 e^{kt}$. Now, $N = 40{,}000$ when $t = 0$, so $N_0 = 40{,}000$. Therefore $N = 40{,}000e^{kt}$. Since $N = 48{,}000$ when $t = 10$, we have $48{,}000 = 40{,}000e^{10k}$, $1.2 = e^{10k}$, $\ln 1.2 = 10k$, $k = \dfrac{\ln 1.2}{10}$

Thus

$$N = 40{,}000e^{\ln(1.2)\left(\frac{t}{10}\right)} \qquad (*)$$

$$N = 40{,}000e^{0.18\left(\frac{t}{10}\right)}$$

$$N = 40{,}000e^{0.018t} \qquad \text{(First form)}$$

From (*), we have $N = 40{,}000\left[e^{\ln 1.2}\right]^{\frac{t}{10}}$, so

$$N = 40{,}000(1.2)^{\frac{t}{10}} \qquad \text{(Second form)}$$

At year 2005, $t = 20$ and so

$$N = 40{,}000(1.2)^{\frac{20}{10}} = 40{,}000(1.2)^2$$
$$= 57{,}600.$$

25. Let N be the population (in billions) at time t, where t is the number of years past 1930. N follows exponential growth, so
$N = N_0 e^{kt}$. When $t = 0$, then $N = 2$, so
$N_0 = 2$. Thus $N = 2e^{kt}$. Since $N = 3$ when $t = 30$, then
$$3 = 2e^{30k}$$
$$\frac{3}{2} = e^{30k}$$
$$30k = \ln\frac{3}{2}$$
$$k = \frac{\ln\frac{3}{2}}{30}$$
Thus $N = 2e^{\frac{t}{30}\ln\frac{3}{2}}$.
In 2015, $t = 85$ and so
$N = 2e^{\frac{85}{30}\ln\frac{3}{2}} \approx 2e^{1.14882}$ billion.

27. Let N be amount of sample that remains after t seconds. Then $N = N_0 e^{-\lambda t}$, where N_0 is the initial amount present. When $t = 100$, then $N = 0.3N_0$. Thus
$$0.3N_0 = N_0 e^{-100\lambda}$$
$$0.3 = e^{-100\lambda}$$
$$-100\lambda = \ln 0.3$$
$$\lambda = -\frac{\ln 0.3}{100}$$
Thus $\lambda \approx 0.01204$. The half-life is
$$\frac{\ln 2}{\lambda} = \frac{\ln 2}{-\frac{\ln 0.3}{100}} = -100\frac{\ln 2}{\ln 0.3} \approx 57.57 \text{ s.}$$

29. Let N be the amount of ^{14}C present in the scroll t years after it was made. Then $N = N_0 e^{-\lambda t}$, where N_0 is amount of ^{14}C present when $t = 0$. We must find t when $N = 0.7N_0$.
$$0.7N_0 = N_0 e^{-\lambda t}$$
$$0.7 = e^{-\lambda t}$$
$$-\lambda t = \ln 0.7$$

so $t = -\dfrac{\ln 0.7}{\lambda}$. By Equation 15 in the text,
$$\lambda = \frac{\ln 2}{5730}, \text{ so}$$
$$t = -\frac{\ln 0.7}{\frac{\ln 2}{5730}} = -\frac{5730\ln 0.7}{\ln 2} \approx 2900 \text{ years.}$$

31. $\dfrac{dN}{dt} = kN$
$$N = Ae^{kt}$$
$$N_0 = Ae^{kt_0}$$
$$A = \frac{N_0}{e^{kt_0}}$$
Thus $N = \dfrac{N_0}{e^{kt_0}}\left(e^{kt}\right) = N_0 e^{kt-kt_0}$, or
$N = N_0 e^{k(t-t_0)}$, where $t \geq t_0$.

33. $N = N_0 e^{-\lambda t}$
When $t = 2$, then $N = 10$. Thus
$10 = N_0 e^{-2\lambda}$, $N_0 = 10e^{2\lambda}$. By Equation 15 in the text,
$$6 = \frac{\ln 2}{\lambda}$$
$$\lambda = \frac{\ln 2}{6}$$
Thus $N_0 = 10e^{2\frac{\ln 2}{6}} = 10e^{\frac{\ln 2}{3}} \approx 12.6$ units.

35. $\dfrac{dA}{dt} = 200 - 0.50A$
$$\int\frac{dA}{200 - 0.50A} = \int dt$$
$$-\frac{1}{0.50}\ln(200 - 0.50A) = t + C_1$$
$$\ln(200 - 0.50A) = -0.50t - 0.50C_1$$
$$= -0.50t + C_2$$
Thus

$$200 - 0.50A = e^{-0.50t + C_2} = e^{-0.50t} e^{C_2}$$

$$200 - \frac{A}{2} = Ce^{-0.50t}$$

Given that $A = 0$ when $t = 0$, we have $C = 200$, so $200 - \frac{A}{2} = 200e^{-\frac{t}{2}}$

$$200 - 200e^{-\frac{t}{2}} = \frac{A}{2}$$

$$200\left(1 - e^{-\frac{t}{2}}\right) = \frac{A}{2}$$

Thus $A = 400\left(1 - e^{-\frac{t}{2}}\right)$. If $t = 1$,

$$A = 400\left(1 - e^{-\frac{1}{2}}\right) \approx 157 \text{ grams per square}$$

meter.

37. a. $\dfrac{dV}{dt} = kV$

$$\int \frac{1}{V} dV = \int k \, dt$$

$$\ln V = kt + C_1$$

$$V = e^{kt} e^{C_1}$$

or $V = Ce^{kt}$. Now $t = 0$ corresponds to July 1, 1996 where
$V = 0.75 \cdot 80{,}000 = 60{,}000$, so
$60{,}000 = C(1)$. Thus $V = 60{,}000e^{kt}$.
Also $V = 38{,}900$ for $t = 9.5$, so

$$38{,}900 = 60{,}000e^{9.5k}$$

$$\frac{389}{600} = e^{9.5k}$$

$$9.5k = \ln\left(\frac{389}{600}\right)$$

$$k = \frac{1}{9.5}\ln\left(\frac{389}{600}\right)$$

Thus $V = 60{,}000e^{\frac{t}{9.5}\ln\left(\frac{389}{600}\right)}$.

b. $14{,}000 = 60{,}000e^{\frac{t}{9.5}\ln\left(\frac{389}{600}\right)}$

$$\frac{7}{30} = e^{\frac{t}{9.5}\ln\left(\frac{389}{600}\right)}$$

$$\ln\left(\frac{7}{30}\right) = \frac{t}{9.5}\ln\left(\frac{389}{600}\right)$$

$$t = \frac{9.5\ln\left(\frac{7}{30}\right)}{\ln\left(\frac{389}{600}\right)} \approx 31.903$$

This corresponds to about 31 years and 11 months after
July 1, 1996 $\Rightarrow$ June 2028.

Problems 15.6

1. $N = \dfrac{M}{1 + be^{-ct}}$

$M = 100{,}000$

Since $N = 50{,}000$ at $t = 0$ (1995), we have

$$50{,}000 = \frac{100{,}000}{1+b}, \text{ so } 1+b = \frac{100{,}000}{50{,}000} = 2,$$

or $b = 1$.

Hence, $N = \dfrac{100{,}000}{1 + e^{-ct}}$. If $t = 5$, then

$N = 60{,}000$, so

$$60{,}000 = \frac{100{,}000}{1 + e^{-5c}}$$

$$1 + e^{-5c} = \frac{100{,}000}{60{,}000} = \frac{5}{3}$$

$$e^{-5c} = \frac{5}{3} - 1 = \frac{2}{3}$$

$$e^{-c} = \left(\frac{2}{3}\right)^{1/5}$$

Hence, $N = \dfrac{100{,}000}{1 + \left(\frac{2}{3}\right)^{t/5}}$. In 2005, $t = 10$, so

$$N = \frac{100{,}000}{1 + \left(\frac{2}{3}\right)^{2}} \approx 69{,}200.$$

3. $N = \dfrac{M}{1+be^{-ct}}$

$M = 40,000$, and $N = 20$ when $t = 0$, so

$20 = \dfrac{40,000}{1+b}$

$1+b = \dfrac{40,000}{20} = 2000$

$b = 1999$

Hence $N = \dfrac{40,000}{1+1999e^{-ct}}$.

Since $N = 100$ when $t = 1$,

$100 = \dfrac{40,000}{1+1999e^{-c}}$,

$1+1999e^{-c} = \dfrac{40,000}{100} = 400$

$e^{-c} = \dfrac{399}{1999}$

Hence $N = \dfrac{40,000}{1+1999\left(\frac{399}{1999}\right)^{t}}$.

If $t = 2$, then $N = \dfrac{40,000}{1+1999\left(\frac{399}{1999}\right)^{2}} \approx 500$.

5. $N = \dfrac{M}{1+be^{-ct}}$

$M = 100,000$, and since $N = 500$ when $t = 0$, we have

$500 = \dfrac{100,000}{1+b}$

$1+b = \dfrac{100,000}{500} = 200$

$b = 199$

Hence $N = \dfrac{100,000}{1+199e^{-ct}}$. If $t = 1$, then

$N = 1000$. Thus

$1000 = \dfrac{100,000}{1+199e^{-c}}$

$1+199e^{-c} = \dfrac{100,000}{1000} = 100$

$199e^{-c} = 99$

$e^{-c} = \dfrac{99}{199}$

Hence $N = \dfrac{100,000}{1+199\left(\frac{99}{199}\right)^{t}}$. If $t = 2$, then

$N = \dfrac{100,000}{1+199\left(\frac{99}{199}\right)^{2}} \approx 1990$.

7. a. $N = \dfrac{375}{1+e^{5.2-2.3t}} = \dfrac{375}{1+e^{5.2}e^{-2.3t}}$

$\approx \dfrac{375}{1+181.27e^{-2.3t}}$

b. $\displaystyle\lim_{t\to\infty} N = \dfrac{375}{1+181.27(0)} = 375$

9. $\dfrac{dT}{dt} = k(T-a)$ where $a = -10$.

$\dfrac{dT}{T+10} = k\,dt$

$\displaystyle\int\dfrac{dT}{T+10} = \int k\,dt$

Thus $\ln(T + 10) = kt + C$. At $t = 0$, we have
$T = 28$, so $\ln(28 + 10) = 0 + C$, $C = \ln 38$, and

$\ln(T + 10) = kt + \ln 38$.

$\ln(T + 10) - \ln 38 = kt$

Hence $\ln\left(\dfrac{T+10}{38}\right) = kt$.

If $t = 1$, then $T = 20$. Thus

$\ln\left(\dfrac{20+10}{38}\right) = k\cdot 1$, so $k = \ln\dfrac{30}{38} = \ln\dfrac{15}{19}$.

Hence $\ln\left(\dfrac{T+10}{38}\right) = \left(\ln\dfrac{15}{19}\right)t$.

If $T = 37$, then $\ln\left(\dfrac{47}{38}\right) = \left(\ln\dfrac{15}{19}\right)t$

$$t = \dfrac{\ln\dfrac{47}{38}}{\ln\dfrac{15}{19}} \approx -0.90 \text{ hr}$$

which corresponds to 54 minutes. Time of murder: 4:15 A.M. $-$ 54 min $=$ 3:21 A.M.

11. $\dfrac{dx}{dt} = k(200,000 - x)$

$\displaystyle\int\dfrac{dx}{200,000 - x} = \int k\ dt$

$-\ln(200,000 - x) = kt + C$

$\ln(200,000 - x) = -kt - C$

$200,000 - x = e^{-kt-C} = e^{-C}e^{-kt} = Ae^{-kt}$,

where $A = e^{-C}$. Thus $x = 200,000 - Ae^{-kt}$.

If $t = 0$, then $x = 50,000$, so

$50,000 = 200,000 - A \Rightarrow A = 150,000$.

Thus $x = 200,000 - 150,000e^{-kt}$. If $t = 1$, then

$x = 100,000$, so

$100,000 = 200,000 - 150,000e^{-k}$

$150,000e^{-k} = 100,000$

$e^{-k} = \dfrac{100,000}{150,000} = \dfrac{2}{3}$

Thus $x = 200,000 - 150,000\left(\dfrac{2}{3}\right)^t$. If $t = 3$,

then

$x = 200,000 - 150,000\left(\dfrac{8}{27}\right) \approx \$155,555.56$.

13. $\dfrac{dN}{dt} = k(M - N)$

$\displaystyle\int\dfrac{dN}{M - N} = \int k\ dt$

$-\ln(M - N) = kt + C$

If $t = 0$, then $N = N_0$, so

$-\ln(M - N_0) = C$. Thus we have

$-\ln(M - N) = kt - \ln(M - N_0)$

$\ln(M - N_0) - \ln(M - N) = kt$

$\ln\dfrac{M - N_0}{M - N} = kt$

$\ln\dfrac{M - N}{M - N_0} = -kt$

$\dfrac{M - N}{M - N_0} = e^{-kt}$

$M - N = (M - N_0)e^{-kt}$

$N = M - (M - N_0)e^{-kt}$

Principles in Practice 15.7

1. $\displaystyle\int_0^\infty\left(3e^{-0.1t} - 3e^{-0.3t}\right)dt$

$= \displaystyle\lim_{r\to\infty}\int_0^r\left(3e^{-0.1t} - 3e^{-0.3t}\right)dt$

$= \displaystyle\lim_{r\to\infty}\left(-30e^{-0.1t} + 10e^{-0.3t}\right)\Big|_0^r$

$= \displaystyle\lim_{r\to\infty}\left[-\dfrac{30}{e^{-0.1r}} + \dfrac{10}{e^{0.3r}} - \left(-30e^0 + 10e^0\right)\right]$

$= \displaystyle\lim_{r\to\infty}\left[-\dfrac{30}{e^{0.1r}} + \dfrac{10}{e^{0.3r}} - (-20)\right]$

$= 0 + 0 + 20 = 20$

The total amount of the drug that is eliminated is approximately 20 milliliters.

Problems 15.7

1. $\displaystyle\int_3^\infty\dfrac{1}{x^3}\,dx = \lim_{r\to\infty}\int_3^r x^{-3}dx$

$= \displaystyle\lim_{r\to\infty}\dfrac{x^{-2}}{-2}\Big|_3^r = -\dfrac{1}{2}\lim_{r\to\infty}\dfrac{1}{x^2}\Big|_3^r$

$= -\dfrac{1}{2}\lim_{r\to\infty}\left(\dfrac{1}{r^2} - \dfrac{1}{9}\right) = -\dfrac{1}{2}\left(0 - \dfrac{1}{9}\right) = \dfrac{1}{18}$

3. $\int_1^\infty \frac{1}{x}\,dx = \lim_{r\to\infty} \int_1^r \frac{1}{x}\,dx = \lim_{r\to\infty} \ln|x| \Big|_1^r$

$= \lim_{r\to\infty} \left(\ln|r| - 0 \right)$

$= \lim_{r\to\infty} \ln|r| = \infty \Rightarrow$ diverges

5. $\int_1^\infty e^{-x}\,dx = \lim_{r\to\infty} -\int_1^r e^{-x}[-dx]$

$= \lim_{r\to\infty} (-e^{-x}) \Big|_1^r$

$= \lim_{r\to\infty} \left(-e^{-r} + e^{-1} \right) = \lim_{r\to\infty} \left(-\frac{1}{e^r} + \frac{1}{e} \right)$

$= 0 + \frac{1}{e} = \frac{1}{e}$

7. $\int_1^\infty \frac{1}{\sqrt{x}}\,dx = \lim_{r\to\infty} \int_1^r x^{-\frac{1}{2}}\,dx = \lim_{r\to\infty} 2x^{\frac{1}{2}} \Big|_1^r$

$= \lim_{r\to\infty} \left(2\sqrt{r} - 2 \right) = \infty \Rightarrow$ diverges

9. $\int_{-\infty}^{-3} \frac{1}{(x+1)^2}\,dx = \lim_{r\to-\infty} \int_r^{-3} (x+1)^{-2}\,dx$

$= \lim_{r\to-\infty} -\frac{1}{x+1} \Big|_r^{-3}$

$= \lim_{r\to-\infty} \left[\frac{1}{2} + \frac{1}{r+1} \right]$

$= \frac{1}{2} + 0$

$= \frac{1}{2}$

11. $\int_{-\infty}^\infty 2xe^{-x^2}\,dx$

$= \int_{-\infty}^0 2xe^{-x^2}\,dx + \int_0^\infty 2xe^{-x^2}\,dx$

$\int_{-\infty}^0 2xe^{-x^2}\,dx = \lim_{r\to-\infty} -\int_r^0 e^{-x^2}[-2x\,dx]$

$= \lim_{r\to-\infty} -e^{-x^2} \Big|_r^0$

$= \lim_{r\to-\infty} \left[-1 + \frac{1}{e^{r^2}} \right]$

$= -1 + 0 = -1$

$\int_0^\infty 2xe^{-x^2}\,dx = \lim_{r\to\infty} -\int_0^r e^{-x^2}[-2x\,dx]$

$= \lim_{r\to\infty} -e^{x^2} \Big|_0^r$

$= \lim_{r\to\infty} \left[-\frac{1}{e^{r^2}} + 1 \right] = 0 + 1 = 1$

Thus $\int_{-\infty}^\infty 2xe^{-x^2}\,dx = -1 + 1 = 0.$

13. a. $\int_{800}^\infty \frac{k}{x^2}\,dx = 1$

$\lim_{r\to\infty} k \int_{800}^r x^{-2}\,dx = 1$

$\lim_{r\to\infty} -\frac{k}{x} \Big|_{800}^r = 1$

$\lim_{r\to\infty} \left(-\frac{k}{r} + \frac{k}{800} \right) = 1$

$0 + \frac{k}{800} = 1$

$k = 800$

b. $\int_{1200}^\infty \frac{800}{x^2}\,dx = \lim_{r\to\infty} 800 \int_{1200}^r x^{-2}\,dx$

$= \lim_{r\to\infty} -\frac{800}{x} \Big|_{1200}^r$

$= \lim_{r\to\infty} \left(-\frac{800}{r} + \frac{800}{1200} \right) = 0 + \frac{2}{3} = \frac{2}{3}$

15. $\displaystyle\int_0^\infty 240,000e^{-0.06t}\,dt = \lim_{r\to\infty} \frac{240,000}{-0.06} \int_0^r e^{-0.06t}[-0.06\,dt]$

$\displaystyle = \lim_{r\to\infty} -\frac{240,000}{0.06} e^{-0.06t}\Big|_0^r$

$\displaystyle = \lim_{r\to\infty} -\frac{240,000}{0.06}\left(\frac{1}{e^{0.06r}} - 1\right)$

$\displaystyle = -\frac{240,000}{0.06}(-1) = 4,000,000$

17. $\displaystyle\text{Area} = -\int_{-\infty}^0 -e^{3x}\,dx = \lim_{r\to-\infty} \frac{1}{3}\int_r^0 e^{3x}[3\,dx]$

$\displaystyle = \lim_{r\to-\infty} \frac{1}{3}\cdot e^{3x}\Big|_r^0 = \lim_{r\to-\infty}\left[\frac{1}{3} - \frac{1}{3}e^r\right]$

$\displaystyle = \frac{1}{3} - 0 = \frac{1}{3}\text{ sq units}$

19. $\displaystyle\int_0^\infty \frac{40,000}{(t+2)^2}\,dt = \lim_{r\to\infty}\int_0^r \frac{40,000}{(t+2)^2}\,dt$

$\displaystyle = \lim_{r\to\infty} -\frac{40,000}{t+2}\Big|_0^r$

$\displaystyle = \lim_{r\to\infty}\left[-\frac{40,000}{r+2} + \frac{40,000}{2}\right]$

$\displaystyle = 0 + \frac{40,000}{2} = 20,000 \text{ increase}$

Chapter 15 Review Problems

1. Use Formula 42 with $u = x$ and $n = 1$. Then $du = dx$.

$\displaystyle\int x\ln x\,dx = \frac{x^2\ln x}{2} - \frac{x^2}{4} + C$

3. Use Formula 23 with $u = 3x$, $a^2 = 16$. Then $du = 3\, dx$.

$$\int_0^2 \sqrt{9x^2 + 16}\, dx = \frac{1}{3}\int_0^2 \sqrt{(3x)^2 + 16}\,(3\, dx)$$

$$= \frac{1}{3}\left[\frac{1}{2}\left((3x)\sqrt{9x^2 + 16} + 16\ln\left|3x + \sqrt{9x^2 + 16}\right|\right)\right]_0^2$$

$$= \left(2\sqrt{13} + \frac{8}{3}\ln\left(6 + 2\sqrt{13}\right)\right) - \left(0 + \frac{8}{3}\ln 4\right)$$

$$= 2\sqrt{13} + \frac{8}{3}\ln\left(\frac{6 + 2\sqrt{13}}{4}\right)$$

$$= 2\sqrt{13} + \frac{8}{3}\ln\left(\frac{3 + \sqrt{13}}{2}\right)$$

5. $\displaystyle\int\frac{15x - 2}{(3x+1)(x-2)}\, dx = \int\left(\frac{15x}{(3x+1)(x-2)} - \frac{2}{(3x+1)(x-2)}\right)dx$

For $\displaystyle\int\frac{15x}{(3x+1)(x-2)}\, dx$, use Formula 12 with $u = x$, $a = 1$, $b = 3$, $c = -2$, and $k = 1$. Then $du = dx$.

$$\int\frac{15x}{(3x+1)(x-2)}\, dx = 15\int\frac{x}{(3x+1)(x-2)}\, dx = 15\left[\frac{1}{-7}\left(-2\ln|x-2| - \frac{1}{3}\ln|3x+1|\right)\right] + C$$

For $\displaystyle\int\frac{2}{(3x+1)(x-2)}\, dx$, use Formula 11 with

$u = x$, $a = 1$, $b = 3$, $c = -2$, and $k = 1$. Then $du = dx$.

$$\int\frac{2}{(3x+1)(x-2)}\, dx = 2\int\frac{dx}{(3x+1)(x-2)} = 2\left(\frac{1}{-7}\ln\left|\frac{3x+1}{x-2}\right|\right) + C$$

Thus, $\displaystyle\int\frac{15x - 2}{(3x+1)(x-2)}\, dx$

$$= \frac{30}{7}\ln|x-2| + \frac{5}{7}\ln|3x+1| + \frac{2}{7}\ln\left|\frac{3x+1}{x-2}\right| + C$$

$$= \frac{30}{7}\ln|x-2| + \frac{5}{7}\ln|3x+1| + \frac{2}{7}\ln|3x+1| - \frac{2}{7}\ln|x-2| + C$$

$$= 4\ln|x-2| + \ln|3x+1| + C$$

7. Use Formula 9 with $u = x$, $a = 2$, and $b = 1$. Then $du = dx$.

$$\int\frac{dx}{x(x+2)^2} = \frac{1}{2(x+2)} + \frac{1}{4}\ln\left|\frac{x}{x+2}\right| + C$$

9. Use Formula 21 with $u = 4x$ and $a^2 = 9$.
Then $du = 4\ dx$.

$$\int \frac{dx}{x^2 \sqrt{9-16x^2}} = 4 \int \frac{(4\ dx)}{(4x)^2 \sqrt{9-(4x)^2}}$$

$$= 4\left[-\frac{\sqrt{9-16x^2}}{9(4x)} \right] + C$$

$$= -\frac{\sqrt{9-16x^2}}{9x} + C$$

11. Use Formula 35 with $u = x$ and $a = 3$. Then $du = dx$.

$$\int \frac{9\,dx}{x^2-9} = 9 \int \frac{dx}{x^2-9} = 9\left[\frac{1}{6} \ln\left| \frac{x-3}{x+3} \right| \right] + C$$

$$= \frac{3}{2} \ln\left| \frac{x-3}{x+3} \right| + C$$

13. Use Formula 38 with $u = x$ and $a = 7$. Then $du = dx$.

$$\int 49xe^{7x}\,dx = 49 \int xe^{7x}\,du$$

$$= 49\left[\frac{e^{7x}}{49}(7x-1) \right] + C = e^{7x}(7x-1) + C$$

15. The integral has the form $\int \frac{1}{u}\,du$.

$$\int \frac{dx}{2x \ln x^2} = \frac{1}{4} \int \frac{1}{\ln x^2}\left[\frac{2}{x}dx \right]$$

$$= \frac{1}{4} \ln\left| \ln x^2 \right| + C$$

17. Long division or Formula 3. For long division,

$$\int \frac{2x}{3+2x}\,dx = \int \left[1 - \frac{3}{3+2x} \right] dx$$

$$= x - 3 \cdot \frac{1}{2} \int \frac{1}{3+2x}[2\ dx]$$

$$= x - \frac{3}{2} \ln|3+2x| + C.$$

For Formula 3, use $u = x$, $a = 3$, and $b = 2$.

Then $du = dx$.

$$\int \frac{2x}{3+2x}\,dx = 2 \int \frac{x}{3+2x}\,dx$$

$$= 2\left(\frac{x}{2} - \frac{3}{4} \ln|3+2x| \right) + C$$

$$= x - \frac{3}{2} \ln|3+2x| + C$$

19. Partial fractions

$$\frac{5x^2+2}{x^3+x} = \frac{5x^2+2}{x(x^2+1)} = \frac{A}{x} + \frac{Bx+C}{x^2+1}$$

$$5x^2+2 = A(x^2+1) + (Bx+C)x$$

$$= (A+B)x^2 + Cx + A$$

Thus, $A + B = 5$, $C = 0$, $A = 2$. This gives
$A = 2$, $B = 3$, $C = 0$.

$$\int \frac{5x^2+2}{x^3+x}\,dx = \int \left[\frac{2}{x} + \frac{3x}{x^2+1} \right] dx$$

$$= 2 \ln|x| + \frac{3}{2} \ln\left(x^2+1 \right) + C$$

21. Integration by parts
$u = \ln(x+1)$

$$dv = (x+1)^{-\frac{1}{2}}\,dx$$

Then $du = \dfrac{1}{x+1}\,dx$ and $v = 2(x+1)^{\frac{1}{2}}$.

$$\int \frac{\ln(x+1)}{\sqrt{x+1}}\,dx$$

$$= 2(x+1)^{\frac{1}{2}} \ln(x+1) - 2 \int (x+1)^{-\frac{1}{2}}\,dx$$

$$= 2(x+1)^{\frac{1}{2}} \ln(x+1) - 4(x+1)^{\frac{1}{2}} + C$$

$$= 2\sqrt{x+1}[\ln(x+1) - 2] + C$$

23. $\bar{f} = \dfrac{1}{4-2} \int_2^4 \left(3x^2 + 2x \right) dx = \dfrac{1}{2}\left(x^3 + x^2 \right) \Big|_2^4$

$$= \frac{1}{2}[(64+16) - (8+4)] = 34$$

25. $y' = 3x^2 y + 2xy$, $y > 0$

$$\frac{dy}{y} = \left(3x^2 + 2x\right) dx$$

$$\int \frac{dy}{y} = \int \left(3x^2 + 2x\right) dx$$

$\ln y = x^3 + x^2 + C_1$, from which

$y = e^{x^3 + x^2 + C_1}$, $y = Ce^{x^3 + x^2}$, where $C > 0$.

27. $\displaystyle\int_1^\infty \frac{1}{x^{2.5}} dx = \lim_{r \to \infty} \int_1^r x^{-2.5} dx$

$$= \lim_{r \to \infty} \frac{x^{-1.5}}{-1.5} \Big|_1^r$$

$$= \lim_{r \to \infty} -\frac{2}{3x^{1.5}} \Big|_1^r$$

$$= \lim_{r \to \infty} \left(-\frac{2}{3r^{1.5}} + \frac{2}{3}\right)$$

$$= 0 + \frac{2}{3}$$

$$= \frac{2}{3}$$

29. $\displaystyle\int_1^\infty \frac{1}{2x} dx = \lim_{r \to \infty} \int_1^r \frac{1}{2x} dx = \lim_{r \to \infty} \frac{1}{2} \ln|x| \Big|_1^r$

$$= \lim_{r \to \infty} \left[\frac{1}{2} \ln|r| - 0\right] = \infty \Rightarrow \text{ diverges}$$

31. $N = N_0 e^{kt}$

Since $N = 100{,}000$ when $t = 0$ (1985),

$N_0 = 100{,}000$. Thus $N = 100{,}000 e^{kt}$.

Since $N = 120{,}000$ when $t = 15$, then

$120{,}000 = 100{,}000 e^{15k}$

$1.2 = e^{15k}$

$\ln 1.2 = 15k$, or $k = \dfrac{\ln 1.2}{15}$. Thus

$N = 100{,}000 e^{t \frac{\ln 1.2}{15}} = 100{,}000 \left(e^{\ln 1.2}\right)^{\frac{t}{15}}$

$= 100{,}000(1.2)^{\frac{t}{15}}$

For the year 2015 we have $t = 30$ and

$N = 100{,}000(1.2)^{\frac{30}{15}} = 100{,}000(1.2)^2$

$= 144{,}000$

33. $N = N_0 e^{-\lambda t}$, where N_0 is the original amount present. When $t = 100$, then $N = 0.95 N_0$, so we have

$0.95 N_0 = N_0 e^{-100\lambda}$

$0.95 = e^{-100\lambda}$

$-100\lambda = \ln 0.95$

$\lambda = -\dfrac{\ln 0.95}{100} \approx 0.0005$ (decay constant).

After 200 years, $N = N_0 e^{-200\lambda}$. Thus

$\dfrac{N}{N_0} = e^{-200\lambda} = e^{-200\left[-\frac{\ln 0.95}{100}\right]} = e^{2\ln 0.95}$

$\approx 0.90 = 90\%$

35. $N = \dfrac{450}{1 + be^{-ct}}$

If $t = 0$, then $N = 2$. Thus $2 = \dfrac{450}{1+b}$,

$1 + b = \dfrac{450}{2} = 225$, $b = 224$, so

$N = \dfrac{450}{1 + 224e^{-ct}}$. If $t = 6$, then

$N = 300 \Rightarrow 300 = \dfrac{450}{1 + 224e^{-6c}}$

$1 + 224e^{-6c} = \dfrac{450}{300} = \dfrac{3}{2}$

$224e^{-6c} = \dfrac{3}{2} - 1 = \dfrac{1}{2}$

$e^{-6c} = \dfrac{1}{448}$

$e^{6c} = 448$

$6c = \ln 448$

$c = \dfrac{\ln 448}{6} \approx 1.02$

Thus $N \approx \dfrac{450}{1 + 224e^{-1.02t}}$.

37. $\dfrac{dT}{dt} = k(T - 25)$

$\dfrac{dT}{T - 25} = k\ dt$

$\displaystyle\int \dfrac{dT}{T - 25} = \int k\ dt$

$\ln(T - 25) = kt + C$

If $t = 0$, then $T = 35$. Thus $\ln 10 = C$, so

$\ln(T - 25) = kt + \ln 10$, or $\ln\left(\dfrac{T - 25}{10}\right) = kt$. If $t = 1$, then $T = 34$ and $\ln\left(\dfrac{9}{10}\right) = k$. Thus

$\ln\left(\dfrac{T - 25}{10}\right) = (\ln 0.9)t$. If $T = 37$,

$\ln\dfrac{12}{10} = (\ln 0.9)t$

$\ln 1.2 = (\ln 0.9)t,$

$t = \dfrac{\ln 1.2}{\ln 0.9} \approx -1.73$

Note that 1.73 hr corresponds approximately to 1 hr 44 min. Thus 6:00 P.M. – 1 hr 44 min = 4:16 P.M.

39. $\displaystyle\int_0^\infty f(x)\,dx$

$= \displaystyle\lim_{r \to \infty} \int_0^r (0.007e^{-0.01x} + 0.00005e^{-0.0002x})\,dx$

$= \displaystyle\lim_{r \to \infty} (-0.7e^{-0.01x} - 0.25e^{-0.0002x})\Big|_0^r$

$= \displaystyle\lim_{r \to \infty} \left[-\dfrac{0.7}{e^{0.01r}} - \dfrac{0.25}{e^{-0.0002r}} - (-0.7 - 0.25) \right]$

$= 0 - 0 + 0.7 + 0.25$

$= 0.95$

41. a. Total revenue $= r(12) - r(0) = \displaystyle\int_0^{12} \dfrac{dr}{dq}\,dq$. $f(q) = \dfrac{dr}{dq}$

$n = 4,\ a = 0,\ b = 12$

$h = \dfrac{b - a}{n} = \dfrac{12 - 0}{4} = 3$

Trapezoidal

$\begin{array}{rcl}
f(0) = & 25 \\
2f(3) = & 44 \\
2f(6) = & 36 \\
2f(9) = & 26 \\
f(12) = & 7 \\
\hline
& 138
\end{array}$

$$TR \approx \frac{3}{2}(138) = 207$$

Simpson's

$$
\begin{array}{rl}
f(0) = & 25 \\
4f(3) = & 88 \\
2f(6) = & 36 \\
4f(9) = & 52 \\
f(12) = & 7 \\
\hline
& 208
\end{array}
$$

$$TR \approx \frac{3}{3}(208) = 208$$

b. Total variable cost

$$c(12) - c(0) = \int_0^{12} \frac{dc}{dq}\, dq$$

$$f(q) = \frac{dc}{dq}$$

$a = 0,\ b = 12$

Using as few data values as possible, we choose $n = 1$ for Trapezoidal and $n = 2$ for Simpson's (n must be even).

Trapezoidal ($n = 1$)

$$h = \frac{b-a}{n} = \frac{12-0}{1} = 12$$

$$
\begin{array}{l}
f(0) = 15 \\
f(12) = 7 \\
\hline
22
\end{array}
$$

$$VC \approx \frac{12}{2}(22) = 132$$

Simpson's ($n = 2$)

$$h = \frac{b-a}{n} = \frac{12-0}{2} = 6$$

$$
\begin{array}{l}
f(0) = 15 \\
4f(6) = 48 \\
f(12) = 7 \\
\hline
70
\end{array}
$$

$$VC \approx \frac{6}{3}(70) = 140$$

To each of our results we must add on the fixed cost of 25 to obtain total cost. Thus for trapezoidal we get $TC \approx 132 + 25 = 157$, and for Simpson's we have $TC \approx 140 + 25 = 165$.

c. We use the relation

$$P(12) = \int_0^{12} \left[\frac{dr}{dq} - \frac{dc}{dq} \right] dq - 25\,.$$ First we

determine variable cost for each rule with

$$n = 4 \text{ and } h = \frac{b-a}{n} = \frac{12-0}{4} = 3\,.$$

Trapezoidal

$$
\begin{array}{l}
f(0) = 15 \\
2f(3) = 28 \\
2f(6) = 24 \\
2f(9) = 20 \\
f(12) = 7 \\
\hline
94
\end{array}
$$

$$VC \approx \frac{3}{2}(94) = 141$$

Simpson's

$$
\begin{array}{rl}
f(0) = & 15 \\
4f(3) = & 56 \\
2f(6) = & 24 \\
4f(12) = & 40 \\
f(12) = & 7 \\
\hline
& 142
\end{array}
$$

$$VC \approx \frac{3}{3}(142) = 142$$

Using these results and those of part (a), we have:

Trapezoidal

$$P(12) \approx 207 - 141 - 25 = 41$$

Simpson's

$$P(12) \approx 208 - 142 - 25 = 41$$

Mathematical Snapshot Chapter 15

1. $C = 2000,\ w_0 = 200$

$$w_{eq} = \frac{C}{17.5} = \frac{2000}{17.5} \approx 114$$

$$w(t) = \frac{C}{17.5} + \left(w_0 - \frac{C}{17.5}\right)e^{-0.005t}$$

$$= \frac{2000}{17.5} + \left(200 - \frac{2000}{17.5}\right)e^{-0.005t}$$

Letting $w(t) = 175$ and solving for t gives

$$175 = \frac{2000}{17.5} + \left(200 - \frac{2000}{17.5}\right)e^{-0.005t}$$

$$175 - \frac{2000}{17.5} = \left(200 - \frac{2000}{17.5}\right)e^{-0.005t}$$

$$\frac{175 - \frac{2000}{17.5}}{200 - \frac{2000}{17.5}} = e^{-0.005t}$$

$$-0.005t = \ln\left[\frac{175 - \frac{2000}{17.5}}{200 - \frac{2000}{17.5}}\right]$$

$$t = \frac{\ln\left[\frac{175 - \frac{2000}{17.5}}{200 - \frac{2000}{17.5}}\right]}{-0.005} \approx 69$$

Thus $w_{eq} = 114$ and $t = 69$ days.

3. $w(t) = \frac{C}{17.5} + \left(w_0 - \frac{C}{17.5}\right)e^{-0.005t}$

Since $\frac{C}{17.5} = w_{eq}$, we have

$$w(t) = w_{eq} + \left(w_0 - w_{eq}\right)e^{-0.005t}.$$

Simplifying the equation

$$w(t + d) = w(t) - \frac{1}{2}\left[w(t) - w_{eq}\right] \text{ gives}$$

$$w(t + d) = \frac{1}{2}\left[w(t) + w_{eq}\right]. \text{ Thus}$$

$$w_{eq} + \left(w_0 - w_{eq}\right)e^{-0.005(t+d)}$$

$$= \frac{1}{2}\left[w_{eq} + \left(w_0 - w_{eq}\right)e^{-0.005t} + w_{eq}\right], \text{ or}$$

$$w_{eq} + \left(w_0 - w_{eq}\right)e^{-0.005(t+d)}$$

$$= w_{eq} + \frac{1}{2}\left(w_0 - w_{eq}\right)e^{-0.005t}$$

Solving for d gives

$$e^{-0.005(t+d)} = \frac{1}{2}e^{-0.005t}$$

$$e^{-0.005t}e^{-0.005d} = \frac{1}{2}e^{-0.005t}$$

$$e^{-0.005d} = \frac{1}{2}$$

$$-0.005d = \ln\frac{1}{2} = -\ln 2$$

$$d = \frac{\ln 2}{0.005}$$

as was to be shown.

5. Answers may vary.

Chapter 16

Principles in Practice 16.1

1. The uniform density function is given by

$$f(x) = \begin{cases} \dfrac{1}{60}, & \text{if } 0 \le x \le 60 \\ 0, & \text{otherwise.} \end{cases}$$

The probability of waiting between 25 and 45 minutes is

$$P(25 \le X \le 45) = \int_{25}^{45} \frac{1}{60}\,dx = \frac{x}{60}\bigg|_{25}^{45}$$

$$= \frac{45-25}{60} = \frac{20}{60} = \frac{1}{3}.$$

3. The exponential density function is given by

$$f(x) = \begin{cases} 0.2e^{-0.2x}, & \text{for } x \ge 0 \\ 0, & \text{otherwise.} \end{cases}$$

The mean is given by $\mu = \dfrac{1}{k} = \dfrac{1}{0.2} = 5.$

The standard deviation is given by

$$\sigma = \frac{1}{k} = \frac{1}{0.2} = 5.$$

Problems 16.1

1. a. $P(1 < X < 2) = \displaystyle\int_{1}^{2} \frac{1}{6}(x+1)\,dx$

$$= \frac{(x+1)^2}{12}\bigg|_{1}^{2}$$

$$= \frac{9}{12} - \frac{4}{12} = \frac{5}{12}$$

b. $P(X < 2.5) = \displaystyle\int_{1}^{2.5} \frac{1}{6}(x+1)\,dx$

$$= \frac{(x+1)^2}{12}\bigg|_{1}^{2.5}$$

$$= \frac{49}{48} - \frac{4}{12} = \frac{11}{16} = 0.6875$$

c. $P\left(X \ge \dfrac{3}{2}\right) = \displaystyle\int_{3/2}^{3} \frac{1}{6}(x+1)\,dx$

$$= \frac{(x+1)^2}{12}\bigg|_{3/2}^{3}$$

$$= \frac{16}{12} - \frac{25}{48} = \frac{13}{16} = 0.8125$$

d. $\displaystyle\int_{1}^{c} \frac{1}{6}(x+1)\,dx = \frac{1}{2}$

$$\frac{(x+1)^2}{12}\bigg|_{1}^{c} = \frac{1}{2}$$

$$\frac{(c+1)^2}{12} - \frac{1}{3} = \frac{1}{2}$$

$$(c+1)^2 - 4 = 6$$

$$(c+1)^2 = 10$$

$$c+1 = \pm\sqrt{10}$$

$$c = -1 \pm \sqrt{10}$$

We choose $c = -1 + \sqrt{10}$ since $1 < c < 3$.

3. a. $f(x) = \begin{cases} \dfrac{1}{3}, & \text{if } 1 \le x \le 4 \\ 0, & \text{otherwise} \end{cases}$

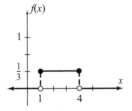

b. $P\left(\dfrac{3}{2} < X < \dfrac{7}{2}\right) = \dfrac{\frac{7}{2} - \frac{3}{2}}{4-1} = \dfrac{2}{3}$

c. $P(0 < X < 1) = \displaystyle\int_{0}^{1} 0\,dx = 0$

d. $P(X \le 3.5) = P(1 \le X \le 3.5)$

$$= \frac{3.5-1}{4-1} = \frac{2.5}{3} = \frac{5}{6}$$

e. $P(X > 3) = P(3 < X \le 4) = \frac{4-3}{4-1} = \frac{1}{3}$

f. $P(X = 2) = 0$

g. $P(X < 5) = P(1 \le X \le 4) = \frac{4-1}{4-1} = 1$

h. $\mu = \int_1^4 x\left(\frac{1}{3}\right)dx = \frac{x^2}{6}\bigg|_1^4 = \frac{16}{6} - \frac{1}{6} = \frac{5}{2}$

i. $\sigma^2 = \int_1^4 x^2\left(\frac{1}{3}\right)dx - \mu^2$

$$= \frac{x^3}{9}\bigg|_1^4 - \left(\frac{5}{2}\right)^2$$

$$= \left[\frac{64}{9} - \frac{1}{9}\right] - \frac{25}{4}$$

$$= 7 - \frac{25}{4}$$

$$= \frac{3}{4}$$

Thus, $\sigma = \sqrt{\frac{3}{4}} = \frac{\sqrt{3}}{2}$.

j. If $1 \le x \le 4$,

$$F(x) = \int_1^x \frac{1}{3}dt = \frac{t}{3}\bigg|_1^x = \frac{x-1}{3}.$$

Thus $F(x) = \begin{cases} 0, & \text{if } x < 1 \\ \dfrac{x-1}{3}, & \text{if } 1 \le x \le 4 \\ 1, & \text{if } x > 4. \end{cases}$

$$P(X < 2) = F(2) = \frac{2-1}{3} = \frac{1}{3}$$

$$P(1 < X < 3) = F(3) - F(1) = \frac{2}{3} - 0 = \frac{2}{3}$$

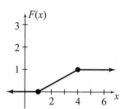

5. a. $f(x) = \begin{cases} \dfrac{1}{b-a}, & \text{if } a \le x \le b \\ 0, & \text{otherwise} \end{cases}$

b. $\mu = \int_a^b x\left(\frac{1}{b-a}\right)dx = \frac{x^2}{2(b-a)}\bigg|_a^b$

$$= \frac{b^2 - a^2}{2(b-a)} = \frac{a+b}{2}$$

c. $\sigma^2 = \int_a^b x^2\left(\frac{1}{b-a}\right)dx - \mu^2$

$$= \frac{x^3}{3(b-a)}\bigg|_a^b - \left(\frac{a+b}{2}\right)^2$$

$$= \frac{b^3 - a^3}{3(b-a)} - \frac{(a+b)^2}{4}$$

$$= \frac{b^2 + ab + a^2}{3} - \frac{a^2 + 2ab + b^2}{4}$$

$$= \frac{b^2 - 2ab + a^2}{12} = \frac{(b-a)^2}{12}$$

Thus $\sigma = \frac{b-a}{\sqrt{12}}$.

7. $f(x) = \begin{cases} 3e^{-3x}, & \text{if } x \ge 0 \\ 0, & \text{if } x < 0 \end{cases}$

a. $P(1 < X < 4) = \int_1^4 3e^{-3x}dx$

$$= -e^{-3x}\bigg|_1^4$$

$$= -e^{-12} + e^{-3}$$

$$\approx 0.04978$$

b. $P(X < 4) = \int_0^4 3e^{-3x} dx$

$$= -e^{-3x} \Big|_0^4$$

$$= -e^{-12} + 1$$

$$\approx 0.99999$$

c. $P(X > 6) = 1 - P(X \le 6)$

$$= 1 - \int_0^6 3e^{-3x} dx$$

$$= 1 - (-e^{-18} + 1)$$

$$= e^{-18} \approx 0.00000$$

d. From the text, $\mu = \sigma = \dfrac{1}{k} = \dfrac{1}{3}$.

$$P(\mu - 2\sigma < X < \mu + 2\sigma)$$

$$= P\left(-\frac{1}{3} < X < 1\right)$$

$$= \int_0^1 3e^{-3x} dx$$

$$= -e^{-3x} \Big|_0^1$$

$$= -e^{-3} + 1$$

$$\approx 0.95021$$

e. $\displaystyle\int_{-\infty}^{\infty} f(x)\, dx = \int_0^{\infty} 3e^{-3x}\, dx$

$$= \lim_{r \to \infty} \int_0^r 3e^{-3x}\, dx$$

$$= \lim_{r \to \infty} (-e^{-3x}) \Big|_0^r$$

$$= \lim_{r \to \infty} (-e^{-3r} + 1)$$

$$= 0 + 1 = 1$$

f. $F(x) = P(X \le x) = \displaystyle\int_{-\infty}^x f(t)\, dt$

If $x \ge 0$,

$$F(x) = \int_0^x 3e^{-3t}\, dt = -e^{-3t} \Big|_0^x$$

$$= -e^{-3x} + 1.$$

Thus $F(x) = \begin{cases} 0 & \text{if } x < 0 \\ -e^{-3x} + 1 & \text{if } x \ge 0 \end{cases}$

9. a. $\displaystyle\int_0^4 kx\, dx = 1$

$$\frac{kx^2}{2} \Big|_0^4 = 1$$

$$8k = 1$$

$$k = \frac{1}{8}$$

b. $P(2 < X < 3) = \displaystyle\int_2^3 \frac{x}{8}\, dx = \frac{x^2}{16} \Big|_2^3 = \frac{9}{16} - \frac{4}{16}$

$$= \frac{5}{16}$$

c. $P(X > 2.5) = \displaystyle\int_{2.5}^4 \frac{x}{8}\, dx = \frac{x^2}{16} \Big|_{2.5}^4$

$$= 1 - \frac{25}{64} = \frac{39}{64} \approx 0.609$$

d. $P(X > 0) = P(0 \le X \le 4) = 1$

e. $\mu = \displaystyle\int_0^4 x\left(\frac{x}{8}\right) dx = \frac{x^3}{24} \Big|_0^4 = \frac{64}{24} - 0 = \frac{8}{3}$

f. $\sigma^2 = \displaystyle\int_0^4 x^2 \left(\frac{x}{8}\right) dx - \mu^2 = \frac{x^4}{32} \Big|_0^4 - \left(\frac{8}{3}\right)^2$

$$= 8 - \frac{64}{9} = \frac{8}{9}$$

Thus $\sigma = \sqrt{\dfrac{8}{9}} = \dfrac{\sqrt{8}}{3} = \dfrac{2\sqrt{2}}{3}$

g. $P(X < c) = \dfrac{1}{2}$

$$\int_0^c \frac{x}{8}\,dx = \frac{1}{2}$$

$$\left.\frac{x^2}{16}\right|_0^c = \frac{1}{2}$$

$$\frac{c^2}{16} = \frac{1}{2}$$

$$c^2 = 8$$

$$c = \pm 2\sqrt{2}$$

We choose $c = 2\sqrt{2}$ since $0 < c < 4$.

h. $P(3 < X < 5) = P(3 < X < 4) = \displaystyle\int_3^4 \frac{x}{8}\,dx$

$$= \left.\frac{x^2}{16}\right|_3^4 = \frac{16}{16} - \frac{9}{16} = \frac{7}{16}$$

11. $P(X \le 7) = \displaystyle\int_0^7 \frac{1}{10}\,dx = \left.\frac{x}{10}\right|_0^7 = \frac{7}{10}$

$$E(X) = \int_0^{10} x\left(\frac{1}{10}\right)dx = \left.\frac{x^2}{20}\right|_0^{10} = 5 \text{ min}$$

13. $P(X > 1) = 1 - P(X \le 1)$

$$= 1 - \int_0^1 3e^{-3x}\,dx = 1 - \left.\left(-e^{-3x}\right)\right|_0^1$$

$$= 1 - \left(-e^{-3} + 1\right) \approx 0.050$$

Problems 16.2

1. a. $P(0 < Z < 1.7) = A(1.7) = 0.4554$

b. $P(0.43 < Z < 2.89) = A(2.89) - A(0.43)$
$$= 0.4981 - 0.1664$$
$$= 0.3317$$

c. $P(Z > -1.23) = 0.5 + A(1.23)$
$$= 0.5 + 0.3907$$
$$= 0.8907$$

d. $P(Z \le 2.91) = 0.5 + A(2.91)$
$$= 0.5 + 0.4982$$
$$= 0.9982$$

e. $P(-2.51 < Z \le 1.3) = A(2.51) + A(1.3)$
$$= 0.4940 + 0.4032$$
$$= 0.8972$$

f. $P(Z > 0.03) = 0.5 - A(0.03)$
$$= 0.5 - 0.0120$$
$$= 0.4880$$

3. $P(Z < z_0) = 0.5517$
$$0.5 + A(z_0) = 0.5517$$
$$A(z_0) = 0.0517$$
$$z_0 = 0.13$$

5. $P(Z > z_0) = 0.8599$
$$0.5 + A(-z_0) = 0.8599$$
$$A(-z_0) = 0.3599$$
$$-z_0 = 1.08$$
$$z_0 = -1.08$$

7. $P(-z_0 < Z < z_0) = 0.2662$
$$2A(z_0) = 0.2662$$
$$A(z_0) = 0.1331$$
$$z_0 = 0.34$$

9. a. $P(X < 27) = P\left(Z < \dfrac{27 - 16}{4}\right)$
$$= P(Z < 2.75) = 0.5 + A(2.75)$$
$$= 0.5 + 0.4970 = 0.9970$$

b. $P(X < 10) = P\left(Z < \dfrac{10 - 16}{4}\right)$
$$= P(Z < -1.5) = 0.5 - A(1.5)$$
$$= 0.5 - 0.4332 = 0.0668$$

c. $P(10.8 < X < 12.4)$

$$= P\left(\frac{10.8-16}{4} < Z < \frac{12.4-16}{4}\right)$$

$$= P(-1.3 < Z < -0.9) = A(1.3) - A(0.9)$$

$$= 0.4032 - 0.3159 = 0.0873$$

11. $P(X > -2) = P\left(Z > \frac{-2-(-3)}{2}\right)$

$$= P\left(Z > \frac{1}{2}\right) = 0.5 - A\left(\frac{1}{2}\right)$$

$$= 0.5 - 0.1915 = 0.3085$$

13. Since $\sigma^2 = 100$, $\sigma = 10$. Thus

$$35 = 65 - 30 = \mu - 3\sigma$$

$$95 = 65 + 30 = \mu + 3\sigma$$

Thus

$$P(35 < X \le 95) = P(\mu - 3\sigma < X \le \mu + 3\sigma)$$

$$= 0.997 \quad (99.7\%)$$

15. $P(X > 54) = P(Z > z_0) = 0.0401$

$$0.5 - A(z_0) = 0.0401$$

$$A(z_0) = 0.4599$$

$$z_0 = 1.75$$

Since $\dfrac{54-40}{\sigma} = 1.75$

$$\frac{14}{\sigma} = 1.75$$

so $\sigma = \dfrac{14}{1.75} = 8$.

17. Let X be score on test. Then the probability that X lies within $2\sigma = 2(100) = 200$ points of 500 is 0.95. Thus, 95% of those who took the test had scores between 300 and 700.

19. Let X be height of an adult. Then

$$P(X < 72) = P\left(Z < \frac{72-68}{3}\right) = P(Z < 1.33)$$

$$= 0.5 + A(1.33) = 0.5 + 0.4082 = 0.9082$$

90.82% are over 6 feet.

21. Let X be IQ of a child in population.

a. $P(X > 125) = P\left(Z > \dfrac{125-100.4}{11.6}\right)$

$$= P(Z > 2.12) = 0.5 - A(2.12)$$

$$= 0.5 - 0.4830 = 0.0170.$$

Thus 1.7% of the children have IQ's greater than 125.

b. If x_0 is the value, then

$$P(X > x_0) = 0.90. \text{ Thus}$$

$$P\left(Z > \frac{x_0 - 100.4}{11.6}\right) = 0.90 \text{ or}$$

$$0.5 + A\left(-\frac{x_0 - 100.4}{11.6}\right) = 0.90. \text{ Hence}$$

$$A\left(-\frac{x_0 - 100.4}{11.6}\right) = 0.4, \text{ so}$$

$$-\frac{x_0 - 100.4}{11.6} = 1.28 \text{ or}$$

$$x_0 = 85.552 \approx 85.6.$$

Principles in Practice 16.3

1. X is the number of winners and X is binomial with $n = 60$ and $p = \dfrac{1}{4}$. To find $P(X = 20)$, use the normal approximation to the binomial distribution with

$$\mu = np = 60\left(\frac{1}{4}\right) = 15 \text{ and}$$

$$\sigma = \sqrt{npq} = \sqrt{60\left(\frac{1}{4}\right)\left(\frac{3}{4}\right)} = \sqrt{\frac{45}{4}} \approx 3.35.$$

Converting the correct X-values 19.5 and 20.5 to Z-values gives

$$z_1 = \frac{19.5-15}{\sqrt{\frac{45}{4}}} \approx 1.34$$

$$z_2 = \frac{20.5-15}{\sqrt{\frac{45}{4}}} \approx 1.64$$

Thus $P(X = 20) \approx P(1.34 \le Z \le 1.64)$
$= A(1.64) - A(1.34) = 0.4495 - 0.4099$
$= 0.0396$
The probability of 20 winners out of 60 contestants is 0.0396.

Problems 16.3

1. $n = 150$, $p = 0.4$, $q = 0.6$,
$\mu = np = 150(0.4) = 60$,
$\sigma = \sqrt{npq} = \sqrt{150(0.4)(0.6)} = \sqrt{36} = 6$
$P(X \le 52) = P(X < 52.5)$

$z = \dfrac{52.5 - 60}{6} = -1.25$

$P(X \le 52) = P(X \le 52.5)$
$\approx P(Z \le -1.25) = 0.5 - A(1.25)$
$= 0.5 - 0.3944 = 0.1056$
$P(X \ge 74) = P(X \ge 73.5)$

$\approx P\left(Z \ge \dfrac{73.5 - 60}{6}\right)$

$= P(Z \ge 2.25)$
$= 0.5 - A(2.25)$
$= 0.5 - 0.4878 = 0.0122$

3. $n = 200$, $p = 0.6$, $q = 0.4$,
$\mu = np = 200(0.6) = 120$
$\sigma = \sqrt{npq} = \sqrt{200(0.6)(0.4)} = \sqrt{48} \approx 6.93$
$P(X = 125) = P(124.5 \le X \le 125.5)$

$\approx P\left(\dfrac{124.5 - 120}{\sqrt{48}} \le Z \le \dfrac{125.5 - 120}{\sqrt{48}}\right)$

$= P(0.65 \le Z \le 0.79)$
$= A(0.79) - A(0.65)$
$= 0.2852 - 0.2422$
$= 0.0430$

$P(110 \le X \le 135)$
$= P(109.5 \le X \le 135.5)$

$\approx P\left(\dfrac{109.5 - 120}{\sqrt{48}} \le Z \le \dfrac{135.5 - 120}{\sqrt{48}}\right)$

$= P(-1.52 \le Z \le 2.24)$
$= A(1.52) + A(2.24)$
$= 0.4357 + 0.4875$
$= 0.9232$

5. Let X = no. of times 5 occurs. Then X is
binomial with $n = 300$, $p = \dfrac{1}{6}$, $q = \dfrac{5}{6}$,

$\mu = np = 50$, $\sigma = \sqrt{npq} = \sqrt{\dfrac{125}{3}} \approx 6.45$.

$P(45 \le X \le 60) = P(44.5 \le X \le 60.5)$

$\approx P\left(\dfrac{44.5 - 50}{\sqrt{\dfrac{125}{3}}} \le Z \le \dfrac{60.5 - 50}{\sqrt{\dfrac{125}{3}}}\right)$

$= P(-0.85 \le Z \le 1.63) = A(0.85) + A(1.63)$
$= 0.3023 + 0.4484 = 0.7507$

7. Let X = no. of trucks out of service. Then X
can be considered binomial with $n = 60$, $p = 0.1$,
$q = 0.9$, $\mu = np = 6$,
$\sigma = \sqrt{npq} = \sqrt{5.4} \approx 2.32$

$P(X \ge 7) = P(X \ge 6.5) \approx P\left(Z \ge \dfrac{6.5 - 6}{\sqrt{5.4}}\right)$

$= P(Z \ge 0.22) = 0.5 - A(0.22)$
$= 0.5 - 0.0871 = 0.4129$

9. Let X = no. of correct answers. Then X is
binomial and $p = 0.5$, $q = 0.5$. If $n = 25$, then
$\mu = np = 25(0.5) = 12.5$,
$\sigma = \sqrt{npq} = \sqrt{25(0.5)(0.5)} = \sqrt{6.25} = 2.5$
and
$P(X \ge 13) = P(X \ge 12.5)$

$\approx P\left(Z \ge \dfrac{12.5 - 12.5}{2.5}\right)$

$= P(Z \ge 0.00)$
$= 0.5 - A(0.00)$
$= 0.5 - 0$
$= 0.5$

If $n = 100$, then $\mu = 100(0.5) = 50$,
$\sigma = \sqrt{npq} = \sqrt{100(0.5)(0.5)} = \sqrt{25} = 5$
$P(X \ge 60) = P(X \ge 59.5)$

$= P\left(Z \ge \dfrac{59.5 - 50}{5}\right)$

$= P(Z \ge 1.9)$
$= 0.5 - A(1.9)$
$= 0.5 - 0.4713$
$= 0.0287$

11. Let X = no. of deals consisting of three cards of one suit and two cards of another suit. Then X is binomial with $n = 100$, $p = 0.1$, $q = 0.9$,
$$\mu = np = 10, \ \sigma = \sqrt{npq} = \sqrt{9} = 3.$$
$$P(X \geq 16) = P(X \geq 15.5)$$
$$\approx P\left(Z \geq \frac{15.5 - 10}{3}\right)$$
$$= P(Z \geq 1.83)$$
$$= 0.5 - A(1.83)$$
$$= 0.5 - 0.4664$$
$$= 0.0336$$

Chapter 16 Review Problems

1. a. $P(0 \leq X \leq 1) = 1$
$$\int_0^1 \left(\frac{1}{3} + kx^2\right) dx = 1$$
$$\left(\frac{x}{3} + \frac{kx^3}{3}\right)\Big|_0^1 = 1$$
$$\frac{1}{3} + \frac{k}{3} = 1$$
$$\frac{k}{3} = \frac{2}{3}$$
$$k = 2$$

b. $P\left(\frac{1}{2} < X < \frac{3}{4}\right) = \int_{1/2}^{3/4}\left(\frac{1}{3} + 2x^2\right) dx$
$$= \left(\frac{x}{3} + \frac{2x^3}{3}\right)\Big|_{1/2}^{3/4} = \frac{1}{3}\left(x + 2x^3\right)\Big|_{1/2}^{3/4}$$
$$= \frac{1}{3}\left[\left(\frac{3}{4} + \frac{27}{32}\right) - \left(\frac{1}{2} + \frac{1}{4}\right)\right] = \frac{9}{32}$$

c. $P\left(X \geq \frac{1}{2}\right) = \int_{1/2}^1 \left(\frac{1}{3} + 2x^2\right) dx$
$$= \left(\frac{x}{3} + \frac{2x^3}{3}\right)\Big|_{1/2}^1 = \frac{1}{3}\left(x + 2x^3\right)\Big|_{1/2}^1$$
$$= \frac{1}{3}\left[(1 + 2) - \left(\frac{1}{2} + \frac{1}{4}\right)\right] = \frac{3}{4}$$

d. If $0 \leq x \leq 1$, $F(x) = \int_0^x \left(\frac{1}{3} + 2t^2\right) dt$
$$= \left(\frac{t}{3} + \frac{2t^3}{3}\right)\Big|_0^x = \frac{x}{3} + \frac{2x^3}{3}$$

Therefore,
$$F(x) = \begin{cases} 0, & \text{if } x < 0 \\ \frac{x}{3} + \frac{2x^3}{3}, & \text{if } 0 \leq x \leq 1 \\ 1, & \text{if } x > 1 \end{cases}$$

3. a. $\mu = \int_0^5 x\left(\frac{2}{25}x\right) dx = \frac{2x^3}{75}\Big|_0^5 = \frac{10}{3}$

b. $\sigma^2 = \int_0^5 x^2\left(\frac{2}{25}x\right) dx - \mu^2$
$$= \frac{x^4}{50}\Big|_0^5 - \left(\frac{10}{3}\right)^2$$
$$= \frac{625}{50} - \frac{100}{9} = \frac{25}{18}$$
Thus $\sigma = \sqrt{\frac{25}{18}} \approx 1.18$.

5. $P(X > 22) = P\left(Z > \frac{22 - 20}{4}\right)$
$$= P(Z > 0.5) = 0.5 - A(0.5)$$
$$= 0.5 - 0.1915 = 0.3085$$

7. $P(14 < X < 18) = P\left(\dfrac{14-20}{4} < Z < \dfrac{18-20}{4}\right)$

   $= P(-1.5 < Z < -0.5)$

   $= A(1.5) - A(0.5) = 0.4332 - 0.1915 =$
   0.2417

9. $P(X < 23) = P\left(Z < \dfrac{23-20}{4}\right)$

   $= P(Z < 0.75) = 0.5 + A(0.75)$

   $= 0.5 + 0.2734 = 0.7734$

11. $n = 100, p = 0.35, q = 0.65, \mu = np = 35,$
    $\sigma = \sqrt{npq} = \sqrt{22.75} \approx 4.77$
    $P(25 \le X \le 47) = P(24.5 \le X \le 47.5)$
    $\approx P\left(\dfrac{24.5-35}{\sqrt{22.75}} \le Z \le \dfrac{47.5-35}{\sqrt{22.75}}\right)$
    $= P(-2.20 \le Z \le 2.62) = A(2.20) + A(2.62)$
    $= 0.4861 + 0.4956 = 0.9817$

13. Let X = height of an individual. X is
    normally distributed with $\mu = 68$ and $\sigma = 2$.

    $P(X > 72) = P\left(Z > \dfrac{72-68}{2}\right)$

    $= P(Z > 2) = 0.5 - A(2)$

    $= 0.5 - 0.4772 = 0.0228$

Mathematical Snapshot Chapter 16

1. The result should correspond to the known distribution function.

3. The list of earthquake magnitudes will appear to have a normal density function. This is surprising, since one would expect something more like an exponential density function, with most earthquakes being very low-magnitude. Presumably, the explanation for the normal density function is that as magnitude declines, the likelihood of a quake's being reported and thus appearing on the list also goes down.

Chapter 17

Principles in Practice 17.1

1. a. $c(500, 700) = 160 + 2(500) + 3(700)$
$= 160 + 1000 + 2100 = 3260$
The cost of manufacturing 500 12-ounce and 700 20-ounce mugs is $3260.

b. $c(1000, 750) = 160 + 2(1000) + 3(750)$
$= 160 + 2000 + 2250 = 4410$
The cost of manufacturing 1000 12-ounce mugs and 750 20-ounce mugs is $4410.

Problems 17.1

1. $f(1, 2) = 4(1) - (2)^2 + 3 = 4 - 4 + 3 = 3$

3. $g(0, 3, -1) = e^{2 \cdot 0}[3(3) + (-1)] = e^0(8) = 8$

5. $h(-3, 3, 5, 4) = \dfrac{-3(3)}{5^2 - 4^2} = \dfrac{-9}{25 - 16} = \dfrac{-9}{9} = -1$

7. $g(4, 8) = 2(4)\left(4^2 - 5\right) = 2(4)(11) = 88$

9. $F(2, 0, -1) = 3$

11. $f(x_0 + h, y_0) = e^{(x_0 + h) + y_0} = e^{x_0 + h + y_0}$

13. $f(400, 400, 80) = \dfrac{400(400)}{80} = 2000$

15. A plane parallel to the x,z-plane has the form $y = $ constant. Because $(0, 2, 0)$ lies on the plane, the equation is $y = 2$.

17. A plane parallel to the x,y-plane has the form $z = $ constant. Because $(2, 7, 6)$ lies on the plane, the equation is $z = 6$.

19. $x + y + z = 1$ can be put in the form $Ax + By + Cz + D = 0$, so the graph is a plane. The intercepts are $(1, 0, 0)$, $(0, 1, 0)$, and $(0, 0, 1)$.

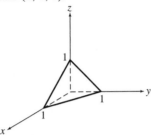

21. $3x + 6y + 2z = 12$ can be put in the form $Ax + By + Cz + D = 0$, so the graph is a plane. The intercepts are $(4, 0, 0)$, $(0, 2, 0)$, and $(0, 0, 6)$.

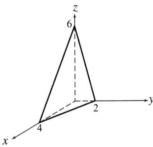

23. $x + 2y = 2$ can be put in the form $Ax + By + Cz + D = 0$, so the graph is a plane. There are only two intercepts: $(2, 0, 0)$ and $(0, 1, 0)$. The x,y-trace is $x + 2y = 2$, which is a line. For any fixed value of z, we obtain the line $x + 2y = 2$.

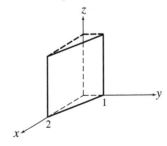

25. $z = 4 - x^2$. The x,z-trace is $z = 4 - x^2$, which is a parabola. For any fixed value of y, we obtain the parabola $z = 4 - x^2$.

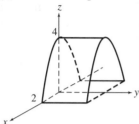

27. $x^2 + y^2 + z^2 = 9$. The x,y-trace is $x^2 + y^2 = 9$, which is a circle. The x,z-trace is $x^2 + z^2 = 9$, which is a circle. The y,z-trace is $y^2 + z^2 = 9$, which is a circle. The surface is a sphere.

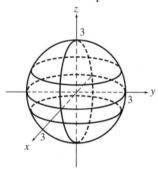

Problems 17.2

1. $f(x, y) = 4x^2 + 3y^2 - 7$
$f_x(x, y) = 4(2x) + 0 + 0 = 8x$
$f_y(x, y) = 0 + 3(2y) + 0 = 6y$

3. $f(x, y) = 2y + 1$
$f_x(x, y) = 0 + 0 = 0$
$f_y(x, y) = 2(1) + 0 = 2$

5. $g(x, y) = 3x^4 y + 2xy^2 - 5xy + 8x - 9y$
$g_x(x, y) = 3(4)x^3 y + 2(1)y^2 - 5(1)y + 8(1)$
$\qquad = 12x^3 y + 2y^2 - 5y + 8$
$g_y(x, y) = 3x^4(1) + 2x(2)y - 5x(1) - 9(1)$
$\qquad = 3x^4 + 4xy - 5x - 9$

7. $g(p, q) = \sqrt{pq} = (pq)^{\frac{1}{2}}$
$g_p(p, q) = \dfrac{1}{2}(pq)^{-\frac{1}{2}} \cdot q = \dfrac{q}{2\sqrt{pq}}$
$g_q(p, q) = \dfrac{1}{2}(pq)^{-\frac{1}{2}} \cdot p = \dfrac{p}{2\sqrt{pq}}$

9. $h(s, t) = \dfrac{s^2 + 4}{t - 3}$
$h_s(s, t) = \dfrac{1}{t - 3}(2s) = \dfrac{2s}{t - 3}$
Rewriting $h(s, t)$ as $\left(s^2 + 4\right)(t - 3)^{-1}$, we have
$h_t(s, t) = \left(s^2 + 4\right)\left[(-1)(t - 3)^{-2}(1)\right]$
$\qquad = -\dfrac{s^2 + 4}{(t - 3)^2}$

11. $u(q_1, q_2) = \dfrac{1}{2}\ln(q_1 + 2) + \dfrac{1}{3}\ln(q_2 + 5)$
$u_{q_1}(q_1, q_2) = \dfrac{1}{2} \cdot \dfrac{1}{q_1 + 2} + 0 = \dfrac{1}{2(q_1 + 2)}$
$u_{q_2}(q_1, q_2) = 0 + \dfrac{1}{3} \cdot \dfrac{1}{q_2 + 5} = \dfrac{1}{3(q_2 + 5)}$

13. $h(x, y) = \dfrac{x^2 + 3xy + y^2}{\sqrt{x^2 + y^2}}$

$$h_x(x, y) = \dfrac{\left(x^2 + y^2\right)^{\frac{1}{2}}[2x + 3y] - \left(x^2 + 3xy + y^2\right)\left[\frac{1}{2}\left(x^2 + y^2\right)^{-\frac{1}{2}}(2x)\right]}{\left[\left(x^2 + y^2\right)^{\frac{1}{2}}\right]^2}$$

$$= \dfrac{\left(x^2 + y^2\right)^{-\frac{1}{2}}\left[\left(x^2 + y^2\right)(2x + 3y) - \left(x^2 + 3xy + y^2\right)x\right]}{x^2 + y^2}$$

$$= \dfrac{2x^3 + 3x^2y + 2xy^2 + 3y^3 - x^3 - 3x^2y - xy^2}{\left(x^2 + y^2\right)^{\frac{3}{2}}} = \dfrac{x^3 + xy^2 + 3y^3}{\left(x^2 + y^2\right)^{\frac{3}{2}}}$$

$$h_y(x, y) = \dfrac{\left(x^2 + y^2\right)^{\frac{1}{2}}[3x + 2y] - \left(x^2 + 3xy + y^2\right)\left[\frac{1}{2}\left(x^2 + y^2\right)^{-\frac{1}{2}}(2y)\right]}{\left[\left(x^2 + y^2\right)^{\frac{1}{2}}\right]^2}$$

$$= \dfrac{\left(x^2 + y^2\right)^{-\frac{1}{2}}\left[\left(x^2 + y^2\right)(3x + 2y) - \left(x^2 + 3xy + y^2\right)y\right]}{x^2 + y^2}$$

$$= \dfrac{3x^3 + 2x^2y + 3xy^2 + 2y^3 - x^2y - 3xy^2 - y^3}{\left(x^2 + y^2\right)^{\frac{3}{2}}} = \dfrac{3x^3 + x^2y + y^3}{\left(x^2 + y^2\right)^{\frac{3}{2}}}$$

15. $z = e^{5xy}$

$$\dfrac{\partial z}{\partial x} = e^{5xy}(5y) = 5ye^{5xy} \; ; \; \dfrac{\partial z}{\partial y} = e^{5xy}(5x) = 5xe^{5xy}$$

17. $z = 5x \ln\left(x^2 + y\right)$

$$\dfrac{\partial z}{\partial x} = 5\left\{x\left[\dfrac{1}{x^2 + y}(2x)\right] + \ln\left(x^2 + y\right)[1]\right\} = 5\left[\dfrac{2x^2}{x^2 + y} + \ln\left(x^2 + y\right)\right]$$

$$\dfrac{\partial z}{\partial y} = 5x\left(\dfrac{1}{x^2 + y}[1]\right) = \dfrac{5x}{x^2 + y}$$

19. $f(r, s) = (r+2s)^{\frac{1}{2}}\left(r^3 - 2rs + s^2\right)$

$f_r(r, s) = (r+2s)^{\frac{1}{2}}\left[3r^2 - 2s\right] + \left(r^3 - 2rs + s^2\right)\left[\frac{1}{2}(r+2s)^{-\frac{1}{2}}(1)\right]$

$\qquad = \sqrt{r+2s}\left(3r^2 - 2s\right) + \dfrac{r^3 - 2rs + s^2}{2\sqrt{r+2s}}$

$f_s(r, s) = (r+2s)^{\frac{1}{2}}\left[-2r + 2s\right] + \left(r^3 - 2rs + s^2\right)\left[\frac{1}{2}(r+2s)^{-\frac{1}{2}}(2)\right]$

$\qquad = 2(s-r)\sqrt{r+2s} + \dfrac{r^3 - 2rs + s^2}{\sqrt{r+2s}}$

21. $f(r, s) = e^{3-r}\ln(7-s)$

$f_r(r, s) = \ln(7-s)\left[e^{3-r}(-1)\right] = -e^{3-r}\ln(7-s)$

$f_s(r, s) = e^{3-r}\left[\dfrac{1}{7-s}(-1)\right] = \dfrac{e^{3-r}}{s-7}$

23. $g(x, y, z) = 2x^3y^2 + 2xy^3z + 4z^2$

$g_x(x, y, z) = 2y^2(3x^2) + 2y^3z(1) + 0 = 6x^2y^2 + 2y^3z$

$g_y(x, y, z) = 2x^3(2y) + 2xz(3y^2) + 0 = 4x^3y + 6xy^2z$

$g_z(x, y, z) = 0 + 2xy^3(1) + 4(2z) = 2xy^3 + 8z$

25. $g(r, s, t) = e^{s+t}(r^2 + 7s^3)$

$g_r(r, s, t) = e^{s+t}[2r + 0] = 2re^{s+t}$

$g_s(r, s, t) = e^{s+t}\left[0 + 21s^2\right] + \left(r^2 + 7s^3\right)\left[e^{s+t}(1)\right]$

$\qquad = \left(7s^3 + 21s^2 + r^2\right)e^{s+t}$

$g_t(r, s, t) = \left(r^2 + 7s^3\right)\left[e^{s+t}(1)\right] = e^{s+t}\left(r^2 + 7s^3\right)$

27. $f(x, y) = x^3y + 7x^2y^2$

$f_x(x, y) = 3x^2y + 14xy^2$

$f_x(1, -2) = 3(1)^2(-2) + 14(1)(-2)^2 = 50$

29. $g(x, y, z) = e^x \sqrt{y + 2z}$

$$g_z(x, y, z) = e^x \left[\frac{1}{2}(y + 2z)^{-\frac{1}{2}}(2) \right] = \frac{e^x}{\sqrt{y + 2z}}$$

$$g_z(0, 6, 4) = \frac{1}{\sqrt{6 + 8}} = \frac{1}{\sqrt{14}}$$

31. $h(r, s, t, u) = \left(s^2 + tu \right) \ln(2r + 7st)$

$$h_s(r, s, t, u) = \frac{7t\left(s^2 + tu \right)}{2r + 7st} + 2s \ln(2r + 7st)$$

$$h_s(1, 0, 0, 1) = 0$$

33. $f(r, s, t) = rst\left(r^2 + s^3 + t^4 \right) = r^3 st + rs^4 t + rst^5$

$$f_s(r, s, t) = r^3(1)t + r\left(4s^3 \right)t + r(1)t^5 = r^3 t + 4rs^3 t + rt^5$$

$$f_s(1, -1, 2) = 2 + (-8) + 32 = 26$$

35. $z = xe^{x-y} + ye^{y-x}$

$$\frac{\partial z}{\partial x} = \left[xe^{x-y} + e^{x-y} \right] + \left[ye^{y-x}(-1) \right]$$

$$\frac{\partial z}{\partial y} = \left[xe^{x-y}(-1) \right] + \left[ye^{y-x} + e^{y-x} \right]$$

Thus $\dfrac{\partial z}{\partial x} + \dfrac{\partial z}{\partial y} = e^{x-y} + e^{y-x}$, as was to be shown.

37. $F(b, C, T, i) = \dfrac{bT}{C} + \dfrac{iC}{2}$

$$\frac{\partial F}{\partial C} = \frac{\partial}{\partial C}\left[\frac{bT}{C} \right] + \frac{\partial}{\partial C}\left[\frac{iC}{2} \right] = -\frac{bT}{C^2} + \frac{i}{2}$$

39. $R = f(r, a, n) = \dfrac{r}{1 + a\left(\frac{n-1}{2} \right)} = r\left[1 + a\left(\frac{n-1}{2} \right) \right]^{-1}$

$$\frac{\partial R}{\partial n} = r(-1)\left[1 + a\left(\frac{n-1}{2} \right) \right]^{-2} \cdot \frac{a}{2} = -\frac{ra}{2\left[1 + a\left(\frac{n-1}{2} \right) \right]^2}$$

Problems 17.3

1. $c = 7x + 0.3y^2 + 2y + 900$

$$\frac{\partial c}{\partial y} = 0.6y + 2$$

When $x = 20$ and $y = 30$, then $\frac{\partial c}{\partial y} = 0.6(30) + 2 = 20$.

3. $c = 0.03(x+y)^3 - 0.6(x+y)^2 + 9.5(x+y) + 7700$

$$\frac{\partial c}{\partial x} = 0.09(x+y)^2 - 1.2(x+y) + 9.5$$

When $x = 50$ and $y = 80$, then $\frac{\partial c}{\partial x} = 0.09(130)^2 - 1.2(130) + 9.5 = 1374.5$.

5. $P = 2.314 l^{0.357} k^{0.643}$

$$\frac{\partial P}{\partial l} = 2.314(0.357) l^{-0.643} k^{0.643}$$

$$= 0.826098 \left(\frac{k}{l} \right)^{0.643}$$

$$\frac{\partial P}{\partial k} = 2.314(0.643) l^{0.357} k^{-0.357}$$

$$= 1.487902 \left(\frac{l}{k} \right)^{0.357}$$

7. $\frac{\partial q_A}{\partial p_A} = -50$, $\frac{\partial q_A}{\partial p_B} = 2$, $\frac{\partial q_B}{\partial q_A} = 4$, $\frac{\partial q_B}{\partial p_B} = -20$

Since $\frac{\partial q_A}{\partial p_B} > 0$ and $\frac{\partial q_B}{\partial p_A} > 0$ the products are competitive.

9. $q_A = 100 p_A^{-1} p_B^{-\frac{1}{2}}$

$q_B = 500 p_B^{-1} p_A^{-\frac{1}{3}}$

$$\frac{\partial q_A}{\partial p_A} = 100(-1) p_A^{-2} p_B^{-\frac{1}{2}} = \frac{-100}{p_A^2 p_B^{\frac{1}{2}}}$$

$$\frac{\partial q_A}{\partial p_B} = 100 \left(-\frac{1}{2} \right) p_A^{-1} p_B^{-\frac{3}{2}} = \frac{-50}{p_A p_B^{\frac{3}{2}}}$$

$$\frac{\partial q_B}{\partial p_A} = 500 \left(-\frac{1}{3} \right) p_B^{-1} p_A^{-\frac{4}{3}} = \frac{-500}{3 p_B p_A^{\frac{4}{3}}}$$

$$\frac{\partial q_B}{\partial p_B} = 500(-1)p_B^{-2}p_A^{-\frac{1}{3}} = -\frac{500}{p_B^2 p_A^{\frac{1}{3}}}$$

Since $\dfrac{\partial q_A}{\partial p_B} < 0$ and $\dfrac{\partial q_B}{\partial p_A} < 0$, the products are complementary.

11. $\dfrac{\partial P}{\partial B} = 0.01A^{0.27}B^{-0.99}C^{0.01}D^{0.23}E^{0.09}F^{0.27}$

$\dfrac{\partial P}{\partial C} = 0.01A^{0.27}B^{0.01}C^{-0.99}D^{0.23}E^{0.09}F^{0.27}$

13. $\dfrac{\partial z}{\partial x} = 4480$. If a staff manager with an M.B.A. degree had an extra year of work experience before the degree, the manager would receive $4480 more per year in extra compensation.

15. a. $\dfrac{\partial R}{\partial w} = -1.015$; $\dfrac{\partial R}{\partial s} = -0.846$

b. One for which $w = w_0$ and $s = s_0$ since increasing w by 1 while holding s fixed decreases the reading ease score.

17. $\dfrac{\partial g}{\partial x} = \dfrac{1}{V_F} > 0$ for $V_F > 0$. Thus if x increases and V_F and V_S are fixed, then g increases.

19. a. $\dfrac{\partial q_A}{\partial p_A} = 10\sqrt{p_B}\left(-\dfrac{1}{2}p_A^{-\frac{3}{2}}\right)$

$\dfrac{\partial q_A}{\partial p_B} = \dfrac{10}{\sqrt{p_A}}\left(\dfrac{1}{2}p_B^{-\frac{1}{2}}\right)$

When $p_A = 9$ and $p_B = 16$, then $\dfrac{\partial q_A}{\partial p_A} = 10(4)\left(-\dfrac{1}{2} \cdot \dfrac{1}{27}\right) = -\dfrac{20}{27}$ and

$\dfrac{\partial q_A}{\partial p_B} = \dfrac{10}{3}\left(\dfrac{1}{2} \cdot \dfrac{1}{4}\right) = \dfrac{5}{12}$.

b. From (a), when $p_A = 9$ and $p_B = 16$, then $\dfrac{\partial q_A}{\partial p_B} = \dfrac{5}{12}$. Hence each $1 reduction in p_B decreases

q_A by approximately $\dfrac{5}{12}$ unit. Thus a $2 reduction in p_B (from $16 to $14) decreases the demand

for A by approximately $\dfrac{5}{12}(2) = \dfrac{5}{6}$ unit.

21. a. $\dfrac{\partial R}{\partial E_r} = 2.5945 - 0.1608 E_r - 0.0277 I_r$

If $E_r = 18.8$ and $I_r = 10$, then

$\dfrac{\partial R}{\partial E_r} = -0.70564$. Since $\dfrac{\partial R}{\partial E_r} < 0$,

such a candidate should not be so advised.

b. $\dfrac{\partial R}{\partial N} = 0.8579 - 0.0122 N$

If $\dfrac{\partial R}{\partial N} < 0$, then $N > 70.3 \approx 70\%$

23. $q_A = 1000 - 50 p_A + 2 p_B$

$\eta_{p_A} = \left(\dfrac{p_A}{q_A}\right)\dfrac{\partial q_A}{\partial p_A} = \left(\dfrac{p_A}{q_A}\right)(-50)$

$\eta_{p_B} = \left(\dfrac{p_B}{q_A}\right)\dfrac{\partial q_A}{\partial p_B} = \left(\dfrac{p_B}{q_A}\right)(2)$

When $p_A = 2$ and $p_B = 10$, then

$q_A = 920$, from which $\eta_{p_A} = -\dfrac{5}{46}$ and

$\eta_{p_B} = \dfrac{1}{46}$

25. $q_A = \dfrac{100}{p_A \sqrt{p_B}}$

$\eta_{p_A} = \left(\dfrac{p_A}{q_A}\right)\dfrac{\partial q_A}{\partial p_A} = \left(\dfrac{p_A}{q_A}\right)\left(\dfrac{-100}{p_A^2 \sqrt{p_B}}\right)$

$\eta_{p_B} = \left(\dfrac{p_B}{q_A}\right)\dfrac{\partial q_A}{\partial p_B} = \left(\dfrac{p_B}{q_A}\right)\left(\dfrac{-50}{p_A \sqrt{p_B^3}}\right)$

When $p_A = 1$ and $p_B = 4$, then $q_A = 50$.

This gives $\eta_{p_A} = -1$ and $\eta_{p_B} = -\dfrac{1}{2}$.

Problems 17.4

1. $4x + 0 + 10z\dfrac{\partial z}{\partial x} = 0$

$\dfrac{\partial z}{\partial x} = -\dfrac{4x}{10z} = -\dfrac{2x}{5z}$

3. $6z^2 \dfrac{\partial z}{\partial y} - 0 - 8y = 0$

$\dfrac{\partial z}{\partial y} = \dfrac{8y}{6z^2} = \dfrac{4y}{3z^2}$

5. $x^2 - 2y - z^2 + y\left(x^2 z^2\right) = 20$

$2x - 0 - 2z\dfrac{\partial z}{\partial x} + y\left[x^2 \cdot 2z\dfrac{\partial z}{\partial x} + z^2 \cdot 2x\right] = 0$

$\left(2x^2 yz - 2z\right)\dfrac{\partial z}{\partial x} = -2x - 2xyz^2$

$\dfrac{\partial z}{\partial x} = \dfrac{-2x\left(1 + yz^2\right)}{2z\left(x^2 y - 1\right)} = \dfrac{x\left(yz^2 + 1\right)}{z\left(1 - x^2 y\right)}$

7. $0 + e^y + e^z \dfrac{\partial z}{\partial y} = 0$

$\dfrac{\partial z}{\partial y} = -\dfrac{e^y}{e^z} = -e^{y-z}$

9. $\dfrac{1}{z}\dfrac{\partial z}{\partial x} + 9\dfrac{\partial z}{\partial x} - y = 0$

$\left(\dfrac{1}{z} + 9\right)\dfrac{\partial z}{\partial x} = y$

$\left(\dfrac{1 + 9z}{z}\right)\dfrac{\partial z}{\partial x} = y$

$\dfrac{\partial z}{\partial x} = \dfrac{yz}{9 + z}$

11. $\left(2z\dfrac{\partial z}{\partial y}+6x\right)\sqrt{x^3+5}=0$

$2z\dfrac{\partial z}{\partial y}+6x=0$

$\dfrac{\partial z}{\partial y}=\dfrac{-6x}{2z}=-\dfrac{3x}{z}$

13. $3x\left(2z\dfrac{\partial z}{\partial x}\right)+3z^2+2y\left(2z\dfrac{\partial z}{\partial x}\right)-7y(4x^3)=0$

$(6xz+4yz)\dfrac{\partial z}{\partial x}=28x^3y-3z^2$

$\dfrac{\partial z}{\partial x}=\dfrac{28x^3y-3z^2}{6xz+4yz}$

$\dfrac{\partial z}{\partial x}\Big|_{(1,\,0,\,1)}=\dfrac{28(1)^3(0)-3(1)}{6(1)(1)-4(0)(1)}=\dfrac{-3}{6}=-\dfrac{1}{2}$

15. $e^{yz}\cdot y\dfrac{\partial z}{\partial x}=-y\left[x\dfrac{\partial z}{\partial x}+z\cdot1\right]$.

$\left(ye^{yz}+xy\right)\dfrac{\partial z}{\partial x}=-yz$

$\dfrac{\partial z}{\partial x}=-\dfrac{yz}{y\left(e^{yz}+x\right)}$

$\dfrac{\partial z}{\partial x}=-\dfrac{z}{e^{yz}+x}$

If $x=-\dfrac{e^2}{2}$, $y=1$, $z=2$, then

$\dfrac{\partial z}{\partial x}=-\dfrac{2}{e^2+\frac{-e^2}{2}}=-\dfrac{2}{\frac{e^2}{2}}=-\dfrac{4}{e^2}$.

17. $\dfrac{1}{z}\dfrac{\partial z}{\partial x}=4+0$

$\dfrac{\partial z}{\partial x}=4z$

If $x=5$, $y=-20$, $z=1$, then $\dfrac{\partial z}{\partial x}=4$.

19. $\dfrac{(rs)\left[2t\frac{\partial t}{\partial r}\right]-\left(s^2+t^2\right)[s]}{(rs)^2}=0$

$2rst\dfrac{\partial t}{\partial r}-s\left(s^2+t^2\right)=0$

$2rst\dfrac{\partial t}{\partial r}=s\left(s^2+t^2\right)$

$\dfrac{\partial t}{\partial r}=\dfrac{s\left(s^2+t^2\right)}{2rst}=\dfrac{s^2+t^2}{2rt}$

If $r=1$, $s=2$, $t=4$, then

$\dfrac{\partial t}{\partial r}=\dfrac{4+16}{2\cdot1\cdot4}=\dfrac{20}{8}=\dfrac{5}{2}$.

21. $c+\sqrt{c}=12+q_A\sqrt{9+q_B^2}$

a. If $q_A=6$ and $q_B=4$, then

$c+\sqrt{c}=12+6(5)=42$, $\sqrt{c}=42-c$,

$c=(42-c)^2=42^2-84c+c^2$,

$c^2-85c+1764=0$,

$c=\dfrac{85\pm\sqrt{(-85)^2-4(1)(1764)}}{2}$

$=\dfrac{85\pm\sqrt{169}}{2}=\dfrac{85\pm13}{2}$. Thus $c=49$ or

$c=36$. However $c=49$ is extraneous but $c=36$ is not. Thus $c=36$.

b. Differentiating with respect to q_A:

$\dfrac{\partial c}{\partial q_A}+\dfrac{1}{2\sqrt{c}}\cdot\dfrac{\partial c}{\partial q_A}=\sqrt{9+q_B^2}$.

$\left(1+\dfrac{1}{2\sqrt{c}}\right)\dfrac{\partial c}{\partial q_A}=\sqrt{9+q_B^2}$.

When $q_A=6$ and $q_B=4$, then $c=36$

and $\left(1+\dfrac{1}{12}\right)\dfrac{\partial c}{\partial q_A}=5$, $\dfrac{13}{12}\cdot\dfrac{\partial c}{\partial q_A}=5$,

or $\dfrac{\partial c}{\partial q_A}=\dfrac{60}{13}$.

Differentiating with respect to q_B:

$$\frac{\partial c}{\partial q_B} + \frac{1}{2\sqrt{c}} \cdot \frac{\partial c}{\partial q_B} = q_A \cdot \frac{q_B}{\sqrt{9+q_B^2}}$$

$$\left(1 + \frac{1}{2\sqrt{c}}\right)\frac{\partial c}{\partial q_B} = \frac{q_A q_B}{\sqrt{9+q_B^2}}$$

When $q_A = 6$ and $q_B = 4$, then $c = 36$ and

$$\left(1+\frac{1}{12}\right)\frac{\partial c}{\partial q_B} = \frac{24}{5}, \quad \frac{13}{12} \cdot \frac{\partial c}{\partial q_B} = \frac{24}{5}, \text{ or } \frac{\partial c}{\partial q_B} = \frac{288}{65}.$$

Problems 17.5

1. $f_x(x, y) = 6(1)y^2 = 6y^2$
$f_{xy}(x, y) = 6(2y) = 12y$
$f_y(x, y) = 6x(2y) = 12xy$
$f_{yx}(x, y) = 12(1)y = 12y$

3. $f_y(x, y) = 3$
$f_{yy}(x, y) = 0$
$f_{yyx}(x, y) = 0$

5. $f_y(x, y) = 9\left[e^{2xy}(2x)\right] = 18xe^{2xy}$

$f_{yx}(x, y) = 18\left[x\left(e^{2xy} \cdot 2y\right) + e^{2xy}(1)\right] = 18e^{2xy}(2xy+1)$

$f_{yxy}(x, y) = 18\left[e^{2xy}(2x) + (2xy+1)\left(e^{2xy} \cdot 2x\right)\right]$

$= 18e^{2xy}(2x)[1+(2xy+1)] = 18e^{2xy}(2x)[2+2xy]$

$= 72x(1+xy)e^{2xy}$

7. $f(x, y) = (x+y)^2(xy) = \left(x^2 + 2xy + y^2\right)(xy) = x^3y + 2x^2y^2 + xy^3$

$f_x(x, y) = 3x^2y + 4xy^2 + y^3$
$f_y(x, y) = x^3 + 4x^2y + 3xy^2$
$f_{xx}(x, y) = 6xy + 4y^2$
$f_{yy}(x, y) = 4x^2 + 6xy$

9. $z = e^{\sqrt{x^2+y^2}}$

$$\frac{\partial z}{\partial y} = e^{\sqrt{x^2+y^2}} \cdot \frac{2y}{2\sqrt{x^2+y^2}} = \frac{ye^{\sqrt{x^2+y^2}}}{\sqrt{x^2+y^2}} = \frac{zy}{\sqrt{x^2+y^2}}$$

$$\frac{\partial^2 z}{\partial y^2} = \frac{\sqrt{x^2+y^2}\left(e^{\sqrt{x^2+y^2}} + y \cdot e^{\sqrt{x^2+y^2}} \cdot \frac{y}{\sqrt{x^2+y^2}}\right) - ye^{\sqrt{x^2+y^2}}\left(\frac{y}{\sqrt{x^2+y^2}}\right)}{\left(\sqrt{x^2+y^2}\right)^2}$$

$$= \frac{(x^2+y^2)\left(e^{\sqrt{x^2+y^2}} + e^{\sqrt{x^2+y^2}} \cdot \frac{y^2}{\sqrt{x^2+y^2}}\right) - y^2 e^{\sqrt{x^2+y^2}}}{(x^2+y^2)^{3/2}}$$

$$= e^{\sqrt{x^2+y^2}} \frac{x^2+y^2+y^2\sqrt{x^2+y^2}-y^2}{(x^2+y^2)^{3/2}}$$

$$= e^{\sqrt{x^2+y^2}} \frac{x^2+y^2\sqrt{x^2+y^2}}{(x^2+y^2)^{3/2}}$$

$$= z \cdot \frac{x^2+y^2\sqrt{x^2+y^2}}{(x^2+y^2)^{3/2}}$$

11. $f_y(x, y, z) = 0$

$f_{yx}(x, y, z) = 0$

$f_{yxx}(x, y, z) = 0$

Thus $f_{yxx}(4, 3, -2) = 0$.

13. $f_k(l, k) = 18l^3k^5 - 14l^2k^6$

$f_{kl}(l, k) = 54l^2k^5 - 28lk^6$

$f_{klk}(l, k) = 270l^2k^4 - 168lk^5$

Thus $f_{klk}(2, 1) = 270(4)(1) - 168(2)(1) = 744$.

15. $f_x(x, y) = y^2e^x + \frac{1}{x}$

$f_{xy}(x, y) = 2ye^x$

$f_{xyy}(x, y) = 2e^x$

Thus $f_{xyy}(1, 1) = 2e$.

17. $\dfrac{\partial c}{\partial q_B} = \dfrac{1}{3}\left(3q_A^2 + q_B^3 + 4\right)^{-\frac{2}{3}}\left(3q_B^2\right)$

$= q_B^2\left(3q_A^2 + q_B^3 + 4\right)^{-\frac{2}{3}}$

$\dfrac{\partial^2 c}{\partial q_A \partial q_B} = -\dfrac{2}{3}q_B^2\left(3q_A^2 + q_B^3 + 4\right)^{-\frac{5}{3}}\left(6q_A\right)$

$= -4q_A q_B^2\left(3q_A^2 + q_B^3 + 4\right)^{-\frac{5}{3}}$

When $p_A = 25$ and $p_B = 4$, then

$q_A = 10 - 25 + 16 = 1$ and

$q_B = 20 + 25 - 44 = 1$, and

$\dfrac{\partial^2 c}{\partial q_A \partial q_B} = -4(8)^{-\frac{5}{3}} = -\dfrac{4}{32} = -\dfrac{1}{8}.$

19. $f_x(x,\ y) = 24x^2 + 4xy^2$

$f_y(x,\ y) = 4x^2 y + 20y^3$

$f_{xy}(x,\ y) = 8xy$

$f_{yx}(x,\ y) = 8xy$

Thus $f_{xy}(x,\ y) = f_{yx}(x,\ y).$

21. $\dfrac{\partial z}{\partial x} = \dfrac{2x}{x^2 + y^2}$

$\dfrac{\partial^2 z}{\partial x^2} = \dfrac{\left(x^2 + y^2\right)(2) - (2x)(2x)}{\left(x^2 + y^2\right)^2}$

$= \dfrac{2\left(y^2 - x^2\right)}{\left(x^2 + y^2\right)^2}$

$\dfrac{\partial z}{\partial y} = \dfrac{2y}{x^2 + y^2}$

$\dfrac{\partial^2 z}{\partial y^2} = \dfrac{\left(x^2 + y^2\right)(2) - (2y)(2y)}{\left(x^2 + y^2\right)^2}$

$= \dfrac{2\left(x^2 - y^2\right)}{\left(x^2 + y^2\right)^2}$

$\dfrac{\partial^2 z}{\partial x^2} + \dfrac{\partial^2 z}{\partial y^2} = \dfrac{2\left(y^2 - x^2\right)}{\left(x^2 + y^2\right)^2} + \dfrac{2\left(x^2 - y^2\right)}{\left(x^2 + y^2\right)^2} = 0$

23. $2z\dfrac{\partial z}{\partial y} + 2y = 0$

$\dfrac{\partial z}{\partial y} = -\dfrac{2y}{2z} = -\dfrac{y}{z}$

$\dfrac{\partial^2 z}{\partial y^2} = -\dfrac{z(1) - y\cdot\frac{\partial z}{\partial y}}{z^2} = -\dfrac{z - y\left(-\frac{y}{z}\right)}{z^2}$

$= -\dfrac{z^2 + y^2}{z^3}$

From the original equation, $z^2 + y^2 = 3x^2$.

Thus $\dfrac{\partial^2 z}{\partial y^2} = -\dfrac{3x^2}{z^3}.$

Problems 17.6

1. $z = 5x + 3y$, $x = 2r + 3s$, $y = r - 2s$

$\dfrac{\partial z}{\partial r} = \dfrac{\partial z}{\partial x}\dfrac{\partial x}{\partial r} + \dfrac{\partial z}{\partial y}\dfrac{\partial y}{\partial r} = (5)(2) + (3)(1) = 13$

$\dfrac{\partial z}{\partial s} = \dfrac{\partial z}{\partial x}\dfrac{\partial x}{\partial s} + \dfrac{\partial z}{\partial y}\dfrac{\partial y}{\partial s} = (5)(3) + (3)(-2) = 9$

3. $z = e^{x+y}$, $x = t^2 + 3$, $y = \sqrt{t^3}$

$\dfrac{dz}{dt} = \dfrac{\partial t}{\partial x}\dfrac{dx}{dt} + \dfrac{\partial z}{\partial y}\dfrac{dy}{dt}$

$= e^{x+y}(2t) + e^{x+y}\left(\dfrac{3}{2}t^{1/2}\right)$

$= e^{x+y}\left(2t + \dfrac{3}{2}\sqrt{t}\right)$

5. $w = x^2 z^2 + xyz + yz^2$, $x = 5t$, $y = 2t + 3$, $z = 6 - t$

$$\frac{dw}{dt} = \frac{\partial w}{\partial x}\frac{dx}{dt} + \frac{\partial w}{\partial y}\frac{dy}{dt} + \frac{\partial w}{\partial t}\frac{dz}{dt}$$

$$= \left(2xz^2 + yz\right)(5) + \left(xz + z^2\right)(2) + \left(2x^2 z + xy + 2yz\right)(-1)$$

$$= 5\left(2xz^2 + yz\right) + 2\left(xz + z^2\right) - \left(2x^2 z + xy + 2yz\right)$$

7. $z = \left(x^2 + xy^2\right)^3$, $x = r + s + t$, $y = 2r - 3s + 8t$

$$\frac{\partial z}{\partial t} = \frac{\partial z}{\partial x}\frac{\partial x}{\partial t} + \frac{\partial z}{\partial y}\frac{\partial y}{\partial t}$$

$$= 3\left(x^2 + xy^2\right)^2 \left(2x + y^2\right)[1] + 3\left(x^2 + xy^2\right)^2 (2xy)[8]$$

$$= 3\left(x^2 + xy^2\right)^2 \left(2x + y^2 + 16xy\right)$$

9. $w = x^2 + xyz + z^2$, $x = r^2 - s^2$, $y = rs$, $z = r^2 + s^2$

$$\frac{\partial w}{\partial s} = \frac{\partial w}{\partial x}\frac{\partial x}{\partial s} + \frac{\partial w}{\partial y}\frac{\partial y}{\partial s} + \frac{\partial w}{\partial z}\frac{\partial z}{\partial s}$$

$$= (2x + yz)(-2s) + (xz)(r) + \left(xy + 2z\right)(2s)$$

$$= -2s(2x + yz) + r(xz) + 2s(xy + 2z)$$

11. $y = x^2 - 7x + 5$, $x = 19rs + 2s^2 t^2$

$$\frac{\partial y}{\partial r} = \frac{dy}{dx}\frac{\partial x}{\partial r} = (2x - 7)(19s) = 19s(2x - 7)$$

13. $z = (4x + 3y)^3$, $x = r^2 s$, $y = r - 2s$; $r = 0$, $s = 1$

$$\frac{\partial z}{\partial r} = \frac{\partial z}{\partial x}\frac{\partial x}{\partial r} + \frac{\partial z}{\partial y}\frac{\partial y}{\partial r}$$

$$= 12(4x + 3y)^2 (2rs) + 9(4x + 3y)^2 (1)$$

$$= 3(4x + 3y)^2 (8rs + 3)$$

When $r = 0$, $s = 1$, then $x = 0$, $y = -2$, and $\dfrac{\partial z}{\partial r} = 324$.

15. $w = e^{2x+3y}(x^2 + 4z^2)$, $x = rs$, $y = 2s - 3r$, and $z = r + s$; $r = 1$, $s = 0$.

$$\frac{\partial w}{\partial s} = \frac{\partial w}{\partial x}\frac{\partial x}{\partial s} + \frac{\partial w}{\partial y}\frac{\partial y}{\partial s} + \frac{\partial w}{\partial z}\frac{\partial z}{\partial s}$$

$$= [2e^{2x+3y}(x^2 + 4z^2) + e^{2x+3y}(2x)](r) + 3e^{2x+3y}(x^2 + 4z^2)(2) + e^{2x+3y}(8z)(1)$$

When $r = 1$, $s = 0$, then $x = 0$, $y = -3$ and $z = 1$.

$$\frac{\partial w}{\partial s} = [2e^{-9}(4) + e^{-9}(0)](1) + 3e^{-9}(4)(2) + e^{-9}(8)(1)$$

$$= e^{-9}[8 + 24 + 8] = \frac{40}{e^9}$$

17. $\dfrac{\partial c}{\partial p_A} = \dfrac{\partial c}{\partial q_A}\dfrac{\partial q_A}{\partial p_A} + \dfrac{\partial c}{\partial q_B}\dfrac{\partial q_B}{\partial p_A}$

$$= \left[\frac{1}{3}\left(3q_A^2 + q_B^3 + 4\right)^{-\frac{2}{3}}(6q_A)\right](-1) + \left[\frac{1}{3}\left(3q_A^2 + q_B^3 + 4\right)^{-\frac{2}{3}}(3q_B^2)\right](1)$$

$$= \left(3q_A^2 + q_B^3 + 4\right)^{-\frac{2}{3}}\left(-2q_A + q_B^2\right)$$

$\dfrac{\partial c}{\partial p_B} = \dfrac{\partial c}{\partial q_A}\dfrac{\partial q_A}{\partial p_B} + \dfrac{\partial c}{\partial q_B}\dfrac{\partial q_B}{\partial p_B}$

$$= \left[\frac{1}{3}\left(3q_A^2 + q_B^3 + 4\right)^{-\frac{2}{3}}(6q_A)\right](2p_B) + \left[\frac{1}{3}\left(3q_A^2 + q_B^3 + 4\right)^{-\frac{2}{3}}(3q_B^2)\right](-11)$$

$$= \left(3q_A^2 + q_B^3 + 4\right)^{-\frac{2}{3}}\left(4q_A p_B - 11q_B^2\right)$$

When $p_A = 25$ and $p_B = 4$, then $q_A = 10 - 25 + 16 = 1$, $q_B = 20 + 25 - 44 = 1$,

and $\dfrac{\partial c}{\partial p_A} = (8)^{-\frac{2}{3}}(-1) = -\dfrac{1}{4}$ and $\dfrac{\partial c}{\partial p_B} = (8)^{-\frac{2}{3}}(5) = \dfrac{5}{4}$.

19. a. $\dfrac{\partial w}{\partial t} = \dfrac{\partial w}{\partial x}\dfrac{\partial x}{\partial t} + \dfrac{\partial w}{\partial y}\dfrac{\partial y}{\partial t}$

b. $w = 2x^2 \ln|3x - 5y|$, $x = s\sqrt{t^2 + 2}$ and $y = t - 3e^{2-s}$.

$$\frac{\partial w}{\partial t} = \left[4x\ln|3x-5y| + \frac{2x^2(3)}{3x-5y}\right]\frac{s(2t)}{2\sqrt{t^2+2}} + \left[\frac{2x^2}{3x-5y}(-5)\right](1)$$

When $s = 1$ and $t = 0$, then $x = \sqrt{2}$ and $y = -3e$.

$$\frac{\partial w}{\partial t} = \left[4\sqrt{2}\ln|3\sqrt{2} - 5(-3e)| + \frac{2(2)(3)}{3\sqrt{2}-5(-3e)}\right](0) + \left[\frac{2(2)}{3\sqrt{2}-5(-3e)}(-5)\right]$$

$$= -\frac{20}{3\sqrt{2}+15e}$$

Problems 17.7

1. $f(x, y) = x^2 + y^2 - 5x + 4y + xy$

$\begin{cases} f_x(x, y) = 2x + y - 5 = 0 \\ f_y(x, y) = x + 2y + 4 = 0 \end{cases}$

Solving the system gives the critical point $\left(\dfrac{14}{3}, -\dfrac{13}{3} \right)$.

3. $f(x, y) = \dfrac{5}{3}x^3 + \dfrac{2}{3}y^3 - \dfrac{15}{2}x^2 + y^2 - 4y + 7$

$\begin{cases} f_x(x, y) = 5x^2 - 15x = 0 \\ f_y(x, y) = 2y^2 + 2y - 4 = 0 \end{cases}$

Both equations are easily solved by factoring. Critical points: $(0, -2), (0, 1), (3, -2), (3, 1)$

5. $f(x, y, z) = 2x^2 + xy + y^2 + 100 - z(x + y - 200)$

$\begin{cases} f_x(x, y, z) = 4x + y - z = 0 \\ f_y(x, y, z) = x + 2y - z = 0 \\ f_z(x, y, z) = -x - y + 200 = 0 \end{cases}$

Solving the system gives the critical point $(50, 150, 350)$.

7. $f(x, y) = x^2 + 3y^2 + 4x - 9y + 3$

$\begin{cases} f_x(x, y) = 2x + 4 = 0 \\ f_y(x, y) = 6y - 9 = 0 \end{cases}$

Critical point $\left(-2, \dfrac{3}{2} \right)$

Second-Derivative Test

$f_{xx}(x, y) = 2$, $f_{yy}(x, y) = 6$, $f_{xy}(x, y) = 0$. At $\left(-2, \dfrac{3}{2} \right)$, $D = (2)(6) - 0^2 = 12 > 0$ and

$f_{xx}(x, y) = 2 > 0$. Thus at $\left(-2, \dfrac{3}{2} \right)$ there is a relative minimum.

9. $f(x, y) = y - y^2 - 3x - 6x^2$

$\begin{cases} f_x(x, y) = -3 - 12x = 0 \\ f_y(x, y) = 1 - 2y = 0 \end{cases}$

Critical point $\left(-\dfrac{1}{4}, \dfrac{1}{2} \right)$

Second-Derivative Test

$f_{xx}(x, y) = -12$, $f_{yy}(x, y) = -2$, $f_{xy}(x, y) = 0$

At $\left(-\dfrac{1}{4}, \dfrac{1}{2}\right)$, $D = (-12)(-2) - 0^2 = 24 > 0$

and $f_{xx}(x, y) = -12 < 0$. Thus at $\left(-\dfrac{1}{4}, \dfrac{1}{2}\right)$

there is a relative maximum.

11. $f(x, y) = x^2 + 3xy + y^2 + x + 3$

$\begin{cases} f_x(x, y) = 2x + 3y + 1 = 0 \\ f_y(x, y) = 3x + 2y = 0 \end{cases}$

Critical point: $\left(\dfrac{2}{5}, -\dfrac{3}{5}\right)$

Second-Derivative Test

$f_{xx}(x, y) = 2$, $f_{yy} = 2$, $f_{xy} = 3$. At

$\left(\dfrac{2}{5}, -\dfrac{3}{5}\right)$, $D = (2)(2) - (3)^2 = -5 < 0$, so

there is no relative extremum at $\left(\dfrac{2}{5}, -\dfrac{3}{5}\right)$.

13. $f(x, y) = \dfrac{1}{3}\left(x^3 + 8y^3\right) - 2\left(x^2 + y^2\right) + 1$

$\begin{cases} f_x(x, y) = x^2 - 4x = 0 \\ f_y(x, y) = 8y^2 - 4y = 0 \end{cases}$

Critical points: $(0, 0)$, $\left(4, \dfrac{1}{2}\right)$, $\left(0, \dfrac{1}{2}\right)$,

$(4, 0)$

Second-Derivative Test

$f_{xx}(x, y) = 2x - 4$, $f_{yy}(x, y) = 16y - 4$,

$f_{xy}(x, y) = 0$. At $(0, 0)$,

$D = (-4)(-4) - 0^2 = 16 > 0$ and

$f_{xx}(x, y) = -4 < 0$; thus a relative

maximum.

At $\left(4, \dfrac{1}{2}\right)$, $D = (4)(4) - 0^2 = 16 > 0$ and

$f_{xx}(x, y) = 4 > 0$; thus a relative minimum.

At $\left(0, \dfrac{1}{2}\right)$, $D = (-4)(4) - 0^2 = -16 < 0$;

thus neither.

At $(4, 0)$, $D = (4)(-4) - 0^2 = -16 < 0$, thus

neither.

15. $f(l, k) = \dfrac{l^2}{2} + 2lk + 3k^2 - 69l - 164k + 17$

$\begin{cases} f_l(l, k) = l + 2k - 69 = 0 \\ f_k(l, k) = 2l + 6k - 164 = 0 \end{cases}$

Critical point: $(43, 13)$

Second-Derivative Test

$f_{ll}(l, k) = 1$, $f_{kk}(l, k) = 6$, $f_{lk}(l, k) = 2$

At $(43, 13)$, $D = (1)(6) - 2^2 = 2 > 0$ and

$f_{ll}(l, k) = 1 > 0$; thus there is a relative

minimum at $(43, 13)$.

17. $f(p, q) = pq - \dfrac{1}{p} - \dfrac{1}{q}$

$\begin{cases} f_p(p, q) = q + \dfrac{1}{p^2} = 0 \\[2mm] f_q(p, q) = p + \dfrac{1}{q^2} = 0 \end{cases}$

Critical point: $(-1, -1)$

Second-Derivative Test

$f_{pp}(p, q) = -\dfrac{2}{p^3}$, $f_{qq}(p, q) = -\dfrac{2}{q^3}$,

$f_{pq}(p, q) = 1$. At $(-1, -1)$,

$D = (2)(2) - 1^2 = 3 > 0$ and

$f_{pp}(p, q) = 2 > 0$; thus there is a relative

minimum at $(-1, -1)$.

19. $f(x, y) = \left(y^2 - 4\right)\left(e^x - 1\right)$

$\begin{cases} f_x(x, y) = e^x\left(y^2 - 4\right) = 0 & (1) \\ f_y(x, y) = 2y\left(e^x - 1\right) = 0 & (2) \end{cases}$

Critical points: $(0, -2)$, $(0, 2)$

[Note that $y = 0$ does not give rise to a

common solution of (1) and (2).]

Second-Derivative Test

$f_{xx}(x, y) = e^x\left(y^2 - 4\right)$,

$f_{yy}(x, y) = 2\left(e^x - 1\right)$, $f_{xy}(x, y) = 2ye^x$.

At $(0, -2)$, $D = (0)(0) - (-4)^2 = -16 < 0$;

thus neither. At $(0, 2)$,

$D = (0)(0) - (4)^2 = -16 < 0$; thus neither.

21. $P = f(l, k)$

$= 1.08l^2 - 0.03l^3 + 1.68k^2 - 0.08k^3$

$\begin{cases} P_l = 2.16l - 0.09l^2 = 0 \\ P_k = 3.36k - 0.24k^2 = 0 \end{cases}$

Critical points: (0, 0), (0, 14), (24, 0), (24, 14)

Second-Derivative Test

$P_{ll} = 2.16 - 0.18l$, $P_{kk} = 3.36 - 0.48k$, $P_{lk} = 0$. At (0, 0), $D = (2.16)(3.36) - 0^2 > 0$ and $P_{ll} = 2.16 > 0$; thus relative minimum.

At (0, 14), $D = (2.16)(-3.36) - 0^2 < 0$; thus no extremum. At (24, 0), $D = (-2.16)(3.36) - 0^2 < 0$; thus no extremum. At (24, 14), $D = (-2.16)(-3.36) - 0^2 > 0$ and $P_{ll} = -2.16 < 0$; thus $l = 24$, $k = 14$ gives a relative maximum.

23. Profit per lb for $A = p_A - 60$.

Profit per lb for $B = p_B - 70$.

Total Profit $= P = (p_A - 60)q_A + (p_B - 70)q_B$

$P = (p_A - 60)\left[5(p_B - p_A) \right] + (p_B - 70)\left[500 + 5(p_A - 2p_B) \right]$

Thus

$\begin{cases} \dfrac{\partial P}{\partial p_A} = -10(p_A - p_B + 5) = 0 \\ \dfrac{\partial P}{\partial p_B} = 10(p_A - 2p_B + 90) = 0 \end{cases}$

Critical point: $p_A = 80$, $p_B = 85$

$\dfrac{\partial^2 P}{\partial p_A^2} = -10$, $\dfrac{\partial^2 P}{\partial p_B^2} = -20$, $\dfrac{\partial^2 P}{\partial p_B \partial p_A} = 10$. When $p_A = 80$ and $p_B = 85$, then

$D = (-10)(-20) - (10)^2 = 100 > 0$ and $\dfrac{\partial^2 P}{\partial p_A^2} = -10 < 0$; thus relative maximum at $p_A = 80$, $p_B = 85$.

25. $p_A = 100 - q_A$, $p_B = 84 - q_B$, $c = 600 + 4(q_A + q_B)$.

Revenue from market A $= r_A = p_A q_A = (100 - q_A)q_A$. Revenue from market B

$= r_B = p_B q_B = (84 - q_B)q_B$.

Total Profit = Total Revenue – Total Cost

$P = (100 - q_A)q_A + (84 - q_B)q_B - \left[600 + 4(q_A + q_B) \right]$

$\begin{cases} \dfrac{\partial P}{\partial q_A} = 96 - 2q_A = 0 \\ \dfrac{\partial P}{\partial q_B} = 80 - 2q_B = 0 \end{cases}$

Critical point: $q_A = 48$, $q_B = 40$

$$\frac{\partial^2 P}{\partial q_A^2} = -2, \quad \frac{\partial^2 P}{\partial q_B^2} = -2, \quad \frac{\partial^2 P}{\partial q_B \partial q_A} = 0.$$

At $q_A = 48$ and $q_B = 40$, then $D = (-2)(-2) - 0^2 = 4 > 0$ and $\dfrac{\partial^2 P}{\partial q_A^2} = -2 < 0$; thus relative maximum at

$q_A = 48$, $q_B = 40$. When $q_A = 48$ and $q_B = 40$, then selling prices are $p_A = 52$, $p_B = 44$, and profit = 3304.

27. $c = \dfrac{3}{2}q_A^2 + 3q_B^2, \quad p_A = 60 - q_A^2, \quad p_B = 72 - 2q_B^2$

Total Profit = Total Revenue − Total Cost

$$P = (p_A q_A + p_B q_B) - c$$

$$P = 60q_A - q_A^3 + 72q_B - 2q_B^3 - \left(\frac{3}{2}q_A^2 + 3q_B^2\right)$$

$$\begin{cases} \dfrac{\partial P}{\partial q_A} = 60 - 3q_A - 3q_A^2 = 3(5 + q_A)(4 - q_A) \\[2mm] \dfrac{\partial P}{\partial q_B} = 72 - 6q_B - 6q_B^2 = 6(4 + q_B)(3 - q_B) \end{cases}$$

Since we want $q_A \geq 0$ and $q_B \geq 0$, the critical point occurs when $q_A = 4$ and $q_B = 3$.

$$\frac{\partial^2 P}{\partial q_A^2} = -3 - 6q_A, \quad \frac{\partial^2 P}{\partial q_B^2} = -6 - 12q_B, \quad \frac{\partial^2 P}{\partial q_B \partial q_A} = 0. \text{ When } q_A = 4 \text{ and } q_B = 3, \text{ then}$$

$$D = (-27)(-42) - 0^2 > 0 \text{ and } \frac{\partial^2 P}{\partial q_A^2} = -27 < 0; \text{ thus relative maximum at } q_A = 4, q_B = 3.$$

29. Refer to the diagram in the text.
$xyz = 6$
$C = 3xy + 2[1(xz)] + 2[0.5(yz)]$

Note that $z = \dfrac{6}{xy}$. Thus

$$C = 3xy + 2xz + yz = 3xy + 2x\left(\frac{6}{xy}\right) + y\left(\frac{6}{xy}\right) = 3xy + \frac{12}{y} + \frac{6}{x}$$

$$\begin{cases} \dfrac{\partial C}{\partial x} = 3y - \dfrac{6}{x^2} = 0 \\[2mm] \dfrac{dC}{dy} = 3x - \dfrac{12}{y^2} = 0 \end{cases}$$

A critical point occurs at $x = 1$ and $y = 2$. Thus $z = 3$.

$$\frac{\partial^2 C}{\partial x^2} = \frac{12}{x^3}, \quad \frac{\partial^2 C}{\partial y^2} = \frac{24}{y^3}, \quad \frac{\partial^2 C}{\partial x \partial y} = 3.$$

When $x = 1$ and $y = 2$, then $d = (12)(3) - (3)^2 = 27 > 0$ and $\dfrac{\partial^2 C}{\partial x^2} = 12 > 0$. Thus we have a minimum.

The dimensions should be 1 ft by 2 ft by 3 ft.

31. $y = \dfrac{3x-7}{2}$

$f(x,\,y) = -2x^2 + 5\left(\dfrac{3x-7}{2}\right)^2 + 7$

Setting the derivative equal to 0 gives $-4x + 5(2)\left(\dfrac{3x-7}{2}\right)\left(\dfrac{3}{2}\right) = 0$, $-4x + \dfrac{15}{2}(3x-7) = 0$,

$-8x + 15(3x-7) = 0$, $37x = 105$, or $x = \dfrac{105}{37}$. The second-derivative is $\dfrac{37}{2} > 0$, so we have a relative

minimum. If $x = \dfrac{105}{37}$, then $y = \dfrac{28}{37}$. Thus there is a relative minimum at $\left(\dfrac{105}{37}, \dfrac{28}{37}\right)$.

33. $c = q_A^2 + 3q_B^2 + 2q_A q_B + aq_A + bq_B + d$

We are given that $(q_A,\,q_B) = (3,\,1)$ is a critical point.

$$\begin{cases} \dfrac{\partial c}{\partial q_A} = 2q_A + 2q_B + a = 0 \\[2mm] \dfrac{\partial c}{\partial q_B} = 6q_B + 2q_A + b = 0 \end{cases}$$

Substituting the given values for q_A and q_B into both equations gives $a = -8$ and $b = -12$. Since $c = 15$ when $q_A = 3$ and $q_B = 1$, from the joint-cost function we have

$15 = 3^2 + 3\left(1^2\right) + 2(3)(1) + (-8)(3) + (-12) + d$,

$15 = -18 + d$, $33 = d$. Thus $a = -8$, $b = -12$, $d = 33$.

35. a. Profit = Total Revenue − Total Cost

$P = p_A q_A + p_B q_B - $ total cost

$= \left(35 - 2q_A^2 + q_B\right) q_A + \left(20 - q_B + q_A\right) q_B - \left(-8 - 2q_A^3 + 3q_A q_B + 30q_A + 12q_B + \dfrac{1}{2}q_A^2\right)$

$P = 5q_A - \dfrac{1}{2}q_A^2 - q_A q_B + 8q_B - q_B^2 + 8$

$$\begin{cases} \dfrac{\partial P}{\partial q_A} = 5 - q_A - q_B = 0 \\[2mm] \dfrac{\partial P}{\partial q_B} = -q_A + 8 - 2q_B = 0 \end{cases}$$

Critical point: $q_A = 2$, $q_B = 3$

$\dfrac{\partial^2 P}{\partial q_A^2} = -1$, $\dfrac{\partial^2 P}{\partial q_B^2} = -2$, $\dfrac{\partial^2 P}{\partial q_B \partial q_A} = -1$

At $q_A = 2$ and $q_B = 3$, then $D = (-1)(-2) - (-1)^2 = 1 > 0$ and $\dfrac{\partial^2 P}{\partial q_A^2} = -1 < 0$; thus there is a

relative maximum profit for 2 units of A and 3 units of B.

   b. Substituting $q_A = 2$ and $q_B = 3$ into the formulas for p_A, p_B, and P gives a selling price for A of 30, a selling price for B of 19, and a relative maximum profit of 25.

37. a. $P = 5T\left(1 - e^{-x}\right) - 20x - 0.1T^2$

   b. $\dfrac{\partial P}{\partial T} = 5\left(1 - e^{-x}\right) - 0.2T$

$\dfrac{\partial P}{\partial x} = 5Te^{-x} - 20$

At the point $(T, x) = (20, \ln 5)$,

$\dfrac{\partial P}{\partial T} = 5\left(1 - e^{-\ln 5}\right) - 0.2(20)$

$= 5\left(1 - \dfrac{1}{5}\right) - 4 = 0$

$\dfrac{\partial P}{\partial x} = 5(20)e^{-\ln 5} - 20 = 100\left(\dfrac{1}{5}\right) - 20 = 0$

Thus $(20, \ln 5)$ is a critical point. In a similar fashion we verify that $\left(5, \ln \dfrac{5}{4}\right)$ is a critical point.

   c. $\dfrac{\partial^2 P}{\partial T^2} = -0.2$, $\dfrac{\partial^2 P}{\partial x^2} = -5Te^{-x}$, $\dfrac{\partial^2 P}{\partial T \partial x} = 5e^{-x}$

At $(20, \ln 5)$,

$D = (-0.2)\left[-5(20)e^{-\ln 5}\right] - \left(5e^{-\ln 5}\right)^2 = 20\left(\dfrac{1}{5}\right) - \left[5\left(\dfrac{1}{5}\right)\right]^2 = 3 > 0$,

and $\dfrac{\partial^2 P}{\partial T^2} = -0.2 < 0$. Thus we get a relative maximum at $(20, \ln 5)$.

At $\left(5, \ln \dfrac{5}{4}\right)$,

$D = (-0.2)\left[-5(5)e^{-\ln\left(\frac{5}{4}\right)}\right] - \left[5e^{-\ln\left(\frac{5}{4}\right)}\right]^2 = 5\left(\dfrac{4}{5}\right) - \left[5\left(\dfrac{4}{5}\right)\right]^2 = -12 < 0$, so there is no relative

extremum at $\left(5, \ln \dfrac{5}{4}\right)$.

Problems 17.8

1. $f(x, y) = x^2 + 4y^2 + 6,\ 2x - 8y = 20$

$F(x, y, \lambda) = x^2 + 4y^2 + 6 - \lambda(2x - 8y - 20)$

$\begin{cases} F_x = 2x - 2\lambda = 0 & (1) \\ F_y = 8y + 8\lambda = 0 & (2) \\ F_\lambda = -2x + 8y + 20 = 0 & (3) \end{cases}$

From (1), $x = \lambda$; from (2), $y = -\lambda$. Substituting $x = \lambda$ and $y = -\lambda$ into (3) gives $-2\lambda - 8\lambda + 20 = 0$, $-10\lambda = -20$, so $\lambda = 2$. Thus $x = 2$ and $y = -2$. Critical point of F: $(2, -2, 2)$. Critical point of f: $(2, -2)$.

3. $f(x, y, z) = x^2 + y^2 + z^2,\ 2x + y - z = 9$

$F(x, y, z, \lambda) = x^2 + y^2 + z^2 - \lambda(2x + y - z - 9)$

$\begin{cases} F_x = 2x - 2\lambda = 0 & (1) \\ F_y = 2y - \lambda = 0 & (2) \\ F_z = 2z + \lambda = 0 & (3) \\ F_\lambda = -2x - y + z + 9 = 0 & (4) \end{cases}$

From (1), $x = \lambda$; from (2), $y = \dfrac{\lambda}{2}$; from (3), $z = -\dfrac{\lambda}{2}$. Substituting into (4) gives

$-2\lambda - \dfrac{\lambda}{2} + \left(\dfrac{-\lambda}{2}\right) + 9 = 0,\ -6\lambda + 18 = 0$, so $\lambda = 3$. Thus $x = 3,\ y = \dfrac{3}{2},\ z = -\dfrac{3}{2}$. Critical point of F:

$\left(3, \dfrac{3}{2}, -\dfrac{3}{2}, 3\right)$. Critical point of f: $\left(3, \dfrac{3}{2}, -\dfrac{3}{2}\right)$.

5. $f(x, y, z) = 2x^2 + xy + y^2 + z,\ x + 2y + 4z = 3$

$F(x, y, z, \lambda) = 2x^2 + xy + y^2 + z - \lambda(x + 2y + 4z - 3)$

$\begin{cases} F_x = 4x + y - \lambda = 0 \\ F_y = x + 2y - 2\lambda = 0 \\ F_z = 1 - 4\lambda = 0 \\ F_\lambda = -x - 2y - 4z - 3 = 0 \end{cases}$

From the third equation we have $\lambda = \dfrac{1}{4}$. Substituting this value into the first two equations and then

eliminating y gives $x = 0$ and $y = \dfrac{1}{4}$. Finally, solving for z in the last equation gives $z = -\dfrac{7}{8}$.

Critical point of F: $\left(0, \dfrac{1}{4}, -\dfrac{7}{8}, \dfrac{1}{4}\right)$

Critical point of f: $\left(0, \dfrac{1}{4}, -\dfrac{7}{8}\right)$

7. $f(x, y, z) = xyz$, $x + 2y + 3z = 18$ $(xyz \neq 0)$

$f(x, y, z, \lambda) = xyz - \lambda(x + 2y + 3z - 18)$

$\begin{cases} F_x = yz - \lambda = 0 & (1) \\ F_y = xz - 2\lambda = 0 & (2) \\ F_z = xy - 3\lambda = 0 & (3) \\ F_\lambda = -x - 2y - 3z + 18 = 0 & (4) \end{cases}$

From (1) and (2), $y = \dfrac{x}{2}$. From (1) and (3), $z = \dfrac{x}{3}$. Hence from (4), $x = 6$, so $y = 3$ and $z = 2$. Critical point of f is (6, 3, 2). Note that it is not necessary to determine λ.

9. $f(x, y, z) = x^2 + 2y - z^2$, $2x - y = 0$, $y + z = 0$

Since there are two constraints, two Lagrange multipliers are used.

$F(x, y, z, \lambda_1, \lambda_2) = x^2 + 2y - z^2 - \lambda_1(2x - y) - \lambda_2(y + z)$

$\begin{cases} F_x = 2x - 2\lambda_1 = 0 & (1) \\ F_y = 2 + \lambda_1 - \lambda_2 = 0 & (2) \\ F_z = -2z - \lambda_2 = 0 & (3) \\ F_{\lambda_1} = -2x + y = 0 & (4) \\ F_{\lambda_2} = -y - z = 0 & (5) \end{cases}$

From (1), $x = \lambda_1$. From (3), $z = -\dfrac{\lambda_2}{2}$. From (4) and (5), $2x = -z$, so $\lambda_1 = \dfrac{\lambda_2}{4}$. Substituting $\lambda_1 = \dfrac{\lambda_2}{4}$ into (2) yields $\lambda_2 = \dfrac{8}{3}$. Thus $\lambda_1 = \dfrac{2}{3}$, $x = \dfrac{2}{3}$, and $z = -\dfrac{4}{3}$. From (5), $y = -z$ and hence $y = \dfrac{4}{3}$. Critical point of f: $\left(\dfrac{2}{3}, \dfrac{4}{3}, -\dfrac{4}{3} \right)$.

11. $f(x, y, z) = xy^2z$, $x + y + z = 1$, $x - y + z = 0$ $(xyz \neq 0)$

Since there are two constraints, two Lagrange multipliers are used.

$F(x, y, z, \lambda_1, \lambda_2) = xy^2z - \lambda_1(x + y + z - 1) - \lambda_2(x - y + z)$

$\begin{cases} F_x = y^2z - \lambda_1 - \lambda_2 = 0 & (1) \\ F_y = 2xyz - \lambda_1 + \lambda_2 = 0 & (2) \\ F_z = xy^2 - \lambda_1 - \lambda_2 = 0 & (3) \\ F_{\lambda_1} = -x - y - z + 1 = 0 & (4) \\ F_{\lambda_2} = -x + y - z = 0 & (5) \end{cases}$

Subtracting (3) from (1) gives $y^2z - xy^2 = 0$, so $x = z$ (since $xy^2z \neq 0$). Subtracting (5) from (4) gives $-2y + 1 = 0$, so $y = \dfrac{1}{2}$. Substituting $z = x$ and $y = \dfrac{1}{2}$ in (5) gives $-2x + \dfrac{1}{2} = 0$, so $x = \dfrac{1}{4}$. Thus, $z = \dfrac{1}{4}$. Critical point of f: $\left(\dfrac{1}{4}, \dfrac{1}{2}, \dfrac{1}{4} \right)$.

13. We minimize $c = f(q_1, q_2) = 0.1q_1^2 + 7q_1 + 15q_2 + 1000$ subject to the constraint $q_1 + q_2 = 100$.

$F(q_1, q_2, \lambda) = 0.1q_1^2 + 7q_1 + 15q_2 + 1000 - \lambda(q_1 + q_2 - 100)$

$$\begin{cases} F_{q_1} = 0.2q_1 + 7 - \lambda = 0 & (1) \\ F_{q_2} = 15 - \lambda = 0 & (2) \\ F_\lambda = -q_1 - q_2 + 100 = 0 & (3) \end{cases}$$

From (2), $\lambda = 15$. Substituting $\lambda = 15$ into (1) gives $0.2q_1 + 7 - 15 = 0$, so $q_1 = 40$. Substituting $q_1 = 40$ into (3) gives $-40 - q_2 + 100 = 0$, so $q_2 = 60$. Thus $\lambda = 15$, $q_1 = 40$, and $q = 60$. Thus plant 1 should produce 40 units and plant 2 should produce 60 units.

15. We maximize $f(l,k) = 12l + 20k - l^2 - 2k^2$ subject to the constraint $4l + 8k = 88$.

$F(l, k, \lambda) = 12l + 20k - l^2 - 2k^2 - \lambda(4l + 8k - 88)$

$$\begin{cases} F_l = 12 - 2l - 4\lambda = 0 & (1) \\ F_k = 20 - 4k - 8\lambda = 0 & (2) \\ F_\lambda = -4l - 8k + 88 = 0 & (3) \end{cases}$$

Eliminating λ from (1) and (2) yields $k = l - 1$. Substituting $k = l - 1$ into (3) yields $l = 8$, so $k = 7$. Therefore the greatest output is $f(8, 7) = 74$ units (when $l = 8$, $k = 7$).

17. We maximize $P(x, y) = 9x^{\frac{1}{4}}y^{\frac{3}{4}} - x - y$ subject to the constraint $x + y = 60{,}000$.

$F(x, y, \lambda) = 9x^{\frac{1}{4}}y^{\frac{3}{4}} - x - y - \lambda(x + y - 60{,}000)$

$$\begin{cases} F_x = \dfrac{9}{4}x^{-\frac{3}{4}}y^{\frac{3}{4}} - 1 - \lambda = 0 & (1) \\ F_y = \dfrac{27}{4}x^{\frac{1}{4}}y^{-\frac{1}{4}} - 1 - \lambda = 0 & (2) \\ F_\lambda = -x - y + 60{,}000 = 0 & (3) \end{cases}$$

Solving (2) for λ and substituting in (1) gives $\dfrac{9}{4}x^{-\frac{3}{4}}y^{\frac{3}{4}} - \dfrac{27}{4}x^{\frac{1}{4}}y^{-\frac{1}{4}} = 0$, $\dfrac{9}{4}x^{-\frac{3}{4}}y^{\frac{3}{4}} = \dfrac{27}{4}x^{\frac{1}{4}}y^{-\frac{1}{4}}$, $y = 3x$. Substituting for y in (3) gives $-4x + 60{,}000 = 0$, so $x = 15{,}000$, from which $y = 45{,}000$. Thus each month \$15,000 should be spent on newspaper advertising and \$45,000 on TV advertising.

19. We minimize $B(x, y, z) = x^2 + y^2 + 2z^2$ subject to $x + y = 20$ and $y + z = 20$.

Since there are two constraints, two Lagrange multipliers are used.

$F(x, y, z, \lambda_1, \lambda_2) = x^2 + y^2 + 2z^2 - \lambda_1(x + y - 20) - \lambda_2(y + z - 20)$

$$\begin{cases} F_x = 2x - \lambda_1 = 0 & (1) \\ F_y = 2y - \lambda_1 - \lambda_2 = 0 & (2) \\ F_z = 4z - \lambda_2 = 0 & (3) \\ F_{\lambda_1} = -x - y + 20 = 0 & (4) \\ F_{\lambda_2} = -y - z + 20 = 0 & (5) \end{cases}$$

Eliminating y from (4) and (5) gives $x = z$. From (1) and (3), $\lambda_1 = 2x$ and $\lambda_2 = 4z$. Substituting in (2) we have $2y - 2x - 4z = 0$, $2y - 2x - 4x = 0$, $2y - 6x = 0$, $y = 3x$. Substituting in (5) gives $-(3x) - x + 20 = 0$, so $x = 5$. Thus $z = 5$ and $y = 15$. Therefore, $x = 5$, $y = 15$, $z = 5$.

21. $U = x^3 y^3$, $p_x = 2$, $p_y = 3$, $I = 48$ $\left(x^3 y^3 \neq 0 \right)$

We want to maximize $U = x^3 y^3$ subject to $2x + 3y = 48$.

$F(x, y, \lambda) = x^3 y^3 - \lambda(2x + 3y - 48)$

$$\begin{cases} F_x = 3x^2 y^3 - 2\lambda = 0 & (1) \\ F_y = 3x^3 y^2 - 3\lambda = 0 & (2) \\ F_\lambda = -2x - 3y + 48 = 0 & (3) \end{cases}$$

From (1), $\lambda = \dfrac{3}{2} x^2 y^3$ and from (2), $\lambda = x^3 y^2$. Thus $\dfrac{3}{2} x^2 y^3 = x^3 y^2$, so $x = \dfrac{3}{2} y$.

Substituting this expression for x into (3) yields $y = 8$. Hence $x = \left(\dfrac{3}{2} \right) 8 = 12$.

23. $U = f(x, y, z) = xyz$

$p_x = p_y = p_z = 1$, $I = 100$

$(xyz \neq 0)$

We want to maximize $U = xyz$ subject to
$x + y + z = 100$.

$F(x, y, z, \lambda) = xyz - \lambda(x + y + z - 100)$

$$\begin{cases} F_x = yz - \lambda = 0 & (1) \\ F_y = xz - \lambda = 0 & (2) \\ F_z = xy - \lambda = 0 & (3) \\ F_\lambda = -x - y - z + 100 = 0 & (4) \end{cases}$$

From (1) and (2), $yz = xz$, so $y = x$. Similarly, from (1) and (3), $z = x$. Substituting $y = x$ and $z = x$ into (4) yields $x = \dfrac{100}{3}$. Thus $y = \dfrac{100}{3}$ and $z = \dfrac{100}{3}$.

Problems 17.9

1. $n = 6$, $\Sigma x_i = 21$, $\Sigma y_i = 18.6$, $\Sigma x_i y_i = 75.7$, $\Sigma x_i^2 = 91$.

$a = 0.98$

$b = 0.61$

Thus $\hat{y} = 0.98 + 0.61x$. When $x = 3.5$, then $\hat{y} = 3.12$.

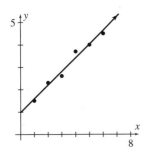

3. $n = 5$, $\Sigma x_i = 22$, $\Sigma y_i = 37$, $\Sigma x_i y_i = 189$,

$\Sigma x_i^2 = 112.5$. $a = 0.057$, $b = 1.67$. Thus

$\hat{y} = 0.057 + 1.67x$. When $x = 3.5$, then

$\hat{y} = 5.90$.

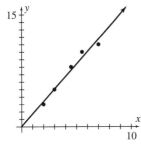

5. $n = 6$, $\Sigma p_i = 250$, $\Sigma q_i = 322$, $\Sigma p_i q_i = 11,690$,

$\Sigma p_i^2 = 13,100$.

$a = 80.5$

$b = -0.643$

Thus $\hat{q} = 80.5 - 0.643p$.

7. $n = 4$,

$\Sigma x_i = 160$, $\Sigma y_i = 420.8$, $\Sigma x_i y_i = 16,915.2$,

$\Sigma x_i^2 = 7040$. $a = 100$, $b = 0.13$. Thus

$\hat{y} = 100 + 0.13x$. When $x = 40$, then

$\hat{y} = 105.2$.

9.

Year (x)	1	2	3	4	5
Production (y)	10	15	16	18	21

$n = 5$,

$\Sigma x_i = 15$, $\Sigma y_i = 80$, $\Sigma x_i y_i = 265$, $\Sigma x_i^2 = 55$.

$a = 8.5$

$b = 2.5$

Thus $\hat{y} = 8.5 + 2.5x$

11. a.

Year (x)	1	2	3	4	5
Quantity (y)	35	31	26	24	26

$n = 5$, $\Sigma x_i = 15$, $\Sigma y_i = 142$, $\Sigma x_i y_i = 401$,

$\Sigma x_i^2 = 55$. $a = 35.9$, $b = -2.5$. Thus

$\hat{y} = 35.9 - 2.5x$.

b.

Year (x)	-2	-1	0	1	2
Quantity (y)	35	31	26	24	26

$n = 5$, $\Sigma x_i = 0$, $\Sigma y_i = 142$, $\Sigma x_i y_i = -25$,

$\Sigma x_i^2 = 10$. $a = \dfrac{\Sigma y_i}{n} = 28.4$ and

$b = \dfrac{\Sigma x_i y_i}{\Sigma x_i^2} = -2.5$. Thus

$\hat{y} = 28.4 - 2.5x$.

Problems 17.10

1. $\displaystyle\int_0^3 \int_0^4 x\,dy\,dx = \int_0^3 xy \Big|_0^4 \, dx = \int_0^3 4x\,dx = 2x^2 \Big|_0^3$

$= 18$

3. $\displaystyle\int_0^1 \int_0^1 xy\,dx\,dy = \int_0^1 \frac{x^2 y}{2}\Big|_0^1 \, dy = \int_0^1 \frac{y}{2}\,dy = \frac{y^2}{4}\Big|_0^1$

$= \dfrac{1}{4}$

5. $\displaystyle\int_1^3 \int_1^2 \left(x^2 - y\right)dx\,dy = \int_1^3 \left(\frac{x^3}{3} - xy\right)\Big|_1^2 \, dy$

$= \displaystyle\int_1^3 \left[\left(\frac{8}{3} - 2y\right) - \left(\frac{1}{3} - y\right)\right]dy = \int_1^3 \left(\frac{7}{3} - y\right)dy$

$= \left(\dfrac{7}{3}y - \dfrac{y^2}{2}\right)\Big|_1^3 = \left(7 - \dfrac{9}{2}\right) - \left(\dfrac{7}{3} - \dfrac{1}{2}\right) = \dfrac{2}{3}$

7. $\int_0^1 \int_0^2 (x+y)\,dy\,dx = \int_0^1 \left(xy + \frac{y^2}{2} \right)\Big|_0^2 dx$

$$= \int_0^1 (2x+2)dx = \left(x^2 + 2x\right)\Big|_0^1 = 3$$

9. $\int_1^4 \int_0^{5x} y\,dy\,dx = \int_1^4 \frac{y^2}{2}\Big|_0^{5x} dx = \int_1^4 \frac{25}{2}x^2\,dx$

$$= \frac{25}{6}x^3\Big|_1^4 = \frac{525}{2}$$

11. $\int_0^1 \int_{3x}^{x^2} 14x^2 y\,dy\,dx = \int_0^1 \left(7x^2 y^2\right)\Big|_{3x}^{x^2} dx$

$$= \int_0^1 \left(7x^6 - 63x^4\right)dx = \left(x^7 - \frac{63x^5}{5}\right)\Big|_0^1$$

$$= -\frac{58}{5}$$

13. $\int_0^3 \int_0^{\sqrt{9-x^2}} y\,dy\,dx = \int_0^3 \frac{y^2}{2}\Big|_0^{\sqrt{9-x^2}} dx$

$$= \int_0^3 \left(\frac{9-x^2}{2} - 0 \right)dx = \frac{1}{2}\int_0^3 (9-x^2)dx$$

$$= \frac{1}{2}\left(9x - \frac{x^3}{3} \right)\Big|_0^3 = \frac{1}{2}(27-9) - 0 = 9$$

15. $\int_{-1}^1 \int_x^{1-x} 3(x+y)\,dy\,dx$

$$= \int_{-1}^1 3\left(xy + \frac{y^2}{2} \right)\Big|_x^{1-x} dx$$

$$= \int_{-1}^1 3\left[x(1-x) + \frac{(1-x)^2}{2} - \left(x^2 + \frac{x^2}{2} \right) \right]dx$$

$$= \int_{-1}^1 3\left[x - \frac{5x^2}{2} + \frac{(1-x)^2}{2} \right]dx$$

$$= 3\left[\frac{x^2}{2} - \frac{5x^3}{6} - \frac{(1-x)^3}{6} \right]\Big|_{-1}^1$$

$$= 3\left[\frac{1}{2} - \frac{5}{6} - 0 \right] - 3\left[\frac{1}{2} + \frac{5}{6} - \frac{4}{3} \right] = -1$$

17. $\int_0^1 \int_0^y e^{x+y}\,dx\,dy = \int_0^1 e^{x+y}\Big|_0^y dy$

$$= \int_0^1 \left(e^{2y} - e^y \right)dy$$

$$= \left[\frac{e^{2y}}{2} - e^y \right]\Big|_0^1 = \frac{e^2}{2} - e - \left(\frac{1}{2} - 1 \right)$$

$$= \frac{e^2}{2} - e + \frac{1}{2}$$

19. $\int_{-1}^0 \int_{-1}^2 \int_1^2 6xy^2 z^3\,dx\,dy\,dz$

$$= \int_{-1}^0 \int_{-1}^2 3x^2 y^2 z^3\Big|_1^2 dy\,dz$$

$$= \int_{-1}^0 \int_{-1}^2 9y^2 z^3\,dy\,dz = \int_{-1}^0 3y^3 z^3\Big|_{-1}^2 dz$$

$$= \int_{-1}^0 27z^3\,dz = \frac{27z^4}{4}\Big|_{-1}^0 = -\frac{27}{4}$$

21. $\int_0^1 \int_{x^2}^x \int_0^{xy} dz\,dy\,dx = \int_0^1 \int_{x^2}^x z\Big|_0^{xy} dy\,dx$

$= \int_0^1 \int_{x^2}^x xy\,dy\,dx = \int_0^1 \frac{xy^2}{2}\Big|_{x^2}^x dx$

$= \int_0^1 \left[\frac{x^3}{2} - \frac{x^5}{2}\right] dx = \left[\frac{x^4}{8} - \frac{x^6}{12}\right]\Big|_0^1 = \frac{1}{24}$

23. $P(0 \le x \le 2, 1 \le y \le 2) = \int_1^2 \int_0^2 e^{-(x+y)}dx\,dy$

$= \int_1^2 -e^{-(x+y)}\Big|_0^2 dy = \int_1^2 \left[-e^{-(2+y)} + e^{-y}\right]dy$

$= \left[e^{-(2+y)} - e^{-y}\right]\Big|_1^2 = e^{-4} - e^{-2} - e^{-3} + e^{-1}$

25. $P\left(x \ge \frac{1}{2},\, y \ge \frac{1}{3}\right) = \int_{1/3}^1 \int_{1/2}^1 1\,dx\,dy$

$= \int_{1/3}^1 x\Big|_{1/2}^1 dy = \int_{1/3}^1 \left(1 - \frac{1}{2}\right)dy$

$= \int_{1/3}^1 \frac{1}{2}dy = \frac{1}{2}y\Big|_{1/3}^1 = \frac{1}{2}\left(1 - \frac{1}{3}\right) = \frac{1}{2}\left(\frac{2}{3}\right) = \frac{1}{3}$

Chapter 17 Review Problems

1. $x + y + z = 1$ can be put in the form $Ax + By + Cz + D = 0$, so the graph is a plane. The intercepts are $(1, 0, 0)$, $(0, 1, 0)$, and $(0, 0, 1)$.

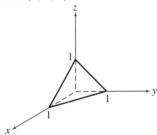

3. $z = y^2$

The *y,z*-trace is $z = y^2$, which is a parabola. For any fixed value of *x*, we obtain the curve $z = y^2$.

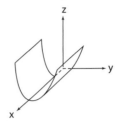

5. $f_x(x, y) = 4(2x) + 6(1)y + 0 - 0 = 8x + 6y$
$f_y(x, y) = 0 + 6x(1) + 2y - 0 = 6x + 2y$

7. $\frac{\partial z}{\partial x} = \frac{(x+y)(1) - x(1)}{(x+y)^2} = \frac{y}{(x+y)^2}$

Because $z = x(x+y)^{-1}$,

$\frac{\partial z}{\partial y} = x\left[(-1)(x+y)^{-2}(1)\right] = -\frac{x}{(x+y)^2}$.

9. $f(x, y) = \ln\sqrt{x^2 + y^2} = \frac{1}{2}\ln\left(x^2 + y^2\right)$

$\frac{\partial}{\partial y}[f(x, y)] = \frac{1}{2} \cdot \frac{1}{x^2 + y^2}(2y) = \frac{y}{x^2 + y^2}$

11. $w_x(x, y, z) = 2xyze^{x^2yz}$

$w_{xy}(x, y, z) = 2xz\left[y\left(e^{x^2yz} \cdot x^2z\right) + e^{x^2yz} \cdot 1\right]$

$= 2xze^{x^2yz}\left(x^2yz + 1\right)$

13. $\dfrac{\partial}{\partial z}[f(x,\ y,\ z)] = (x+y+z)(2z)+(x^2+y^2+z^2)(1)=3z^2+2z(x+y)+x^2+y^2$

$\dfrac{\partial^2}{\partial z^2}[f(x,\ y,\ z)] = 6z+2(x+y)=2x+2y+6z$

15. $w=e^{x+y+z}\ln xyz = e^{x+y+z}(\ln x+\ln y+\ln z)$

$\dfrac{\partial w}{\partial y} = e^{x+y+z}(\ln x+\ln y+\ln z)+e^{x+y+z}\left(\dfrac{1}{y}\right)$

$\qquad = e^{x+y+z}\left[\ln xyz + \dfrac{1}{y}\right]$

By symmetry, $\dfrac{\partial w}{\partial x} = e^{x+y+z}\left[\ln xyz + \dfrac{1}{x}\right]$.

$\dfrac{\partial^2 w}{\partial z\partial x} = e^{x+y+z}\left[\ln xyz + \dfrac{1}{x}\right]+e^{x+y+z}\left[\dfrac{1}{z}\right]$

$\qquad = e^{x+y+z}\left[\ln xyz + \dfrac{1}{x}+\dfrac{1}{z}\right]$.

17. $f(x,\ y,\ z) = \dfrac{x+y}{xz} = \dfrac{1}{z}+\dfrac{y}{xz}$

$f_x(x,\ y,\ z) = -\dfrac{y}{x^2 z}$

$f_{xy}(x,\ y,\ z) = -\dfrac{1}{x^2 z}$

$f_{xyz}(x,\ y,\ z) = \dfrac{1}{x^2 z^2}$

$f_{xyz}(2,\ 7,\ 4) = \dfrac{1}{2^2\cdot 4^2} = \dfrac{1}{64}$

19. $\dfrac{\partial w}{\partial r} = \dfrac{\partial w}{\partial x}\dfrac{\partial x}{\partial r}+\dfrac{\partial w}{\partial y}\dfrac{\partial y}{\partial r} = (2x+2y)\left(e^r\right)+(2x+6y)\left(\dfrac{1}{r+s}\right)$

$\qquad = 2(x+y)e^r + \dfrac{2(x+3y)}{r+s}$

$\dfrac{\partial w}{\partial s} = \dfrac{\partial w}{\partial x}\dfrac{\partial x}{\partial s}+\dfrac{\partial w}{\partial y}\dfrac{\partial y}{\partial s} = (2x+2y)(0)+(2x+6y)\left(\dfrac{1}{r+s}\right)$

$\qquad = \dfrac{2(x+3y)}{r+s}$

21. $2x + 2y - 4z\dfrac{\partial z}{\partial x} + \left[x\dfrac{\partial z}{\partial x} + z(1) \right] + 0 = 0$

$(-4z + x)\dfrac{\partial z}{\partial x} = -(2x + 2y + z)$

$\dfrac{\partial z}{\partial x} = \dfrac{-(2x + 2y + z)}{-4z + x} = \dfrac{2x + 2y + z}{4z - x}$

23. $P = 20l^{0.7}k^{0.3}$. Marginal productivity functions are given by $\dfrac{\partial P}{\partial l} = 20(0.7)l^{-0.3}k^{0.3}$ and

$\dfrac{\partial P}{\partial k} = 20(0.3)l^{0.7}k^{-0.7}$. Thus $\dfrac{\partial P}{\partial l} = 14l^{-0.3}k^{0.3}$ and $\dfrac{\partial P}{\partial k} = 6l^{0.7}k^{-0.7}$.

25. $q_A = 100 - p_A + 2p_B$, $q_B = 150 + 3p_A - 2p_B$. Since $\dfrac{\partial q_A}{\partial p_B} = 2 > 0$ and $\dfrac{\partial q_B}{\partial p_A} = 3 > 0$, A and B are

competitive products.

27. $f(x, y) = x^2 + 2y^2 - 2xy - 4y + 3$

$\begin{cases} f_x(x, y) = 2x - 2y = 0 \\ f_y(x, y) = 4y - 2x - 4 = 0 \end{cases}$

Critical point: (2, 2)

$f_{xx}(x, y) = 2, \ f_{yy}(x, y) = 4, \ f_{xy}(x, y) = -2$

At (2, 2), $D = (2)(4) - (-2)^2 = 4 > 0$ and $f_{xx}(x, y) = 2 > 0$; thus relative minimum at (2, 2).

29.

$xyz = 32 \ (xyz \neq 0)$. Let S be the amount of cardboard used.

$S = xy + 2yz + 2xz$

$= xy + 2y\left[\dfrac{32}{xy} \right] + 2x\left[\dfrac{32}{xy} \right]$

$= xy + \dfrac{64}{x} + \dfrac{64}{y}$

$\dfrac{\partial S}{\partial x} = y - \dfrac{64}{x^2}, \ \dfrac{\partial S}{\partial y} = x - \dfrac{64}{y^2}$

The critical point occurs when $x = 4$, $y = 4$, and
$z = 2$, which gives a minimum. The dimensions are 4 ft by 4 ft by 2 ft.

31. Profit $= P = (p_A - 50)q_A + (p_B - 60)q_B$

$$P = (p_A - 50)\left[250(p_B - p_A)\right] + (p_B - 60)\left[32,000 + 250(p_A - 2p_B)\right].$$

$$\frac{\partial P}{\partial p_A} = (p_A - 50)(-250) + \left[250(p_B - p_A)\right](1) + 250(p_B - 60)$$

$$= -500p_A + 500p_B - 250(10) = 500(-p_A + p_B - 5)$$

Also, $\dfrac{\partial P}{\partial p_B} = (p_A - 50)(250) + (p_B - 60)(-500) + \left[32,000 + 250(p_A - 2p_B)\right](1)$

$$= 500p_A - 1000p_B + 49,500 = 500(p_A - 2p_B + 99)$$

Setting $\dfrac{\partial P}{\partial p_A} = 0$ and $\dfrac{\partial P}{\partial p_B} = 0$ gives

$$-p_A + p_B - 5 = 0 \qquad (1)$$

and

$$p_A - 2p_B + 99 = 0 \qquad (2)$$

Adding Equations (1) and (2) gives $-p_B + 94 = 0$. So $p_B = 94$. From Equation (1), $p_A = p_B - 5$, so

$p_A = 94 - 5 = 89$. At $p_A = 89$ and $p_B = 94$, $D = \dfrac{\partial^2 P}{\partial p_A^2}\dfrac{\partial^2 P}{\partial p_B^2} - \dfrac{\partial^2 P}{\partial p_B \partial p_A} = (-500)(-1000) - (500)^2 > 0$

and $\dfrac{\partial^2 P}{\partial p_A^2} = -500 < 0$.

Thus there is a relative maximum profit when the price of A is 89 cents per pound and the price of B is 94 cents per pound.

33. $f(x, y, z) = x^2 + y^2 + z^2$, $3x + 2y + z = 14$

$F(x, y, z, \lambda)$

$= x^2 + y^2 + z^2 - \lambda(3x + 2y + z - 14)$

$$\begin{cases} F_x = 2x - 3\lambda = 0 & (1) \\ F_y = 2y - 2\lambda = 0 & (2) \\ F_z = 2z - \lambda = 0 & (3) \\ F_\lambda = -3x - 2y - z + 14 = 0 & (4) \end{cases}$$

From (1), $x = \dfrac{3\lambda}{2}$; from (2), $y = \lambda$, from (3), $z = \dfrac{\lambda}{2}$. Substituting into (4) gives

$-3\left(\dfrac{3\lambda}{2}\right) - 2\lambda - \dfrac{\lambda}{2} + 14 = 0$, from which $\lambda = 2$. Thus $x = 3$, $y = 2$, and $z = 1$.

Critical point of F: (3, 2, 1, 2), so the critical point of f is (3, 2, 1).

35.

Year (x)	1	2	3	4	5	6
Expenditures (y)	15	22	21	26	27	34

$n = 6$, $\Sigma x_i = 21$, $\Sigma y_i = 145$, $\Sigma x_i y_i = 565$,

$\Sigma x_i^2 = 91$

$$a = \frac{\left(\Sigma x_i^2\right)\left(\Sigma y_i\right) - \left(\Sigma x_i\right)\left(\Sigma x_i y_i\right)}{n\Sigma x_i^2 - \left(\Sigma x_i\right)^2} = 12.67$$

$$b = \frac{n\Sigma x_i y_i - \left(\Sigma x_i\right)\left(\Sigma y_i\right)}{n\Sigma x_i^2 - \left(\Sigma x_i\right)^2} = 3.29$$

Thus $\hat{y} = 12.67 + 3.29x$

37.

$$\int_0^1 \int_0^{y^2} xy\, dx\, dy = \int_0^1 \frac{x^2 y}{2}\bigg|_0^{y^2} dy$$

$$= \int_0^1 \left(\frac{y^4 \cdot y}{2} - 0\right) dy = \frac{1}{2}\int_0^1 y^5 dy$$

$$= \frac{1}{2}\cdot\frac{y^6}{6}\bigg|_0^1 = \frac{1}{2}\cdot\frac{1}{6} - 0 = \frac{1}{12}$$

39.

$$\int_0^1 \int_{\sqrt{x}}^{x^2} 7\left(x^2 + 2xy - 3y^2\right) dy\, dx$$

$$= 7\int_0^1 \left(x^2 y + xy^2 - y^3\right)\bigg|_{\sqrt{x}}^{x^2} dx$$

$$= 7\int_0^1 \left[\left(x^4 + x^5 - x^6\right) - \left(x^{\frac{5}{2}} + x^2 - x^{\frac{3}{2}}\right)\right] dx$$

$$= 7\left[\frac{x^5}{5} + \frac{x^6}{6} - \frac{x^7}{7} - \frac{2x^{\frac{7}{2}}}{7} - \frac{x^3}{3} + \frac{2x^{\frac{5}{2}}}{5}\right]\Bigg|_0^1$$

$$= 7\left[\frac{1}{5} + \frac{1}{6} - \frac{1}{7} - \frac{2}{7} - \frac{1}{3} + \frac{2}{5}\right] - 0 = \frac{1}{30}$$

Mathematical Snapshot Chapter 17

1.

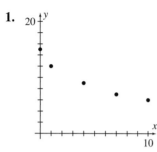

$y = Ce^{ax} + 5$, $y - 5 = Ce^{ax}$, $\ln(y - 5) = ax + \ln C$

x	y	$y - 5$	$\ln(y-5)$
0	15	10	2.30259
1	12	7	1.94591
4	9	4	1.38629
7	7	2	0.69315
10	6	1	0.00000

$n = 5$, $\Sigma x_i = 22$, $\Sigma \ln(y_i - 5) = 6.32794$, $\Sigma\left[x_i \ln(y_i - 5)\right] = 12.34312$, $\Sigma x_i^2 = 166$

$$a = \frac{n\Sigma\left[x_i \ln(y_i - 5)\right] - (\Sigma x_i)\left[\Sigma \ln(y_i - 5)\right]}{n\left(\Sigma x_i^2\right) - (\Sigma x_i)^2} = \frac{5(12.34312) - 22(6.32794)}{5(166) - (22)^2} \approx -0.22399$$

$$\ln C = \frac{\left(\Sigma x_i^2\right)\left[\Sigma \ln(y_i - 5)\right] - (\Sigma x_i)\left\{\Sigma\left[x_i \ln(y_i - 5)\right]\right\}}{n\left(\Sigma x_i^2\right) - (\Sigma x_i)^2} = \frac{166(6.32794) - 22(12.34312)}{5(166) - (22)^2} \approx 2.25112$$

$C \approx e^{2.25112} \approx 9.50$

Thus $y = 9.50e^{-0.22399x} + 5$.

3. Newton's law of cooling: $\dfrac{dT}{dt} = k(T - a)$, where $a = 45$. Thus $\dfrac{dT}{dt} = k(T - 45)$, $\dfrac{dT}{T - 45} = k\, dt$,

$\displaystyle\int \frac{dT}{T - 45} = \int k\, dt$, $\ln|T - 45| = kt + C$. Because $T - 45 > 0$, $\ln(T - 45) = kt + C$. Thus $T - 45 = e^{kt+C}$, or

$T = e^{kt+C} + 45 = e^C e^{kt} + 45 = C_1 e^{kt} + 45$, where $C_1 = e^C$. So $T = C_1 e^{kt} + 45$. When $t = 0$, then $T = 124$.

Hence $124 = C_1 + 45$, or $C_1 = 79$. Thus $T = 79e^{kt} + 45$. When $t = 128$, then $T = 64$, so

$64 = 79e^{128k} + 45$, $19 = 79e^{128k}$, $e^{128k} = \dfrac{19}{79}$, $128k = \ln\dfrac{19}{79}$, $k = \dfrac{\ln\left(\frac{19}{79}\right)}{128} \approx -0.01113$. Thus

$T = 79e^{-0.01113t} + 45$.